An Introduction to Theoretical Chemistry

In this unique textbook Jack Simons goes back to basics and focuses on the foundations that lie at the heart of modern day theoretical chemistry. Emphasis is on the concepts, tools and equations that govern the three main theoretical chemistry sub-disciplines: electronic structure, statistical mechanics and reaction dynamics.

Part I provides the foundations of quantum mechanics and molecular spectroscopy as applied to chemistry today. This section can be used either as stand alone material in a junior level physical chemistry class or to provide the reader with the tools and background needed to cover the second part of the book. Part II starts with a general overview of theoretical chemistry and then gives a very accessible introduction to each of the three main sub-disciplines in the subject.

Highly illustrated with numerous exercises and worked solutions, this book provides a concise, up-to-date treatise on the underpinnings of modern theoretical chemistry.

Born April 2, 1945 in Ohio, JACK SIMONS earned a B.S. degree (Magna Cum Laude) in Chemistry from Case Institute of Technology in 1967. His Ph.D. degree, in 1970 as an NSF Fellow, is from the University of Wisconsin, Madison. After serving an NSF Postdoctoral Fellowship at MIT, he joined the University of Utah Chemistry Faculty in 1971, where he was appointed to the Henry Eyring Chair in 1989. He is the author of more than 270 scientific papers and four books, and has mentored more than sixty Ph.D. and postdoctoral students. He is the recipient of many awards including Sloan, Dreyfus, and Guggenheim Fellowships, the Medal of the International Academy of Quantum Molecular Science and the R. W. Parry Teaching Award.

An Introduction to Theoretical Chemistry

Jack Simons

University of Utah, Salt Lake City, Utah

PUBLISHED BY THE PRESS SYNDICATE OF THE UNIVERSITY OF CAMBRIDGE
The Pitt Building, Trumpington Street, Cambridge, United Kingdom

CAMBRIDGE UNIVERSITY PRESS
The Edinburgh Building, Cambridge CB2 2RU, UK
40 West 20th Street, New York, NY 10011-4211, USA
477 Williamstown Road, Port Melbourne, VIC 3207, Australia
Ruiz de Alarcón 13, 28014 Madrid, Spain
Dock House, The Waterfront, Cape Town 8001, South Africa

http://www.cambridge.org

First published 2003

Printed in the United Kingdom at the University Press, Cambridge

Typefaces Times New Roman 10/13 pt and Univers *System* LaTeX 2_ε [TB]

A catalogue record for this book is available from the British Library

Library of Congress Cataloging in Publication data

Simons, Jack.
An introduction to theoretical chemistry / Jack Simons.
p. cm.
Includes bibliographical references and index.
ISBN 0 521 82360 9 – ISBN 0 521 53047 4 (pbk.)
1. Chemistry, Physical and theoretical. I. Title.
QD453.3 .S56 2003
541.2–dc21 2002073610

ISBN 0 521 82360 9 hardback
ISBN 0 521 53047 4 paperback

Contents

Introductory remarks

Let's begin by discussing what the discipline of theoretical chemistry is about. I think most young students of chemistry think that theory deals with using computers to model or simulate molecular behaviors. This is only partly true. Theory indeed contains under its broad umbrella the field of computational simulations, and it is such applications of theory that have gained much recent attention especially as powerful computers and user-friendly software packages have become widely available.

However, this discipline also involves analytical theory, which deals with how the fundamental equations used to perform the simulations are derived from the Schrödinger equation or from classical mechanics, among other things. The discipline also has to do with obtaining the equations that relate laboratory data (e.g., spectra, heat capacities, reaction cross-sections) to molecular properties (e.g., geometries, bond energies, activation energies). This analytical side of theory is also where the equations of statistical mechanics that relate macroscopic properties of matter to the microscopic properties of the constituent molecules are obtained.

So, theory is a diverse field of chemistry that uses physics, mathematics and computers to help us understand molecular behavior, to simulate molecular phenomena, and to predict the properties of new molecules. It is common to hear this discipline referred to as theoretical and computational chemistry. This text is focused more on the theory than on the computation. That is, I deal primarily with the basic ideas upon which theoretical chemistry is centered, and I discuss the equations and tools that enter into the three main sub-disciplines of theory – electronic structure, statistical mechanics, and reaction dynamics. I have chosen to emphasize these elements rather than to stress computational applications of theory because there are already many good sources available that deal with computational chemistry.

Now, let me address the issue of "who does theory?" It is common for chemists whose primary research activities involve laboratory work to also use theoretical concepts, equations, simulations and methods to assist in interpreting their data. Sometimes, these researchers also come up with new concepts or new models in terms of which to understand their laboratory findings. These experimental chemists are using theory in the former case and doing new theory in the latter. Many of my experimental chemistry colleagues have evolved into using theory in this manner.

However, for several decades now there have also been chemists who do not carry out laboratory experiments but whose research focus lies in developing new theory (new analytical equations, new computational tools, new concepts) and applying theory to understanding chemical processes as a full-time endeavor. These people are what we call theoretical chemists. I am proud to say that I am a member of this community of theorists and that I believe this discipline offers the most powerful background for understanding essentially all other areas of chemistry.

Where does one learn about theoretical chemistry? Most chemistry students in colleges and universities take classes in introductory chemistry, organic, analytical, inorganic, physical, and biochemistry. It is extremely rare for students to encounter a class that has "theoretical chemistry" in its title. This book is intended to illustrate to students that the subject of theoretical chemistry pervades most if not all of the other classes they take in an undergraduate chemistry curriculum. It is also intended to offer students a modern introduction to the field of theoretical chemistry and to illustrate how it has evolved into a discipline of its own and now stands shoulder-to-shoulder with the traditional experimental sub-disciplines of chemical science.

I have tried to write this book so it could be used in any of several ways:

(i) As a textbook that could be used to learn the quantum mechanics and molecular spectroscopy components of a typical junior-level physical chemistry class, including an overview of point group symmetry. This would involve covering Part I (the Background Material covered in Chapters 1–4) and then Chapter 5 and Sections 5–7 of Chapter 6. It would also be wise to solve many of the problems that I offer. Certainly, any student who has not yet taken an undergraduate class in physical chemistry should follow this route.

(ii) As a first-year graduate text in which selected topics in the areas of introductory quantum chemistry, spectroscopy, statistical mechanics, and reaction dynamics are surveyed. This would involve covering Chapters 5–8 and solving many of the problems. Although the Background Material of Chapters 1–4 should have been learned by such students in an undergraduate physical chemistry class, it would be wise to read this material to refresh one's memory. It is likely that full-semester classes in statistical mechanics and in reaction dynamics will require more material than offered in Chapters 7 and 8, but these chapters should suffice for briefer classes and for gaining an introduction to these fields.

(iii) As an introductory survey source for experimental chemists interested in learning about the central concepts and many of the most common tools of theoretical chemistry. To pursue this avenue, the reader should focus on Chapters 5–8 because the Background Material of Chapters 1–4 covers what such readers probably already know.

Because of the flexibility in how this text can be used, some duplication of material occurs. However, it has been my experience that students benefit from encountering subjects more than one time, especially if each subsequent encounter is at a deeper level. I believe this is the case for subjects that are covered in more than one place in this text.

I have also offered many exercises (small problems) and problems to be solved by the reader, as well as detailed solutions. Most of these problems deal with topics contained in Chapters 1–4 because it is these subjects that are likely to be studied in an undergraduate classroom setting where homework assignments are common. Chapters 5–8 are designed to give the reader an introduction to electronic structure theory, statistical mechanics, and reaction dynamics at the graduate and beginning-research level. In such settings, it is my experience that individual instructors prefer to construct their own problems, so I offer fewer exercises and problems associated with these chapters. Most, if not all, of the problems presented here require many steps to solve, so the reader is encouraged not to despair when attempting them; they may be difficult, but they teach valuable lessons.

The reader will notice that I do not provide many references, nor do I mention many names of theoretical chemists who work on the subjects I discuss. I avoided such citations because most of the people who have pioneered or are now actively working on the theories I discuss are friends of mine. I felt that mentioning any of them and not citing others would risk offending many colleagues, so I decided to severely limit such references. I hope the readers and my friends in the world theory community will appreciate my decision.

Before launching into the subject of theoretical chemistry, allow me to mention other sources that can be used to obtain information at a somewhat more advanced level than is presented in this text. Keep in mind that this is a text intended to offer an introduction to the field of theoretical chemistry, and is directed primarily at advanced undergraduate- and beginning graduate-level readerships. It is my hope that such readers will, from time to time, want to learn more about certain topics that are especially appealing to them. For this purpose, I suggest two sources that I have been instrumental in developing. First, a World Wide Web site that I created can be accessed at simons.hec.utah.edu/TheoryPage. This site provides a wealth of information including:

(i) web links to home pages of a multitude of practicing theoretical chemists who specialize in many of the topics discussed in this text;
(ii) numerous education-site web links that allow students ranging from fresh-persons to advanced graduate students to seek out a variety of information;

(iii) textual information much of which covers at a deeper level those subjects discussed in this text at an introductory level.

Another major source of information at a more advanced level is my textbook *Quantum Mechanics in Chemistry* (*QMIC*) written with Dr. Jeff Nichols (Past Director of the High Performance Computing Group at the Pacific Northwest National Laboratory and now Director of Mathematics and Computational Science at Oak Ridge National Laboratory). The full content of that book can be accessed in .pdf file format through the TheoryPage web link mentioned above. In several locations within the present introductory text, I specifically refer the reader either to my TheoryPage or *QMIC* textbook, but I urge you to also use these two sources whenever you want a more in-depth treatment of a subject.

To the readers who want to access up-to-date research-level treatments of many of the topics we introduce in this text, I suggest several recent monographs to which I refer throughout this text:

Molecular Electronic Structure Theory, T. Helgaker, P. Jørgensen, and J. Olsen, J. Wiley, New York (2000),
Modern Electronic Structure Theory, D. R. Yarkony, Ed., World Scientific Publishing, Singapore (1999),
Theory of Chemical Reaction Dynamics, M. Baer, Ed., Vols. 1–4; CRC Press, Boca Raton, Fla. (1985),
Essentials of Computational Chemistry, C. J. Cramer, Wiley, Chichester (2002),
An Introduction to Computational Chemistry, F. Jensen, John Wiley, New York (1998),
Molecular Modeling, 2nd edn., A. R. Leach, Prentice Hall, Englewood Cliffs (2001).
Molecular Reaction Dynamics and Chemical Reactivity, R. D. Levine and R. B. Bernstein, Oxford University Press, New York (1997),
Computer Simulations of Liquids, M. P. Allen and D. J. Tildesley, Oxford University Press, New York (1997),

as well as a few longer-standing texts in areas covered in this work:

Statistical Mechanics, D. A. McQuarrie, Harper and Row, New York (1977),
Quantum Chemistry, H. Eyring, J. Walter, and G. E. Kimball, John Wiley, New York (1944),
Introduction to Quantum Mechanics, L. Pauling and E. B. Wilson, Dover, New York (1963),
Molecular Quantum Mechanics, 3rd edn., P. W. Atkins and R. S. Friedman, Oxford University Press, New York (1997),
Modern Quantum Chemistry, A. Szabo and N. S. Ostlund, McGraw-Hill, New York (1989).

Because the science of theoretical chemistry makes much use of high-speed computers, it is essential that we appreciate to what extent the computer revolution has impacted this field. Primarily, the advent of modern computers has revolutionized the range of problems to which theoretical chemistry can be applied. Before this revolution, the classical Newton or quantum Schrödinger equations in terms of which theory expresses the behavior of atoms and molecules simply could not be solved for any but the simplest species, and then often only by making rather crude approximations. However, present-day computers, which routinely perform 10^9operations per second, have 10^9 bytes of memory and 50 times this much hard disk storage, have made it possible to solve these equations for large collections of molecules and for molecules containing hundreds of atoms and electrons. Moreover, the vast improvement in computing power has inspired many scientists to develop better (more accurate and more efficient) approximations to use in solving these equations. Because this text is intended for both an undergraduate and beginning graduate audience and is designed to offer an introduction to the field of theoretical chemistry, it does not devote much time to describing the computer implementation of this subject. Nevertheless, I will attempt to introduce some of the more basic computational aspects of theory especially when doing so will help the reader understand the basic principles. In addition, the TheoryPage web site contains a large number of links to scientists and to commercial software providers that can give the reader more detail about the computational aspects of theoretical chemistry.

Let's now begin the journey that I hope will give the reader a basic understanding of what theoretical chemistry is and how it fits into the amazing broad discipline of modern chemistry.

Acknowledgements

The following figures are reprinted with the permission of Oxford University Press.

(A) Figures 13.12, 13.17, 14.50, 13.22, 17.11, 17.12, 16.32, 13.23, 14.46, 14.47, 18.4, 16.49, 14.51, 16.15, 15.3, 14.30, 17.9, 18.3, 12.4, 12.7, 14.27, 16.17, 18.1, 22.9, 13.18 & 13.1 from *Physical Chemistry* by P. W. Atkins (Sixth Edition, 1998) © P. W. Atkins, 1998.

(B) Figures 7.18 (p.265), 7.12 (p.256), 11.24 (p.448), 9.9 (p.351), 11.29 (p.452), 2 (p.258), 9.7 (p.350) & 11.8 (p.434) from *Elements of Physical Chemistry* by P. W. Atkins (1992) © P. W. Atkins, 1992.

(C) Figures 12.2, 14.5, 18.5, 18.8, 18.9, 21.7 & 28.3 from *Physical Chemistry* by P. W. Atkins (Fourth edition, 1990) © P. W. Atkins, 1990.

Part I

Background material

In this portion of the text, most of the topics that are appropriate to an undergraduate reader are covered. Many of these subjects are subsequently discussed again in Chapter 5, where a broad perspective of what theoretical chemistry is about is offered. They are treated again in greater detail in Chapters 6–8 where the three main disciplines of theory are covered in depth appropriate to a graduate-student reader.

Chapter 1
The basics of quantum mechanics

1.1 Why quantum mechanics is necessary for describing molecular properties

We know that all molecules are made of atoms which, in turn, contain nuclei and electrons. As I discuss in this introductory section, the equations that govern the motions of electrons and of nuclei are not the familiar Newton equations,

$$\mathbf{F} = m\mathbf{a}, \tag{1.1}$$

but a new set of equations called Schrödinger equations. When scientists first studied the behavior of electrons and nuclei, they tried to interpret their experimental findings in terms of classical Newtonian motions, but such attempts eventually failed. They found that such small light particles behaved in a way that simply is not consistent with the Newton equations. Let me now illustrate some of the experimental data that gave rise to these paradoxes and show you how the scientists of those early times then used these data to suggest new equations that these particles might obey. I want to stress that the Schrödinger equation was not derived but postulated by these scientists. In fact, to date, no one has been able to derive the Schrödinger equation.

From the pioneering work of Bragg on diffraction of x-rays from planes of atoms or ions in crystals, it was known that peaks in the intensity of diffracted x-rays having wavelength λ would occur at scattering angles θ determined by the famous Bragg equation:

$$n\lambda = 2d \sin\theta, \tag{1.2}$$

where d is the spacing between neighboring planes of atoms or ions. These quantities are illustrated in Fig. 1.1. There are many such diffraction peaks, each labeled by a different value of the integer n ($n = 1, 2, 3, \ldots$). The Bragg formula can be derived by considering when two photons, one scattering from the second plane in the figure and the second scattering from the third plane, will undergo constructive interference. This condition is met when the "extra path length"

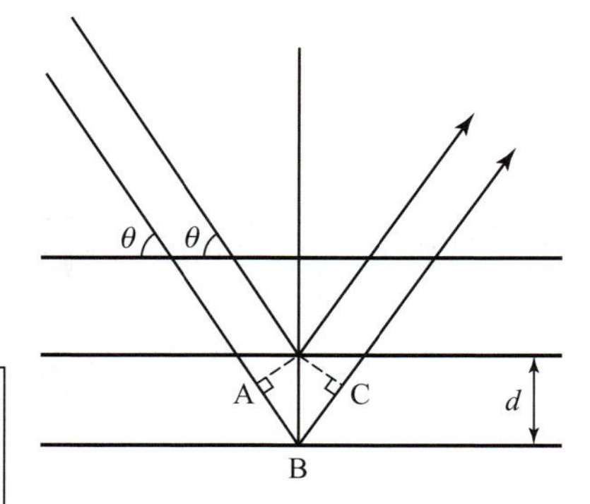

Figure 1.1 Scattering of two beams at angle θ from two planes in a crystal spaced by d.

covered by the second photon (i.e., the length from points A to B to C) is an integer multiple of the wavelength of the photons.

The importance of these x-ray scattering experiments to the study of electrons and nuclei appears in the experiments of Davisson and Germer, in 1927, who scattered electrons of (reasonably) fixed kinetic energy E from metallic crystals. These workers found that plots of the number of scattered electrons as a function of scattering angle θ displayed "peaks" at angles θ that obeyed a Bragg-like equation. The startling thing about this observation is that electrons are particles, yet the Bragg equation is based on the properties of waves. An important observation derived from the Davisson–Germer experiments was that the scattering angles θ observed for electrons of kinetic energy E could be fit to the Bragg $n\lambda = 2d \sin\theta$ equation if a wavelength were ascribed to these electrons that was defined by

$$\lambda = h/(2m_e E)^{1/2}, \tag{1.3}$$

where m_e is the mass of the electron and h is the constant introduced by Max Planck and Albert Einstein in the early 1900s to relate a photon's energy E to its frequency ν via $E = h\nu$. These amazing findings were among the earliest to suggest that electrons, which had always been viewed as particles, might have some properties usually ascribed to waves. That is, as de Broglie suggested in 1925, an electron seems to have a wavelength inversely related to its momentum, and to display wave-type diffraction. I should mention that analogous diffraction was also observed when other small light particles (e.g., protons, neutrons, nuclei, and small atomic ions) were scattered from crystal planes. In all such cases, Bragg-like diffraction is observed and the Bragg equation is found to govern the scattering angles if one assigns a wavelength to the scattering particle according to

$$\lambda = h/(2mE)^{1/2}, \tag{1.4}$$

where m is the mass of the scattered particle and h is Planck's constant (6.62×10^{-27} erg s).

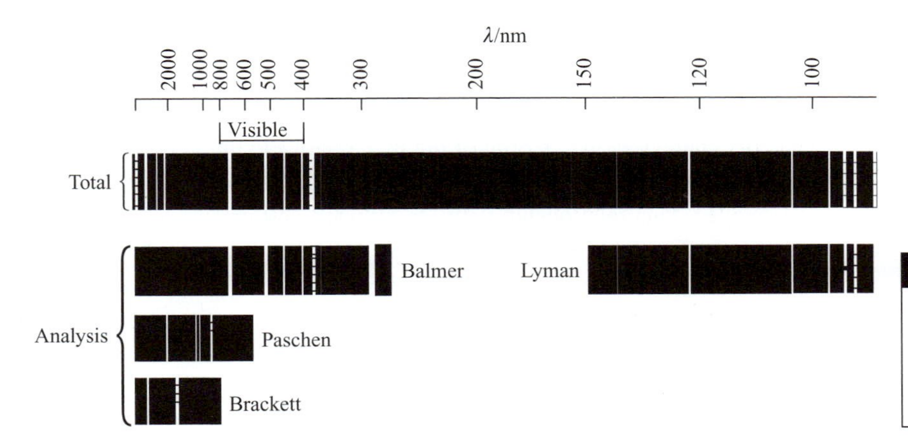

Figure 1.2 Emission spectrum of atomic hydrogen with some lines repeated below to illustrate the series to which they belong.

The observation that electrons and other small light particles display wave-like behavior was important because these particles are what all atoms and molecules are made of. So, if we want to fully understand the motions and behavior of molecules, we must be sure that we can adequately describe such properties for their constituents. Because the classical Newton equations do not contain factors that suggest wave properties for electrons or nuclei moving freely in space, the above behaviors presented significant challenges.

Another problem that arose in early studies of atoms and molecules resulted from the study of the photons emitted from atoms and ions that had been heated or otherwise excited (e.g., by electric discharge). It was found that each kind of atom (i.e., H or C or O) emitted photons whose frequencies ν were of very characteristic values. An example of such emission spectra is shown in Fig. 1.2 for hydrogen atoms. In the top panel, we see all of the lines emitted with their wavelengths indicated in nanometers. The other panels show how these lines have been analyzed (by scientists whose names are associated) into patterns that relate to the specific energy levels between which transitions occur to emit the corresponding photons.

In the early attempts to rationalize such spectra in terms of electronic motions, one described an electron as moving about the atomic nuclei in circular orbits such as shown in Fig. 1.3. A circular orbit was thought to be stable when the outward centrifugal force characterized by radius r and speed v ($m_e v^2/r$) on the electron perfectly counterbalanced the inward attractive Coulomb force (Ze^2/r^2) exerted by the nucleus of charge Z:

$$m_e v^2/r = Ze^2/r^2. \tag{1.5}$$

This equation, in turn, allows one to relate the kinetic energy $\frac{1}{2}m_e v^2$ to the Coulombic energy Ze^2/r, and thus to express the total energy E of an orbit in

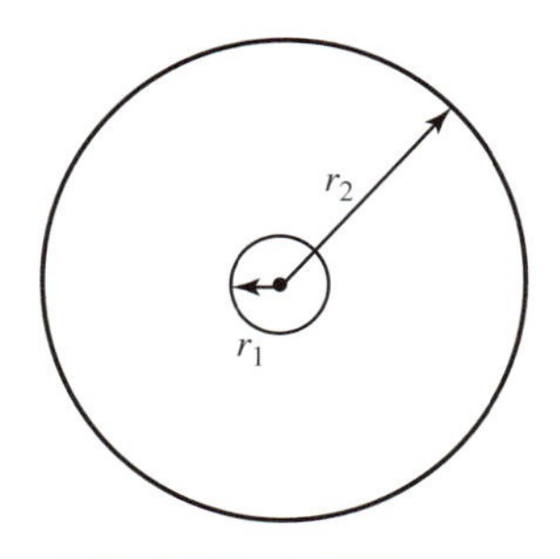

Figure 1.3
Characterization of small and large stable orbits of radii r_1 and r_2 for an electron moving around a nucleus.

terms of the radius of the orbit:

$$E = \frac{1}{2} m_e v^2 - Ze^2/r = \frac{-1}{2} Ze^2/r. \tag{1.6}$$

The energy characterizing an orbit of radius r, relative to the $E = 0$ reference of energy at $r \to \infty$, becomes more and more negative (i.e., lower and lower) as r becomes smaller. This relationship between outward and inward forces allows one to conclude that the electron should move faster as it moves closer to the nucleus since $v^2 = Ze^2/(rm_e)$. However, nowhere in this model is a concept that relates to the experimental fact that each atom emits only certain kinds of photons. It was believed that photon emission occurred when an electron moving in a larger circular orbit lost energy and moved to a smaller circular orbit. However, the Newtonian dynamics that produced the above equation would allow orbits of any radius, and hence any energy, to be followed. Thus, it would appear that the electron should be able to emit photons of any energy as it moved from orbit to orbit.

The breakthrough that allowed scientists such as Niels Bohr to apply the circular-orbit model to the observed spectral data involved first introducing the idea that the electron has a wavelength and that this wavelength λ is related to its momentum by the de Broglie equation $\lambda = h/p$. The key step in the Bohr model was to also specify that the radius of the circular orbit be such that the circumference of the circle $2\pi r$ equal an integer (n) multiple of the wavelength λ. Only in this way will the electron's wave experience constructive interference as the electron orbits the nucleus. Thus, the Bohr relationship that is analogous to the Bragg equation that determines at what angles constructive interference can occur is

$$2\pi r = n\lambda. \tag{1.7}$$

Both this equation and the analogous Bragg equation are illustrations of what we call boundary conditions; they are extra conditions placed on the wavelength to produce some desired character in the resultant wave (in these cases, constructive interference). Of course, there remains the question of why one must impose these extra conditions when the Newtonian dynamics do not require them. The resolution of this paradox is one of the things that quantum mechanics does.

Returning to the above analysis and using $\lambda = h/p = h/(mv)$, $2\pi r = n\lambda$, as well as the force-balance equation $m_e v^2/r = Ze^2/r^2$, one can then solve for the radii that stable Bohr orbits obey:

$$r = (nh/2\pi)^2/(m_e Ze^2) \tag{1.8}$$

and, in turn, for the velocities of electrons in these orbits,

$$v = Ze^2/(nh/2\pi). \tag{1.9}$$

These two results then allow one to express the sum of the kinetic ($\frac{1}{2}m_e v^2$) and Coulomb potential ($-Ze^2/r$) energies as

$$E = -\frac{1}{2}m_e Z^2 e^4/(nh/2\pi)^2. \quad (1.10)$$

Just as in the Bragg diffraction result, which specified at what angles special high intensities occurred in the scattering, there are many stable Bohr orbits, each labeled by a value of the integer n. Those with small n have small radii, high velocities and more negative total energies (n.b., the reference zero of energy corresponds to the electron at $r = \infty$, and with $v = 0$). So, it is the result that only certain orbits are "allowed" that causes only certain energies to occur and thus only certain energies to be observed in the emitted photons.

It turned out that the Bohr formula for the energy levels (labeled by n) of an electron moving about a nucleus could be used to explain the discrete line emission spectra of all one-electron atoms and ions (i.e., H, He^+, Li^{+2}, etc.) to very high precision. In such an interpretation of the experimental data, one claims that a photon of energy

$$h\nu = R\left(1/n_f^2 - 1/n_i^2\right) \quad (1.11)$$

is emitted when the atom or ion undergoes a transition from an orbit having quantum number n_i to a lower-energy orbit having n_f. Here the symbol R is used to denote the following collection of factors:

$$R = \frac{1}{2}m_e Z^2 e^4/(h/2\pi)^2. \quad (1.12)$$

The Bohr formula for energy levels did not agree as well with the observed pattern of emission spectra for species containing more than a single electron. However, it does give a reasonable fit, for example, to the Na atom spectra if one examines only transitions involving the single valence electron. The primary reason for the breakdown of the Bohr formula is the neglect of electron–electron Coulomb repulsions in its derivation. Nevertheless, the success of this model made it clear that discrete emission spectra could only be explained by introducing the concept that not all orbits were "allowed". Only special orbits that obeyed a constructive-interference condition were really accessible to the electron's motions. This idea that not all energies were allowed, but only certain "quantized" energies could occur was essential to achieving even a qualitative sense of agreement with the experimental fact that emission spectra were discrete.

In summary, two experimental observations on the behavior of electrons that were crucial to the abandonment of Newtonian dynamics were the observations of electron diffraction and of discrete emission spectra. Both of these findings seem to suggest that electrons have some wave characteristics and that these waves have only certain allowed (i.e., quantized) wavelengths.

So, now we have some idea why the Newton equations fail to account for the dynamical motions of light and small particles such as electrons and nuclei. We see that extra conditions (e.g., the Bragg condition or constraints on the de Broglie wavelength) could be imposed to achieve some degree of agreement with experimental observation. However, we still are left wondering what the equations are that can be applied to properly describe such motions and why the extra conditions are needed. It turns out that a new kind of equation based on combining wave and particle properties needed to be developed to address such issues. These are the so-called Schrödinger equations to which we now turn our attention.

As I said earlier, no one has yet shown that the Schrödiger equation follows deductively from some more fundamental theory. That is, scientists did not derive this equation; they postulated it. Some idea of how the scientists of that era "dreamed up" the Schrödinger equation can be had by examining the time and spatial dependence that characterizes so-called traveling waves. It should be noted that the people who worked on these problems knew a great deal about waves (e.g., sound waves and water waves) and the equations they obeyed. Moreover, they knew that waves could sometimes display the characteristic of quantized wavelengths or frequencies (e.g., fundamentals and overtones in sound waves). They knew, for example, that waves in one dimension that are constrained at two points (e.g., a violin string held fixed at two ends) undergo oscillatory motion in space and time with characteristic frequencies and wavelengths. For example, the motion of the violin string just mentioned can be described as having an amplitude $A(x, t)$ at a position x along its length at time t given by

$$A(x, t) = A(x, 0)\cos(2\pi\nu t), \tag{1.13}$$

where ν is its oscillation frequency. The amplitude's spatial dependence also has a sinusoidal dependence given by

$$A(x, 0) = A\sin(2\pi x/\lambda), \tag{1.14}$$

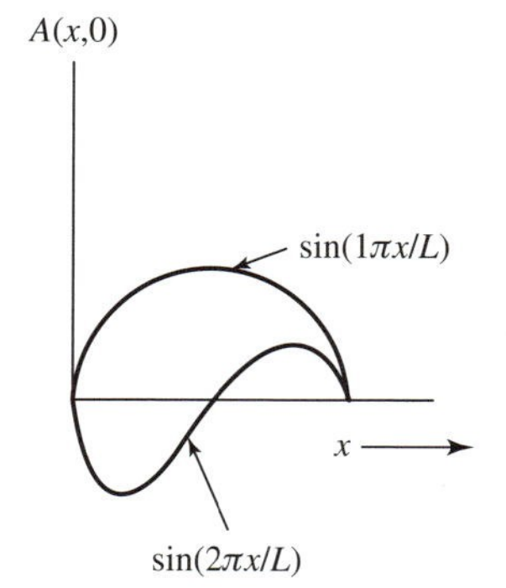

Figure 1.4 Fundamental and first overtone notes of a violin string.

where λ is the crest-to-crest length of the wave. Two examples of such waves in one dimension are shown in Fig. 1.4. In these cases, the string is fixed at $x = 0$ and at $x = L$, so the wavelengths belonging to the two waves shown are $\lambda = 2L$ and $\lambda = L$. If the violin string were not clamped at $x = L$, the waves could have any value of λ. However, because the string is attached at $x = L$, the allowed wavelengths are quantized to obey

$$\lambda = L/n, \tag{1.15}$$

where $n = 1, 2, 3, 4, \ldots$ The equation that such waves obey, called the wave equation, reads

$$\frac{d^2 A(x, t)}{dt^2} = c^2 \frac{d^2 A}{dx^2}, \tag{1.16}$$

where c is the speed at which the wave travels. This speed depends on the composition of the material from which the violin string is made. Using the earlier expressions for the x- and t-dependences of the wave, $A(x, t)$, we find that the wave's frequency and wavelength are related by the so-called dispersion equation:

$$\nu^2 = (c/\lambda)^2, \tag{1.17}$$

or

$$c = \lambda \nu. \tag{1.18}$$

This relationship implies, for example, that an instrument string made of a very stiff material (large c) will produce a higher frequency tone for a given wavelength (i.e., a given value of n) than will a string made of a softer material (smaller c).

For waves moving on the surface of, for example, a rectangular two-dimensional surface of lengths L_x and L_y, one finds

$$A(x, y, t) = \sin(n_x \pi x/L_x)\sin(n_y \pi y/L_y)\cos(2\pi \nu t). \tag{1.19}$$

Hence, the waves are quantized in two dimensions because their wavelengths must be constrained to cause $A(x, y, t)$ to vanish at $x = 0$ and $x = L_x$ as well as at $y = 0$ and $y = L_y$ for all times t. Let us now return to the issue of waves that describe electrons moving.

The pioneers of quantum mechanics examined functional forms similar to those shown above. For example, forms such as $A = \exp[\pm 2\pi \mathrm{i}(\nu t - x/\lambda)]$ were considered because they correspond to periodic waves that evolve in x and t under no external x- or t-dependent forces. Noticing that

$$\frac{d^2 A}{dx^2} = -\left(\frac{2\pi}{\lambda}\right)^2 A \tag{1.20}$$

and using the de Broglie hypothesis $\lambda = h/p$ in the above equation, one finds

$$\frac{d^2 A}{dx^2} = -p^2\left(\frac{2\pi}{h}\right)^2 A. \tag{1.21}$$

If A is supposed to relate to the motion of a particle of momentum p under no external forces (since the waveform corresponds to this case), p^2 can be related to the energy E of the particle by $E = p^2/2m$. So, the equation for A can be rewritten as

$$\frac{d^2 A}{dx^2} = -2mE\left(\frac{2\pi}{h}\right)^2 A, \tag{1.22}$$

or, alternatively,

$$-\left(\frac{h}{2\pi}\right)^2 \frac{d^2 A}{dx^2} = EA. \tag{1.23}$$

Returning to the time-dependence of $A(x, t)$ and using $\nu = E/h$, one can also show that

$$i\left(\frac{h}{2\pi}\right)\frac{dA}{dt} = EA, \tag{1.24}$$

which, using the first result, suggests that

$$i\left(\frac{h}{2\pi}\right)\frac{dA}{dt} = -\left(\frac{h}{2\pi}\right)^2\frac{d^2A}{dx^2}. \tag{1.25}$$

This is a primitive form of the Schrödinger equation that we will address in much more detail below. Briefly, what is important to keep in mind is that the use of the de Broglie and Planck/Einstein connections ($\lambda = h/p$ and $E = h\nu$), both of which involve the constant h, produces suggestive connections between

$$i\left(\frac{h}{2\pi}\right)\frac{d}{dt} \quad \text{and} \quad E \tag{1.26}$$

and between

$$p^2 \quad \text{and} \quad -\left(\frac{h}{2\pi}\right)^2\frac{d^2}{dx^2} \tag{1.27}$$

or, alternatively, between

$$p \quad \text{and} \quad -i\left(\frac{h}{2\pi}\right)\frac{d}{dx}. \tag{1.28}$$

These connections between physical properties (energy E and momentum p) and differential operators are some of the unusual features of quantum mechanics.

The above discussion about waves and quantized wavelengths as well as the observations about the wave equation and differential operators are not meant to provide or even suggest a derivation of the Schrödinger equation. Again the scientists who invented quantum mechanics did not derive its working equations. Instead, the equations and rules of quantum mechanics have been postulated and designed to be consistent with laboratory observations. My students often find this to be disconcerting because they are hoping and searching for an underlying fundamental basis from which the basic laws of quantum mechanics follow logically. I try to remind them that this is not how theory works. Instead, one uses experimental observation to postulate a rule or equation or theory, and one then tests the theory by making predictions that can be tested by further experiments. If the theory fails, it must be "refined", and this process continues until one has a better and better theory. In this sense, quantum mechanics, with all of its unusual mathematical constructs and rules, should be viewed as arising from the imaginations of scientists who tried to invent a theory that was consistent with experimental data and which could be used to predict things that could then be tested in the laboratory. Thus far, this theory has proven reliable, but, of course, we are always searching for a "new and improved" theory that describes how small light particles move.

If it helps you to be more accepting of quantum theory, I should point out that the quantum description of particles will reduce to the classical Newton description under certain circumstances. In particular, when treating heavy particles (e.g., macroscopic masses and even heavier atoms), it is often possible to use Newton dynamics. Briefly, we will discuss in more detail how the quantum and classical dynamics sometimes coincide (in which case one is free to use the simpler Newton dynamics). So, let us now move on to look at this strange Schrödinger equation that we have been digressing about for so long.

1.2 The Schrödinger equation and its components

It has been well established that electrons moving in atoms and molecules do not obey the classical Newton equations of motion. People long ago tried to treat electronic motion classically, and found that features observed clearly in experimental measurements simply were not consistent with such a treatment. Attempts were made to supplement the classical equations with conditions that could be used to rationalize such observations. For example, early workers required that the angular momentum $\mathbf{L} = \mathbf{r} \times \mathbf{p}$ be allowed to assume only integer multiples of $h/2\pi$ (which is often abbreviated as $\hbar$), which can be shown to be equivalent to the Bohr postulate $n\lambda = 2\pi r$. However, until scientists realized that a new set of laws, those of quantum mechanics, applied to light microscopic particles, a wide gulf existed between laboratory observations of molecule-level phenomena and the equations used to describe such behavior.

Quantum mechanics is cast in a language that is not familiar to most students of chemistry who are examining the subject for the first time. Its mathematical content and how it relates to experimental measurements both require a great deal of effort to master. With these thoughts in mind, I have organized this material in a manner that first provides a brief introduction to the two primary constructs of quantum mechanics – operators and wave functions that obey a Schrödinger equation. Next, I demonstrate the application of these constructs to several chemically relevant model problems. By learning the solutions of the Schrödinger equation for a few model systems, the student can better appreciate the treatment of the fundamental postulates of quantum mechanics as well as their relation to experimental measurement for which the wave functions of the known model problems offer important interpretations.

1.2.1 Operators

Each physically measurable quantity has a corresponding operator. The eigenvalues of the operator tell the only values of the corresponding physical property that can be observed.

Any experimentally measurable physical quantity F (e.g., energy, dipole moment, orbital angular momentum, spin angular momentum, linear momentum, kinetic energy) has a classical mechanical expression in terms of the Cartesian positions $\{q_i\}$ and momenta $\{p_i\}$ of the particles that comprise the system of interest. Each such classical expression is assigned a corresponding quantum mechanical operator **F** formed by replacing the $\{p_i\}$ in the classical form by the differential operator $-i\hbar\,\partial/\partial q_j$ and leaving the coordinates q_j that appear in F untouched. For example, the classical kinetic energy of N particles (with masses m_l) moving in a potential field containing both quadratic and linear coordinate-dependence can be written as

$$F = \sum_{l=1,N} \left[p_l^2/2m_l + 1/2k\left(q_l - q_l^0\right)^2 + L\left(q_l - q_l^0\right) \right]. \tag{1.29}$$

The quantum mechanical operator associated with this F is

$$\mathbf{F} = \sum_{l=1,N} \left[\frac{-\hbar^2}{2m_l}\frac{\partial^2}{\partial q_l^2} + \frac{1}{2}k\left(q_l - q_l^0\right)^2 + \mathrm{L}\left(q_l - q_l^0\right) \right]. \tag{1.30}$$

Such an operator would occur when, for example, one describes the sum of the kinetic energies of a collection of particles (the $\sum_{l=1,N}(p_l^2/2m_l)$ term), plus the sum of "Hooke's Law" parabolic potentials (the $1/2\sum_{l=1,N} k(q_l - q_l^0)^2$), and (the last term in F) the interactions of the particles with an externally applied field whose potential energy varies linearly as the particles move away from their equilibrium positions $\{q_l^0\}$.

Let us try more examples. The sum of the z-components of angular momenta (recall that vector angular momentum **L** is defined as $\mathbf{L} = \mathbf{r} \times \mathbf{p}$) of a collection of N particles has the following classical expression:

$$F = \sum_{j=1,N} (x_j p_{yj} - y_j p_{xj}), \tag{1.31}$$

and the corresponding operator is

$$\mathbf{F} = -i\hbar \sum_{j=1,N} \left(x_j \frac{\partial}{\partial y_j} - y_j \frac{\partial}{\partial x_j} \right). \tag{1.32}$$

If one transforms these Cartesian coordinates and derivatives into polar coordinates, the above expression reduces to

$$\mathbf{F} = -i\hbar \sum_{j=1,N} \frac{\partial}{\partial \phi_j}. \tag{1.33}$$

The x-component of the dipole moment for a collection of N particles has a classical form of

$$F = \sum_{j=1,N} Z_j e x_j, \tag{1.34}$$

for which the quantum operator is

$$\mathbf{F} = \sum_{j=1,N} Z_j e x_j, \tag{1.35}$$

where $Z_j e$ is the charge on the jth particle. Notice that in this case, classical and quantum forms are identical because F contains no momentum operators.

The mapping from F to $\mathbf{F}$ is straightforward only in terms of Cartesian coordinates. To map a classical function F, given in terms of curvilinear coordinates (even if they are orthogonal), into its quantum operator is not at all straightforward. The mapping can always be done in terms of Cartesian coordinates after which a transformation of the resulting coordinates and differential operators to a curvilinear system can be performed.

The relationship of these quantum mechanical operators to experimental measurement lies in the eigenvalues of the quantum operators. Each such operator has a corresponding eigenvalue equation

$$\mathbf{F}\chi_j = \alpha_j \chi_j \tag{1.36}$$

in which the χ_j are called eigenfunctions and the (scalar numbers) α_j are called eigenvalues. All such eigenvalue equations are posed in terms of a given operator ($\mathbf{F}$ in this case) and those functions $\{\chi_j\}$ that $\mathbf{F}$ acts on to produce the function back again but multiplied by a constant (the eigenvalue). Because the operator $\mathbf{F}$ usually contains differential operators (coming from the momentum), these equations are differential equations. Their solutions χ_j depend on the coordinates that $\mathbf{F}$ contains as differential operators. An example will help clarify these points. The differential operator d/dy acts on what functions (of y) to generate the same function back again but multiplied by a constant? The answer is functions of the form $\exp(ay)$ since

$$\frac{d(\exp(ay))}{dy} = a\exp(ay). \tag{1.37}$$

So, we say that $\exp(ay)$ is an eigenfunction of d/dy and a is the corresponding eigenvalue.

As I will discuss in more detail shortly, the eigenvalues of the operator $\mathbf{F}$ tell us the *only* values of the physical property corresponding to the operator $\mathbf{F}$ that can be observed in a laboratory measurement. Some $\mathbf{F}$ operators that we encounter possess eigenvalues that are discrete or quantized. For such properties, laboratory measurement will result in only those discrete values. Other $\mathbf{F}$ operators have eigenvalues that can take on a continuous range of values; for these properties, laboratory measurement can give any value in this continuous range.

1.2.2 Wave functions

The eigenfunctions of a quantum mechanical operator depend on the coordinates upon which the operator acts. The particular operator that corresponds to the total energy of the system is called the Hamiltonian operator. The eigenfunctions of this particular operator are called wave functions.

A special case of an operator corresponding to a physically measurable quantity is the Hamiltonian operator H that relates to the total energy of the system. The energy eigenstates of the system Ψ are functions of the coordinates $\{q_j\}$ that H depends on and of time t. The function $|\Psi(q_j, t)|^2 = \Psi^*\Psi$ gives the probability density for observing the coordinates at the values q_j at time t. For a many-particle system such as the H_2O molecule, the wave function depends on many coordinates. For H_2O, it depends on the x, y, and z (or r, θ, and ϕ) coordinates of the ten electrons and the x, y, and z (or r, θ, and ϕ) coordinates of the oxygen nucleus and of the two protons; a total of 39 coordinates appear in Ψ.

In classical mechanics, the coordinates q_j and their corresponding momenta p_j are functions of time. The state of the system is then described by specifying $q_j(t)$ and $p_j(t)$. In quantum mechanics, the concept that q_j is known as a function of time is replaced by the concept of the probability density for finding q_j at a particular value at a particular time $|\Psi(q_j, t)|^2$. Knowledge of the corresponding momenta as functions of time is also relinquished in quantum mechanics; again, only knowledge of the probability density for finding p_j with any particular value at a particular time t remains.

The Hamiltonian eigenstates are especially important in chemistry because many of the tools that chemists use to study molecules probe the energy states of the molecule. For example, most spectroscopic methods are designed to determine which energy state a molecule is in. However, there are other experimental methods that measure other properties (e.g., the z-component of angular momentum or the total angular momentum).

As stated earlier, if the state of some molecular system is characterized by a wave function Ψ that happens to be an eigenfunction of a quantum mechanical operator **F**, one can immediately say something about what the outcome will be if the physical property F corresponding to the operator **F** is measured. In particular, since

$$\mathbf{F}\chi_j = \lambda_j \chi_j, \tag{1.38}$$

where λ_j is one of the eigenvalues of **F**, we know that the value λ_j will be observed if the property F is measured while the molecule is described by the wave function $\Psi = \chi_j$. In fact, once a measurement of a physical quantity F has been carried out and a particular eigenvalue λ_j has been observed, the system's wave function Ψ becomes the eigenfunction χ_j that corresponds to that eigenvalue. That is, the act of making the measurement causes the system's wave function to become the eigenfunction of the property that was measured.

What happens if some other property G, whose quantum mechanical operator is $\mathbf{G}$ is measured in such a case? We know from what was said earlier that some eigenvalue μ_k of the operator $\mathbf{G}$ will be observed in the measurement. But, will the molecule's wave function remain, after G is measured, the eigenfunction of F, or will the measurement of G cause Ψ to be altered in a way that makes the molecule's state no longer an eigenfunction of F? It turns out that if the two operators $\mathbf{F}$ and $\mathbf{G}$ obey the condition

$$\mathbf{F}\,\mathbf{G} = \mathbf{G}\,\mathbf{F}, \tag{1.39}$$

then, when the property G is measured, the wave function $\Psi = \chi_j$ will remain unchanged. This property, that the order of application of the two operators does not matter, is called commutation; that is, we say the two operators commute if they obey this property. Let us see how this property leads to the conclusion about Ψ remaining unchanged if the two operators commute. In particular, we apply the $\mathbf{G}$ operator to the above eigenvalue equation:

$$\mathbf{G}\,\mathbf{F}\chi_j = \mathbf{G}\lambda_j\chi_j. \tag{1.40}$$

Next, we use the commutation to re-write the left-hand side of this equation, and use the fact that λ_j is a scalar number to thus obtain

$$\mathbf{F}\,\mathbf{G}\chi_j = \lambda_j\mathbf{G}\chi_j. \tag{1.41}$$

So, now we see that $(\mathbf{G}\chi_j)$ itself is an eigenfunction of $\mathbf{F}$ having eigenvalue λ_j. So, unless there are more than one eigenfunctions of $\mathbf{F}$ corresponding to the eigenvalue λ_j (i.e., unless this eigenvalue is degenerate), $\mathbf{G}\chi_j$ must itself be proportional to χ_j. We write this proportionality conclusion as

$$\mathbf{G}\chi_j = \mu_j\chi_j, \tag{1.42}$$

which means that χ_j is also an eigenfunction of $\mathbf{G}$. This, in turn, means that measuring the property G while the system is described by the wave function $\Psi = \chi_j$ does not change the wave function; it remains χ_j.

So, when the operators corresponding to two physical properties commute, once one measures one of the properties (and thus causes the system to be an eigenfunction of that operator), subsequent measurement of the second operator will (if the eigenvalue of the first operator is not degenerate) produce a unique eigenvalue of the second operator and will not change the system wave function.

If the two operators do not commute, one simply can not reach the above conclusions. In such cases, measurement of the property corresponding to the first operator will lead to one of the eigenvalues of that operator and cause the system wave function to become the corresponding eigenfunction. However, subsequent measurement of the second operator will produce an eigenvalue of that operator, but the system wave function will be changed to become an eigenfunction of the second operator and thus no longer the eigenfunction of the first.

1.2.3 The Schrödinger equation

This equation is an eigenvalue equation for the energy or Hamiltonian operator; its eigenvalues provide the only allowed energy levels of the system.

The time-dependent equation

If the Hamiltonian operator contains the time variable explicitly, one must solve the time-dependent Schrödinger equation.

Before moving deeper into understanding what quantum mechanics "means", it is useful to learn how the wave functions Ψ are found by applying the basic equation of quantum mechanics, the Schrödinger equation, to a few exactly soluble model problems. Knowing the solutions to these "easy" yet chemically very relevant models will then facilitate learning more of the details about the structure of quantum mechanics.

The Schrödinger equation is a differential equation depending on time and on all of the spatial coordinates necessary to describe the system at hand (thirty-nine for the H_2O example cited above). It is usually written

$$\mathbf{H}\Psi = i\hbar\,\partial\Psi/\partial t, \tag{1.43}$$

where $\Psi(q_j, t)$ is the unknown wave function and $\mathbf{H}$ is the operator corresponding to the total energy of the system. This operator is called the Hamiltonian and is formed, as stated above, by first writing down the classical mechanical expression for the total energy (kinetic plus potential) in Cartesian coordinates and momenta and then replacing all classical momenta p_j by their quantum mechanical operators $p_j = -i\hbar\,\partial/\partial q_j$.

For the H_2O example used above, the classical mechanical energy of all thirteen particles is

$$E = \sum_i \left\{ \frac{p_i^2}{2m_e} + 1/2 \sum_j \frac{e^2}{r_{i,j}} - \sum_a \frac{Z_a e^2}{r_{i,a}} \right\} + \sum_a \left\{ \frac{p_a^2}{2m_a} + 1/2 \sum_b \frac{Z_a Z_b e^2}{r_{a,b}} \right\}, \tag{1.44}$$

where the indices i and j are used to label the 10 electrons whose 30 Cartesian coordinates are $\{q_i\}$ and a and b label the three nuclei whose charges are denoted $\{Z_a\}$, and whose nine Cartesian coordinates are $\{q_a\}$. The electron and nuclear masses are denoted m_e and $\{m_a\}$, respectively. The corresponding Hamiltonian operator is

$$\mathbf{H} = \sum_i \left\{ -\frac{\hbar^2}{2m_e} \frac{\partial^2}{\partial q_i^2} + \frac{1}{2} \sum_j \frac{e^2}{r_{i,j}} - \sum_a \frac{Z_a e^2}{r_{i,a}} \right\} + \sum_a \left\{ -\frac{\hbar^2}{2m_a} \frac{\partial^2}{\partial q_a^2} + \frac{1}{2} \sum_b \frac{Z_a Z_b e^2}{r_{a,b}} \right\}. \tag{1.45}$$

Notice that $\mathbf{H}$ is a second order differential operator in the space of the 39 Cartesian coordinates that describe the positions of the ten electrons and three nuclei. It is a second order operator because the momenta appear in the kinetic energy as p_j^2 and p_a^2, and the quantum mechanical operator for each momentum $p = -i\hbar\partial/\partial q$ is of first order.The Schrödinger equation for the H_2O example at hand then reads:

$$\sum_i \left\{ -\frac{\hbar^2}{2m_e} \frac{\partial^2}{\partial q_i^2} + \frac{1}{2} \sum_j \frac{e^2}{r_{i,j}} - \sum_a \frac{Z_a e^2}{r_{i,a}} \right\} \Psi + \sum_a \left\{ -\frac{\hbar^2}{2m_a} \frac{\partial^2}{\partial q_a^2} + \frac{1}{2} \sum_b \frac{Z_a Z_b e^2}{r_{a,b}} \right\} \Psi = i\hbar \frac{\partial \Psi}{\partial t}. \tag{1.46}$$

The Hamiltonian in this case contains t nowhere. An example of a case where H does contain t occurs when an oscillating electric field $\mathbf{E}\cos(\omega t)$ along the x-axis interacts with the electrons and nuclei and a term

$$\sum_a Z_a e X_a \mathbf{E} \cos(\omega t) - \sum_j e x_j \mathbf{E} \cos(\omega t) \tag{1.47}$$

is added to the Hamiltonian. Here, X_a and x_j denote the x coordinates of the ath nucleus and the jth electron, respectively.

The time-independent equation

If the Hamiltonian operator does not contain the time variable explicitly, one can solve the time-independent Schrödinger equation.

In cases where the classical energy, and hence the quantum Hamiltonian, do not contain terms that are explicitly time dependent (e.g., interactions with time varying external electric or magnetic fields would add to the above classical energy expression time dependent terms), the separations of variables techniques can be used to reduce the Schrödinger equation to a time-independent equation.

In such cases, $\mathbf{H}$ is not explicitly time dependent, so one can assume that $\Psi(q_j, t)$ is of the form (n.b., this step is an example of the use of the separations of variables method to solve a differential equation)

$$\Psi(q_j, t) = \Psi(q_j) F(t). \tag{1.48}$$

Substituting this “ansatz” into the time-dependent Schrödinger equation gives

$$\Psi(q_j)\, i\hbar\, \partial F/\partial t = F(t)\mathbf{H}\Psi(q_j). \tag{1.49}$$

Dividing by $\Psi(q_j)F(t)$ then gives

$$F^{-1}\,(i\hbar\,\partial F/\partial t) = \Psi^{-1}[\mathbf{H}\Psi(q_j)]. \tag{1.50}$$

Since $F(t)$ is only a function of time t, and $\Psi(q_j)$ is only a function of the spatial coordinates $\{q_j\}$, and because the left-hand and right-hand sides must be equal for all values of t and of $\{q_j\}$, both the left- and right-hand sides must equal a constant. If this constant is called E, the two equations that are embodied in this separated Schrödinger equation read as follows:

$$\mathbf{H}\Psi(q_j) = E\Psi(q_j), \tag{1.51}$$

$$i\hbar\, dF(t)/dt = EF(t). \tag{1.52}$$

The first of these equations is called the time-independent Schrödinger equation; it is a so-called eigenvalue equation in which one is asked to find functions that yield a constant multiple of themselves when acted on by the Hamiltonian operator. Such functions are called eigenfunctions of **H** and the corresponding constants are called eigenvalues of **H**. For example, if **H** were of the form $(-\hbar^2/2M)\partial^2/\partial\phi^2 = \mathbf{H}$, then functions of the form $\exp(im\phi)$ would be eigenfunctions because

$$\left\{-\frac{\hbar^2}{2M}\frac{\partial^2}{\partial\phi^2}\right\}\exp(im\phi) = \left\{\frac{m^2\hbar^2}{2M}\right\}\exp(im\phi). \tag{1.53}$$

In this case, $m^2\hbar^2/2M$ is the eigenvalue. In this example, the Hamiltonian contains the square of an angular momentum operator (recall earlier that we showed the z-component of angular momentum is to equal $-i\hbar\, d/d\phi$).

When the Schrödinger equation can be separated to generate a time-independent equation describing the spatial coordinate dependence of the wave function, the eigenvalue E must be returned to the equation determining $F(t)$ to find the time-dependent part of the wave function. By solving

$$i\hbar\, dF(t)/dt = EF(t) \tag{1.54}$$

once E is known, one obtains

$$F(t) = \exp(-iEt/\hbar), \tag{1.55}$$

and the full wave function can be written as

$$\Psi(q_j, t) = \Psi(q_j)\exp(-iEt/\hbar). \tag{1.56}$$

For the above example, the time dependence is expressed by

$$F(t) = \exp(-it\{m^2\hbar^2/2M\}/\hbar). \tag{1.57}$$

In summary, whenever the Hamiltonian does not depend on time explicitly, one can solve the time-independent Schrödinger equation first and then obtain the time dependence as $\exp(-iEt/\hbar)$ once the energy E is known. In the case of molecular structure theory, it is a quite daunting task even to approximately solve the full Schrödinger equation because it is a partial differential equation depending on all of the coordinates of the electrons and nuclei in the molecule. For this reason, there are various approximations that one usually implements when attempting to study molecular structure using quantum mechanics.

The Born–Oppenheimer approximation

One of the most important approximations relating to applying quantum mechanics to molecules is known as the Born–Oppenheimer (BO) approximation. The basic idea behind this approximation involves realizing that in the full electrons-plus-nuclei Hamiltonian operator introduced above,

$$\mathbf{H} = \sum_i \left\{ -\frac{\hbar^2}{2m_e}\frac{\partial^2}{\partial q_i^2} + \frac{1}{2}\sum_j \frac{e^2}{r_{i,j}} - \sum_a \frac{Z_a e^2}{r_{i,a}} \right\} + \sum_a \left\{ -\frac{\hbar^2}{2m_a}\frac{\partial^2}{\partial q_a^2} + \frac{1}{2}\sum_b \frac{Z_a Z_b e^2}{r_{a,b}} \right\}, \tag{1.58}$$

the time scales with which the electrons and nuclei move are generally quite different. In particular, the heavy nuclei (i.e., even a H nucleus weighs nearly 2000 times what an electron weighs) move (i.e., vibrate and rotate) more slowly than do the lighter electrons. Thus, we expect the electrons to be able to "adjust" their motions to the much more slowly moving nuclei. This observation motivates us to solve the Schrödinger equation for the movement of the electrons in the presence of fixed nuclei as a way to represent the fully adjusted state of the electrons at any fixed positions of the nuclei.

The electronic Hamiltonian that pertains to the motions of the electrons in the presence of so-called clamped nuclei,

$$\mathbf{H} = \sum_i \left\{ -\frac{\hbar^2}{2m_e}\frac{\partial^2}{\partial q_i^2} + \frac{1}{2}\sum_j \frac{e^2}{r_{i,j}} - \sum_a \frac{Z_a e^2}{r_{i,a}} \right\}, \tag{1.59}$$

produces as its eigenvalues, through the equation

$$\mathbf{H}\psi_J(q_j \mid q_a) = E_J(q_a)\psi_J(q_j \mid q_a), \tag{1.60}$$

energies $E_K(q_a)$ that depend on where the nuclei are located (i.e., the $\{q_a\}$ coordinates). As its eigenfunctions, one obtains what are called electronic wave functions $\{\psi_K(q_i \mid q_a)\}$ which also depend on where the nuclei are located. The energies $E_K(q_a)$ are what we usually call potential energy surfaces. An example of such a surface is shown in Fig. 1.5. This surface depends on two geometrical coordinates $\{q_a\}$ and is a plot of one particular eigenvalue $E_J(q_a)$ vs. these two coordinates.

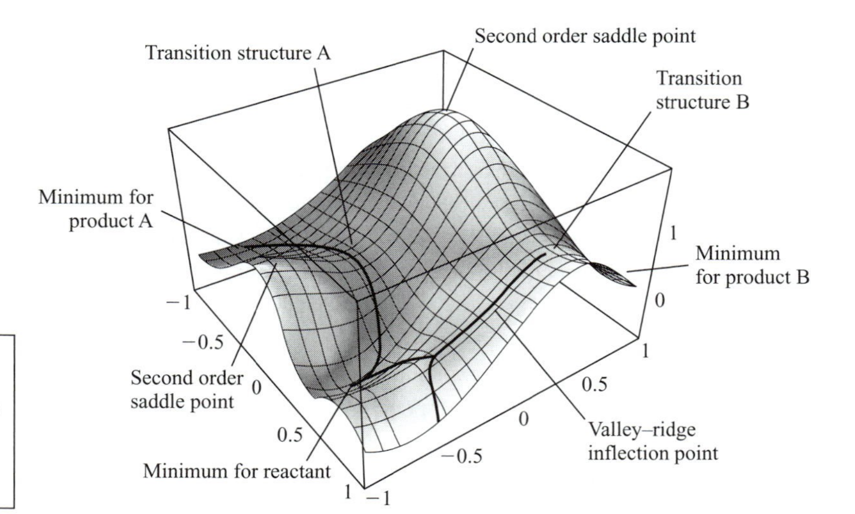

Figure 1.5 Two-dimensional potential energy surface showing local minima, transition states and paths connecting them.

Although this plot has more information on it than we shall discuss now, a few features are worth noting. There appear to be three minima (i.e., points where the derivatives of E_J with respect to both coordinates vanish and where the surface has positive curvature). These points correspond, as we will see toward the end of this introductory material, to geometries of stable molecular structures. The surface also displays two first order saddle points (labeled transition structures A and B) that connect the three minima. These points have zero first derivative of E_J with respect to both coordinates but have one direction of negative curvature. As we will show later, these points describe transition states that play crucial roles in the kinetics of transitions among the three stable geometries.

Keep in mind that Fig. 1.5 shows just one of the E_J surfaces; each molecule has a ground-state surface (i.e., the one that is lowest in energy) as well as an infinite number of excited-state surfaces. Let's now return to our discussion of the BO model and ask what one does once one has such an energy surface in hand.

The motions of the nuclei are subsequently, within the BO model, assumed to obey a Schrödinger equation in which $\sum_a\{-(\hbar^2/2m_a)\partial^2/\partial q_a^2 + 1/2\sum_b Z_a Z_b e^2/r_{a,b}\} + E_K(q_a)$ defines a rotation–vibration Hamiltonian for the particular energy state E_K of interest. The rotational and vibrational energies and wave functions belonging to each electronic state (i.e., for each value of the index K in $E_K(q_a)$) are then found by solving a Schrödinger equation with such a Hamiltonian.

This BO model forms the basis of much of how chemists view molecular structure and molecular spectroscopy. For example, as applied to formaldehyde $H_2C{=}O$, we speak of the singlet ground electronic state (with all electrons spin paired and occupying the lowest energy orbitals) and its vibrational states as well as the $n \rightarrow \pi^*$ and $\pi \rightarrow \pi^*$ electronic states and their vibrational levels. Although much more will be said about these concepts later in this text, the student should be aware of the concepts of electronic energy surfaces (i.e., the $\{E_K(q_a)\}$) and the vibration–rotation states that belong to each such surface.

Having been introduced to the concepts of operators, wave functions, the Hamiltonian and its Schrödinger equation, it is important to now consider several examples of the applications of these concepts. The examples treated below were chosen to provide the reader with valuable experience in solving the Schrödinger equation; they were also chosen because they form the most elementary chemical models of electronic motions in conjugated molecules and in atoms, rotations of linear molecules, and vibrations of chemical bonds.

1.3 Your first application of quantum mechanics – motion of a particle in one dimension

This is a very important problem whose solutions chemists use to model a wide variety of phenomena.

Let's begin by examining the motion of a single particle of mass m in one direction which we will call x while under the influence of a potential denoted $V(x)$. The classical expression for the total energy of such a system is $E = p^2/2m + V(x)$, where p is the momentum of the particle along the x-axis. To focus on specific examples, consider how this particle would move if $V(x)$ were of the forms shown in Fig. 1.6, where the total energy E is denoted by the position of the horizontal line.

1.3.1 Classical probability density

I would like you to imagine what the probability density would be for this particle moving with total energy E and with $V(x)$ varying as the three plots in Fig. 1.6

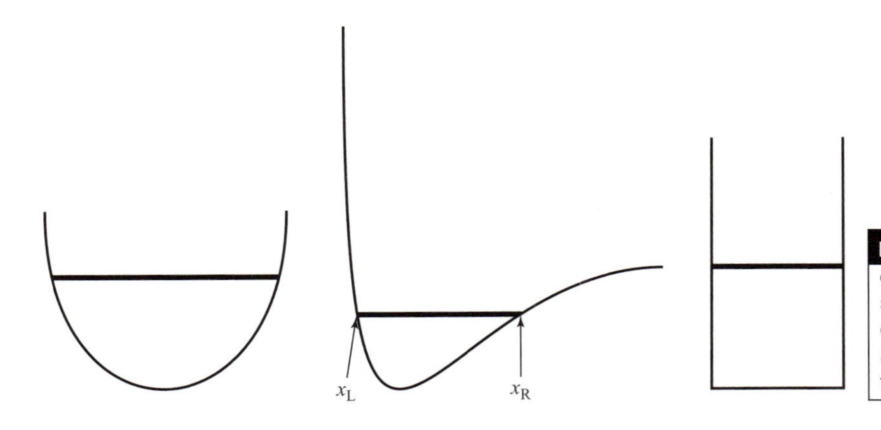

Figure 1.6 Three characteristic potentials showing left and right classical turning points at energies denoted by the horizontal lines.

illustrate. To conceptualize the probability density, imagine the particle to have a blinking lamp attached to it and think of this lamp blinking say 100 times for each time interval it takes for the particle to complete a full transit from its left turning point to its right turning point and back to the former. The turning points x_L and x_R are the positions at which the particle, if it were moving under Newton's laws, would reverse direction (as the momentum changes sign) and turn around. These positions can be found by asking where the momentum goes to zero:

$$0 = p = (2m(E - V(x))^{1/2}. \tag{1.61}$$

These are the positions where all of the energy appears as potential energy $E = V(x)$ and correspond in the above figures to the points where the dark horizontal lines touch the $V(x)$ plots as shown in the central plot.

The probability density at any value of x represents the fraction of time the particle spends at this value of x (i.e., within x and $x + dx$). Think of forming this density by allowing the blinking lamp attached to the particle to shed light on a photographic plate that is exposed to this light for many oscillations of the particle between x_L and x_R. Alternatively, one can express this probability amplitude $P(x)$ by dividing the spatial distance dx by the velocity of the particle at the point x:

$$P(x) = (2m(E - V(x))^{-1/2} m\, dx. \tag{1.62}$$

Because E is constant throughout the particle's motion, $P(x)$ will be small at x values where the particle is moving quickly (i.e., where V is low) and will be high where the particle moves slowly (where V is high). So, the photographic plate will show a bright region where V is high (because the particle moves slowly in such regions) and less brightness where V is low.

The bottom line is that the probability densities anticipated by analyzing the classical Newtonian dynamics of this one particle would appear as the histogram plots shown in Fig. 1.7 illustrate. Where the particle has high kinetic energy (and thus lower $V(x)$), it spends less time and $P(x)$ is small. Where the particle moves

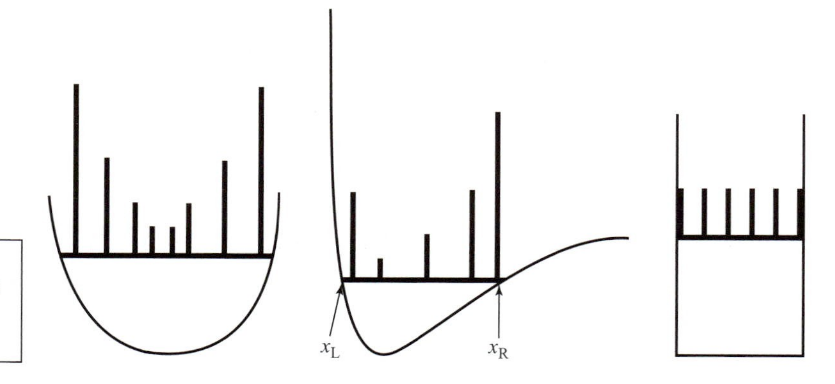

Figure 1.7 Classical probability plots for the three potentials shown in Fig. 1.6.

slowly, it spends more time and $P(x)$ is larger. For the plot on the right, $V(x)$ is constant within the "box", so the speed is constant, hence $P(x)$ is constant for all x values within this one-dimensional box. I ask that you keep these plots in mind because they are very different from what one finds when one solves the Schrödinger equation for this same problem. Also please keep in mind that these plots represent what one expects if the particle were moving according to classical Newtonian dynamics (which we know it is not!).

1.3.2 Quantum treatment

To solve for the quantum mechanical wave functions and energies of this same problem, we first write the Hamiltonian operator as discussed above by replacing p by $-i\hbar d/dx$:

$$H = -\frac{\hbar^2}{2m}\frac{d^2}{dx^2} + V(x). \tag{1.63}$$

We then try to find solutions $\psi(x)$ to $H\psi = E\psi$ that obey certain conditions. These conditions are related to the fact that $|\psi(x)|^2$ is supposed to be the probability density for finding the particle between x and $x + dx$. To keep things as simple as possible, let's focus on the "box" potential V shown in the right side of Fig. 1.7. This potential, expressed as a function of x is: $V(x) = \infty$ for $x < 0$ and for $x > L$; $V(x) = 0$ for x between 0 and L.

The fact that V is infinite for $x < 0$ and for $x > L$, and that the total energy E must be finite, says that ψ must vanish in these two regions ($\psi = 0$ for $x < 0$ and for $x > L$). This condition means that the particle can not access these regions where the potential is infinite. The second condition that we make use of is that $\psi(x)$ must be continuous; this means that the probability of the particle being at x can not be discontinuously related to the probability of it being at a nearby point.

1.3.3 Energies and wave functions

The second order differential equation

$$-\frac{\hbar^2}{2m}\frac{d^2\psi}{dx^2} + V(x)\psi = E\psi \tag{1.64}$$

has two solutions (because it is a second order equation) in the region between $x = 0$ and $x = L$:

$$\psi = \sin(kx) \quad \text{and} \quad \psi = \cos(kx), \quad \text{where } k \text{ is defined as} \quad k = (2mE/\hbar^2)^{1/2}. \tag{1.65}$$

Hence, the most general solution is some combination of these two:

$$\psi = A\sin(kx) + B\cos(kx). \tag{1.66}$$

The fact that ψ must vanish at $x = 0$ (n.b., ψ vanishes for $x < 0$ and is continuous, so it must vanish at the point $x = 0$) means that the weighting amplitude of the $\cos(kx)$ term must vanish because $\cos(kx) = 1$ at $x = 0$. That is,

$$B = 0. \tag{1.67}$$

The amplitude of the $\sin(kx)$ term is not affected by the condition that ψ vanish at $x = 0$, since $\sin(kx)$ itself vanishes at $x = 0$. So, now we know that ψ is really of the form:

$$\psi(x) = A\sin(kx). \tag{1.68}$$

The condition that ψ also vanish at $x = L$ has two possible implications. Either $A = 0$ or k must be such that $\sin(kL) = 0$. The option $A = 0$ would lead to an answer ψ that vanishes at all values of x and thus a probability that vanishes everywhere. This is unacceptable because it would imply that the particle is never observed anywhere.

The other possibility is that $\sin(kL) = 0$. Let's explore this answer because it offers the first example of energy quantization that you have probably encountered. As you know, the sine function vanishes at integral multiples of π. Hence kL must be some multiple of π; let's call the integer n and write $Lk = n\pi$ (using the definition of k) in the form:

$$L(2mE/\hbar^2)^{1/2} = n\pi. \tag{1.69}$$

Solving this equation for the energy E, we obtain:

$$E = n^2\pi^2\hbar^2/(2mL^2). \tag{1.70}$$

This result says that the only energy values that are capable of giving a wave function $\psi(x)$ that will obey the above conditions are these specific E values. In other words, not all energy values are "allowed" in the sense that they can produce ψ functions that are continuous and vanish in regions where $V(x)$ is infinite. If one uses an energy E that is not one of the allowed values and substitutes this E into $\sin(kx)$, the resultant function will not vanish at $x = L$. I hope the solution to this problem reminds you of the violin string that we discussed earlier. Recall that the violin string being tied down at $x = 0$ and at $x = L$ gave rise to quantization of the wavelength just as the conditions that ψ be continuous at $x = 0$ and $x = L$ gave energy quantization.

Substituting $k = n\pi/L$ into $\psi = A\sin(kx)$ gives

$$\psi(x) = A\sin(n\pi x/L). \tag{1.71}$$

The value of A can be found by remembering that $|\Psi|^2$ is supposed to represent the probability density for finding the particle at x. Such probability densities are supposed to be normalized, meaning that their integral over all x values should amount to unity. So, we can find A by requiring that

$$1 = \int |\psi(x)|^2 dx = |A|^2 \int \sin^2(n\pi x/L)\, dx, \quad (1.72)$$

where the integral ranges from $x = 0$ to $x = L$. Looking up the integral of $\sin^2(ax)$ and solving the above equation for the so-called normalization constant A gives

$$A = (2/L)^{1/2} \quad (1.73)$$

and so

$$\psi(x) = (2/L)^{1/2} \sin(n\pi x/L). \quad (1.74)$$

The values that n can take on are $n = 1, 2, 3, \ldots$; the choice $n = 0$ is unacceptable because it would produce a wave function $\psi(x)$ that vanishes at all x.

The full x- and t-dependent wave functions are then given as

$$\Psi(x, t) = \left(\frac{2}{L}\right)^{1/2} \sin\frac{n\pi x}{L} \exp\left[\frac{-itn^2\pi^2\hbar^2}{2mL^2/\hbar}\right]. \quad (1.75)$$

Notice that the spatial probability density $|\Psi(x, t)|^2$ is not dependent on time and is equal to $|\psi(x)|^2$ because the complex exponential disappears when $\Psi^*\Psi$ is formed. This means that the probability of finding the particle at various values of x is time-independent.

Another thing I want you to notice is that, unlike the classical dynamics case, not all energy values E are allowed. In the Newtonian dynamics situation, E could be specified and the particle's momentum at any x value was then determined to within a sign. In contrast, in quantum mechanics one must determine, by solving the Schrödinger equation, what the allowed values of E are. These E values are quantized, meaning that they occur only for discrete values $E = n^2\pi^2\hbar^2/(2mL^2)$ determined by a quantum number n, by the mass of the particle m, and by characteristics of the potential (L in this case).

1.3.4 Probability densities

Let's now look at some of the wave functions $\Psi(x)$ and compare the probability densities $|\Psi(x)|^2$ that they represent to the classical probability densities discussed earlier. The $n = 1$ and $n = 2$ wave functions are shown in the top of Fig. 1.8. The corresponding probability densities are shown below the wave functions in two formats (as x–y plots and shaded plots that could relate to the flashing light way of monitoring the particle's location that we discussed earlier). A more complete set of wave functions (for n ranging from 1 to 7) are shown in Fig. 1.9.

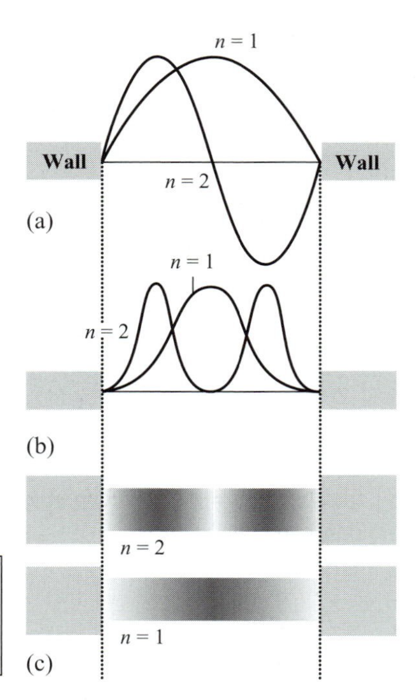

Figure 1.8 The two lowest wave functions and probability densities.

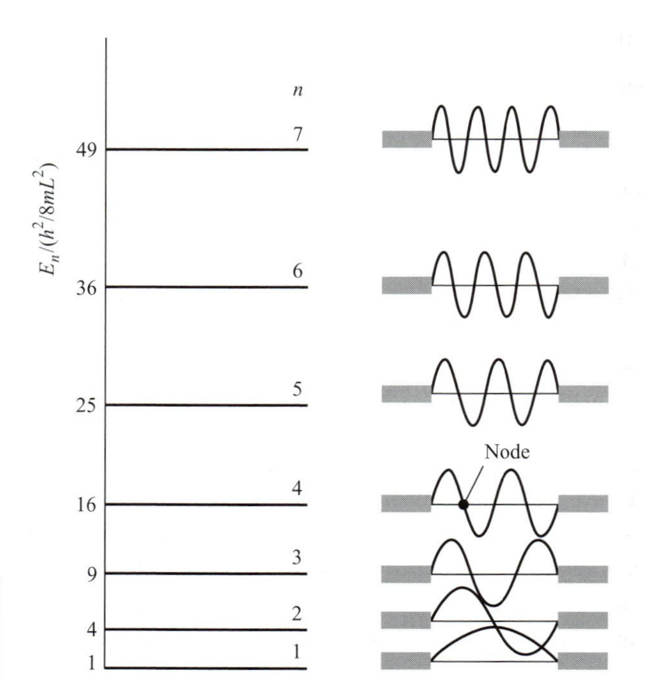

Figure 1.9 Seven lowest wave functions and energies.

Notice that as the quantum number n increases, the energy E also increases (quadratically with n in this case) and the number of nodes in Ψ also increases. Also notice that the probability densities are very different from what we encountered earlier for the classical case. For example, look at the $n = 1$ and $n = 2$ densities and compare them to the classical density illustrated in Fig. 1.10. The classical density is easy to understand because we are familiar with classical dynamics. In this case, we say that $P(x)$ is constant within the box because the fact that $V(x)$ is constant causes the kinetic energy and hence the speed of the particle to remain constant. In contrast, the $n = 1$ quantum wave function's $P(x)$ plot is peaked in the middle of the box and falls to zero at the walls. The $n = 2$ density has two peaks $P(x)$ (one to the left of the box midpoint, and one to the right), a node at the box midpoint, and falls to zero at the walls. One thing that students often ask me is "how does the particle get from being in the left peak to being in the right peak if it has zero chance of ever being at the midpoint where the node is?" The difficulty with this question is that it is posed in a terminology that asks for a classical dynamics answer. That is, by asking "how does the particle get . . ." one is demanding an answer that involves describing its motion (i.e., it moves from here at time t_1 to there at time t_2). Unfortunately, quantum mechanics does not deal with issues such as a particle's trajectory (i.e., where it is at various times) but only with its probabilty of being somewhere (i.e., $|\Psi|^2$). The next section will treat such paradoxical issues even further.

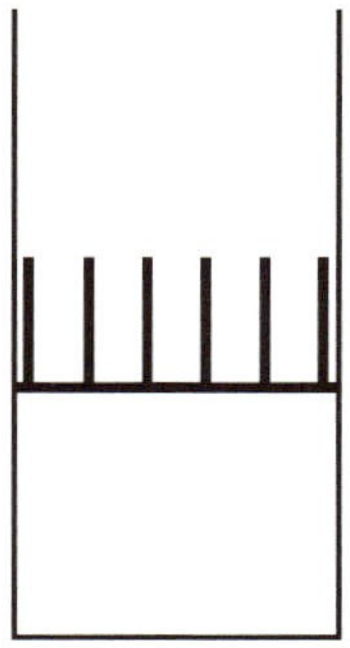

Figure 1.10 Classical probability density for potential shown.

1.3.5 Classical and quantum probability densities

As just noted, it is tempting for most beginning students of quantum mechanics to attempt to interpret the quantum behavior of a particle in classical terms. However, this adventure is full of danger and bound to fail because small, light particles simply do not move according to Newton's laws. To illustrate, let's try to "understand" what kind of (classical) motion would be consistent with the $n = 1$ or $n = 2$ quantum $P(x)$ plots shown in Fig. 1.8. However, as I hope you anticipate, this attempt at gaining classical understanding of a quantum result will not "work" in that it will lead to nonsensical results. My point in leading you to attempt such a classical understanding is to teach you that classical and quantum results are simply different and that you must resist the urge to impose a classical understanding on quantum results.

For the $n = 1$ case, we note that $P(x)$ is highest at the box midpoint and vanishes at $x = 0$ and $x = L$. In a classical mechanics world, this would mean that the particle moves slowly near $x = L/2$ and more quickly near $x = 0$ and $x = L$. Because the particle's total energy E must remain constant as it moves, in regions where it moves slowly, the potential it experiences must be high, and where it moves quickly, V must be small. This analysis (n.b., based on classical concepts) would lead us to conclude that the $n = 1$ $P(x)$ arises from the particle moving in a potential that is high near $x = L/2$ and low as x approaches 0 or L.

A similar analysis of the $n = 2$ $P(x)$ plot would lead us to conclude that the particle for which this is the correct $P(x)$ must experience a potential that is high midway between $x = 0$ and $x = L/2$, high midway between $x = L/2$ and $x = L$, and very low near $x = L/2$ and near $x = 0$ and $x = L$. These conclusions are "crazy" because we know that the potential $V(x)$ for which we solved the Schrödinger equation to generate both of the wave functions (and both probability densities) is constant between $x = 0$ and $x = L$. That is, we know the same $V(x)$ applies to the particle moving in the $n = 1$ and $n = 2$ states, whereas the classical motion analysis offered above suggests that $V(x)$ is different for these two cases.

What is wrong with our attempt to understand the quantum $P(x)$ plots? The mistake we made was in attempting to apply the equations and concepts of classical dynamics to a $P(x)$ plot that did not arise from classical motion. Simply put, one can not ask how the particle is moving (i.e., what is its speed at various positions) when the particle is undergoing quantum dynamics. Most students, when first experiencing quantum wave functions and quantum probabilities, try to think of the particle moving in a classical way that is consistent with the quantum $P(x)$. This attempt to retain a degree of classical understanding of the particle's movement is always met with frustration, as I illustrated with the above example and will illustrate later in other cases.

Continuing with this first example of how one solves the Schrödinger equation and how one thinks of the quantized E values and wave functions Ψ, let me offer a little more optimistic note than offered in the preceding discussion. If we examine the $\Psi(x)$ plot shown in Fig. 1.9 for $n = 7$, and think of the corresponding $P(x) = |\Psi(x)|^2$, we note that the $P(x)$ plot would look something like that shown in Fig. 1.11. It would have seven maxima separated by six nodes. If we were to plot

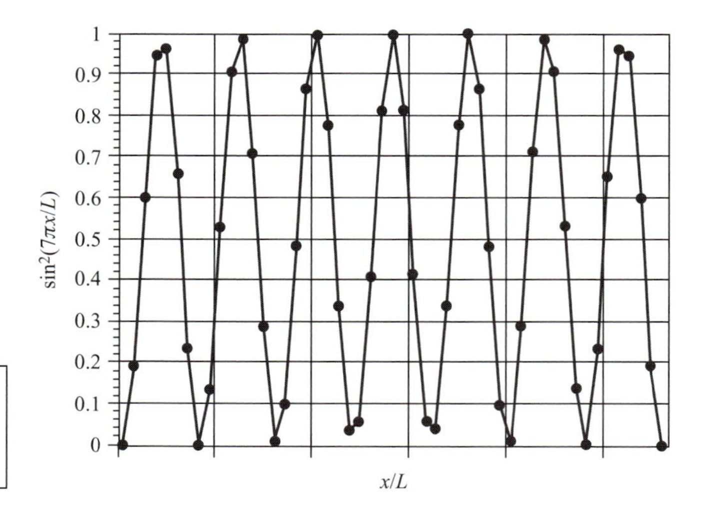

Figure 1.11 Quantum probability density for $n = 7$ showing seven peaks and six nodes.

$|\Psi(x)|^2$ for a very large n value such as $n = 55$, we would find a $P(x)$ plot having 55 maxima separated by 54 nodes, with the maxima separated approximately by distances of $(1/55L)$. Such a plot, when viewed in a "coarse grained" sense (i.e., focusing with somewhat blurred vision on the positions and heights of the maxima) looks very much like the classical $P(x)$ plot in which $P(x)$ is constant for all x. In fact, it is a general result of quantum mechanics that the quantum $P(x)$ distributions for large quantum numbers take on the form of the classical $P(x)$ for the same potential V that was used to solve the Schrödinger equation. It is also true that classical and quantum results agree when one is dealing with heavy particles. For example, a given particle-in-a-box energy $E_n = n^2\hbar^2/(2mL^2)$ would be achieved for a heavier particle at higher n-values than for a lighter particle. Hence, heavier particles, moving with a given energy E, have higher n and thus more classical probability distributions.

We will encounter this so-called quantum–classical correspondence principle again when we examine other model problems. It is an important property of solutions to the Schrödinger equation because it is what allows us to bridge the "gap" between using the Schrödinger equation to treat small, light particles and the Newton equations for macroscopic (big, heavy) systems.

Another thing I would like you to be aware of concerning the solutions ψ and E to this Schrödinger equation is that each pair of wave functions ψ_n and $\psi_{n'}$ belonging to different quantum numbers n and n' (and to different energies) display a property termed orthonormality. This property means that not only are ψ_n and $\psi_{n'}$ each normalized

$$1 = \int |\psi_n|^2\, dx = \int |\psi_{n'}|^2\, dx, \tag{1.76}$$

but they are also orthogonal to each other

$$0 = \int (\psi_n)^* \psi_{n'}\, dx, \tag{1.77}$$

where the complex conjugate * of the first function appears only when the ψ solutions contain imaginary components (you have only seen one such case thus far – the $\exp(im\phi)$ eigenfunctions of the z-component of angular momentum). It is common to write the integrals displaying the normalization and orthogonality conditions in the following so-called Dirac notation:

$$1 = \langle \psi_n \mid \psi_n \rangle \qquad 0 = \langle \psi_n \mid \psi_{n'} \rangle, \tag{1.78}$$

where the $|\rangle$ and $\langle|$ symbols represent ψ and ψ^*, respectively, and putting the two together in the $\langle|\rangle$ construct implies the integration over the variable that ψ depends upon.

The orthogonality condition can be viewed as similar to the condition of two vectors $\mathbf{v}_1$ and $\mathbf{v}_2$ being perpendicular, in which case their scalar (sometimes called "dot") product vanishes $\mathbf{v}_1 \cdot \mathbf{v}_2 = 0$. I want you to keep this property in mind because you will soon see that it is a characteristic not only of these particle-in-a-box wave functions but of all wave functions obtained from any Schrödinger equation.

In fact, the orthogonality property is even broader than the above discussion suggests. It turns out that all quantum mechanical operators formed as discussed earlier (replacing Cartesian momenta p by the corresponding $-i\hbar\partial/\partial q$ operator and leaving all Cartesian coordinates as they are) can be shown to be so-called Hermitian operators. This means that they form Hermitian matrices when they are placed between pairs of functions and the coordinates are integrated over. For example, the matrix representation of an operator **F** when acting on a set of functions denoted $\{\phi_J\}$ is

$$F_{I,J} = \langle\phi_I|\mathbf{F}|\phi_J\rangle = \int \phi_I^* \mathbf{F} \phi_J \, dq. \tag{1.79}$$

For all of the operators formed following the rules stated earlier, one finds that these matrices have the following property:

$$F_{I,J} = F_{J,I}^*, \tag{1.80}$$

which makes the matrices what we call Hermitian. If the functions upon which **F** acts and **F** itself have no imaginary parts (i.e., are real), then the matrices turn out to be symmetric:

$$F_{I,J} = F_{J,I}. \tag{1.81}$$

The importance of the Hermiticity or symmetry of these matrices lies in the fact that it can be shown that such matrices have all real (i.e., not complex) eigenvalues and have eigenvectors that are orthogonal.

So, all quantum mechanical operators, not just the Hamiltonian, have real eigenvalues (this is good since these eigenvalues are what can be measured in any experimental observation of that property) and orthogonal eigenfunctions. It is important to keep these facts in mind because we make use of them many times throughout this text.

1.3.6 Time propagation of wave functions

For a system that exists in an eigenstate $\Psi(x) = (2/L)^{1/2}\sin(n\pi x/L)$ having an energy $E_n = n^2\pi^2\hbar^2/(2mL^2)$, the time-dependent wave function is

$$\Psi(x, t) = \left(\frac{2}{L}\right)^{1/2} \sin\frac{n\pi x}{L} \exp\left(-\frac{itE_n}{\hbar}\right), \tag{1.82}$$

which can be generated by applying the so-called time evolution operator $U(t, 0)$ to the wave function at $t = 0$:

$$\Psi(x, t) = U(t, 0)\Psi(x, 0), \tag{1.83}$$

where an explicit form for $U(t, t')$ is

$$U(t, t') = \exp\left[-\frac{i(t - t')H}{\hbar}\right]. \tag{1.84}$$

The function $\Psi(x, t)$ has a spatial probability density that does not depend on time because

$$\Psi^*(x, t)\Psi(x, t) = \left(\frac{2}{L}\right)\sin^2\left(\frac{n\pi x}{L}\right), \tag{1.85}$$

since $\exp(-itE_n/\hbar)\exp(itE_n/\hbar) = 1$. However, it is possible to prepare systems (even in real laboratory settings) in states that are not single eigenstates; we call such states superposition states. For example, consider a particle moving along the x-axis within the "box" potential but in a state whose wave function at some initial time $t = 0$ is

$$\Psi(x, 0) = 2^{-1/2}\left(\frac{2}{L}\right)^{1/2}\sin\left(\frac{1\pi x}{L}\right) - 2^{-1/2}\left(\frac{2}{L}\right)^{1/2}\sin\left(\frac{2\pi x}{L}\right). \tag{1.86}$$

This is a superposition of the $n = 1$ and $n = 2$ eigenstates. The probability density associated with this function is

$$|\Psi|^2 = \frac{1}{2}\left\{\frac{2}{L}\sin^2\left(\frac{1\pi x}{L}\right) + \frac{2}{L}\sin^2\left(\frac{2\pi x}{L}\right) - 2\left(\frac{2}{L}\right) \times \sin\left(\frac{1\pi x}{L}\right)\sin\left(\frac{2\pi x}{L}\right)\right\}. \tag{1.87}$$

The $n = 1$ and $n = 2$ components, the superposition Ψ, and the probability density at $t = 0|\Psi|^2$ are shown in the first three panels of Fig. 1.12. It should be noted that the probability density associated with this superposition state is not symmetric about the $x = L/2$ midpoint even though the $n = 1$ and $n = 2$ component wave functions and densities are. Such a density describes the particle localized more strongly in the large-x region of the box than in the small-x region.

Now, let's consider the superposition wave function and its density at later times. Applying the time evolution operator $\exp(-itH/\hbar)$ to $\Psi(x, 0)$ generates this time-evolved function at time t:

$$\begin{aligned}\Psi(x, t) &= \exp\left(-\frac{itH}{\hbar}\right)\left\{2^{-1/2}\left(\frac{2}{L}\right)^{1/2}\sin\left(\frac{1\pi x}{L}\right) - 2^{-1/2}\left(\frac{2}{L}\right)^{1/2}\sin\left(\frac{2\pi x}{L}\right)\right\}\\ &= \left[2^{-1/2}\left(\frac{2}{L}\right)^{1/2}\sin\left(\frac{1\pi x}{L}\right)\exp\left(-\frac{itE_1}{\hbar}\right) - 2^{-1/2}\left(\frac{2}{L}\right)^{1/2}\right.\\ &\quad\left.\times \sin\left(\frac{2\pi x}{L}\right)\exp\left(-\frac{itE_2}{\hbar}\right)\right].\end{aligned} \tag{1.88}$$

The spatial probability density associated with this Ψ is

$$\begin{aligned}|\Psi(x, t)|^2 = \frac{1}{2}&\left\{\left(\frac{2}{L}\right)\sin^2\left(\frac{1\pi x}{L}\right) + \left(\frac{2}{L}\right)\sin^2\left(\frac{2\pi x}{L}\right)\right.\\ &\left. - 2\left(\frac{2}{L}\right)\cos\left[(E_2 - E_1)\frac{t}{\hbar}\right]\sin\left(\frac{1\pi x}{L}\right)\sin\left(\frac{2\pi x}{L}\right)\right\}.\end{aligned} \tag{1.89}$$

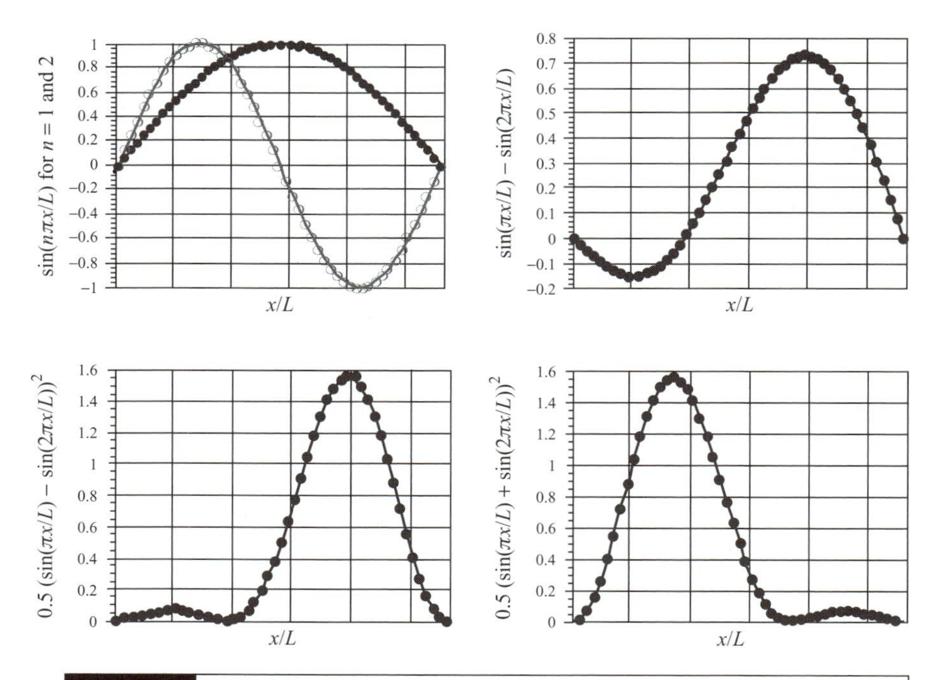

Figure 1.12 The $n = 1$ and $n = 2$ wave functions, their superposition, and the $t = 0$ and time-evolved probability densities of the superposition.

At $t = 0$, this function clearly reduces to that written earlier for $\Psi(x, 0)$. Notice that as time evolves, this density changes because of the $\cos[(E_2 - E_1)t/\hbar]$ factor it contains. In particular, note that as t moves through a period of length $\delta t = \pi\hbar/(E_2 - E_1)$, the cos factor changes sign. That is, for $t = 0$, the cos factor is $+1$; for $t = \pi\hbar/(E_2 - E_1)$, the cos factor is -1; for $t = 2\pi\hbar/(E_2 - E_1)$, it returns to $+1$. The result of this time variation in the cos factor is that $|\Psi|^2$ changes in form from that shown in the bottom left panel of Fig. 1.12 to that shown in the bottom right panel (at $t = \pi\hbar/(E_2 - E_1)$) and then back to the form in the bottom left panel (at $t = 2\pi\hbar/(E_2 - E_1)$). One can interpret this time variation as describing the particle's probability density (not its classical position!), initially localized toward the right side of the box, moving to the left and then back to the right. Of course, this time evolution will continue over more and more cycles as time evolves.

This example illustrates once again the difficulty with attempting to localize particles that are being described by quantum wave functions. For example, a particle that is characterized by the eigenstate $(2/L)^{1/2} \sin(1\pi x/L)$ is more likely to be detected near $x = L/2$ than near $x = 0$ or $x = L$ because the square of this function is large near $x = L/2$. A particle in the state $(2/L)^{1/2} \sin(2\pi x/L)$ is most likely to be found near $x = L/4$ and $x = 3L/4$, but not near $x = 0$, $x = L/2$, or

$x = L$. The issue of how the particle in the latter state moves from being near $x = L/4$ to $x = 3L/4$ is not something quantum mechanics deals with. Quantum mechanics does not allow us to follow the particle's trajectory, which is what we need to know when we ask how it moves from one place to another. Nevertheless, superposition wave functions can offer, to some extent, the opportunity to follow the motion of the particle. For example, the superposition state written above as $2^{-1/2}(2/L)^{1/2}\sin(1\pi x/L) - 2^{-1/2}(2/L)^{1/2}\sin(2\pi x/L)$ has a probability amplitude that changes with time as shown in the figure. Moreover, this amplitude's major peak does move from side to side within the box as time evolves. So, in this case, we can say with what frequency the major peak moves back and forth. In a sense, this allows us to "follow" the particle's movements, but only to the extent that we are satisfied with ascribing its location to the position of the major peak in its probability distribution. That is, we can not really follow its "precise" location, but we can follow the location of where it is very likely to be found. This is an important observation that I hope the student will keep fresh in mind. It is also an important ingredient in modern quantum dynamics in which localized wave packets, similar to superposed eigenstates, are used to detail the position and speed of a particle's main probability density peak.

The above example illustrates how one time-evolves a wave function that can be expressed as a linear combination (i.e., superposition) of eigenstates of the problem at hand. There is a large amount of current effort in the theoretical chemistry community aimed at developing efficient approximations to the $\exp(-itH/\hbar)$ evolution operator that do not require $\Psi(x, 0)$ to be explicitly written as a sum of eigenstates. This is important because, for most systems of direct relevance to molecules, one can not solve for the eigenstates; it is simply too difficult to do so. You can find a significantly more detailed treatment of this subject at the research-level in my TheoryPage web site and my *QMIC* textbook. However, let's spend a little time on a brief introduction to what is involved.

The problem is to express $\exp(-itH/\hbar)\Psi(q_j)$, where $\Psi(q_j)$ is some initial wave function but not an eigenstate, in a manner that does not require one to first find the eigenstates $\{\Psi_J\}$ of H and to expand Ψ in terms of these eigenstates:

$$\Psi = \sum_J C_J \Psi_J \tag{1.90}$$

after which the desired function is written as

$$\exp\left(-\frac{itH}{\hbar}\right)\Psi(q_j) = \sum_J C_J \Psi_J \exp\left(-\frac{itE_J}{\hbar}\right). \tag{1.91}$$

The basic idea is to break H into its kinetic T and potential V energy components and to realize that the differential operators appear in T only. The importance of this observation lies in the fact that T and V do not commute, which means that TV is not equal to VT (n.b., for two quantities to commute means that their order of appearance does not matter). Why do they not commute? Because T contains second derivatives with respect to the coordinates $\{q_j\}$ that V depends on, so, for example, $d^2/dq^2[V(q)\Psi(q)]$ is not equal to $V(q)\, d^2/dq^2\ \Psi(q)$. The fact that T and V do not commute is important because the most common approach to approximating $\exp(-itH/\hbar)$ is to write this single exponential in terms of $\exp(-itT/\hbar)$ and $\exp(-itV/\hbar)$. However, the identity

$$\exp\left(-\frac{itH}{\hbar}\right) = \exp\left(-\frac{itV}{\hbar}\right)\exp\left(-\frac{itT}{\hbar}\right) \tag{1.92}$$

is not fully valid as one can see by expanding all three of the above exponential factors as $\exp(x) = 1 + x + x^2/2! + \cdots$, and noting that the two sides of the above equation only agree if one can assume that $TV = VT$, which, as we noted, is not true.

In most modern approaches to time propagation, one divides the time interval t into many (i.e., P of them) small time "slices" $\tau = t/P$. One then expresses the evolution operator as a product of P short-time propagators:

$$\begin{aligned}\exp\left(-\frac{itH}{\hbar}\right) &= \exp\left(-\frac{i\tau H}{\hbar}\right)\exp\left(-\frac{i\tau H}{\hbar}\right)\exp\left(-\frac{i\tau H}{\hbar}\right)\cdots \\ &= \left[\exp\left(-\frac{i\tau H}{\hbar}\right)\right]^P.\end{aligned} \tag{1.93}$$

If one can develop an efficient means of propagating for a short time τ, one can then do so over and over again P times to achieve the desired full-time propagation.

It can be shown that the exponential operator involving H can better be approximated in terms of the T and V exponential operators as follows:

$$\exp\left(-\frac{i\tau H}{\hbar}\right) \approx \exp\left(-\tau^2\frac{(TV-VT)}{\hbar^2}\right)\exp\left(-\frac{i\tau V}{\hbar}\right)\exp\left(-\frac{i\tau T}{\hbar}\right). \tag{1.94}$$

So, if one can be satisfied with propagating for very short time intervals (so that the τ^2 term can be neglected), one can indeed use

$$\exp\left(-\frac{i\tau H}{\hbar}\right) \approx \exp\left(-\frac{i\tau V}{\hbar}\right)\exp\left(-\frac{i\tau T}{\hbar}\right) \tag{1.95}$$

as an approximation for the propagator $U(\tau, 0)$.

To progress further, one then expresses $\exp(-i\tau T/\hbar)$ acting on the initial function $\Psi(q)$ in terms of the eigenfunctions of the kinetic energy operator T. Note that these eigenfunctions do not depend on the nature of the potential V, so this step is valid for any and all potentials. The eigenfunctions of $T = -\hbar^2/2md^2/dq^2$ are

$$\psi_p(q) = \left(\frac{1}{2\pi}\right)^{1/2}\exp\left(\frac{ipq}{\hbar}\right) \tag{1.96}$$

and they obey the following orthogonality

$$\int \psi_{p'}^*(q)\psi_p(q)dq = \delta(p' - p) \tag{1.97}$$

and completeness relations

$$\int \psi_p(q)\psi_p^*(q')dp = \delta(q - q'). \tag{1.98}$$

Writing $\Psi(q)$ as

$$\Psi(q) = \int \delta(q - q')\Psi(q')dq', \tag{1.99}$$

and using the above expression for $\delta(q - q')$ gives

$$\Psi(q) = \iint \psi_p(q)\psi_p^*(q')\Psi(q')dq'dp. \tag{1.100}$$

Then inserting the explicit expressions for $\psi_p(q)$ and $\psi_p^*(q')$ in terms of $\psi_p(q) = (1/2\pi)^{1/2}\exp(ipq/\hbar)$ gives

$$\Psi(q) = \iint \left(\frac{1}{2\pi}\right)^{1/2} \exp\left(\frac{ipq}{\hbar}\right)\left(\frac{1}{2\pi}\right)^{1/2} \exp\left(-\frac{ipq'}{\hbar}\right)\Psi(q')dq'dp. \tag{1.101}$$

Now, allowing $\exp(-i\tau T/\hbar)$ to act on this form for $\Psi(q)$ produces

$$\begin{aligned}\exp\left(-\frac{i\tau T}{\hbar}\right)\Psi(q) = \iint \exp\left(-\frac{i\tau p^2\hbar^2}{2m\hbar}\right)\left(\frac{1}{2\pi}\right)^{1/2} \\ \times \exp\left[\frac{ip(q-q')}{\hbar}\right]\left(\frac{1}{2\pi}\right)^{1/2}\Psi(q')\,dq'dp.\end{aligned} \tag{1.102}$$

The integral over p above can be carried out analytically and gives

$$\exp\left(-\frac{i\tau T}{\hbar}\right)\Psi(q) = \left(\frac{m}{2\pi\tau\hbar}\right)^{1/2}\int \exp\left[\frac{im(q-q')^2}{2\tau\hbar}\right]\Psi(q')\,dq'. \tag{1.103}$$

So, the final expression for the short-time propagated wave function is

$$\Psi(q,\tau) = \exp\left[-\frac{i\tau V(q)}{\hbar}\right]\left(\frac{m}{2\pi\tau\hbar}\right)^{1/2}\int \exp\left[\frac{im(q-q')^2}{2\tau\hbar}\right]\Psi(q')\,dq', \tag{1.104}$$

which is the working equation one uses to compute $\Psi(q,\tau)$ knowing $\Psi(q)$. Notice that all one needs to know to apply this formula is the potential $V(q)$ at each point in space. One does not need to know any of the eigenfunctions of the Hamiltonian to apply this method. However, one does have to use this formula over and over again to propagate the initial wave function through many small time steps τ to achieve full propagation for the desired time interval $t = P\tau$.

Because this type of time propagation technique is a very active area of research in the theory community, it is likely to continue to be refined and improved. Further discussion of it is beyond the scope of this book, so I will not go further in this direction.

1.4 Free particle motions in more dimensions

The number of dimensions depends on the number of particles and the number of spatial (and other) dimensions needed to characterize the position and motion of each particle.

1.4.1 The Schrödinger equation

Consider an electron of mass m and charge e moving on a two-dimensional surface that defines the x, y plane (e.g., perhaps an electron is constrained to the surface of a solid by a potential that binds it tightly to a narrow region in the z-direction), and assume that the electron experiences a constant and not time-varying potential V_0 at all points in this plane. The pertinent time-independent Schrödinger equation is

$$-\frac{\hbar^2}{2m}\left(\frac{\partial^2}{\partial x^2}+\frac{\partial^2}{\partial y^2}\right)\psi(x,y)+V_0\psi(x,y)=\mathrm{E}\psi(x,y). \quad (1.105)$$

The task at hand is to solve the above eigenvalue equation to determine the "allowed" energy states for this electron. Because there are no terms in this equation that couple motion in the x and y directions (e.g., no terms of the form $x^a y^b$ or $\partial/\partial x\ \partial/\partial y$ or $x\ \partial/\partial y$), separation of variables can be used to write ψ as a product $\psi(x,y)=A(x)B(y)$. Substitution of this form into the Schrödinger equation, followed by collecting together all x-dependent and all y-dependent terms, gives

$$-\frac{\hbar^2}{2m}\frac{1}{A}\frac{\partial^2 A}{\partial x^2}-\frac{\hbar^2}{2m}\frac{1}{B}\frac{\partial^2 B}{\partial y^2}=E-V_0. \quad (1.106)$$

Since the first term contains no y-dependence and the second contains no x-dependence, and because the right side of the equation is independent of both x and y, both terms on the left must actually be constant (these two constants are denoted E_x and E_y, respectively). This observation allows two separate Schrödinger equations to be written:

$$-\frac{\hbar^2}{2m}A^{-1}\frac{\partial^2 A}{\partial x^2}=Ex, \quad (1.107)$$

and

$$-\frac{\hbar^2}{2m}B^{-1}\frac{\partial^2 B}{\partial y^2}=Ey. \quad (1.108)$$

The total energy E can then be expressed in terms of these separate energies E_x and E_y as $E_x+E_y=E-V_0$. Solutions to the x- and y-Schrödinger equations are easily seen to be:

$$A(x)=\exp\left[ix\left(\frac{2mE_x}{\hbar^2}\right)^{1/2}\right] \quad \text{and} \quad \exp\left[-ix\left(\frac{2mE_x}{\hbar^2}\right)^{1/2}\right], \quad (1.109)$$

$$B(y)=\exp\left[iy\left(\frac{2mE_y}{\hbar^2}\right)^{1/2}\right] \quad \text{and} \quad \exp\left[-iy\left(\frac{2mE_y}{\hbar^2}\right)^{1/2}\right]. \quad (1.110)$$

Two independent solutions are obtained for each equation because the x- and y-space Schrödinger equations are both second order differential equations (i.e., a second order differential equation has two independent solutions).

1.4.2 Boundary conditions

The boundary conditions, not the Schrödinger equation, determine whether the eigenvalues will be discrete or continuous.

If the electron is entirely unconstrained within the x, y plane, the energies E_x and E_y can assume any values; this means that the experimenter can "inject" the electron onto the x, y plane with any total energy E and any components E_x and E_y along the two axes as long as $E_x + E_y = E$. In such a situation, one speaks of the energies along both coordinates as being "in the continuum" or "not quantized".

In contrast, if the electron is constrained to remain within a fixed area in the x, y plane (e.g., a rectangular or circular region), then the situation is qualitatively different. Constraining the electron to any such specified area gives rise to boundary conditions that impose additional requirements on the above A and B functions. These constraints can arise, for example, if the potential $V_0(x, y)$ becomes very large for x, y values outside the region, in which case the probability of finding the electron outside the region is very small. Such a case might represent, for example, a situation in which the molecular structure of the solid surface changes outside the enclosed region in a way that is highly repulsive to the electron (e.g., as in the case of molecular corrals on metal surfaces). This case could then represent a simple model of so-called "corrals" in which the particle is constrained to a finite region of space.

For example, if motion is constrained to take place within a rectangular region defined by $0 \le x \le L_x; 0 \le y \le L_y$, then the continuity property that all wave functions must obey (because of their interpretation as probability densities, which must be continuous) causes $A(x)$ to vanish at 0 and at L_x. That is, because A must vanish for $x < 0$ and must vanish for $x > L_x$, and because A is continuous, it must vanish at $x = 0$ and at $x = L_x$. Likewise, $B(y)$ must vanish at 0 and at L_y. To implement these constraints for $A(x)$, one must linearly combine the above two solutions $\exp[ix(2mE_x/\hbar^2)^{1/2}]$ and $\exp[-ix(2mE_x/\hbar^2)^{1/2}]$ to achieve a function that vanishes at $x = 0$:

$$A(x) = \exp\left[ix\left(\frac{2mE_x}{\hbar^2}\right)^{1/2}\right] - \exp\left[-ix\left(\frac{2mE_x}{\hbar^2}\right)^{1/2}\right]. \tag{1.111}$$

One is allowed to linearly combine solutions of the Schrödinger equation that have the same energy (i.e., are degenerate) because Schrödinger equations are linear differential equations. An analogous process must be applied to $B(y)$ to achieve a function that vanishes at $y = 0$:

$$B(y) = \exp\left[iy\left(\frac{2mE_y}{\hbar^2}\right)^{1/2}\right] - \exp\left[-iy\left(\frac{2mE_y}{\hbar^2}\right)^{1/2}\right]. \tag{1.112}$$

Further requiring $A(x)$ and $B(y)$ to vanish, respectively, at $x = L_x$ and $y = L_y$, gives equations that can be obeyed only if E_x and E_y assume particular values:

$$\exp\left[iL_x\left(\frac{2mE_x}{\hbar^2}\right)^{1/2}\right] - \exp\left[-iL_x\left(\frac{2mE_x}{\hbar^2}\right)^{1/2}\right] = 0, \tag{1.113}$$

and

$$\exp\left[iL_y\left(\frac{2mE_y}{\hbar^2}\right)^{1/2}\right] - \exp\left[-iL_y\left(\frac{2mE_y}{\hbar^2}\right)^{1/2}\right] = 0. \tag{1.114}$$

These equations are equivalent (i.e., using $\exp(ix) = \cos x + i \sin x$) to

$$\sin\left[L_x\left(\frac{2mE_x}{\hbar^2}\right)^{1/2}\right] = \sin\left[L_y\left(\frac{2mE_y}{\hbar^2}\right)^{1/2}\right] = 0. \tag{1.115}$$

Knowing that $\sin\theta$ vanishes at $\theta = n\pi$, for $n = 1, 2, 3, \ldots$ (although the $\sin(n\pi)$ function vanishes for $n = 0$, this function vanishes for all x or y, and is therefore unacceptable because it represents zero probability density at all points in space), one concludes that the energies E_x and E_y can assume only values that obey

$$L_x\left(\frac{2mE_x}{\hbar^2}\right)^{1/2} = n_x\pi, \tag{1.116}$$

$$L_y\left(\frac{2mE_y}{\hbar^2}\right)^{1/2} = n_y\pi, \tag{1.117}$$

$$\text{or} \quad E_x = \frac{n_x^2\pi^2\hbar^2}{2mL_x^2}, \tag{1.118}$$

$$\text{and} \quad E_y = \frac{n_y^2\pi^2\hbar^2}{2mL_y^2}, \quad \text{with } n_x \text{ and } n_y = 1, 2, 3, \ldots \tag{1.119}$$

It is important to stress that it is the imposition of boundary conditions, expressing the fact that the electron is spatially constrained, that gives rise to quantized energies. In the absence of spatial confinement, or with confinement only at $x = 0$ or L_x or only at $y = 0$ or L_y, quantized energies would *not* be realized.

In this example, confinement of the electron to a finite interval along both the x and y coordinates yields energies that are quantized along both axes. If the electron is confined along one coordinate (e.g., between $0 \le x \le L_x$) but not along the other (i.e., $B(y)$ is either restricted to vanish at $y = 0$ or at $y = L_y$ or at neither point), then the total energy E lies in the continuum; its E_x component is quantized but E_y is not. Analogs of such cases arise, for example, when a linear triatomic molecule has more than enough energy in one of its bonds to rupture it but not much energy in the other bond; the first bond's energy lies in the continuum, but the second bond's energy is quantized.

Perhaps more interesting is the case in which the bond with the higher dissociation energy is excited to a level that is not enough to break it but that is in excess of the dissociation energy of the weaker bond. In this case, one has two degenerate states: (i) the strong bond having high internal energy and the weak bond having low energy (ψ_1), and (ii) the strong bond having little energy and the weak bond having more than enough energy to rupture it (ψ_2). Although an experiment may prepare the molecule in a state that contains only the former component (i.e., $\psi = C_1\psi_1 + C_2\psi_2$ with $C_1 = 1, C_2 = 0$), coupling between the two degenerate functions (induced by terms in the Hamiltonian $\mathbf{H}$ that have been ignored in defining ψ_1 and ψ_2) usually causes the true wave function $\Psi = \exp(-it\,\mathbf{H}/\hbar)\psi$ to acquire a component of the second function as time evolves. In such a case, one speaks of internal vibrational energy relaxation (IVR) giving rise to unimolecular decomposition of the molecule.

1.4.3 Energies and wave functions for bound states

For discrete energy levels, the energies are specified functions that depend on quantum numbers, one for each degree of freedom that is quantized.

Returning to the situation in which motion is constrained along both axes, the resultant total energies and wave functions (obtained by inserting the quantum energy levels into the expressions for $A(x)B(y)$) are as follows:

$$E_x = \frac{n_x^2\pi^2\hbar^2}{2mL_x^2}, \tag{1.120}$$

and

$$E_y = \frac{n_y^2\pi^2\hbar^2}{2mL_y^2}, \tag{1.121}$$

$$E = E_x + E_y + V_0, \tag{1.122}$$

$$\psi(x, y) = \left(\frac{1}{2L_x}\right)^{\frac{1}{2}}\left(\frac{1}{2L_y}\right)^{\frac{1}{2}}\left[\exp\left(\frac{in_x\pi x}{L_x}\right) - \exp\left(\frac{-in_x\pi x}{L_x}\right)\right] \times\left[\exp\left(\frac{in_y\pi y}{L_y}\right) - \exp\left(-\frac{in_y\pi y}{L_y}\right)\right],$$
$$\text{with } n_x \text{ and } n_y = 1, 2, 3, \ldots \tag{1.123}$$

The two $(1/2L)^{1/2}$ factors are included to guarantee that ψ is normalized:

$$\int |\psi(x, y)|^2 dx\, dy = 1. \tag{1.124}$$

Normalization allows $|\psi(x, y)|^2$ to be properly identified as a probability density for finding the electron at a point x, y.

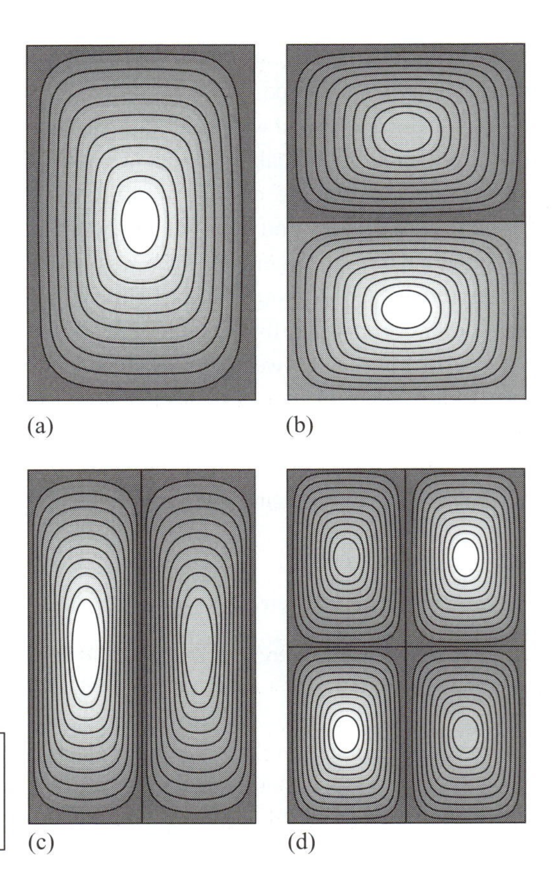

Figure 1.13 Plots of the (a) (1,1), (b) (2,1), (c) (1,2) and (d) (2,2) wave functions.

Shown in Fig. 1.13 are plots of four such two-dimensional wave functions for n_x and n_y values of (1,1), (2,1), (1,2) and (2,2), respectively. Note that the functions vanish on the boundaries of the box, and notice how the number of nodes (i.e., zeroes encountered as the wave function oscillates from positive to negative) is related to the n_x and n_y quantum numbers and to the energy. This pattern of more nodes signifying higher energy is one that we encounter again and again in quantum mechanics and is something the student should be able to use to "guess" the relative energies of wave functions when their plots are at hand. Finally, you should also notice that, as in the one-dimensional box case, any attempt to classically interpret the probabilities $P(x, y)$ corresponding to the above quantum wave functions will result in failure. As in the one-dimensional case, the classical $P(x, y)$ would be constant along slices of fixed x and varying y or slices of fixed y and varying x within the box because the speed is constant there. However, the quantum $P(x, y)$ plots, at least for small quantum numbers, are not constant. For large n_x and n_y values, the quantum $P(x, y)$ plots will again, via the quantum–classical correspondence principle, approach the (constant) classical $P(x, y)$ form.

1.4.4 Quantized action can also be used to derive energy levels

There is another approach that can be used to find energy levels and is especially straightforward to use for systems whose Schrödinger equations are separable. The so-called classical action (denoted S) of a particle moving with momentum $\mathbf{p}$ along a path leading from initial coordinate $\mathbf{q}_\mathrm{i}$ at initial time t_i to a final coordinate $\mathbf{q}_\mathrm{f}$ at time t_f is defined by

$$S = \int_{\mathbf{q}_\mathrm{i};t_\mathrm{i}}^{\mathbf{q}_\mathrm{f};t_\mathrm{f}} \mathbf{p}\cdot\mathbf{dq}. \tag{1.125}$$

Here, the momentum vector $\mathbf{p}$ contains the momenta along all coordinates of the system, and the coordinate vector $\mathbf{q}$ likewise contains the coordinates along all such degrees of freedom. For example, in the two-dimensional particle-in-a-box problem considered above, $\mathbf{q} = (x, y)$ has two components as does $\mathbf{p} = (p_x, p_y)$, and the action integral is

$$S = \int_{x_\mathrm{i};y_\mathrm{i};t_\mathrm{i}}^{x_\mathrm{f};y_\mathrm{f};t_\mathrm{f}} (p_x dx + p_y d_y). \tag{1.126}$$

In computing such actions, it is essential to keep in mind the sign of the momentum as the particle moves from its initial to its final positions. An example will help clarify these matters.

For systems such as the above particle-in-a-box example for which the Hamiltonian is separable, the action integral decomposes into a sum of such integrals, one for each degree of freedom. In this two-dimensional example, the additivity of H,

$$\begin{aligned} H = H_x + H_y &= \frac{p_x^2}{2m} + \frac{p_y^2}{2m} + V(x) + V(y) \\ &= -\frac{\hbar^2}{2m}\frac{\partial^2}{\partial x^2} + V(x) - \frac{\hbar^2}{2m}\frac{\partial^2}{\partial y^2} + V(y), \end{aligned} \tag{1.127}$$

means that p_x and p_y can be independently solved for in terms of the potentials $V(x)$ and $V(y)$ as well as the energies E_x and E_y associated with each separate degree of freedom:

$$p_x = \pm\sqrt{2m(E_x - V(x))}, \tag{1.128}$$

$$p_y = \pm\sqrt{2m(E_y - V(y))}; \tag{1.129}$$

the signs on p_x and p_y must be chosen to properly reflect the motion that the particle is actually undergoing. Substituting these expressions into the action integral yields

$$S = S_x + S_y \tag{1.130}$$

$$= \int_{x_\mathrm{i};t_\mathrm{i}}^{x_\mathrm{f};t_\mathrm{f}} \pm\sqrt{2m(E_x - V(x))}\,dx + \int_{y_\mathrm{i};t_\mathrm{i}}^{y_\mathrm{f};t_\mathrm{f}} \pm\sqrt{2m(E_y - V(y))}\,dy. \tag{1.131}$$

The relationship between these classical action integrals and the existence of quantized energy levels has been shown to involve equating the classical action for motion on a closed path to an integral multiple of Planck's constant:

$$S_{\text{closed}} = \int_{q_i;t_i}^{q_f=q_i;t_f} \mathbf{p}\cdot\mathbf{dq} = nh \qquad (n = 1, 2, 3, 4, \ldots). \tag{1.132}$$

Applied to each of the independent coordinates of the two-dimensional particle-in-a-box problem, this expression reads

$$n_x h = \int_{x=0}^{x=L_x} \sqrt{2m(E_x - V(x))}\, dx + \int_{x=L_x}^{x=0} -\sqrt{2m(E_x - V(x))}\, dx, \tag{1.133}$$

$$n_y h = \int_{y=0}^{y=L_y} \sqrt{2m(E_y - V(y))}\, dy + \int_{y=L_y}^{y=0} -\sqrt{2m(E_y - V(y))}\, dy. \tag{1.134}$$

Notice that the signs of the momenta are positive in each of the first integrals appearing above (because the particle is moving from $x = 0$ to $x = L_x$, and analogously for y-motion, and thus has positive momentum) and negative in each of the second integrals (because the motion is from $x = L_x$ to $x = 0$ (and analogously for y-motion) and thus the particle has negative momentum). Within the region bounded by $0 \le x \le L_x; 0 \le y \le L_y$, the potential vanishes, so $V(x) = V(y) = 0$. Using this fact, and reversing the upper and lower limits, and thus the sign, in the second integrals above, one obtains

$$n_x h = 2\int_{x=0}^{x=L_x} \sqrt{2mE_x}\, dx = 2\sqrt{2mE_x}\, L_x, \tag{1.135}$$

$$n_y h = 2\int_{y=0}^{y=L_y} \sqrt{2mE_y}\, dy = 2\sqrt{2mE_y}\, L_y. \tag{1.136}$$

Solving for E_x and E_y, one finds

$$E_x = \frac{(n_x h)^2}{8mL_x^2}, \tag{1.137}$$

$$E_y = \frac{(n_y h)^2}{8mL_y^2}. \tag{1.138}$$

These are the same quantized energy levels that arose when the wave function boundary conditions were matched at $x = 0, x = L_x$ and $y = 0, y = L_y$. In this case, one says that the Bohr–Sommerfeld quantization condition,

$$nh = \int_{\mathbf{q}_i;t_i}^{\mathbf{q}_f=\mathbf{q}_i;t_f} \mathbf{p}\cdot\mathbf{dq}, \tag{1.139}$$

has been used to obtain the result.

The use of action quantization as illustrated above has become a very important tool. It has allowed scientists to make great progress toward bridging the gap between classical and quantum descriptions of molecular dynamics. In particular, by using classical concepts such as trajectories and then appending quantal action conditions, people have been able to develop so-called semi-classical models of molecular dynamics. In such models, one is able to retain a great deal of classical understanding while building in quantum effects such as energy quantization, zero-point energies, and interferences.

1.4.5 Quantized action does not always work

Unfortunately, the approach of quantizing the action does not always yield the correct expression for the quantized energies. For example, when applied to the so-called harmonic oscillator problem that we will study in quantum form later, which serves as the simplest reasonable model for vibration of a diatomic molecule AB, one expresses the total energy as

$$E = \frac{p^2}{2\mu} + \frac{k}{2}x^2 \tag{1.140}$$

where $\mu = m_A m_B/(m_A + m_B)$ is the reduced mass of the AB diatom, k is the force constant describing the bond between A and B, x is the bond-length displacement, and p is the momentum along the bond length. The quantized action requirement then reads

$$nh = \int p\, dx = \int \left[2\mu\left(E - \frac{k}{2}x^2\right)\right]^{1/2} dx. \tag{1.141}$$

This integral is carried out between $x = -(2E/k)^{1/2}$ and $(2E/k)^{1/2}$, the left and right turning points of the oscillatory motion, and back again to form a closed path. Carrying out this integral and equating it to nh gives the following expression for the energy E:

$$E = n\frac{h}{2\pi}\left(\frac{k}{\mu}\right)^{1/2}, \tag{1.142}$$

where the quantum number n is allowed to assume integer values ranging from $n = 0, 1, 2,$ to infinity. The problem with this result is that it is wrong! As experimental data clearly show, the lowest energy levels for the vibrations of a molecule do not have $E = 0$; they have a "zero-point" energy that is approximately equal to $1/2(h/2\pi)(k/\mu)^{1/2}$. So, although the action quantization condition yields energies whose spacings are reasonably in agreement with laboratory data for low-energy states (e.g., such states have approximately constant spacings), it fails to predict the zero-point energy content of such vibrations. As we will see later, a proper quantum mechanical treatment of the harmonic oscillator yields energies of the form

$$E = \left(n + \frac{1}{2}\right)\left(\frac{h}{2\pi}\right)\left(\frac{k}{\mu}\right)^{1/2} \tag{1.143}$$

which differs from the action-based result by the proper zero-point energy.

Even with such difficulties known, much progress has been made in extending the most elementary action-based methods to more and more systems by introducing, for example, rules that allow the quantum number n to assume half-integer as well as integer values. Clearly, if n were allowed to equal $1/2, 3/2, 5/2, \ldots$, the earlier action integral would have produced the correct result. However, how does one know when to allow n to assume only integer or only half-integer or both integer and half-integer values? The answers to this question are beyond the scope of this text and constitute an active area of research. For now, it is enough for the student to be aware that one can often find energy levels by using action integrals, but one must be careful in doing so because sometimes the answers are wrong.

Before leaving this section, it is worth noting that the appearance of half-integer quantum numbers does not only occur in the harmonic oscillator case. To illustrate, let us consider the L_z angular momentum operator discussed earlier. As we showed, this operator, when computed as the z-component of $\mathbf{r} \times \mathbf{p}$, can be written in polar (r, θ, ϕ) coordinates as

$$L_z = -i\hbar d/d\phi. \tag{1.144}$$

The eigenfunctions of this operator have the form $\exp(ia\phi)$, and the eigenvalues are $a\hbar$. Because geometries with azimuthal angles equal to ϕ or equal to $\phi + 2\pi$ are exactly the same geometries, the function $\exp(ia\phi)$ should be exactly the same as $\exp(ia(\phi + 2\pi))$. This can only be the case if a is an integer. Thus, one concludes that only integral multiples of $\hbar$ can be "allowed" values of the z-component of angular momentum. Experimentally, one measures the z-component of an angular momentum by placing the system possessing the angular momentum in a magnetic field of strength B and observing how many z-component energy states arise. For example, a boron atom with its 2p orbital has one unit of orbital angular momentum, so one finds three separate z-component values which are usually denoted $m = -1$, $m = 0$, and $m = 1$. Another example is offered by the scandium atom with one unpaired electron in a d orbital; this atom's states split into five ($m = -2, -1, 0, 1, 2$) z-component states. In each case, one finds $2L + 1$ values of the m quantum number, and, because L is an integer, $2L + 1$ is an odd integer. Both of these observations are consistent with the expectation that only integer values can occur for L_z eigenvalues.

However, it has been observed that some species do not possess 3 or 5 or 7 or 9 z-component states but an even number of such states. In particular, when electrons, protons, or neutrons are subjected to the kind of magnetic field experiment mentioned above, these particles are observed to have only two z-component eigenvalues. Because, as we discuss later in this text, all angular momenta have z-component eigenvalues that are separated from one another by unit multiples of $\hbar$, one is forced to conclude that these three fundamental building-block particles have z-component eigenvalues of $1/2\hbar$ and $-1/2\hbar$. The appearance of half-integer angular momenta is not consistent with the observation made earlier that ϕ and $\phi + 2\pi$ correspond to exactly the same physical point in coordinate space, which, in turn, implies that only full-integer angular momenta are possible.

The resolution of the above paradox (i.e., how can half-integer angular momenta exist?) involved realizing that some angular momenta correspond not to the $\mathbf{r} \times \mathbf{p}$ angular momenta of a physical mass rotating, but, instead, are intrinsic properties of certain particles. That is, the intrinsic angular momenta of electrons, protons, and neutrons can not be viewed as arising from rotation of some mass that comprises these particles. Instead, such intrinsic angular momenta are fundamental "built in" characteristics of these particles. For example, the two $1/2\hbar$ and $-1/2\hbar$ angular momentum states of an electron, usually denoted α and β, respectively, are two internal states of the electron that are degenerate in the absence of a magnetic field but which represent two distinct states of the electron. Analogously, a proton has $1/2\hbar$ and $-1/2\hbar$ states, as do neutrons. All such half-integer angular momentum states can not be accounted for using classical mechanics but are known to arise in quantum mechanics.

Chapter 2
Model problems that form important starting points

The model problems discussed in this chapter form the basis for chemists' understanding of the electronic states of atoms, molecules, clusters, and solids as well as the rotational and vibrational motions of molecules.

2.1 Free electron model of polyenes

The particle-in-a-box problem provides an important model for several relevant chemical situations.

The "particle-in-a-box" model for motion in two dimensions discussed earlier can obviously be extended to three dimensions or to one. For two and three dimensions, it provides a crude but useful picture for electronic states on surfaces or in metallic crystals, respectively. I say metallic crystals because it is in such systems that the outermost valence electrons are reasonably well treated as moving freely. Free motion within a spherical volume gives rise to eigenfunctions that are used in nuclear physics to describe the motions of neutrons and protons in nuclei. In the so-called shell model of nuclei, the neutrons and protons fill separate s, p, d, etc. orbitals with each type of nucleon forced to obey the Pauli principle (i.e., to have no more than two nucleons in each orbital because protons and neutrons are fermions). To remind you, I display in Fig. 2.1 the angular shapes that characterize s, p, and d orbitals.

This same spherical box model has also been used to describe the orbitals of valence electrons in clusters of metal atoms such as Cs_n, Cu_n, Na_n and their positive and negative ions. Because of the metallic nature of these species, their valence electrons are essentially free to roam over the entire spherical volume of the cluster, which renders this simple model rather effective. In this model, one thinks of each electron being free to roam within a sphere of radius R (i.e., having a potential that is uniform within the sphere and infinite outside the sphere). Finally, as noted above, this same spherical box model forms the basis of the so-called shell model of nuclear structure. In this model, one assumes that the protons and neutrons that make up a nucleus, both of which are fermions, occupy

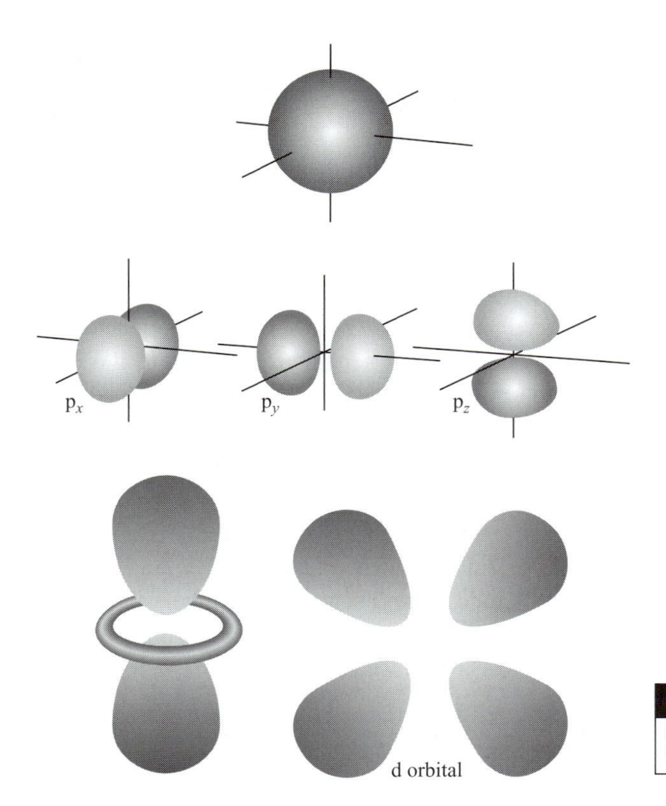

Figure 2.1 The angular shapes of s, p, and d functions.

spherical-box orbitals (one set of orbitals for protons, another set for neutrons because they are distinguishable from one another). By placing the protons and neutrons into these orbitals, two to an orbital, one achieves a description of the energy levels of the nucleus. Excited states are achieved by promoting a neutron or proton from an occupied orbital to a virtual (i.e., previously unoccupied) orbital. In such a model, especially stable nuclei are achieved when "closed-shell" configurations such as $1s^2$ or $1s^22s^22p^6$ are realized (e.g., ^{4}He has both neutrons and protons in $1s^2$ configurations).

The orbitals that solve the Schrödinger equation inside such a spherical box are not the same in their radial "shapes" as the s, p, d, etc. orbitals of atoms because, in atoms, there is an additional radial potential $V(r) = -Ze^2/r$ present. However, their angular shapes are the same as in atomic structure because, in both cases, the potential is independent of θ and ϕ. As the orbital plots shown above indicate, the angular shapes of s, p, and d orbitals display varying numbers of nodal surfaces. The s orbitals have none, p orbitals have one, and d orbitals have two. Analogous to how the number of nodes related to the total energy of the particle constrained to the x, y plane, the number of nodes in the angular wave functions indicates the amount of angular or rotational energy. Orbitals of s shape have no angular energy, those of p shape have less then do d orbitals, etc.

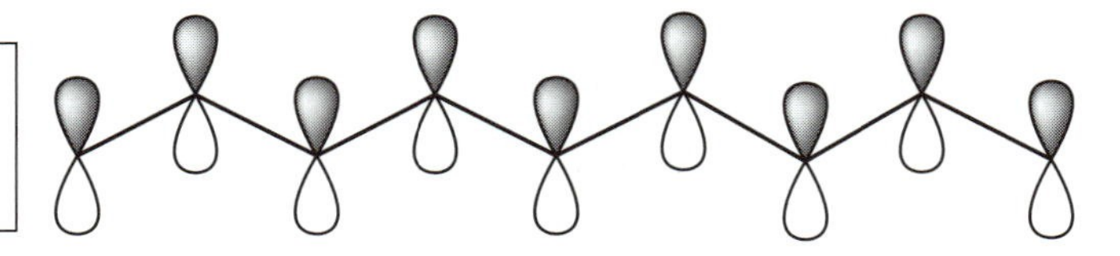

Figure 2.2 The π atomic orbitals of a conjugated chain of nine carbon atoms.

One-dimensional free particle motion provides a qualitatively correct picture for π-electron motion along the p_π orbitals of delocalized polyenes. The one Cartesian dimension then corresponds to motion along the delocalized chain. In such a model, the box length L is related to the carbon–carbon bond length R and the number N of carbon centers involved in the delocalized network $L = (N - 1)R$. In Fig. 2.2, such a conjugated network involving nine centers is depicted. In this example, the box length would be eight times the C—C bond length. The eigenstates $\psi_n(x)$ and their energies E_n represent orbitals into which electrons are placed. In the example case, if nine π electrons are present (e.g., as in the 1,3,5,7-nonatetraene radical), the ground electronic state would be represented by a total wave function consisting of a product in which the lowest four ψs are doubly occupied and the fifth ψ is singly occupied:

$$\Psi = \psi_1\alpha\psi_1\beta\psi_2\alpha\psi_2\beta\psi_3\alpha\psi_3\beta\psi_4\alpha\psi_4\beta\psi_5\alpha. \tag{2.1}$$

The z-component angular momentum states of the electrons are labeled α and β as discussed earlier.

A product wave function is appropriate because the total Hamiltonian involves the kinetic plus potential energies of nine electrons. To the extent that this total energy can be represented as the sum of nine separate energies, one for each electron, the Hamiltonian allows a separation of variables

$$H \cong \sum_j H(j) \tag{2.2}$$

in which each $H(j)$ describes the kinetic and potential energy of an individual electron. Recall that when a partial differential equation has no operators that couple its different independent variables (i.e., when it is separable), one can use separation of variables methods to decompose its solutions into products. Thus, the (approximate) additivity of H implies that solutions of $H\Psi = E\Psi$ are products of solutions to

$$H(j)\psi(\mathbf{r}_j) = E_j\psi(\mathbf{r}_j). \tag{2.3}$$

The two lowest π-excited states would correspond to states of the form

$$\Psi^* = \psi_1\alpha\psi_1\beta\psi_2\alpha\psi_2\beta\psi_3\alpha\psi_3\beta\psi_4\alpha\psi_5\beta\psi_5\alpha, \tag{2.4}$$

$$\text{and} \quad \Psi'^* = \psi_1\alpha\psi_1\beta\psi_2\alpha\psi_2\beta\psi_3\alpha\psi_3\beta\psi_4\alpha\psi_4\beta\psi_6\alpha, \tag{2.5}$$

where the spin-orbitals (orbitals multiplied by α or β) appearing in the above products depend on the coordinates of the various electrons. For example,

$$\psi_1\alpha\psi_1\beta\psi_2\alpha\psi_2\beta\psi_3\alpha\psi_3\beta\psi_4\alpha\psi_5\beta\psi_5\alpha \quad (2.6)$$

denotes

$$\psi_1\alpha(\mathbf{r}_1)\psi_1\beta(\mathbf{r}_2)\psi_2\alpha(\mathbf{r}_3)\psi_2\beta(\mathbf{r}_4)\psi_3\alpha(\mathbf{r}_5)\psi_3\beta(\mathbf{r}_6)\psi_4\alpha(\mathbf{r}_7)\psi_5\beta(\mathbf{r}_8)\psi_5\alpha(\mathbf{r}_9). \quad (2.7)$$

The electronic excitation energies from the ground state to each of the above excited states within this model would be

$$\begin{aligned} \Delta E^* &= \pi^2 \frac{\hbar^2}{2m}\left[\frac{5^2}{L^2} - \frac{4^2}{L^2}\right] \\ \text{and} \quad \Delta E'^* &= \pi^2 \frac{\hbar^2}{2m}\left[\frac{6^2}{L^2} - \frac{5^2}{L^2}\right]. \end{aligned} \quad (2.8)$$

It turns out that this simple model of π-electron energies provides a qualitatively correct picture of such excitation energies. Its simplicity allows one, for example, to easily suggest how a molecule's color (as reflected in the complementary color of the light the molecule absorbs) varies as the conjugation length L of the molecule varies. That is, longer conjugated molecules have lower-energy orbitals because L^2 appears in the denominator of the energy expression. As a result, longer conjugated molecules absorb light of lower energy than do shorter molecules.

This simple particle-in-a-box model does not yield orbital energies that relate to ionization energies unless the potential "inside the box" is specified. Choosing the value of this potential V_0 such that $V_0 + [\pi^2\hbar^2/2m][5^2/L^2]$ is equal to minus the lowest ionization energy of the 1,3,5,7-nonatetraene radical, gives energy levels (as $E = V_0 + [\pi^2\hbar^2/2m][n^2/L^2]$) which can then be used as approximations to ionization energies.

The individual π-molecular orbitals

$$\psi_n = \left(\frac{2}{L}\right)^{1/2} \sin\left(\frac{n\pi x}{L}\right) \quad (2.9)$$

are depicted in Fig. 2.3 for a model of the 1,3,5-hexatriene π-orbital system for which the "box length" L is five times the distance R_{CC} between neighboring pairs of carbon atoms. The magnitude of the kth C-atom centered atomic orbital in the nth π-molecular orbital is given by $(2/L)^{1/2}\sin(n\pi k R_{CC}/L)$. In this figure, positive amplitude is denoted by the clear spheres, and negative amplitude is shown by the darkened spheres. Where two spheres of like shading overlap, the wave function has enhanced amplitude; where two spheres of different shading overlap, a node occurs. Once again, we note that the number of nodes increases as one ranges from the lowest energy orbital to higher energy orbitals. The reader is once again encouraged to keep in mind this ubiquitous characteristic of quantum mechanical wave functions.

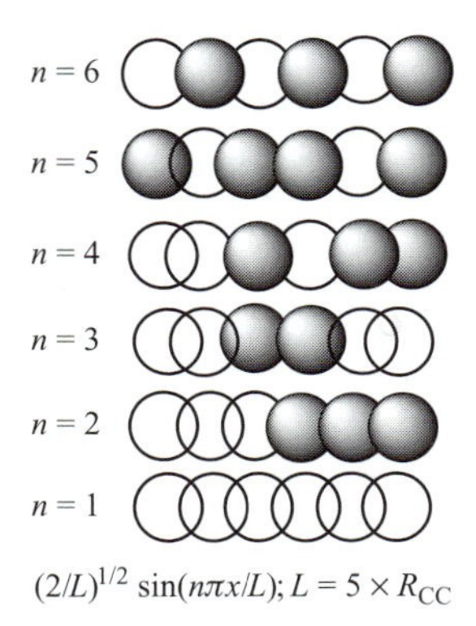

Figure 2.3 The phases of the six molecular orbitals of a chain containing six atoms.

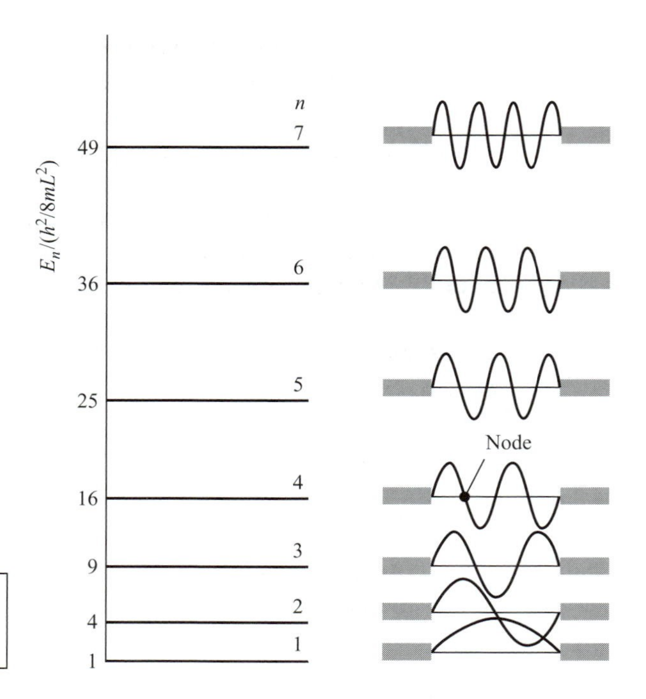

Figure 2.4 The nodal pattern for a chain containing seven atoms.

This simple model allows one to estimate spin densities at each carbon center and provides insight into which centers should be most amenable to electrophilic or nucleophilic attack. For example, radical attack at the C_5 carbon of the nine-atom nonatetraene system described earlier would be more facile for the ground state Ψ than for either Ψ^* or Ψ'^*. In the former, the unpaired spin density resides in ψ_5, which has non-zero amplitude at the C_5 site $x = L/2$. In Ψ^* and Ψ'^*, the unpaired density is in ψ_4 and ψ_6, respectively, both of which have zero density at C_5. These densities reflect the values $(2/L)^{1/2} \sin(n\pi k R_{CC}/L)$ of the amplitudes for this case in which $L = 8 \times R_{CC}$ for $n = 5$, 4, and 6, respectively. Plots of the wave functions for n ranging from 1 to 7 are shown in another format in Fig. 2.4 where the nodal pattern is emphasized. I hope that by now the student is not tempted to ask how the electron "gets" from one region of high amplitude, through a node, to another high-amplitude region. Remember, such questions are cast in classical Newtonian language and are not appropriate when addressing the wave-like properties of quantum mechanics.

2.2 Bands of orbitals in solids

Not only does the particle-in-a-box model offer a useful conceptual representation of electrons moving in polyenes, but it also is the zeroth-order model of band structures in solids. Let us consider a simple one-dimensional "crystal" consisting

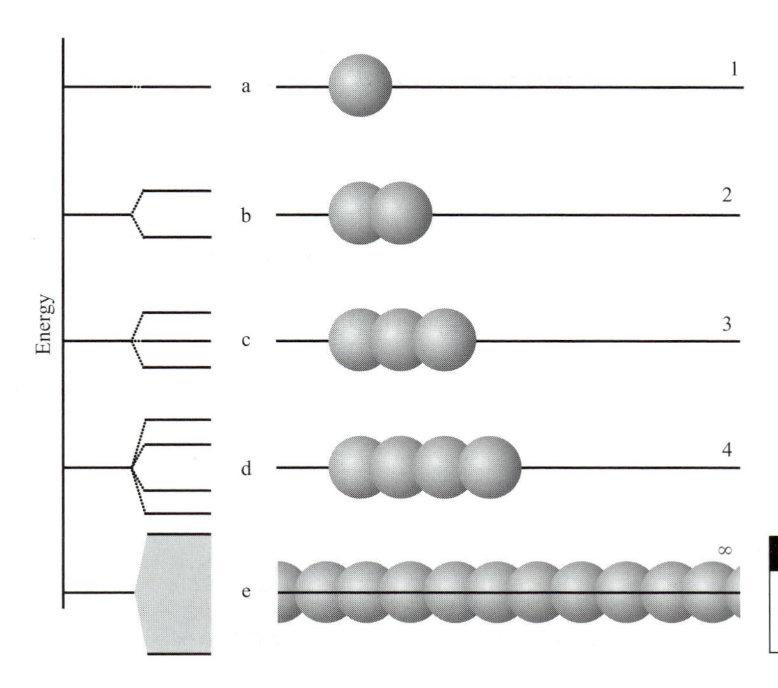

Figure 2.5 The energy levels arising from 1, 2, 3, 4, and an infinite number of orbitals.

of a large number of atoms or molecules, each with a single orbital (the spheres shown) that it contributes to the bonding. Let us arrange these building blocks in a regular "lattice" as shown in Fig. 2.5. In the top four rows of this figure we show the case with 1, 2, 3, and 4 building blocks. To the left of each row, we display the energy splitting pattern into which the building blocks' orbitals evolve as they overlap and form delocalized molecular orbitals. Not surprisingly, for $n = 2$, one finds a bonding and an antibonding orbital. For $n = 3$, one has one bonding, one non-bonding, and one antibonding orbital. Finally, in the bottom row, we attempt to show what happens for an infinitely long chain. The key point is that the discrete number of molecular orbitals appearing in the 1–4 orbital cases evolves into a continuum of orbitals called a band. This band of orbital energies ranges from its bottom (whose orbital consists of a fully in-phase bonding combination of the building block orbitals) to its top (whose orbital is a fully out-of-phase antibonding combination). In Fig. 2.6 we illustrate these fully bonding and fully antibonding band orbitals for two cases – the bottom involving s-type building block orbitals, and the top involving p-type orbitals. Notice that when the energy gap between the building block s and p orbitals is larger than is the dispersion (spread) in energy within the band of s or band of p orbitals, a band gap occurs between the highest member of the s band and the lowest member of the p band. The splitting between the s and p orbitals is a property of the individual atoms comprising the solid and varies among the elements of the periodic table. The dispersion in energies that a given band of orbitals is split into as these atomic

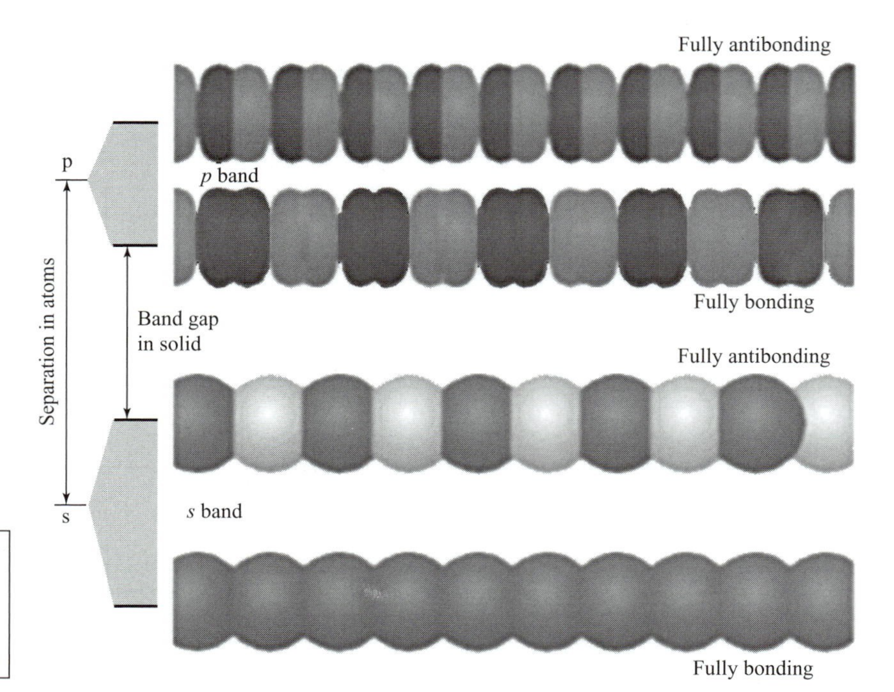

Figure 2.6 The bonding through antibonding energies and band orbitals arising from s and from p orbitals.

orbitals combine to form a band is determined by how strongly the orbitals on neighboring atoms overlap. Small overlap produces small dispersion, and large overlap yields a broad band.

Depending on how many valence electrons each building block contributes, the various bands formed by overlapping the building block orbitals of the constituent atoms will be filled to various levels. For example, if each orbital shown above has a single valence electron in an s orbital (e.g., as in the case of the alkali metals), the s-band will be half filled in the ground state with α and β paired electrons. Such systems produce very good conductors because their partially filled bands allow electrons to move with very little (e.g., only thermal) excitation among other orbitals in this same band. On the other hand, for alkaline earth systems with two s electrons per atom, the s band will be completely filled. In such cases, conduction requires excitation to the lowest members of the nearby p-orbital band. Finally, if each building block were an Al ($3s^2$ $3p^1$) atom, the s band would be full and the p band would be half-filled. Systems whose highest energy occupied band is completely filled and for which the gap in energy to the lowest unfilled band is large are called insulators because they have no way to easily (i.e., with little energy requirement) promote some of their higher energy electrons from orbital to orbital and thus effect conduction. If the band gap between a filled band and an unfilled band is small, it may be possible for thermal excitation (i.e., collisions with neighboring atoms or molecules) to cause excitation of electrons

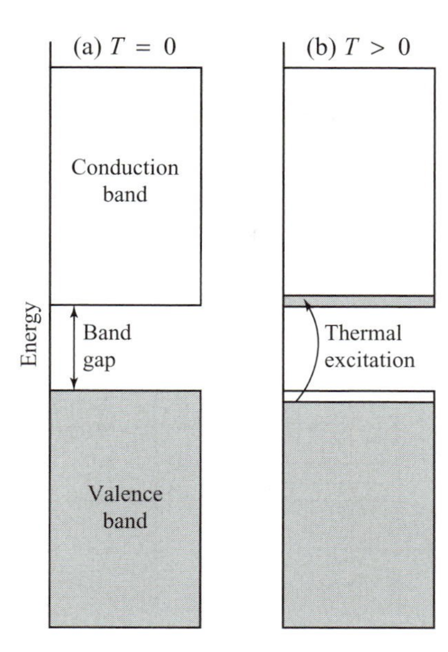

Figure 2.7 The valence and conduction bands and the band gap.

from the former to the latter thereby inducing conductive behavior. An example of such a case is illustrated in Fig. 2.7. In contrast, systems whose highest energy occupied band is partially filled are conductors because they have little spacing among their occupied and unoccupied orbitals.

To form a semiconductor, one starts with an insulator as shown in Fig. 2.8 with its filled (dark) band and a band gap between this band and its empty (clear) upper band. If this insulator material were synthesized with a small amount of "dopant" whose valence orbitals have energies between the filled and empty bands of the insulator, one may generate a semiconductor. If the dopant species has no valence electrons (i.e., has an empty valence orbital), it gives rise to an empty band lying between the filled and empty bands of the insulator as shown in Fig. 2.8a. In this case, the dopant band can act as an electron acceptor for electrons excited (either thermally or by light) from the filled band into the dopant band. Once electrons enter the dopant band, charge can flow and the system becomes a conductor. Another case is illustrated in Fig. 2.8b. Here, the dopant has its own band filled but lies close to the empty band of the insulator. Hence excitation of electrons from the dopant band to the empty band can induce current to flow.

2.3 Densities of states in one, two, and three dimensions

When a large number of neighboring orbitals overlap, bands are formed. However, the nature of these bands is very different in different dimensions.

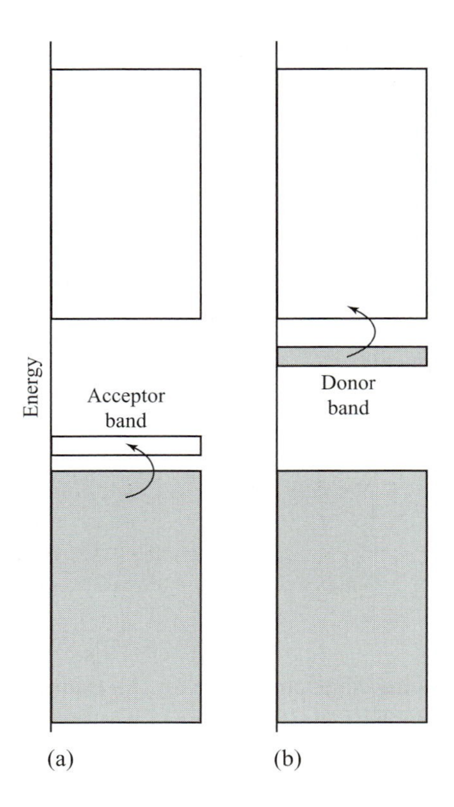

Figure 2.8 The filled and empty bands, the band gap, and empty acceptor or filled donor bands.

Before leaving our discussion of bands of orbitals and orbital energies in solids, I want to address the issue of the density of electronic states and the issue of what determines the energy range into which orbitals of a given band will split. First, let's recall the energy expression for the one- and two-dimensional electron in a box case, and let's generalize it to three dimensions. The general result is

$$E = \sum_j n_j^2 \pi^2 \hbar^2 / \left(2mL_j^2\right), \tag{2.10}$$

where the sum over j runs over the number of dimensions (one, two, or three), and L_j is the length of the box along the jth direction. For one dimension, one observes a pattern of energy levels that grows with increasing n, and whose spacing between neighboring energy levels also grows. However, in two and three dimensions, the pattern of energy level spacing displays a qualitatively different character at high quantum number.

Consider first the three-dimensional case and, for simplicity, let's use a "box" that has equal length sides L. In this case, the total energy E is $(\hbar^2\pi^2/2mL^2)$ times $(n_x^2 + n_y^2 + n_z^2)$. The latter quantity can be thought of as the square of the length of a vector **R** having three components n_x, n_y, n_z. Now think of three Cartesian axes labeled n_x, n_y, and n_z and view a sphere of radius R in this space. The volume of the 1/8 sphere having positive values of n_x, n_y, and n_z and having radius R is $1/8(4/3\pi R^3)$. Because each cube having unit length along the n_x, n_y, and n_z axes corresponds to a single quantum wave function and its energy, the total number $N_{\rm tot}(E)$ of quantum states with positive n_x, n_y, and n_z and with energy between zero and $E = (\hbar^2\pi^2/2mL^2)R^2$ is

$$N_{\rm tot} = \frac{1}{8}\left(\frac{4}{3}\pi R^3\right) = \frac{1}{8}\left(\frac{4}{3}\pi\left[\frac{2mEL^2}{\hbar^2\pi^2}\right]^{3/2}\right). \quad (2.11)$$

The number of quantum states with energies between E and $E + dE$ is $(dN_{\rm tot}/dE)dE$, which is the density $\Omega(E)$ of states near energy E:

$$\Omega(E) = \frac{1}{8}\left(\frac{4}{3}\pi\left[\frac{2mL^2}{\hbar^2\pi^2}\right]^{3/2}\frac{3}{2}E^{1/2}\right). \quad (2.12)$$

Notice that this state density increases as E increases. This means that, in the three-dimensional case, the number of quantum states per unit energy grows; in other words, the spacing between neighboring state energies decreases, very unlike the one-dimensional case where the spacing between neighboring states grows as n and thus E grows. This growth in state density in the three-dimensional case is a result of the degeneracies and near-degeneracies that occur. For example, the states with n_x, n_y, $n_z = 2, 1, 1$ and $1, 1, 2$, and $1, 2, 1$ are degenerate, and those with n_x, n_y, $n_z = 5, 3, 1$ or $5, 1, 3$ or $1, 3, 5$ or $1, 5, 3$ or $3, 1, 5$ or $3, 5, 1$ are degenerate and nearly degenerate to those having quantum numbers 4, 4, 1 or 1, 4, 4, or 4, 1, 4.

In the two-dimensional case, degereracies also occur and cause the density of states to possess an interesting E dependence. In this case, we think of states having energy $E = (\hbar^2\pi^2/2mL^2)R^2$, but with $R^2 = n_x^2 + n_y^2$. The total number of states having energy between zero and E is

$$N_{\rm total} = 4\pi R^2 = 4\pi E\left(\frac{2mL^2}{\hbar^2\pi^2}\right). \quad (2.13)$$

So, the density of states between E and $E + dE$ is

$$\Omega(E) = \frac{dN_{\rm total}}{dE} = 4\pi\left(\frac{2mL^2}{\hbar^2\pi^2}\right). \quad (2.14)$$

That is, in this two-dimensional case, the number of states per unit energy is constant for high E values (where the analysis above applies best).

This kind of analysis for the one-dimensional case gives

$$N_{\rm total} = R = \left(\frac{2mEL^2}{\hbar^2\pi^2}\right)^{1/2}, \quad (2.15)$$

so the state density between E and $E + dE$ is

$$\Omega(E) = 1/2\left(\frac{2mL^2}{\hbar^2\pi^2}\right)^{1/2} E^{-1/2}, \quad (2.16)$$

which clearly shows the widening spacing, and thus lower density, as one goes to higher energies.

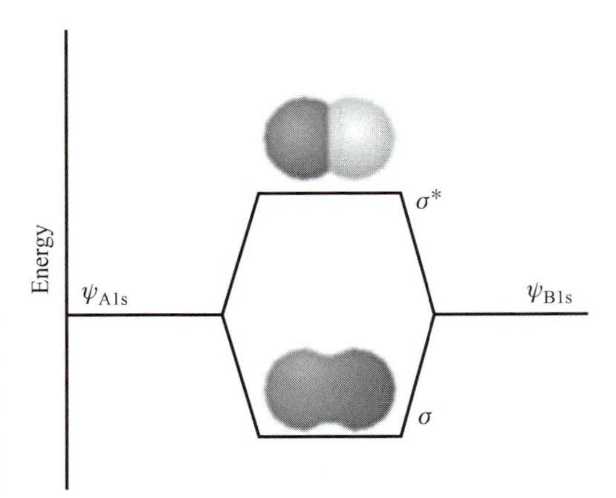

Figure 2.9 Two 1s orbitals combine to produce a σ bonding and a σ^* antibonding molecular orbital.

These findings about densities of states in one, two, and three dimensions are important because, in various problems one encounters in studying electronic states of extended systems such as solids and surfaces, one needs to know how the number of states available at a given total energy E varies with E. Clearly, the answer to this question depends upon the dimensionality of the problem, and this fact is what I want the students reading this text to keep in mind.

2.4 The most elementary model of orbital energy splittings: Hückel or tight-binding theory

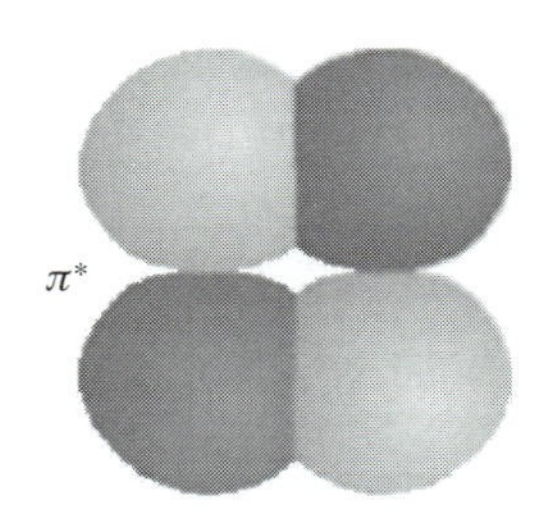

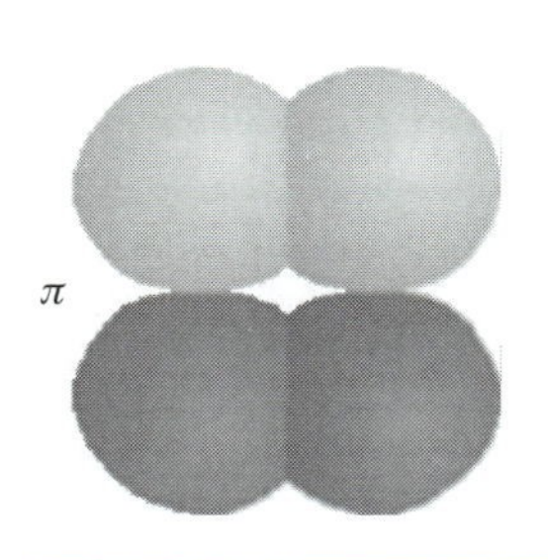

Figure 2.10 Two atomic p_π orbitals form a bonding π and antibonding π^* molecular orbital.

Now let's examine what determines the energy range into which orbitals (e.g., p_π orbitals in polyenes or metal s or p orbitals in a solid) split. To begin, consider two orbitals, one on an atom labeled A and another on a neighboring atom labeled B; these orbitals could be, for example, the 1s orbitals of two hydrogen atoms, such as Fig. 2.9 illustrates. However, the two orbitals could instead be two p_π orbitals on neighboring carbon atoms such as are shown in Fig. 2.10 as they form bonding and π^* antibonding orbitals. In both of these cases, we think of forming the molecular orbitals (MOs) ϕ_K as linear combinations of the atomic orbitals (AOs) χ_a on the constituent atoms, and we express this mathematically as follows:

$$\phi_K = \sum_a C_{K,a}\chi_a, \tag{2.17}$$

where the $C_{K,a}$ are called linear combination of atomic orbitals to form molecular orbital (LCAO-MO) coefficients. The MOs are supposed to be solutions to the Schrödinger equation in which the Hamiltonian H involves the kinetic energy of the electron as well as the potentials V_{L} and V_{R} detailing its attraction to the left and right atomic centers:

$$H = -\hbar^2/2m\nabla^2 + V_{\mathrm{L}} + V_{\mathrm{R}}. \tag{2.18}$$

In contrast, the AOs centered on the left atom A are supposed to be solutions of the Schrödinger equation whose Hamiltonian is $H = -\hbar^2/2m\nabla^2 + V_\mathrm{L}$, and the AOs on the right atom B have $H = -\hbar^2/2m\nabla^2 + V_R$. Substituting $\phi_K = \sum_a C_{K,a}\chi_a$ into the MO's Schrödinger equation $H\phi_K = \varepsilon_K\phi_K$ and then multiplying on the left by the complex conjugate of χ_b and integrating over the r, θ and ϕ coordinates of the electron produces

$$\sum_a \langle\chi_b|-\hbar^2/2m\nabla^2 + V_\mathrm{L} + V_\mathrm{R}|\chi_a\rangle C_{K,a} = \varepsilon_K \sum_a \langle\chi_b \mid \chi_a\rangle C_{K,a}. \tag{2.19}$$

Recall that the Dirac notation $\langle a \mid b\rangle$ denotes the integral of a^* and b, and $\langle a|\mathrm{op}|b\rangle$ denotes the integral of a^* and the operator op acting on b.

In what is known as the Hückel model in organic chemistry or the tight-binding model in solid-state theory, one approximates the integrals entering into the above set of linear equations as follows:

(i) The diagonal integral $\langle\chi_b|-\hbar^2/2m\nabla^2 + V_\mathrm{L} + V_\mathrm{R}|\chi_b\rangle$ involving the AO centered on the right atom and labeled χ_b is assumed to be equivalent to $\langle\chi_b|-\hbar^2/2m\nabla^2 + V_\mathrm{R}|\chi_\mathrm{b}\rangle$, which means that net attraction of this orbital to the left atomic center is neglected. Moreover, this integral is approximated in terms of the binding energy (denoted α, not to be confused with the electron spin function α) for an electron that occupies the χ_b orbital: $\langle\chi_b|-\hbar^2/2m\nabla^2 + V_\mathrm{R}|\chi_b\rangle = \alpha_b$. The physical meaning of α_b is the kinetic energy of the electron in χ_b plus the attraction of this electron to the right atomic center while it resides in χ_b. Of course, an analogous approximation is made for the diagonal integral involving χ_a; $\langle\chi_a|-\hbar^2/2m\nabla^2 + V_\mathrm{L}|\chi_\mathrm{a}\rangle = \alpha_a$.

(ii) The off-diagonal integrals $\langle\chi_b| - \hbar^2/2m\nabla^2 + V_\mathrm{L} + V_\mathrm{R}|\chi_a\rangle$ are expressed in terms of a parameter $\beta_{a,b}$ which relates to the kinetic and potential energy of the electron while it resides in the "overlap region" in which both χ_a and χ_b are non-vanishing. This region is shown pictorially in Fig. 2.10 as the region where the left and right orbitals touch or overlap. The magnitude of β is assumed to be proportional to the overlap $S_{a,b}$ between the two AOs: $S_{a,b} = \langle\chi_a \mid \chi_b\rangle$. It turns out that β is usually a negative quantity, which can be seen by writing it as $\langle\chi_b|-\hbar^2/2m\nabla^2 + V_\mathrm{R}|\chi_a\rangle + \langle\chi_b|V_\mathrm{L}|\chi_a\rangle$. Since χ_a is an eigenfunction of $-\hbar^2/2m\nabla^2 + V_\mathrm{R}$ having the eigenvalue α_a, the first term is equal to α_a (a negative quantity) times $\langle\chi_b \mid \chi_a\rangle$, the overlap S. The second quantity $\langle\chi_b|V_\mathrm{L}|\chi_\mathrm{a}\rangle$ is equal to the integral of the overlap density $\chi_b(r)\chi_a(r)$ multiplied by the (negative) Coulomb potential for attractive interaction of the electron with the left atomic center. So, whenever $\chi_b(r)$ and $\chi_a(r)$ have positive overlap, β will turn out negative.

(iii) Finally, in the most elementary Hückel or tight-binding model, the overlap integrals $\langle\chi_a \mid \chi_b\rangle = S_{a,b}$ are neglected and set equal to zero on the right side of the matrix eigenvalue equation. However, in some Hückel models, overlap between neighboring orbitals is explictly treated, so in some of the discussion below we will retain $S_{a,b}$.

With these Hückel approximations, the set of equations that determine the orbital energies ε_K and the corresponding LCAO-MO coefficients $C_{K,a}$ are written for the two-orbital case at hand as in the first 2×2 matrix equations shown below:

$$\begin{bmatrix} \alpha & \beta \\ \beta & \alpha \end{bmatrix} \begin{bmatrix} C_\mathrm{L} \\ C_\mathrm{R} \end{bmatrix} = \varepsilon \begin{bmatrix} 1 & S \\ S & 1 \end{bmatrix} \begin{bmatrix} C_\mathrm{L} \\ C_\mathrm{R} \end{bmatrix}, \tag{2.20}$$

which is sometimes written as

$$\begin{bmatrix} \alpha - \varepsilon & \beta - \varepsilon S \\ \beta - \varepsilon S & \alpha - \varepsilon \end{bmatrix} \begin{bmatrix} C_\mathrm{L} \\ C_\mathrm{R} \end{bmatrix} = \begin{bmatrix} 0 \\ 0 \end{bmatrix}. \tag{2.21}$$

These equations reduce with the assumption of zero overlap to

$$\begin{bmatrix} \alpha & \beta \\ \beta & \alpha \end{bmatrix} \begin{bmatrix} C_\mathrm{L} \\ C_\mathrm{R} \end{bmatrix} = \varepsilon \begin{bmatrix} 1 & 0 \\ 0 & 1 \end{bmatrix} \begin{bmatrix} C_\mathrm{L} \\ C_\mathrm{R} \end{bmatrix}. \tag{2.22}$$

The α parameters are identical if the two AOs χ_a and χ_b are identical, as would be the case for bonding between the two 1s orbitals of two H atoms or two $2p_\pi$ orbitals of two C atoms or two 3s orbitals of two Na atoms. If the left and right orbitals were not identical (e.g., for bonding in HeH^+ or for the π bonding in a C–O group), their α values would be different and the Hückel matrix problem would look like:

$$\begin{bmatrix} \alpha & \beta \\ \beta & \alpha' \end{bmatrix} \begin{bmatrix} C_\mathrm{L} \\ C_\mathrm{R} \end{bmatrix} = \varepsilon \begin{bmatrix} 1 & S \\ S & 1 \end{bmatrix} \begin{bmatrix} C_\mathrm{L} \\ CR \end{bmatrix}. \tag{2.23}$$

To find the MO energies that result from combining the AOs, one must find the values of ε for which the above equations are valid. Taking the 2×2 matrix consisting of ε times the overlap matrix to the left-hand side, the above set of equations reduces to Eq. (2.21). It is known from matrix algebra that such a set of linear homogeneous equations (i.e., having zeroes on the right-hand sides) can have non-trivial solutions (i.e., values of C that are not simply zero) only if the determinant of the matrix on the left side vanishes. Setting this determinant equal to zero gives a quadratic equation in which the ε values are the unknowns:

$$(\alpha - \varepsilon)^2 - (\beta - \varepsilon S)^2 = 0. \tag{2.24}$$

This quadratic equation can be factored into a product

$$(\alpha - \beta - \varepsilon + \varepsilon S)(\alpha + \beta - \varepsilon - \varepsilon S) = 0, \tag{2.25}$$

which has two solutions

$$\varepsilon = (\alpha + \beta)/(1 + S), \quad \text{and} \quad \varepsilon = (\alpha - \beta)/(1 - S). \tag{2.26}$$

As discussed earlier, it turns out that the β values are usually negative, so the lowest energy such solution is the $\varepsilon = (\alpha + \beta)/(1 + S)$ solution, which gives the energy of the bonding MO. Notice that the energies of the bonding and antibonding MOs are not symmetrically displaced from the value α within this version of the Hückel model that retains orbital overlap. In fact, the bonding orbital lies less than β below α, and the antibonding MO lies more than β above α because of the $1 + S$ and $1 - S$ factors in the respective denominators. This asymmetric lowering and raising of the MOs relative to the energies of the constituent AOs is commonly observed in chemical bonds; that is, the antibonding orbital is more antibonding than the bonding orbital is bonding. This is another important thing to keep in mind because its effects pervade chemical bonding and spectroscopy.

Having noted the effect of inclusion of AO overlap effects in the Hückel model, I should admit that it is far more common to utilize the simplified version of the Hückel model in which the S factors are ignored. In so doing, one obtains patterns of MO orbital energies that do not reflect the asymmetric splitting in bonding and antibonding orbitals noted above. However, this simplified approach is easier to use and offers qualitatively correct MO energy orderings. So, let's proceed with our discussion of the Hückel model in its simplified version.

To obtain the LCAO-MO coefficients corresponding to the bonding and antibonding MOs, one substitutes the corresponding α values into the linear equations

$$\begin{bmatrix} \alpha - \varepsilon & \beta \\ \beta & \alpha - \varepsilon \end{bmatrix} \begin{bmatrix} C_{\mathrm{L}} \\ C_{\mathrm{R}} \end{bmatrix} = \begin{bmatrix} 0 \\ 0 \end{bmatrix} \tag{2.27}$$

and solves for the C_a coefficients (actually, one can solve for all but one C_a, and then use normalization of the MO to determine the final C_a). For example, for the bonding MO, we substitute $\varepsilon = \alpha + \beta$ into the above matrix equation and obtain two equations for C_{L} and C_{R}:

$$-\beta C_{\mathrm{L}} + \beta C_{\mathrm{R}} = 0, \tag{2.28}$$

$$\beta C_{\mathrm{L}} - \beta C_{\mathrm{R}} = 0. \tag{2.29}$$

These two equations are clearly not independent; either one can be solved for one C in terms of the other C to give

$$C_{\mathrm{L}} = C_{\mathrm{R}}, \tag{2.30}$$

which means that the bonding MO is

$$\phi = C_{\mathrm{L}}(\chi_{\mathrm{L}} + \chi_{\mathrm{R}}). \tag{2.31}$$

The final unknown, C_{L}, is obtained by noting that ϕ is supposed to be a normalized function $\langle \phi \mid \phi \rangle = 1$. Within this version of the Hückel model, in which the overlap S is neglected, the normalization of ϕ leads to the following condition:

$$1 = \langle \phi \mid \phi \rangle = C_{\mathrm{L}}^2(\langle \chi_{\mathrm{L}} \mid \chi_{\mathrm{L}} \rangle + \langle \chi_{\mathrm{R}} \chi_{\mathrm{R}} \rangle) = 2C_{\mathrm{L}}^2, \tag{2.32}$$

with the final result depending on assuming that each χ is itself also normalized. So, finally, we know that $C_L = (1/2)^{1/2}$, and hence the bonding MO is

$$\phi = (1/2)^{1/2}(\chi_L + \chi_R). \tag{2.33}$$

Actually, the solution of $1 = 2C_L^2$ could also have yielded $C_L = -(1/2)^{1/2}$ and then we would have

$$\phi = -(1/2)^{1/2}(\chi_L + \chi_R). \tag{2.34}$$

These two solutions are not independent (one is just -1 times the other), so only one should be included in the list of MOs. However, either one is just as good as the other because, as shown very early in this text, all of the physical properties that one computes from a wave function depend not on ψ but on $\psi^*\psi$. So, two wave functions that differ from one another by an overall sign factor, as we have here, have exactly the same $\psi^*\psi$ and thus are equivalent.

In like fashion, we can substitute $\varepsilon = \alpha - \beta$ into the matrix equation and solve for the C_L and C_R values that are appropriate for the antibonding MO. Doing so gives us

$$\phi^* = (1/2)^{1/2}(\chi_L - \chi_R) \tag{2.35}$$

or, alternatively,

$$\phi^* = (1/2)^{1/2}(\chi_R - \chi_L). \tag{2.36}$$

Again, the fact that either expression for ϕ^* is acceptable shows a property of all solutions to any Schrödinger equations; any multiple of a solution is also a solution. In the above example, the two "answers" for ϕ^* differ by a multiplicative factor of (-1).

Let's try another example to practice using Hückel or tight-binding theory. In particular, I'd like you to imagine two possible structures for a cluster of three Na atoms (i.e., pretend that someone came to you and asked what geometry you think such a cluster would assume in its ground electronic state), one linear and one an equilateral triangle. Further, assume that the Na–Na distances in both such clusters are equal (i.e., that the person asking for your theoretical help is willing to assume that variations in bond lengths are not the crucial factor in determining which structure is favored). In Fig. 2.11, I show the two candidate clusters and their 3s orbitals.

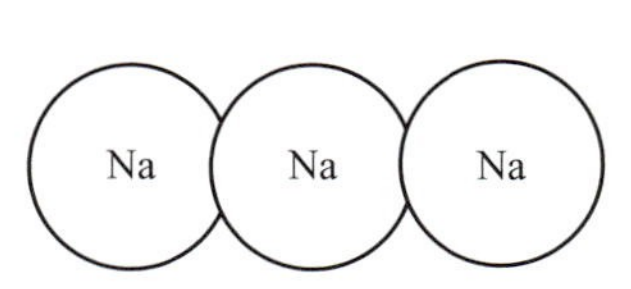

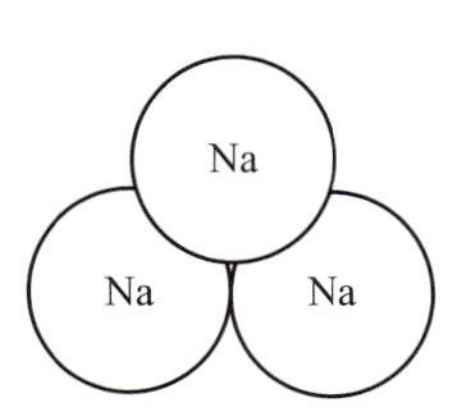

Figure 2.11 Linear and equilateral triangle structures of sodium trimer.

Numbering the three Na atoms' valence 3s orbitals χ_1, χ_2, and χ_3, we then set up the 3×3 Hückel matrix appropriate to the two candidate structures:

$$\begin{bmatrix} \alpha & \beta & 0 \\ \beta & \alpha & \beta \\ 0 & \beta & \alpha \end{bmatrix} \tag{2.37}$$

for the linear structure (n.b., the zeroes arise because χ_1 and χ_3 do not overlap and thus have no β coupling matrix element). Alternatively, for the triangular structure, we find

$$\begin{bmatrix} \alpha & \beta & \beta \\ \beta & \alpha & \beta \\ \beta & \beta & \alpha \end{bmatrix} \tag{2.38}$$

as the Hückel matrix. Each of these 3×3 matrices will have three eigenvalues that we obtain by subtracting ε from their diagonals and setting the determinants of the resulting matrices to zero. For the linear case, doing so generates

$$(\alpha - \varepsilon)^3 - 2\beta^2(\alpha - \varepsilon) = 0, \tag{2.39}$$

and for the triangle case it produces

$$(\alpha - \varepsilon)^3 - 3\beta^2(\alpha - \varepsilon) + 2\beta^2 = 0. \tag{2.40}$$

The first cubic equation has three solutions that give the MO energies:

$$\varepsilon = \alpha + (2)^{1/2}\beta, \qquad \varepsilon = \alpha, \qquad \text{and} \quad \varepsilon = \alpha - (2)^{1/2}\beta, \tag{2.41}$$

for the bonding, non-bonding and antibonding MOs, respectively. The second cubic equation also has three solutions

$$\varepsilon = \alpha + 2\beta, \qquad \varepsilon = \alpha - \beta, \qquad \text{and} \quad \varepsilon = \alpha - \beta. \tag{2.42}$$

So, for the linear and triangular structures, the MO energy patterns are as shown in Fig. 2.12.

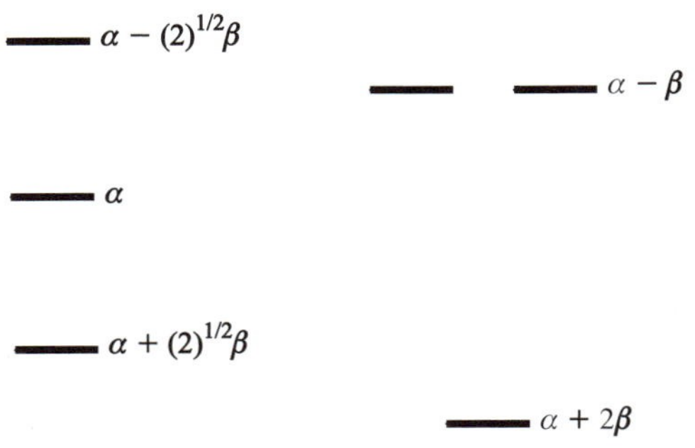

$\alpha - (2)^{1/2}\beta$

$\alpha - \beta$

α

$\alpha + (2)^{1/2}\beta$

$\alpha + 2\beta$

Figure 2.12 Energy orderings of molecular orbitals of linear (left) and triangular (right) sodium trimers.

For the neutral Na_3 cluster about which you were asked, you have three valence electrons to distribute among the lowest available orbitals. In the linear case, we place two electrons into the lowest orbital and one into the second orbital. Doing so produces a three-electron state with a total energy of $E = 2(\alpha + 2^{1/2}\beta) + \alpha = 3\alpha + 2\,2^{1/2}\beta$. Alternatively, for the triangular species, we put two electrons into the lowest MO and one into either of the degenerate MOs resulting in a three-electron state with total energy $E = 3\alpha + 3\beta$. Because β is a negative quantity, the total energy of the triangular structure is lower than that of the linear structure since $3 > 2\,2^{1/2}$.

The above example illustrates how we can use Hückel/tight-binding theory to make qualitative predictions (e.g., which of two "shapes" is likely to be of lower energy). Notice that all one needs to know to apply such a model to any set of atomic orbitals that overlap to form MOs is:

(i) the individual AO energies α (which relate to the electronegativity of the AOs) and
(ii) the degree to which the AOs couple (the β parameters which relate to AO overlaps).

Let's see if you can do some of this on your own. Using the above results, would you expect the cation Na_3^+ to be linear or triangular? What about the anion Na_3^-? Next, I want you to substitute the MO energies back into the 3×3 matrix and find the C_1, C_2, and C_3 coefficients appropriate to each of the three MOs of the linear and of the triangular structure. See if doing so leads you to solutions that can be depicted as shown in Fig. 2.13, and see if you can place each set of MOs in the proper energy ordering.

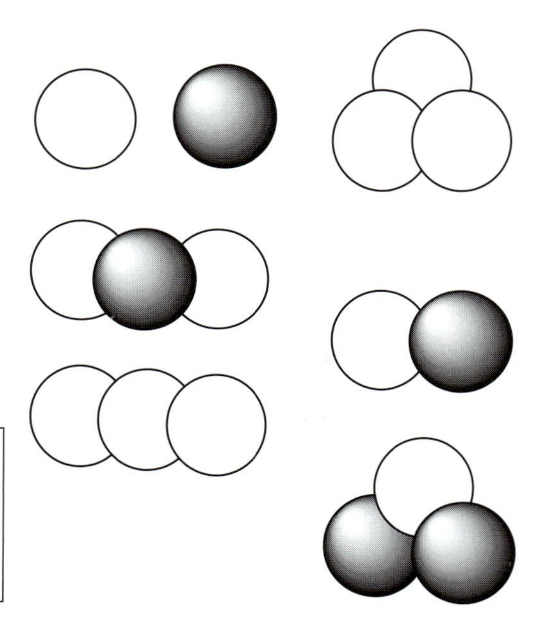

Figure 2.13 The molecular orbitals of linear and triangular sodium trimers (note, they are not energy ordered).

Figure 2.14 Ethylene molecule with four C—H bonds, one C—C σ bond, and one C—C π bond.

Now, I want to show you how to broaden your horizons and use tight-binding theory to describe all of the bonds in a more complicated molecule such as ethylene shown in Fig. 2.14. Within the model described above, each pair of orbitals that "touch" or overlap gives rise to a 2×2 matrix. More correctly, all n of the constituent AOs form an $n \times n$ matrix, but this matrix is broken up into 2×2 blocks whenever each AO touches only one other AO. Notice that this did not happen in the triangular Na_3 case where each AO touched two other AOs. For the ethylene case, the valence AOs consist of (a) four equivalent C sp^2 orbitals that are directed toward the four H atoms, (b) four H 1s orbitals, (c) two C sp^2 orbitals directed toward one another to form the C—C σ bond, and (d) two C p_π orbitals that will form the C—C π bond. This total of 12 AOs generates six Hückel matrices as shown below. We obtain one 2×2 matrix for the C—C σ bond of the form

$$\begin{bmatrix} \alpha_{sp^2} & \beta_{sp^2,sp^2} \\ \beta_{sp^2,sp^2} & \alpha_{sp^2} \end{bmatrix} \tag{2.43}$$

and one 2×2 matrix for the C—C π bond of the form

$$\begin{bmatrix} \alpha_{p\pi} & \beta_{p\pi,p\pi} \\ \beta_{p\pi,p\pi} & \alpha_{p\pi} \end{bmatrix}. \tag{2.44}$$

Finally, we also obtain four identical 2×2 matrices for the C—H bonds:

$$\begin{bmatrix} \alpha_{sp^2} & \beta_{sp^2,H} \\ \beta_{sp^2,H} & \alpha_{H} \end{bmatrix}. \tag{2.45}$$

The above matrices will then produce (i) four identical C—H bonding MOs having energies $\varepsilon = 1/2\{(\alpha_H + \alpha_C) - [(\alpha_H - \alpha_C)^2 + 4\beta^2]^{1/2}\}$, (ii) four identical C—H antibonding MOs having energies $\varepsilon^* = 1/2\{(\alpha_H + \alpha_C) + [(\alpha_H - \alpha_C)^2 + 4\beta^2]^{1/2}\}$, (iii) one bonding C—C π orbital with $\varepsilon = \alpha_{p\pi} + \beta$, (iv) a partner antibonding C—C orbital with $\varepsilon^* = \alpha_{p\pi} - \beta$, (v) a C—C σ bonding MO with $\varepsilon = \alpha_{sp^2} + \beta$, and (vi) its antibonding partner with $\varepsilon^* = \alpha_{sp^2} - \beta$. In all of these expressions, the β parameter is supposed to be that appropriate to the specific orbitals that overlap as shown in the matrices.

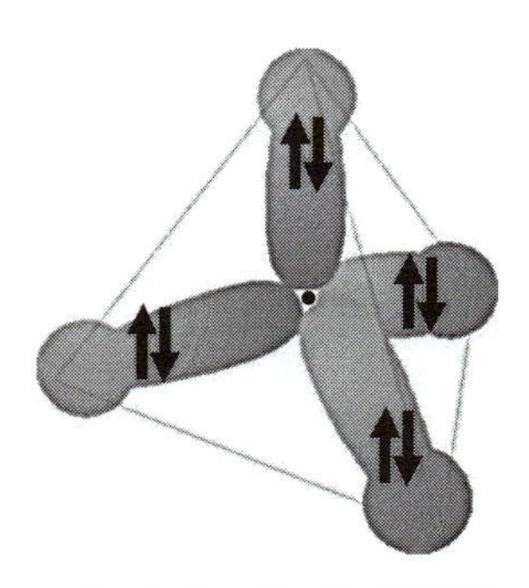

Figure 2.15 Methane molecule with four C—H bonds.

If you wish to practice this exercise of breaking a large molecule down into sets of interacting AOs, try to see what Hückel matrices you obtain and what bonding and antibonding MO energies you obtain for the valence orbitals of methane shown in Fig. 2.15.

Before leaving this discussion of the Hückel/tight-binding model, I need to stress that it has its flaws (because it is based on approximations and involves neglecting certain terms in the Schrödinger equation). For example, it predicts (see above) that ethylene has four energetically identical C—H bonding MOs (and four degenerate C—H antibonding MOs). However, this is not what is seen when photoelectron spectra are used to probe the energies of these MOs. Likewise, it suggests that methane has four equivalent C—H bonding and antibonding orbitals, which again is not true. It turns out that, in each of these two cases (ethylene and methane), the experiments indicate a grouping of four *nearly* iso-energetic bonding MOs and four nearly iso-energetic antibonding MOs. However, there is some "splitting" among these clusters of four MOs. The splittings can be interpreted, within the Hückel model, as arising from couplings or interactions among, for example, one sp^2 or sp^3 orbital on a given C atom and another such orbital on the same atom. Such couplings cause the $n \times n$ Hückel matrix to not block-partition into groups of 2×2 sub-matrices because now there exist off-diagonal β factors that couple one pair of directed AOs to another. When such couplings are included in the analysis, one finds that the clusters of MOs expected to be degenerate are not, but are split just as the photoelectron data suggest.

2.5 Hydrogenic orbitals

The hydrogenic atom problem forms the basis of much of our thinking about atomic structure. To solve the corresponding Schrödinger equation requires separation of the r, θ, and ϕ variables.

The Schrödinger equation for a single particle of mass μ moving in a central potential (one that depends only on the radial coordinate r) can be written as

$$-\frac{\hbar^2}{2\mu}\left(\frac{\partial^2}{\partial x^2}+\frac{\partial^2}{\partial y^2}+\frac{\partial^2}{\partial z^2}\right)\psi + V\left(\sqrt{x^2+y^2+z^2}\right)\psi = E\psi, \tag{2.46}$$

or, introducing the short-hand notation ∇^2:

$$-\hbar^2/2m\nabla^2\psi + V\psi = E\psi. \tag{2.47}$$

This equation is not separable in Cartesian coordinates (x, y, z) because of the way x, y, and z appear together in the square root. However, it is separable in spherical coordinates where it has the form

$$-\frac{\hbar^2}{2\mu r^2}\left[\frac{\partial}{\partial r}\left(r^2\frac{\partial\psi}{\partial r}\right)\right] + \frac{1}{r^2\sin\theta}\frac{\partial}{\partial\theta}\left(\sin\theta\frac{\partial\psi}{\partial\theta}\right) + \frac{1}{r^2\sin^2\theta}\frac{\partial^2\psi}{\partial\phi^2} + V(r)\psi$$

$$= -\hbar^2/2m\nabla^2\psi + V\psi = E\psi. \tag{2.48}$$

Subtracting $V(r)\psi$ from both sides of the equation and multiplying by $-\frac{2\mu r^2}{\hbar^2}$ then moving the derivatives with respect to r to the right-hand side, one obtains

$$\frac{1}{\sin\theta}\frac{\partial}{\partial\theta}\left(\sin\theta\frac{\partial\psi}{\partial\theta}\right)+\frac{1}{\sin^2\theta}\frac{\partial^2\psi}{\partial\phi^2}=-\frac{2\mu r^2}{\hbar^2}[E-V(r)]\psi-\frac{\partial}{\partial r}\left(r^2\frac{\partial\psi}{\partial r}\right). \tag{2.49}$$

Notice that, except for ψ itself, the right-hand side of this equation is a function of r only; it contains no θ or ϕ dependence. Let's call the entire right-hand side $F(r)\psi$ to emphasize this fact.

To further separate the θ and ϕ dependence, we multiply by $\sin^2\theta$ and subtract the θ derivative terms from both sides to obtain

$$\frac{\partial^2\psi}{\partial\phi^2}=F(r)\psi\sin^2\theta-\sin\theta\frac{\partial}{\partial\theta}\left(\sin\theta\frac{\partial\psi}{\partial\theta}\right). \tag{2.50}$$

Now we have separated the ϕ dependence from the θ and r dependence. We now introduce the procedure used to separate variables in differential equations and assume ψ can be written as a function of ϕ times a function of r and θ: $\psi=\Phi(\phi)Q(r,\theta)$. Dividing by ΦQ, we obtain

$$\frac{1}{\Phi}\frac{\partial^2\Phi}{\partial\phi^2}=\frac{1}{Q}\left(F(r)\sin^2\theta\, Q-\sin\theta\frac{\partial}{\partial\theta}\left(\sin\theta\frac{\partial Q}{\partial\theta}\right)\right). \tag{2.51}$$

Now all of the ϕ dependence is isolated on the left-hand side; the right-hand side contains only r and θ dependence.

Whenever one has isolated the entire dependence on one variable as we have done above for the ϕ dependence, one can easily see that the left- and right-hand sides of the equation must equal a constant. For the above example, the left-hand side contains no r or θ dependence and the right-hand side contains no ϕ dependence. Because the two sides are equal, they both must actually contain no r, θ, or ϕ dependence; that is, they are constant.

For the above example, we therefore can set both sides equal to a so-called separation constant that we call $-m^2$. It will become clear shortly why we have chosen to express the constant in the form of minus the square of an integer. You may recall that we studied this same ϕ equation earlier and learned how the integer m arises via the boundary condition that ϕ and $\phi+2\pi$ represent identical geometries.

2.5.1 The Φ equation

The resulting Φ equation reads (the $''$symbol is used to represent second derivative)

$$\Phi''+m^2\Phi=0. \tag{2.52}$$

This equation should be familiar because it is the equation that we treated much earlier when we discussed the z-component of angular momentum. So, its further analysis should also be familiar, but, for completeness, I repeat much of it. The above equation has as its most general solution

$$\Phi = Ae^{im\phi} + Be^{-im\phi}. \tag{2.53}$$

Because the wave functions of quantum mechanics represent probability densities, they must be continuous and single-valued. The latter condition, applied to our Φ function, means that

$$\Phi(\phi) = \Phi(2\pi + \phi) \tag{2.54}$$

or

$$Ae^{im\phi}(1 - e^{2im\pi}) + Be^{-im\phi}(1 - e^{-2im\pi}) = 0. \tag{2.55}$$

This condition is satisfied only when the separation constant is equal to an integer $m = 0, \pm 1, \pm 2, \ldots$ and provides another example of the rule that quantization comes from the boundary conditions on the wave function. Here m is restricted to certain discrete values because the wave function must be such that when you rotate through 2π about the z-axis, you must get back what you started with.

2.5.2 The Θ equation

Now returning to the equation in which the ϕ dependence was isolated from the r and θ dependence and rearranging the θ terms to the left-hand side, we have

$$\frac{1}{\sin\theta}\frac{\partial}{\partial\theta}\left(\sin\theta\frac{\partial Q}{\partial\theta}\right) - \frac{m^2 Q}{\sin^2\theta} = F(r)Q. \tag{2.56}$$

In this equation we have separated θ and r variations so we can further decompose the wave function by introducing $Q = \Theta(\theta)R(r)$, which yields

$$\frac{1}{\Theta}\frac{1}{\sin\theta}\frac{\partial}{\partial\theta}\left(\sin\theta\frac{\partial\Theta}{\partial\theta}\right) - \frac{m^2}{\sin^2\theta} = \frac{F(r)R}{R} = -\lambda, \tag{2.57}$$

where a second separation constant, $-\lambda$, has been introduced once the r and θ dependent terms have been separated onto the right- and left-hand sides, respectively.

We now can write the θ equation as

$$\frac{1}{\sin\theta}\frac{\partial}{\partial\theta}\left(\sin\theta\frac{\partial\Theta}{\partial\theta}\right) - \frac{m^2\Theta}{\sin^2\theta} = -\lambda\Theta, \tag{2.58}$$

where m is the integer introduced earlier. To solve this equation for Θ, we make the substitutions $z = \cos\theta$ and $P(z) = \Theta(\theta)$, so $\sqrt{1 - z^2} = \sin\theta$, and

$$\frac{\partial}{\partial\theta} = \frac{\partial z}{\partial\theta}\frac{\partial}{\partial z} = -\sin\theta\frac{\partial}{\partial z}. \tag{2.59}$$

The range of values for θ was $0 \leq \theta < \pi$, so the range for z is $-1 < z < 1$. The equation for Θ, when expressed in terms of P and z, becomes

$$\frac{d}{dz}\left((1-z^2)\frac{dP}{dz}\right) - \frac{m^2 P}{1-z^2} + \lambda P = 0. \tag{2.60}$$

Now we can look for polynomial solutions for P, because z is restricted to be less than unity in magnitude. If $m = 0$, we first let

$$P = \sum_{k=0}^{\infty} a_k z^k, \tag{2.61}$$

and substitute into the differential equation to obtain

$$\sum_{k=0}^{\infty}(k+2)(k+1)a_{k+2}z^k - \sum_{k=0}^{\infty}(k+1)ka_k z^k + \lambda \sum_{k=0}^{\infty} a_k z^k = 0. \tag{2.62}$$

Equating like powers of z gives

$$a_{k+2} = \frac{a_k(k(k+1)-\lambda)}{(k+2)(k+1)}. \tag{2.63}$$

Note that for large values of k

$$\frac{a_{k+2}}{a_k} \to \frac{k^2\left(1+\frac{1}{k}\right)}{k^2\left(1+\frac{2}{k}\right)\left(1+\frac{1}{k}\right)} = 1. \tag{2.64}$$

Since the coefficients do not decrease with k for large k, this series will diverge for $z = \pm 1$ unless it truncates at finite order. This truncation only happens if the separation constant λ obeys $\lambda = l(l+1)$, where l is an integer. So, once again, we see that a boundary condition (i.e., that the wave function not diverge and thus be normalizable in this case) gives rise to quantization. In this case, the values of λ are restricted to $l(l+1)$; before, we saw that m is restricted to $0, \pm 1, \pm 2, \ldots$

Since the above recursion relation links every other coefficient, we can choose to solve for the even and odd functions separately. Choosing a_0 and then determining all of the even a_k in terms of this a_0, followed by rescaling all of these a_k to make the function normalized, generates an even solution. Choosing a_l and determining all of the odd a_k in like manner generates an odd solution.

For $l = 0$, the series truncates after one term and results in $P_0(z) = 1$. For $l = 1$ the same thing applies and $P_1(z) = z$. For $l = 2$, $a_2 = -6\frac{a_0}{2} = -3a_0$, so one obtains $P_2 = 3z^2 - 1$, and so on. These polynomials are called Legendre polynomials.

For the more general case where $m \neq 0$, one can proceed as above to generate a polynomial solution for the Θ function. Doing so results in the following solutions:

$$P_l^m(z) = (1-z^2)^{\frac{|m|}{2}} \frac{d^{|m|} P_l(z)}{dz^{|m|}}. \tag{2.65}$$

These functions are called associated Legendre polynomials, and they constitute the solutions to the Θ problem for non-zero m values.

The above P and $e^{im\phi}$ functions, when re-expressed in terms of θ and ϕ, yield the full angular part of the wave function for any centrosymmetric potential. These solutions are usually written as $Y_{l,m}(\theta, \phi) = P_l^m(\cos\theta)(2\pi)^{-1/2}\exp(im\phi)$, and are called spherical harmonics. They provide the angular solution of the (r, θ, ϕ) Schrödinger equation for *any* problem in which the potential depends only on the radial coordinate. Such situations include all one-electron atoms and ions (e.g., H, He^+, Li^{++}, etc.), the rotational motion of a diatomic molecule (where the potential depends only on bond length r), the motion of a nucleon in a spherically symmetrical "box" (as occurs in the shell model of nuclei), and the scattering of two atoms (where the potential depends only on interatomic distance).The $Y_{l,m}$ functions possess varying numbers of angular nodes, which, as noted earlier, give clear signatures of the angular or rotational energy content of the wave function. These angular nodes originate in the oscillatory nature of the Legendre and associated Legendre polynomials $P_l^m(\cos\theta)$; the higher l is, the more sign changes occur within the polynomial.

2.5.3 The R equation

Let us now turn our attention to the radial equation, which is the only place that the explicit form of the potential appears. Using our earlier results for the equation obeyed by the $R(r)$ function and specifying $V(r)$ to be the Coulomb potential appropriate for an electron in the field of a nucleus of charge $+Ze$, yields

$$\frac{1}{r^2}\frac{d}{dr}\left(r^2\frac{dR}{dr}\right) + \left[\frac{2\mu}{\hbar^2}\left(E + \frac{Ze^2}{r}\right) - \frac{l(l+1)}{r^2}\right]R = 0. \quad (2.66)$$

We can simplify things considerably if we choose rescaled length and energy units because doing so removes the factors that depend on μ, $\hbar$, and e. We introduce a new radial coordinate ρ and a quantity σ as follows:

$$\rho = \left(\frac{-8\mu E}{\hbar^2}\right)^{\frac{1}{2}} r, \quad \text{and} \quad \sigma^2 = -\frac{\mu Z^2 e^4}{2E\hbar^2}. \quad (2.67)$$

Notice that if E is negative, as it will be for bound states (i.e., those states with energy below that of a free electron infinitely far from the nucleus and with zero kinetic energy), ρ is real. On the other hand, if E is positive, as it will be for states that lie in the continuum, ρ will be imaginary. These two cases will give rise to qualitatively different behavior in the solutions of the radial equation developed below.

We now define a function S such that $S(\rho) = R(r)$ and substitute S for R to obtain

$$\frac{1}{\rho^2}\frac{d}{d\rho}\left(\rho^2\frac{dS}{d\rho}\right) + \left(-\frac{1}{4} - \frac{l(l+1)}{\rho^2} + \frac{\sigma}{\rho}\right)S = 0. \quad (2.68)$$

The differential operator terms can be recast in several ways using

$$\frac{1}{\rho^2}\frac{d}{d\rho}\left(\rho^2\frac{dS}{d\rho}\right) = \frac{d^2S}{d\rho^2} + \frac{2}{\rho}\frac{dS}{d\rho} = \frac{1}{\rho}\frac{d^2}{d\rho^2}(\rho S). \tag{2.69}$$

The strategy that we now follow is characteristic of solving second order differential equations. We will examine the equation for S at large and small ρ values. Having found solutions at these limits, we will use a power series in ρ to "interpolate" between these two limits.

Let us begin by examining the solution of the above equation at small values of ρ to see how the radial functions behave at small r. As $\rho \to 0$, the second term in the brackets in Eq. (2.68) will dominate. Neglecting the other two terms in the brackets, we find that, for small values of ρ (or r), the solution should behave like ρ^L and because the function must be normalizable, we must have $L \geq 0$. Since L can be any non-negative integer, this suggests the following more general form for $S(\rho)$:

$$S(\rho) \approx \rho^L e^{-a\rho}. \tag{2.70}$$

This form will insure that the function is normalizable since $S(\rho) \to 0$ as $r \to \infty$ for all L, as long as ρ is a real quantity. If ρ is imaginary, such a form may not be normalized (see below for further consequences).

Turning now to the behavior of S for large ρ, we make the substitution of $S(\rho)$ into the above equation and keep only the terms with the largest power of ρ (e.g., the first term in brackets in Eq. (2.68)). Upon so doing, we obtain the equation

$$a^2\rho^L e^{-a\rho} = \frac{1}{4}\rho^L e^{-a\rho}, \tag{2.71}$$

which leads us to conclude that the exponent in the large-ρ behavior of S is $a = \frac{1}{2}$. Having found the small- and large-ρ behaviors of $S(\rho)$, we can take S to have the following form to interpolate between large and small ρ-values:

$$S(\rho) = \rho^L e^{-\rho/2} P(\rho), \tag{2.72}$$

where the function P is expanded in an infinite power series in ρ as $P(\rho) = \sum a_k \rho^k$. Again substituting this expression for S into the above equation we obtain

$$P''\rho + P'(2L + 2 - \rho) + P(\sigma - L - 1) = 0, \tag{2.73}$$

and then substituting the power series expansion of P and solving for the a_ks we arrive at a recursion relation for the a_k coefficients:

$$a_{k+1} = \frac{(k - \sigma + L + 1)a_k}{(k + 1)(k + 2L + 2)}. \tag{2.74}$$

For large k, the ratio of expansion coefficients reaches the limit $a_{k+1}/a_k = 1/k$, which has the same behavior as the power series expansion of e^{ρ}. Because the power series expansion of P describes a function that behaves like e^ρ for large ρ, the resulting $S(\rho)$ function would not be normalizable because the $e^{-\rho/2}$ factor would be overwhelmed by this e^ρ dependence. Hence, the series expansion of P must truncate in order to achieve a normalizable S function. Notice that if ρ is imaginary, as it will be if E is in the continuum, the argument that the series must truncate to avoid an exponentially diverging function no longer applies. Thus, we see a key difference between bound (with ρ real) and continuum (with ρ imaginary) states. In the former case, the boundary condition of non-divergence arises; in the latter, it does not because $\exp(\rho/2)$ does not diverge if ρ is imaginary.

To truncate at a polynomial of order n', we must have $n' - \sigma + L + 1 = 0$. This implies that the quantity σ introduced previously is restricted to $\sigma = n' + L + 1$, which is certainly an integer; let us call this integer n. If we label states in order of increasing $n = 1, 2, 3, \ldots$, we see that doing so is consistent with specifying a maximum order (n') in the $P(\rho)$ polynomial $n' = 0, 1, 2, \ldots$ after which the L-value can run from $L = 0$, in steps of unity, up to $L = n - 1$.

Substituting the integer n for σ, we find that the energy levels are quantized because σ is quantized (equal to n):

$$E = -\frac{\mu Z^2 e^4}{2\hbar^2 n^2} \tag{2.75}$$

and the scaled distance turns out to be

$$\rho = \frac{Zr}{a_0 n}. \tag{2.76}$$

Here, the length a_0 is the so-called Bohr radius ($a_0 = \hbar^2/\mu e^2$); it appears once the above expression for E is substituted into the equation for ρ. Using the recursion equation to solve for the polynomial's coefficients a_k for any choice of n and l quantum numbers generates a so-called Laguerre polynomial; $P_{n-L-1}(\rho)$. They contain powers of ρ from zero through $n - L - 1$, and they have $n - L - 1$ sign changes as the radial coordinate ranges from zero to infinity. It is these sign changes in the Laguerre polynomials that cause the radial parts of the hydrogenic wave functions to have $n - L - 1$ nodes. For example, 3d orbitals have no radial nodes, but 4d orbitals have one; and, as shown in Fig. 2.16, 3p orbitals have one while 3s orbitals have two. Once again, the higher the number of nodes, the higher the energy in the radial direction.

Let me again remind you about the danger of trying to understand quantum wave functions or probabilities in terms of classical dynamics. What kind of potential $V(r)$ would give rise to, for example, the 3s $P(r)$ plot shown in Fig. 2.16? Classical mechanics suggests that P should be large where the particle moves slowly and small where it moves quickly. So, the 3s $P(r)$ plot suggests that the radial speed of the electron has three regions where it is low (i.e., where the peaks in P are) and two regions where it is very large (i.e., where the nodes are). This, in

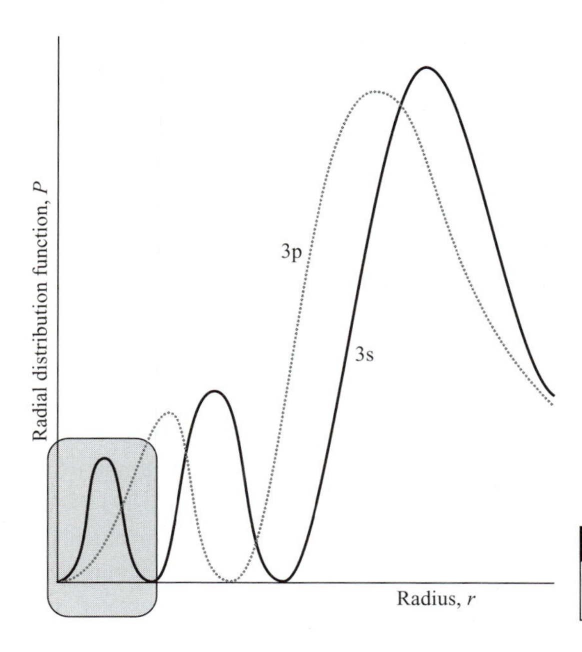

Figure 2.16 Plots of the radial parts of the 3s and 3p orbitals.

turn, suggests that the radial potential $V(r)$ experienced by the 3s electron is high in three regions (near peaks in P) and low in two regions (and at the nucleus). Of course, this conclusion about the form of $V(r)$ is nonsense and again illustrates how one must not be drawn into trying to think of the classical motion of the particle, especially for quantum states with small quantum number. In fact, the low quantum number states of such one-electron atoms and ions have their radial $P(r)$ plots focused in regions of r-space where the potential $-Ze^2/r$ is most attractive (i.e., largest in magnitude).

Finally, we note that the energy quantization does not arise for states lying in the continuum because the condition that the expansion of $P(\rho)$ terminate does not arise. The solutions of the radial equation appropriate to these scattering states (which relate to the scattering motion of an electron in the field of a nucleus of charge Z) are a bit outside the scope of this text, so we will not treat them further here. For the interested student, they are treated on p. 90 of the text by Eyring, Walter, and Kimball.

To review, separation of variables has been used to solve the full (r, θ, ϕ) Schrödinger equation for one electron moving about a nucleus of charge Z. The θ and ϕ solutions are the spherical harmonics $Y_{L,m}(\theta, \phi)$. The bound-state radial solutions

$$R_{n,L}(r) = S(\rho) = \rho^L e^{-\rho/2} P_{n-L-1}(\rho) \tag{2.77}$$

depend on the n and l quantum numbers and are given in terms of the Laguerre polynomials.

2.5.4 Summary

To summarize, the quantum numbers L and m arise through boundary conditions requiring that $\psi(\theta)$ be normalizable (i.e., not diverge) and $\psi(\phi) = \psi(\phi + 2\pi)$. The radial equation, which is the only place the potential energy enters, is found to possess both bound states (i.e., states whose energies lie below the asymptote at which the potential vanishes and the kinetic energy is zero) and continuum states lying energetically above this asymptote. The resulting hydrogenic wave functions (angular and radial) and energies are summarized on pp. 133–136 in the text by Pauling and Wilson for n up to and including 6 and L up to 5.

There are both bound and continuum solutions to the radial Schrödinger equation for the attractive Coulomb potential because, at energies below the asymptote, the potential confines the particle between $r = 0$ and an outer turning point, whereas at energies above the asymptote, the particle is no longer confined by an outer turning point (see Fig. 2.17). The solutions of this one-electron problem form the qualitative basis for much of atomic and molecular orbital theory. For this reason, the reader is encouraged to gain a firmer understanding of the nature of the radial and angular parts of these wave functions. The orbitals that result are labeled by n, L, and m quantum numbers for the bound states and by L and m quantum numbers and the energy E for the continuum states. Much as the particle-in-a-box orbitals are used to qualitatively describe π-electrons in conjugated polyenes, these so-called hydrogen-like orbitals provide qualitative descriptions of orbitals of atoms with more than a single electron. By introducing the concept of screening as a way to represent the repulsive interactions among the electrons of an atom, an effective nuclear charge Z_{eff} can be used in place of Z in the $\psi_{n,L,m}$ and E_n to generate approximate atomic orbitals to be filled by electrons in a many-electron atom. For example, in the crudest approximation of a carbon atom, the two 1s electrons experience the full nuclear attraction so $Z_{\text{eff}} = 6$ for them, whereas the 2s and 2p electrons are screened by the two 1s

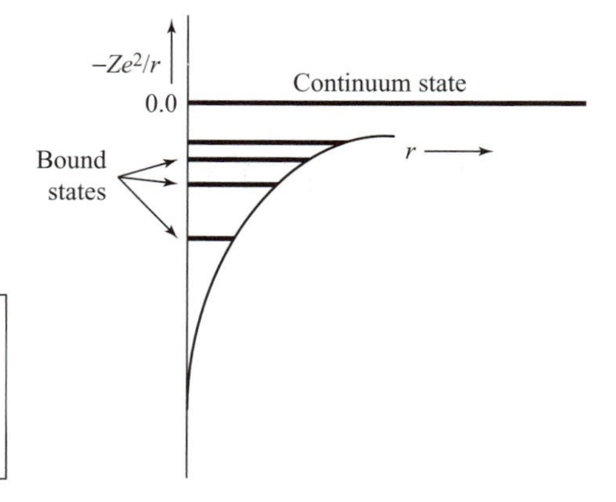

Figure 2.17 Radial potential for hydrogenic atoms and bound and continuum orbital energies.

electrons, so $Z_{eff} = 4$ for them. Within this approximation, one then occupies two 1s orbitals with $Z = 6$, two 2s orbitals with $Z = 4$ and two 2p orbitals with $Z = 4$ in forming the full six-electron wave function of the lowest energy state of carbon.

2.6 Electron tunneling

Tunneling is a phenomenon of quantum mechanics, not classical mechanics. It is an extremely important subject that occurs in a wide variety of chemical species.

Solutions to the Schrödinger equation display several properties that are very different from what one experiences in Newtonian dynamics. One of the most unusual and important is that the particles one describes using quantum mechanics can move into regions of space where they would not be "allowed" to go if they obeyed classical equations. Let us consider an example to illustrate this so-called tunneling phenomenon. Specifically, we think of an electron (a particle that we likely would use quantum mechanics to describe) moving in a direction we will call R under the influence of a potential that is:

(i) Infinite for $R < 0$ (this could, for example, represent a region of space within a solid material where the electron experiences very repulsive interactions with other electrons);
(ii) Constant and negative for some range of R between $R = 0$ and R_{max} (this could represent the attractive interaction of the electrons with those atoms or molecules in a finite region of a solid);
(iii) Constant and repulsive by an amount $\delta V + D_e$ for another finite region from R_{max} to $R_{max} + \delta$ (this could represent the repulsive interactions between the electrons and a layer of molecules of thickness δ lying on the surface of the solid at R_{max});
(iv) Constant and equal to D_e from $R_{maz} + \delta$ to infinity (this could represent the electron being removed from the solid, but with a work function energy cost of D_e, and moving freely in the vacuum above the surface and the ad-layer). Such a potential is shown in Fig. 2.18.

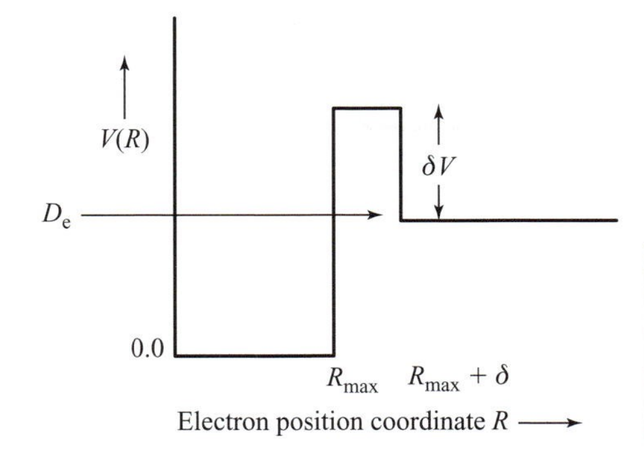

Figure 2.18 One-dimensional potential showing a well, a barrier, and the asymptotic region.

The piecewise nature of this potential allows the one-dimensional Schrödinger equation to be solved analytically. For energies lying in the range $D_e < E < D_e + \delta V$, an especially interesting class of solutions exists. These so-called resonance states occur at energies that are determined by the condition that the amplitude of the wave function within the barrier (i.e., for $0 \le R \le R_{max}$) be large. Let us now turn our attention to this specific energy regime, which also serves to introduce the tunneling phenomenon.

The piecewise solutions to the Schrödinger equation appropriate to the resonance case are easily written down in terms of sin and cos or exponential functions, using the following three definitions:

$$k = \sqrt{2m_e E/\hbar^2}, \qquad k' = \sqrt{2m_e(E - D_e)/\hbar^2}, \qquad \kappa' = \sqrt{2m_e(D_e + \delta V - E)/\hbar^2}. \tag{2.78}$$

The combinations of $\sin(kR)$ and $\cos(kR)$ that solve the Schrödinger equation in the inner region and that vanish at $R = 0$ (because the function must vanish within the region where V is infinite and because it must be continuous, it must vanish at $R = 0$) is

$$\Psi = A\sin(kR) \qquad (\text{for } 0 \le R \le R_{max}). \tag{2.79}$$

Between R_{max} and $R_{max} + \delta$, there are two solutions that obey the Schrödinger equation, so the most general solution is a combination of these two:

$$\Psi = B^+\exp(\kappa' R) + B^-\exp(-\kappa' R) \qquad (\text{for } R_{max} \le R \le R_{max} + \delta). \tag{2.80}$$

Finally, in the region beyond $R_{max} + \delta$, we can use a combination of either $\sin(k'R)$ and $\cos(k'R)$ or $\exp(ik'R)$ and $\exp(-ik'R)$ to express the solution. Unlike the region near $R = 0$, where it was most convenient to use the sin and cos functions because one of them could be "thrown away" since it could not meet the boundary condition of vanishing at $R = 0$, in this large-R region, either set is acceptable. We choose to use the $\exp(ik'R)$ and $\exp(-ik'R)$ set because each of these functions is an eigenfunction of the momentum operator $-i\hbar\partial/\partial R$. This allows us to discuss amplitudes for electrons moving with positive momentum and with negative momentum. So, in this region, the most general solution is

$$\Psi = C\exp(ik'R) + D\exp(-ik'R) \qquad (\text{for } R_{max} + \delta \le R < \infty). \tag{2.81}$$

There are four amplitudes (A, B^+, B^-, and C) that can be expressed in terms of the specified amplitude D of the incoming flux (e.g., pretend that we know the flux of electrons that our experimental apparatus "shoots" at the surface). Four equations that can be used to achieve this goal result when Ψ and $d\Psi/dR$ are matched at R_{max} and at $R_{max} + \delta$ (one of the essential properties of solutions to the Schrödinger equation is that they and their first derivative are continuous; these properties relate to Ψ being a probability and $-ih\partial/\partial R$ being a momentum operator). These four equations are

$$A\sin(kR_{\max}) = B^{+}\exp(\kappa' R_{\max}) + B^{-}\exp(-\kappa' R_{\max}), \tag{2.82}$$

$$Ak\cos(kR_{\max}) = \kappa' B^{+}\exp(\kappa' R_{\max}) - \kappa' B^{-}\exp(-\kappa' R_{\max}), \tag{2.83}$$

$$\begin{aligned} &B^{+}\exp(\kappa'(R_{\max}+\delta)) + B^{-}\exp(-\kappa'(R_{\max}+\delta)) \\ &\quad = C\exp(ik'(R_{\max}+\delta) + D\exp(-ik'(R_{\max}+\delta), \end{aligned} \tag{2.84}$$

$$\begin{aligned} &\kappa' B^{+}\exp(\kappa'(R_{\max}+\delta)) - \kappa' B^{-}\exp(-\kappa'(R_{\max}+\delta)) \\ &\quad = ik'C\exp(ik'(R_{\max}+\delta)) - ik'D\exp(-ik'(R_{\max}+\delta)). \end{aligned} \tag{2.85}$$

It is especially instructive to consider the value of A/D that results from solving this set of four equations in four unknowns because the modulus of this ratio provides information about the relative amount of amplitude that exists inside the barrier in the attractive region of the potential compared to that existing in the asymptotic region as incoming flux.

The result of solving for A/D is

$$\begin{aligned} A/D = {} & 4\kappa'\exp[-ik'(R_{\max}+\delta)]\{\exp(\kappa'\delta)(ik'-\kappa')[\kappa'\sin(kR_{\max}) \\ & + k\cos(kR_{\max})]/ik' + \exp(-\kappa'\delta)(ik'+\kappa')[\kappa'\sin(kR_{\max}) \\ & - k\cos(kR_{\max})]/ik'\}^{-1}. \end{aligned} \tag{2.86}$$

Further, it is instructive to consider this result under conditions of a high (large $D_e + \delta V - E$) and thick (large δ) barrier. In such a case, the "tunneling factor" $\exp(-\kappa'\delta)$ will be very small compared to its counterpart $\exp(\kappa'\delta)$, and so

$$\begin{aligned} A/D = {} & 4\frac{ik'\kappa'}{ik'-\kappa'}\exp[-ik'(R_{\max}+\delta)]\exp(-\kappa'\delta)[\kappa'\sin(kR_{\max}) \\ & + k\cos(kR_{\max})]^{-1}. \end{aligned} \tag{2.87}$$

The $\exp(-\kappa'\delta)$ factor in A/D causes the magnitude of the wave function inside the barrier to be small in most circumstances; we say that incident flux must tunnel through the barrier to reach the inner region and that $\exp(-\kappa'\delta)$ gives the probability of this tunneling.

Keep in mind that, in the energy range we are considering ($E < D_e + \delta$), a classical particle could not even enter the region $R_{\max} < R < R_{\max} + \delta$; this is why we call this the classically forbidden or tunneling region. A classical particle starting in the large-R region can not enter, let alone penetrate, this region, so such a particle could never end up in the $0 < R < R_{\max}$ inner region. Likewise, a classical particle that begins in the inner region can never penetrate the tunneling region and escape into the large-R region. Were it not for the fact that electrons obey a Schrödinger equation rather than Newtonian dynamics, tunneling would not occur and, for example, scanning tunneling microscopy (STM), which has proven to be a wonderful and powerful tool for imaging molecules on and near surfaces, would not exist. Likewise, many of the devices that appear in our modern electronic tools and games, which depend on currents induced by tunneling through various junctions, would not be available. But, of course, tunneling does occur and it can have remarkable effects.

Let us examine an especially important (in chemistry) phenomenon that takes place because of tunneling and that occurs when the energy E assumes very special values. The magnitude of the A/D factor in the above solutions of the Schrödinger equation can become large if the energy E is such that

$$\kappa' \sin(kR_{\max}) + k\cos(kR_{\max}) \tag{2.88}$$

is small. In fact, if

$$\tan(kR_{\max}) = -k/\kappa' \tag{2.89}$$

the denominator factor in A/D will vanish and A/D will become infinite. It can be shown that the above condition is similar to the energy quantization condition

$$\tan(kR_{\max}) = -k/\kappa \tag{2.90}$$

that arises when bound states of a finite potential well are examined. There is, however, a difference. In the bound-state situation, two energy-related parameters occur

$$k = \sqrt{2\mu E/\hbar^2} \tag{2.91}$$

and

$$\kappa = \sqrt{2\mu(D_e - E)/\hbar^2}. \tag{2.92}$$

In the case we are now considering, k is the same, but

$$\kappa' = \sqrt{2\mu(D_e + \delta V - E)/\hbar^2)} \tag{2.93}$$

rather than κ occurs, so the two $\tan(kR_{\max})$ equations are not identical, but they are quite similar.

Another observation that is useful to make about the situations in which A/D becomes very large can be made by considering the case of a very high barrier (so that κ' is much larger than k). In this case, the denominator that appears in A/D,

$$\kappa' \sin(kR_{\max}) + k\cos(kR_{\max}) \cong \kappa' \sin(kR_{\max}), \tag{2.94}$$

can become small if

$$\sin(kR_{\max}) \cong 0. \tag{2.95}$$

This condition is nothing but the energy quantization condition that occurs for the particle-in-a-box potential shown in Fig. 2.19. This potential is identical to the potential that we were examining for $0 \le R \le R_{\max}$, but extends to infinity

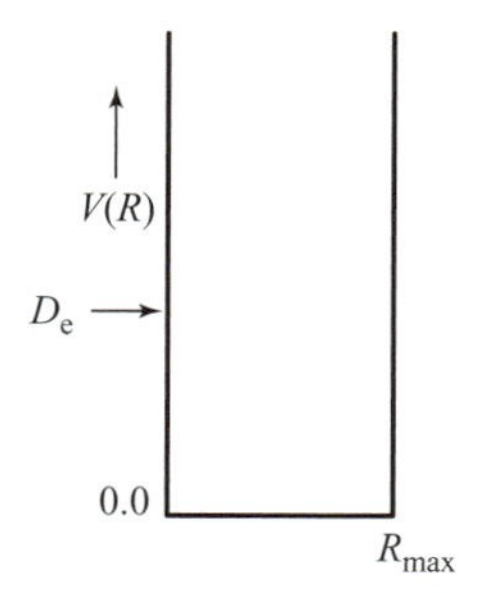

Figure 2.19 One-dimensional potential similar to the tunneling potential but without the barrier and asymptotic region.

beyond $R_{\max}$; the barrier and the dissociation asymptote displayed by our potential are absent.

Let's consider what this tunneling problem has taught us. First, it showed us that quantum particles penetrate into classically forbidden regions. It showed that, at certain so-called resonance energies, tunneling is much more likely than at energies than are "off resonance". In our model problem, this means that electrons impinging on the surface with resonance energies will have a very high probability of tunneling to produce an electron that is trapped in the $0 < R < R_{\max}$ region.

By the way, we could have solved the four equations for the amplitude C of the outgoing wave in the $R > R_{\max}$ region in terms of the A amplitude. We might want to take this approach if wanted to model an experiment in which the electron began in the $0 < R < R_{\max}$ region and we wanted to compute the relative amplitude for the electron to escape. However, if we were to solve for C/A and then examine under what conditions the amplitude of this ratio would become small (so the electron can not escape), we would find the same $\tan(kR_{\max}) = -k/\kappa'$ resonance condition as we found from the other point of view. This means that the resonance energies tell us for what collision energies the electron will tunnel inward and produce a trapped electron and, at these same energies, an electron that is trapped will not escape quickly.

Whenever one has a barrier on a potential energy surface, at energies above the dissociation asymptote D_e but below the top of the barrier ($D_e + \delta V$ here), one can expect resonance states to occur at "special" scattering energies E. As we illustrated with the model problem, these so-called resonance energies can often be approximated by the bound-state energies of a potential that is identical to the potential of interest in the inner region ($0 \le R \le R_{\max}$) but that extends to infinity beyond the top of the barrier (i.e., beyond the barrier, it does not fall back to values below E).

The chemical significance of resonances is great. Highly rotationally excited molecules may have more than enough total energy to dissociate (D_e), but this energy may be "stored" in the rotational motion, and the vibrational energy may be less than D_e. In terms of the above model, high angular momentum may produce a significant centrifugal barrier in the effective potential that characterizes the molecule's vibration, but the system's vibrational energy may lie significantly below D_e. In such a case, and when viewed in terms of motion on an angular-momentum-modified effective potential such as I show in Fig. 2.20, the lifetime of the molecule with respect to dissociation is determined by the rate of tunneling through the barrier.

In that case, one speaks of "rotational predissociation" of the molecule. The lifetime τ can be estimated by computing the frequency ν at which flux that exists inside $R_{\max}$ strikes the barrier at $R_{\max}$,

$$\nu = \frac{\hbar k}{2\mu R_{\max}} \ (\mathrm{s}^{-1}) \tag{2.96}$$

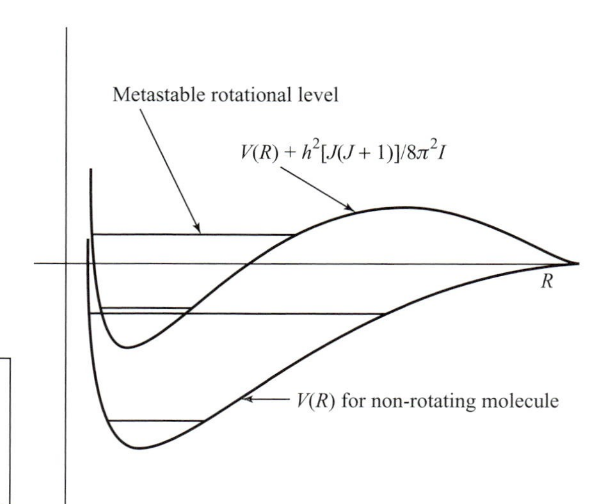

Figure 2.20 Radial potential for non-rotating ($J = 0$) molecule and for rotating molecule.

and then multiplying by the probability P that flux tunnels through the barrier from $R_{\rm max}$ to $R_{\rm max} + \delta$:

$$P = \exp(-2\kappa'\delta). \tag{2.97}$$

The result is that

$$\tau^{-1} = \frac{\hbar k}{2\mu R_{\rm max}} \exp(-2\kappa'\delta) \tag{2.98}$$

with the energy E entering into k and κ' being determined by the resonance condition: $(\kappa' \sin(kR_{\rm max}) + k\cos(kR_{\rm max})) =$ minimum. By looking back at the defintion of κ', we note that the probability of tunneling falls off exponentially with a factor depending on the width δ of the barrier through which the particle must tunnel multiplied by κ', which depends on the height of the barrier $D_{\rm e} + \delta$ above the energy E available. This exponential dependence on thickness and height of the barriers is something you should keep in mind because it appears in all tunneling rate expressions.

Another important case in which tunneling occurs is in electronically metastable states of anions. In so-called shape resonance states, the anion's "extra" electron experiences:

(i) an attractive potential due to its interaction with the underlying neutral molecule's dipole, quadrupole, and induced electrostatic moments, as well as
(ii) a centrifugal potential of the form $L(L+1)h^2/8\pi^2 m_{\rm e} R^2$ whose magnitude depends on the angular character of the orbital the extra electron occupies.

When combined, the above attractive and centrifugal potentials produce an effective radial potential of the form shown in Fig. 2.21 for the N_2^- case in which the added electron occupies the π^* orbital which has $L = 2$ character when viewed

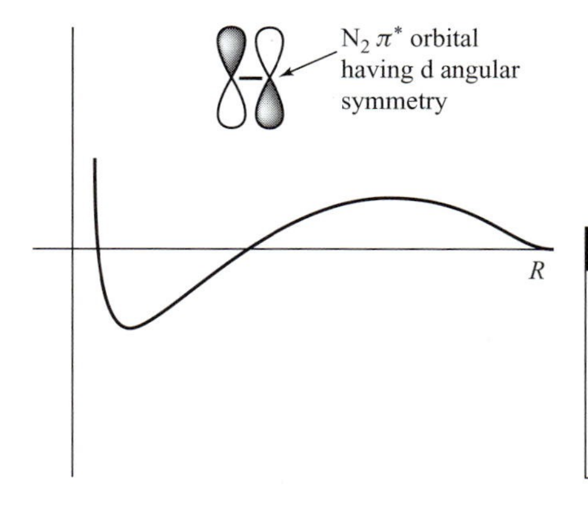

Figure 2.21 Effective radial potential for the excess electron in N_2^- occupying the π^* orbital which has a dominant $L = 2$ component, $L(L+1)h^2/8\pi^2 mR^2$.

from the center of the N—N bond. Again, tunneling through the barrier in this potential determines the lifetimes of such shape resonance states.

Although the examples treated above involved piecewise constant potentials (so the Schrödinger equation and the boundary matching conditions could be solved exactly), many of the characteristics observed carry over to more chemically realistic situations. In fact, one can often model chemical reaction processes in terms of motion along a "reaction coordinate" (s) from a region characteristic of reactant materials where the potential surface is positively curved in all direction and all forces (i.e., gradients of the potential along all internal coordinates) vanish, to a transition state at which the potential surface's curvature along s is negative while all other curvatures are positive and all forces vanish; onward to product materials where again all curvatures are positive and all forces vanish. A prototypical trace of the energy variation along such a reaction coordinate is shown in Fig. 2.22. Near the transition state at the top of the barrier on this surface, tunneling through the barrier plays an important role if the masses of the particles moving in this region are sufficiently light. Specifically, if H or D atoms are involved in the bond breaking and forming in this region of the energy surface, tunneling must usually be considered in treating the dynamics.

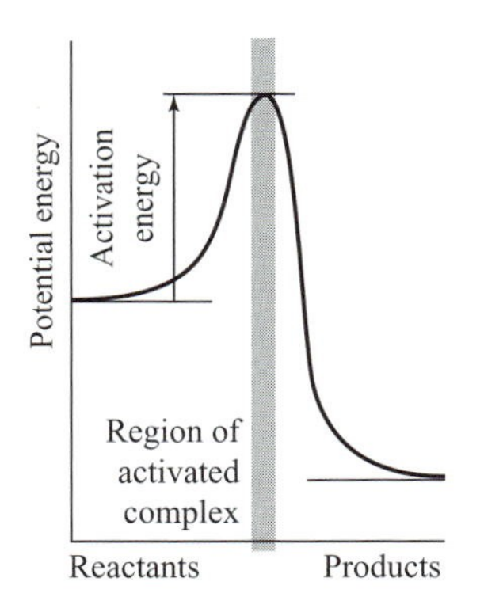

Figure 2.22 Energy profile along a reaction path showing the barrier through which tunneling may occur.

Within the above "reaction path" point of view, motion transverse to the reaction coordinate s is often modeled in terms of local harmonic motion although more sophisticated treatments of the dynamics are possible. This picture leads one to consider motion along a single degree of freedom (s), with respect to which much of the above treatment can be carried over, coupled to transverse motion along all other internal degrees of freedom taking place under an entirely positively curved potential (which therefore produces restoring forces to movement away from the "streambed" traced out by the reaction path s). This point of view constitutes one of the most widely used and successful models of molecular reaction dynamics and is treated in more detail in Chapter 8 of this text.

2.7 Angular momentum

2.7.1 Orbital angular momentum

A particle moving with momentum **p** at a position **r** relative to some coordinate origin has so-called orbital angular momentum equal to $\mathbf{L} = \mathbf{r} \times \mathbf{p}$. The three components of this angular momentum vector in a Cartesian coordinate system located at the origin mentioned above are given in terms of the Cartesian coordinates of **r** and **p** as follows:

$$L_z = xp_y - yp_x, \quad (2.99)$$

$$L_x = yp_z - zp_y, \quad (2.100)$$

$$L_y = zp_x - xp_z. \quad (2.101)$$

Using the fundamental commutation relations among the Cartesian coordinates and the Cartesian momenta:

$$[q_k, p_j] = q_k p_j - p_j q_k = i\hbar\delta_{j,k}(j, k = x, y, z), \quad (2.102)$$

it can be shown that the above angular momentum operators obey the following set of commutation relations:

$$[L_x, L_y] = i\hbar L_z, \quad (2.103)$$

$$[L_y, L_z] = i\hbar L_x, \quad (2.104)$$

$$[L_z, L_x] = i\hbar L_y. \quad (2.105)$$

Although the components of **L** do not commute with one another, they can be shown to commute with the operator L^2 defined by

$$L^2 = L_x^2 + L_y^2 + L_z^2. \quad (2.106)$$

This new operator is referred to as the square of the total angular momentum operator.

The commutation properties of the components of **L** allow us to conclude that complete sets of functions can be found that are eigenfunctions of L^2 and of one, but not more than one, component of **L**. It is convention to select this one component as L_z, and to label the resulting simultaneous eigenstates of L^2 and L_z as $|l, m\rangle$ according to the corresponding eigenvalues:

$$L^2|l, m\rangle = \hbar^2 l(l + 1)|l, m\rangle, \qquad l = 0, 1, 2, 3, \ldots, \quad (2.107)$$

$$L_z|l, m\rangle = \hbar m|l, m\rangle, \qquad m = \pm l, \pm(l - 1), \pm(l - 2), \ldots \pm (l - (l - 1)), 0. \quad (2.108)$$

These eigenfunctions of L^2 and of L_z will not, in general, be eigenfunctions of either L_x or of L_y. This means that any measurement of L_x or L_y will necessarily change the wave function if it begins as an eigenfunction of L_z.

The above expressions for L_x, L_y, and L_z can be mapped into quantum mechanical operators by substituting x, y, and z as the corresponding coordinate operators and $-i\hbar\partial/\partial x$, $-i\hbar\partial/\partial y$, and $-i\hbar\partial/\partial z$ for p_x, p_y, and p_z, respectively. The resulting operators can then be transformed into spherical coordinates the results of which are

$$L_z = -i\hbar\partial/\partial\phi, \quad (2.109)$$
$$L_x = i\hbar\{\sin\phi\ \partial/\partial\theta + \cot\theta\cos\phi\ \partial/\partial\phi\}, \quad (2.110)$$
$$L_y = -i\hbar\{\cos\phi\ \partial/\partial\theta - \cot\theta\sin\phi\partial/\partial\phi\}, \quad (2.111)$$
$$L^2 = -\hbar^2\{(1/\sin\theta)\ \partial/\partial\theta(\sin\theta\ \partial/\partial\theta) + (1/\sin^2\theta)\ \partial^2/\partial\phi^2\}. \quad (2.112)$$

2.7.2 Properties of general angular momenta

There are many types of angular momenta that one encounters in chemistry. Orbital angular momenta, such as introduced above, arise in electronic motion in atoms, in atom–atom and electron–atom collisions, and in rotational motion in molecules. Intrinsic spin angular momentum is present in electrons, H^1, H^2, C^{13}, and many other nuclei. In this section, we will deal with the behavior of any and all angular momenta and their corresponding eigenfunctions.

At times, an atom or molecule contains more than one type of angular momentum. The Hamiltonian's interaction potentials present in a particular species may or may not cause these individual angular momenta to be coupled to an appreciable extent (i.e., the Hamiltonian may or may not contain terms that refer simultaneously to two or more of these angular momenta). For example, the NH^- ion, which has a $^2\Pi$ ground electronic state (its electronic configuration is $1s_N^2 2\sigma^2 3\sigma^2 2p_{\pi x}^2 2p_{\pi y}^1$) has electronic spin, electronic orbital, and molecular rotational angular momenta. The full Hamiltonian H contains terms that couple the electronic spin and orbital angular momenta, thereby causing them individually to not commute with H.

In such cases, the eigenstates of the system can be labeled rigorously only by angular momentum quantum numbers j and m belonging to the total angular momentum $\mathbf{J}$. The total angular momentum of a collection of individual angular momenta is defined, component-by-component, as follows:

$$J_k = \sum_i J_k(i), \quad (2.113)$$

where k labels x, y, and z, and i labels the constituents whose angular momenta couple to produce $\mathbf{J}$.

For the remainder of this section, we will study eigenfunction–eigenvalue relationships that are characteristic of all angular momenta and which are consequences of the commutation relations among the angular momentum vector's three components. We will also study how one combines eigenfunctions of two or more angular momenta $\{\mathbf{J}(i)\}$ to produce eigenfunctions of the total $\mathbf{J}$.

Consequences of the commutation relations

Any set of three operators that obey

$$[J_x, J_y] = i\hbar J_z, \tag{2.114}$$
$$[J_y, J_z] = i\hbar J_x, \tag{2.115}$$
$$[J_z, J_x] = i\hbar J_y, \tag{2.116}$$

will be taken to define an angular momentum **J**, whose square $J^2 = J_x^2 + J_y^2 + J_z^2$ commutes with all three of its components. It is useful to also introduce two combinations of the three fundamental operators:

$$J_\pm = J_x \pm i J_y, \tag{2.117}$$

and to refer to them as raising and lowering operators for reasons that will be made clear below. These new operators can be shown to obey the following commutation relations:

$$[J^2, J_\pm] = 0, \tag{2.118}$$
$$[J_z, J_\pm] = \pm\hbar J_\pm. \tag{2.119}$$

Using only the above commutation properties, it is possible to prove important properties of the eigenfunctions and eigenvalues of J^2 and J_z. Let us assume that we have found a set of simultaneous eigenfunctions of J^2 and J_z; the fact that these two operators commute tells us that this is possible. Let us label the eigenvalues belonging to these functions:

$$J^2|j, m\rangle = \hbar^2 f(j, m)|j, m\rangle, \tag{2.120}$$
$$J_z|j, m\rangle = \hbar m|j, m\rangle, \tag{2.121}$$

in terms of the quantities m and $f(j, m)$. Although we certainly "hint" that these quantities must be related to certain j and m quantum numbers, we have not yet proven this, although we will soon do so. For now, we view $f(j, m)$ and m simply as symbols that represent the respective eigenvalues. Because both J^2 and J_z are Hermitian, eigenfunctions belonging to different $f(j, m)$ or m quantum numbers must be orthogonal:

$$\langle j, m \mid j', m'\rangle = \delta_{m,m'}\delta_{j,j'}. \tag{2.122}$$

We now prove several identities that are needed to discover the information about the eigenvalues and eigenfunctions of general angular momenta that we are after. Later in this section, the essential results are summarized.

(i) There is a maximum and a minimum eigenvalue for J_z

Because all of the components of **J** are Hermitian, and because the scalar product of any function with itself is positive semi-definite, the following identity holds:

$$\langle j, m|J_x^2 + J_y^2|j, m\rangle = \langle J_x\langle j, m|J_x|j, m\rangle + \langle J_y\langle j, m|J_y|j, m\rangle \geq 0. \tag{2.123}$$

However, $J_x^2 + J_y^2$ is equal to $J^2 - J_z^2$, so this inequality implies that

$$\langle j, m|J^2 - J_z^2|j, m\rangle = \hbar^2\{f(j, m) - m^2\} \geq 0, \tag{2.124}$$

which, in turn, implies that m^2 must be less than or equal to $f(j, m)$. Hence, for any value of the total angular momentum eigenvalue f, the z-projection eigenvalue (m) must have a maximum and a minimum value and both of these must be less than or equal to the total angular momentum squared eigenvalue f.

(ii) The raising and lowering operators change the J_z eigenvalue but not the J^2 eigenvalue when acting on $|j, m\rangle$

Applying the commutation relations obeyed by $J_\pm$ to $|j, m\rangle$ yields another useful result:

$$J_z J_\pm|j, m\rangle - J_\pm J_z|j, m\rangle = \pm\hbar J_\pm|j, m\rangle, \tag{2.125}$$

$$J^2 J_\pm|j, m\rangle - J_\pm J^2|j, m\rangle = 0. \tag{2.126}$$

Now, using the fact that $|j, m\rangle$ is an eigenstate of J^2 and of J_z, these identities give

$$J_z J_\pm|j, m\rangle = (m\hbar \pm \hbar)\, J_\pm|j, m\rangle = \hbar(m \pm 1)|j, m\rangle, \tag{2.127}$$

$$J^2 J_\pm|j, m\rangle = \hbar^2 f(j, m)\, J_\pm|j, m\rangle. \tag{2.128}$$

These equations prove that the functions $J_\pm|j, m\rangle$ must either themselves be eigenfunctions of J^2 and J_z, with eigenvalues $\hbar^2 f(j, m)$ and $\hbar(m + 1)$ or $J_\pm|j, m\rangle$ must equal zero. In the former case, we see that $J_\pm$ acting on $|j, m\rangle$ generates a new eigenstate with the same J^2 eigenvalue as $|j, m\rangle$ but with one unit of $\hbar$ higher or lower in J_z eigenvalue. It is for this reason that we call $J_\pm$ raising and lowering operators. Notice that, although $J_\pm|j, m\rangle$ is indeed an eigenfunction of J_z with eigenvalue $(m \pm 1)\hbar$, $J_\pm|j, m\rangle$ is not identical to $|j, m \pm 1\rangle$; it is only proportional to $|j, m \pm 1\rangle$:

$$J_\pm|j, m\rangle = C^\pm_{j,m}|j, m \pm 1\rangle. \tag{2.129}$$

Explicit expressions for these $C^\pm_{j,m}$ coefficients will be obtained below. Notice also that because the $J_\pm|j, m\rangle$, and hence $|j, m \pm 1\rangle$, have the same J^2 eigenvalue as $|j, m\rangle$ (in fact, sequential application of $J_\pm$ can be used to show that all $|j, m'\rangle$, for all m', have this same J^2 eigenvalue), the J^2 eigenvalue $f(j, m)$ must be independent of m. For this reason, f can be labeled by one quantum number j.

(iii) The J^2 eigenvalues are related to the maximum and minimum J_z eigenvalues which are related to one another

Earlier, we showed that there exists a maximum and a minimum value for m, for any given total angular momentum. It is when one reaches these limiting cases that $J_{\pm}|j, m\rangle = 0$ applies. In particular,

$$J_{+}|j, m_{\max}\rangle = 0, \tag{2.130}$$

$$J_{-}|j, m_{\min}\rangle = 0. \tag{2.131}$$

Applying the following identities:

$$J_{-}J_{+} = J^2 - J_z^2 - \hbar J_z, \tag{2.132}$$

$$J_{+}J_{-} = J^2 - J_z^2 + \hbar J_z, \tag{2.133}$$

respectively, to $|j, m_{\max}\rangle$ and $|j, m_{\min}\rangle$ gives

$$\hbar^2 \left\{ f(j, m_{\max}) - m_{\max}^2 - m_{\max} \right\} = 0, \tag{2.134}$$

$$\hbar^2 \left\{ f(j, m_{\min}) - m_{\min}^2 + m_{\min} \right\} = 0, \tag{2.135}$$

which immediately gives the J^2 eigenvalues $f(j, m_{\max})$ and $f(j, m_{\min})$ in terms of $m_{\max}$ or $m_{\min}$:

$$f(j, m_{\max}) = m_{\max}(m_{\max} + 1), \tag{2.136}$$

$$f(j, m_{\min}) = m_{\min}(m_{\min} - 1). \tag{2.137}$$

So, we now know the J^2 eigenvalues for $|j, m_{\max}\rangle$ and $|j, m_{\min}\rangle$. However, we earlier showed that $|j, m\rangle$ and $|j, m-1\rangle$ have the same J^2 eigenvalue (when we treated the effect of $J_{\pm}$ on $|j, m\rangle$) and that the J^2 eigenvalue is independent of m. If we therefore define the quantum number j to be $m_{\max}$, we see that the J^2 eigenvalues are given by

$$J^2|j, m\rangle = \hbar^2 j(j+1)|j, m\rangle. \tag{2.138}$$

We also see that

$$f(j, m) = j(j+1) = m_{\max}(m_{\max} + 1) = m_{\min}(m_{\min} - 1), \tag{2.139}$$

from which it follows that

$$m_{\min} = -m_{\max}. \tag{2.140}$$

(iv) The j quantum number can be integer or half-integer

The fact that the m-values run from j to $-j$ in unit steps (because of the property of the $J_{\pm}$ operators) means that there clearly can be only integer or half-integer values for j. In the former case, the m quantum number runs over $-j, -j+1, -j+2, \ldots, -j+(j-1), 0, 1, 2, \ldots, j$; in the latter, m runs over $-j, -j+1, -j+2, \ldots, -j+(j-1/2), 1/2, 3/2, \ldots, j$. Only integer and half-integer values can range from j to $-j$ in steps of unity. Species with integer spin are known as bosons and those with half-integer spin are called fermions.

(v) More on $J_\pm|j,m\rangle$

Using the above results for the effect of $J_\pm$ acting on $|j, m\rangle$ and the fact that J_+ and J_- are adjoints of one another, allows us to write:

$$\begin{aligned}\langle j, m|J_- J_+|j, m\rangle &= \langle j, m|\left(J^2 - J_z^2 - \hbar J_z\right)|j, m\rangle \\ &= \hbar^2\{j(j+1) - m(m+1)\} \\ &= \langle J_+\langle j, m|J_+|j, m\rangle = (C_{j,m}^+)^2,\end{aligned} \tag{2.141}$$

where $C_{j,m}^+$ is the proportionality constant between $J_+|j, m\rangle$ and the normalized function $|j, m+1\rangle$. Likewise, the effect of J_- can be expressed as

$$\begin{aligned}\langle j, m|J_+ J_-|j, m\rangle &= \langle j, m|\left(J^2 - J_z^2 + \hbar J_z\right)|j, m\rangle \\ &= \hbar^2\{j(j+1) - m(m-1)\} \\ &= \langle J_-\langle j, m|J_-|j, m\rangle = (C_{j,m}^-)^2,\end{aligned} \tag{2.142}$$

where $C_{j,m}^-$ is the proportionality constant between $J_-|j, m\rangle$ and the normalized $|j, m-1\rangle$.

Thus, we can solve for $C_{j,m}^\pm$ after which the effect of $J_\pm$ on $|j, m\rangle$ is given by

$$J_\pm|j, m\rangle = \hbar\{j(j+1) - m(m \pm 1)\}^{1/2}|j, m \pm 1\rangle. \tag{2.143}$$

2.7.3 Summary

The above results apply to any angular momentum operators. The essential findings can be summarized as follows:

(i) J^2 and J_z have complete sets of simultaneous eigenfunctions. We label these eigenfunctions $|j, m\rangle$; they are orthonormal in both their m- and j-type indices: $\langle j, m \mid j', m'\rangle = \delta_{m,m'}\delta_{j,j'}$.

(ii) These $|j, m\rangle$ eigenfunctions obey

$$J^2|j, m\rangle = \hbar^2 j(j+1)|j, m\rangle, \qquad j = \text{integer or half-integer}, \tag{2.144}$$

$$J_z|j, m\rangle = \hbar m|j, m\rangle, \qquad m = -j, \text{ in steps of 1 to } +j. \tag{2.145}$$

(iii) The raising and lowering operators $J_\pm$ act on $|j, m\rangle$ to yield functions that are eigenfunctions of J^2 with the same eigenvalues as $|j, m\rangle$ and eigenfunctions of J_z with eigenvalues of $(m \pm 1)\hbar$:

$$J_\pm|j, m\rangle = \hbar\{j(j+1) - m(m \pm 1)\}^{1/2}|j, m \pm 1\rangle. \tag{2.146}$$

(iv) When $J_\pm$ acts on the "extremal" states $|j, j\rangle$ or $|j, -j\rangle$, respectively, the result is zero.

The results given above are, as stated, general. Any and all angular momenta have quantum mechanical operators that obey these equations. It is convention to designate specific kinds of angular momenta by specific letters; however, it should be kept in mind that no matter what letters are used, there are operators corresponding to J^2, J_z, and $J_\pm$ that obey relations as specified above, and there are eigenfunctions and eigenvalues that have all of the properties obtained above. For electronic or collisional orbital angular momenta, it is common to use L^2 and L_z; for electron spin, S^2 and S_z are used; for nuclear spin I^2 and I_z are most common; and for molecular rotational angular momentum, N^2 and N_z are most common (although sometimes J^2 and J_z may be used). Whenever two or more angular momenta are combined or coupled to produce a "total" angular momentum, the latter is designated by J^2 and J_z.

2.7.4 Coupling of angular momenta

If the Hamiltonian under study contains terms that couple two or more angular momenta $\mathbf{J}(i)$, then only the components of the total angular momentum $\mathbf{J} = \sum_i \mathbf{J}(i)$ and $\mathbf{J}^2$ will commute with H. It is therefore essential to label the quantum states of the system by the eigenvalues of $\mathbf{J}_z$ and $\mathbf{J}^2$ and to construct variational trial or model wavefunctions that are eigenfunctions of these total angular momentum operators. The problem of angular momentum coupling has to do with how to combine eigenfunctions of the uncoupled angular momentum operators, which are given as simple products of the eigenfunctions of the individual angular momenta $\Pi_i |j_i, m_i\rangle$, to form eigenfunctions of J^2 and J_z.

Eigenfunctions of J_z

Because the individual elements of $\mathbf{J}$ are formed additively, but J^2 is not, it is straightforward to form eigenstates of

$$J_z = \sum_i J_z(i); \tag{2.147}$$

simple products of the form $\prod_i |j_i, m_i\rangle$ are eigenfunctions of J_z:

$$J_z \prod_i |j_i, m_i\rangle = \sum_k J_z(k) \prod_i |j_i, m_i\rangle = \sum_k \hbar m_k \prod_i |j_i, m_i\rangle, \tag{2.148}$$

and have J_z eigenvalues equal to the sum of the individual $m_k\hbar$ eigenvalues. Hence, to form an eigenfunction with specified J and M eigenvalues, one must combine only those product states $\prod_i |j_i, m_i\rangle$ whose $m_i\hbar$ sum is equal to the specified M value.

Eigenfunctions of J^2: the Clebsch–Gordon series

The task is then reduced to forming eigenfunctions $|J, M\rangle$, given particular values for the $\{j_i\}$ quantum numbers. When coupling pairs of angular momenta $\{|j, m\rangle$ and $|j', m'\rangle\}$, the total angular momentum states can be written, according to what we determined above, as

$$|J, M\rangle = \sum_{m,m'} C^{J,M}_{j,m;j',m'} |j, m\rangle |j', m'\rangle, \tag{2.149}$$

where the coefficients $C^{J,M}_{j,m;j',m'}$ are called vector coupling coefficients (because angular momentum coupling is viewed much like adding two vectors $\mathbf{j}$ and $\mathbf{j}'$ to produce another vector $\mathbf{J}$), and where the sum over m and m' is restricted to those terms for which $m + m' = M$. It is more common to express the vector coupling or so-called Clebsch–Gordon (CG) coefficients as $\langle j, m; j', m' \mid J, M\rangle$ and to view them as elements of a "matrix" whose columns are labeled by the coupled-state J, M quantum numbers and whose rows are labeled by the quantum numbers characterizing the uncoupled "product basis" $j, m; j', m'$. It turns out that this matrix can be shown to be unitary so that the CG coefficients obey

$$\sum_{m,m'} \langle j, m; j', m' \mid J, M\rangle^* \langle j, m; j', m' \mid J', M'\rangle = \delta_{J,J'}\delta_{M,M'} \tag{2.150}$$

and

$$\sum_{J,M} \langle j, n; j', n' \mid J, M\rangle \langle j, m; j', m' \mid J, M\rangle^* = \delta_{n,m}\delta_{n',m'}. \tag{2.151}$$

This unitarity of the CG coefficient matrix allows the inverse of the relation giving coupled functions in terms of the product functions:

$$|J, M\rangle = \sum_{m,m'} \langle j, m; j', m' \mid J, M\rangle |j, m\rangle |j', m'\rangle \tag{2.152}$$

to be written as

$$\begin{aligned} |j, m\rangle |j', m'\rangle &= \sum_{J,M} \langle j, m; j', m' \mid J, M\rangle^* |J, M\rangle \\ &= \sum_{J,M} \langle J, M \mid j, m; j', m'\rangle |J, M\rangle. \end{aligned} \tag{2.153}$$

This result expresses the product functions in terms of the coupled angular momentum functions.

Generation of the CG coefficients

The CG coefficients can be generated in a systematic manner; however, they can also be looked up in books where they have been tabulated (e.g., see Table 2.4 of R. N. Zare, *Angular Momentum*, John Wiley, New York (1988)). Here, we will demonstrate the technique by which the CG coefficients can be obtained, but we will do so for rather limited cases and refer the reader to more extensive tabulations.

The strategy we take is to generate the $|J, J\rangle$ state (i.e., the state with maximum M-value) and to then use J_- to generate $|J, J-1\rangle$, after which the state $|J-1, J-1\rangle$ (i.e., the state with one lower J-value) is constructed by finding a combination of the product states in terms of which $|J, J-1\rangle$ is expressed (because both $|J, J-1\rangle$ and $|J-1, J-1\rangle$ have the same M-value $M = J-1$) which is orthogonal to $|J, J-1\rangle$ (because $|J-1, J-1\rangle$ and $|J, J-1\rangle$ are eigenfunctions of the Hermitian operator J^2 corresponding to different eigenvalues, they must be orthogonal). This same process is then used to generate $|J, J-2\rangle|J-1, J-2\rangle$ and (by orthogonality construction) $|J-2, J-2\rangle$, and so on.

(i) The states with maximum and minimum M-values

We begin with the state $|J, J\rangle$ having the highest M-value. This state must be formed by taking the highest m and the highest m' values (i.e., $m = j$ and $m' = j'$), and is given by

$$|J, J\rangle = |j, j\rangle|j', j'\rangle. \tag{2.154}$$

Only this one product is needed because only the one term with $m = j$ and $m' = j'$ contributes to the sum in the above CG series. The state

$$|J, -J\rangle = |j, -j\rangle|j', -j'\rangle \tag{2.155}$$

with the minimum M-value is also given as a single product state. Notice that these states have M-values given as $\pm(j + j')$; since this is the maximum M-value, it must be that the J-value corresponding to this state is $J = j + j'$.

(ii) States with one lower M-value but the same J-value

Applying J_- to $|J, J\rangle$, and expressing J_- as the sum of lowering operators for the two individual angular momenta:

$$J_- = J_-(1) + J_-(2) \tag{2.156}$$

gives

$$\begin{aligned} J_-|J, J\rangle &= \hbar\{J(J+1) - J(J-1)\}^{1/2}|J, J-1\rangle \\ &= (J_-(1) + J_-(2))|j, j\rangle|j'j'\rangle \\ &= \hbar\{j(j+1) - j(j-1)\}^{1/2}|j, j-1\rangle|j', j'\rangle + \hbar\{j'(j'+1) \\ &\quad - j'(j'-1)\}^{1/2}|j, j\rangle|j', j'-1\rangle. \end{aligned} \tag{2.157}$$

This result expresses $|J, J-1\rangle$ as follows:

$$\begin{aligned} |J, J-1\rangle = {} & [\{j(j+1) - j(j-1)\}^{1/2}|j, j-1\rangle|j', j'\rangle \\ & + \{j'(j'+1) - j'(j'-1)\}^{1/2}|j, j\rangle|j', j'-1\rangle] \\ & \{J(J+1) - J(J-1)\}^{-1/2}; \end{aligned} \tag{2.158}$$

that is, the $|J, J-1\rangle$ state, which has $M = J-1$, is formed from the two product states $|j, j-1\rangle|j', j'\rangle$ and $|j, j\rangle|j', j'-1\rangle$ that have this same M-value.

(iii) States with one lower J-value

To find the state $|J-1, J-1\rangle$ that has the same M-value as the one found above but one lower J-value, we must construct another combination of the two product states with $M = J-1$ (i.e., $|j, j-1\rangle|j', j'\rangle$ and $|j, j\rangle|j', j'-1\rangle$) that is orthogonal to the combination representing $|J, J-1\rangle$; after doing so, we must scale the resulting function so it is properly normalized. In this case, the desired function is

$$|J-1, J-1\rangle = [\{j(j+1)-j(j-1)\}^{1/2}|j,j\rangle|j',j'-1\rangle \\ - \{j'(j'+1)-j'(j'-1)\}^{1/2}|j,j-1\rangle|j',j'\rangle] \\ \{J(J+1)-J(J-1)\}^{-1/2}. \tag{2.159}$$

It is straightforward to show that this function is indeed orthogonal to $|J, J-1\rangle$.

(iv) States with even one lower J-value

Having expressed $|J, J-1\rangle$ and $|J-1, J-1\rangle$ in terms of $|j, j-1\rangle|j', j'\rangle$ and $|j, j\rangle|j', j'-1\rangle$, we are now prepared to carry on with this stepwise process to generate the states $|J, J-2\rangle$, $|J-1, J-2\rangle$ and $|J-2, J-2\rangle$ as combinations of the product states with $M = J-2$. These product states are $|j, j-2\rangle|j', j'\rangle$, $|j, j\rangle|j', j'-2\rangle$, and $|j, j-1\rangle|j', j'-1\rangle$. Notice that there are precisely as many product states whose $m + m'$ values add up to the desired M-value as there are total angular momentum states that must be constructed (there are three of each in this case).

The steps needed to find the state $|J-2, J-2\rangle$ are analogous to those taken above:

(i) One first applies J_- to $|J-1, J-1\rangle$ and to $|J, J-1\rangle$ to obtain $|J-1, J-2\rangle$ and $|J, J-2\rangle$, respectively, as combinations of $|j, j-2\rangle|j', j'\rangle$, $|j, j\rangle|j', j'-2\rangle$, and $|j, j-1\rangle|j', j'-1\rangle$.
(ii) One then constructs $|J-2, J-2\rangle$ as a linear combination of the $|j, j-2\rangle|j', j'\rangle$, $|j, j\rangle|j', j'-2\rangle$, and $|j, j-1\rangle|j', j'-1\rangle$ that is orthogonal to the combinations found for $|J-1, J-2\rangle$ and $|J, J-2\rangle$.

Once $|J-2, J-2\rangle$ is obtained, it is then possible to move on to form $|J, J-3\rangle$, $|J-1, J-3\rangle$, and $|J-2, J-3\rangle$ by applying J_- to the three states obtained in the preceding application of the process, and to then form $|J-3, J-3\rangle$ as the combination of $|j, j-3\rangle|j', j'\rangle$, $|j, j\rangle|j', j'-3\rangle$, $|j, j-2\rangle|j', j'-1\rangle$, $|j, j-1\rangle|j', j'-2\rangle$ that is orthogonal to the combinations obtained for $|J, J-3\rangle$, $|J-1, J-3\rangle$, and $|J-2, J-3\rangle$.

Again notice that there are precisely the correct number of product states (four here) as there are total angular momentum states to be formed. In fact, the product states and the total angular momentum states are equal in number and are both members of orthonormal function sets (because $J^2(1)$, $J_z(1)$, $J^2(2)$, and $J_z(2)$ as well as J^2 and J_z are Hermitian operators). This is why the CG coefficient matrix is unitary; because it maps one set of orthonormal functions to another, with both sets containing the same number of functions.

An example

Let us consider an example in which the spin and orbital angular momenta of the Si atom in its ^{3}P ground state can be coupled to produce various ^3P_J states. In this case, the specific values for j and j' are $j = S = 1$ and $j' = L = 1$. We could, of course, take $j = L = 1$ and $j' = S = 1$, but the final wave functions obtained would span the same space as those we are about to determine.

The state with highest M-value is the $^3P(M_S = 1, M_L = 1)$ state, which can be represented by the product of an $\alpha\alpha$ spin function (representing $S = 1, M_S = 1$) and a $3p_1 3p_0$ spatial function (representing $L = 1, M_L = 1$), where the first function corresponds to the first open-shell orbital and the second function to the second open-shell orbital. Thus, the maximum M-value is $M = 2$ and corresponds to a state with $J = 2$:

$$|J = 2, M = 2\rangle = |2, 2\rangle = \alpha\alpha 3p_1 3p_0. \tag{2.160}$$

Clearly, the state $|2, -2\rangle$ would be given as $\beta\beta\ 3p_{-1}3p_0$.

The states $|2, 1\rangle$ and $|1, 1\rangle$ with one lower M-value are obtained by applying $J_- = S_- + L_-$ to $|2, 2\rangle$ as follows:

$$\begin{aligned} J_-|2, 2\rangle &= \hbar\{2(3) - 2(1)\}^{1/2}|2, 1\rangle \\ &= (S_- + L_-)\alpha\alpha 3p_1 3p_0. \end{aligned} \tag{2.161}$$

To apply S_- or L_- to $\alpha\alpha\ 3p_1 3p_0$, one must realize that each of these operators is, in turn, a sum of lowering operators for each of the two open-shell electrons:

$$S_- = S_-(1) + S_-(2), \tag{2.162}$$

$$L_- = L_-(1) + L_-(2). \tag{2.163}$$

The result above can therefore be continued as

$$\begin{aligned} (S_- + L_-)\alpha\alpha 3p_1 3p_0 = {} & \hbar\{1/2(3/2) - 1/2(-1/2)\}^{1/2}\ \beta\alpha 3p_1 3p_0 \\ & + \hbar\{1/2(3/2) - 1/2(-1/2)\}^{1/2}\alpha\beta 3p_1 3p_0 \\ & + \hbar\{1(2) - 1(0)\}^{1/2}\alpha\alpha 3p_0 3p_0 \\ & + \hbar\{1(2) - 0(-1)\}^{1/2}\alpha\alpha 3p_1 3p_{-1}. \end{aligned} \tag{2.164}$$

So, the function $|2, 1\rangle$ is given by ($\alpha\alpha 3p_0 3p_0$ violates the Pauli principle, so it is removed)

$$|2, 1\rangle = \left[\beta\alpha 3p_1 3p_0 + \alpha\beta 3p_1 3p_0 + \{2\}^{1/2}\alpha\alpha 3p_1 3p_{-1}\right]/2, \tag{2.165}$$

which can be rewritten as

$$|2, 1\rangle = \left[(\beta\alpha + \alpha\beta)3p_1 3p_0 + \{2\}^{1/2}\alpha\alpha 3p_1 3p_{-1}\right]/2. \tag{2.166}$$

Writing the result in this way makes it clear that $|2, 1\rangle$ is a combination of the product states $|S = 1, M_S = 0\rangle|L = 1, M_L = 1\rangle$ (the terms containing $|S = 1,$

$M_S = 0\rangle = 2^{-1/2}(\alpha\beta + \beta\alpha)$) and $|S = 1, M_S = 1\rangle|L = 1, M_L = 0\rangle$ (the terms containing $|S = 1, M_S = 1\rangle = \alpha\alpha$).

To form the other function with $M = 1$, the $|1, 1\rangle$ state, we must find another combination of $|S = 1, M_S = 0\rangle|L = 1, M_L = 1\rangle$ and $|S = 1, M_S = 1\rangle|L = 1, M_L = 0\rangle$ that is orthogonal to $|2, 1\rangle$ and is normalized. Since

$$|2, 1\rangle = 2^{-1/2}[|S = 1, M_S = 0\rangle|L = 1, M_L = 1\rangle + |S = 1, M_S = 1\rangle \times |L = 1, M_L = 0\rangle], \quad (2.167)$$

we immediately see that the requisite function is

$$|1, 1\rangle = 2^{-1/2}[|S = 1, M_S = 0\rangle|L = 1, M_L = 1\rangle - |S = 1, M_S = 1\rangle \times |L = 1, M_L = 0\rangle]. \quad (2.168)$$

In the spin-orbital notation used above, this state is

$$|1, 1\rangle = \left[(\beta\alpha + \alpha\beta)3\mathrm{p}_1 3\mathrm{p}_0 - \{2\}^{1/2}\alpha\alpha 3\mathrm{p}_1 3\mathrm{p}_{-1}\right]/2. \quad (2.169)$$

Thus far, we have found the $^3\mathrm{P}_J$ states with $J = 2, M = 2$; $J = 2, M = 1$; and $J = 1, M = 1$.

To find the $^3\mathrm{P}_\mathrm{J}$ states with $J = 2, M = 0$; $\mathrm{J} = 1, \mathrm{M} = 0$; and $J = 0, \mathrm{M} = 0$, we must once again apply the J_- tool. In particular, we apply J_- to $|2, 1\rangle$ to obtain $|2, 0\rangle$ and we apply J_- to $|1, 1\rangle$ to obtain $|1, 0\rangle$, each of which will be expressed in terms of $|S = 1, M_S = 0\rangle|L = 1, M_L = 0\rangle$, $|S = 1, M_S = 1\rangle|L = 1, M_L = -1\rangle$, and $|S = 1, M_S = -1\rangle|L = 1, M_L = 1\rangle$. The $|0, 0\rangle$ state is then constructed to be a combination of these same product states which is orthogonal to $|2, 0\rangle$ and to $|1, 0\rangle$. The results are as follows:

$$|J = 2, M = 0\rangle = 6^{-1/2}[2|1, 0\rangle|1, 0\rangle + |1, 1\rangle|1, -1\rangle + |1, -1\rangle|1, 1\rangle], \quad (2.170)$$
$$|J = 1, M = 0\rangle = 2^{-1/2}[|1, 1\rangle|1, -1\rangle - |1, -1\rangle|1, 1\rangle], \quad (2.171)$$
$$|J = 0, M = 0\rangle = 3^{-1/2}[|1, 0\rangle|1, 0\rangle - |1, 1\rangle|1, -1\rangle - |1, -1\rangle|1, 1\rangle], \quad (2.172)$$

where, in all cases, a shorthand notation has been used in which the $|S, M_S\rangle|L, M_L\rangle$ products stated have been represented by their quantum numbers with the spin function always appearing first in the product. To finally express all three of these new functions in terms of spin-orbital products it is necessary to give the $|S, M_S\rangle|L, M_L\rangle$ products with $M = 0$ in terms of these products. For the spin functions, we have

$$|S = 1, M_S = 1\rangle = \alpha\alpha, \quad (2.173)$$
$$|S = 1, M_S = 0\rangle = 2^{-1/2}(\alpha\beta + \beta\alpha) \quad (2.174)$$
$$|S = 1, M_S = -1\rangle = \beta\beta. \quad (2.175)$$

For the orbital product function, we have

$$|L = 1, M_L = 1\rangle = 3\mathrm{p}_1 3\mathrm{p}_0, \quad (2.176)$$

$$|L = 1, M_L = 0\rangle = 2^{-1/2}(3p_0 3p_0 + 3p_1 3p_{-1}), \quad (2.177)$$
$$|L = 1, M_L = -1\rangle = 3p_0 3p_{-1}. \quad (2.178)$$

Coupling angular momenta of equivalent electrons

If equivalent angular momenta are coupled (e.g., to couple the orbital angular momenta of a p^2 or d^3 configuration), one must use the following "box" method to determine which of the term symbols violate the Pauli principle. To carry out this step, one forms all possible unique (determinental) product states with non-negative M_L and M_S values and arranges them into groups according to their M_L and M_S values. For example, the boxes appropriate to the p^2 orbital occupancy are shown below:

	M_L	2	1	0
M_s	1		$\|p_1\alpha p_0\alpha\|$	$\|p_1\alpha p_{-1}\alpha\|$
	0	$\|p_1\alpha p_1\beta\|$	$\|p_1\alpha p_0\beta\|$, $\|p_0\alpha p_1\beta\|$	$\|p_1\alpha p_{-1}\beta\|$, $\|p_{-1}\alpha p_1\beta\|$, $\|p_0\alpha p_0\beta\|$

There is no need to form the corresponding states with negative M_L or negative M_S values because they are simply "mirror images" of those listed above. For example, the state with $M_L = -1$ and $M_S = -1$ is $|p_{-1}\beta p_0\beta|$, which can be obtained from the $M_L = 1$, $M_S = 1$ state $|p_1\alpha p_0\alpha|$ by replacing α by β and replacing p_1 by p_{-1}.

Given the box entries, one can identify those term symbols that arise by applying the following procedure over and over until all entries have been accounted for:

(i) One identifies the highest M_S value (this gives a value of the total spin quantum number that arises, S) in the box. For the above example, the answer is $S = 1$.

(ii) For all product states of *this* M_S value, one identifies the highest M_L value (this gives a value of the total orbital angular momentum, L, that can arise *for this* S). For the above example, the highest M_L within the $M_S = 1$ states is $M_L = 1$ (not $M_L = 2$), hence $L = 1$.

(iii) Knowing an S, L combination, one knows the first term symbol that arises from this configuration. In the p^2 example, this is ^3P.

(iv) Because the level with these L and S quantum numbers contains $(2L + 1)(2S + 1)$ states with M_L and M_S quantum numbers running from $-L$ to L and from $-S$ to S, respectively, one must remove from the original box this number of product states. To do so, one simply erases from the box one entry with each such M_L and M_S value. Actually, since the box need only show those entries with non-negative M_L and M_S values, only these entries need be explicitly deleted. In the ^3P example, this amounts to deleting nine product states with M_L, M_S values of $1, 1; 1, 0; 1, -1; 0, 1; 0, 0; 0, -1; -1, 1; -1, 0; -1, -1$.

(v) After deleting these entries, one returns to step 1 and carries out the process again. For the p^2 example, the box after deleting the first nine product states looks as follows (those that appear in italics should be viewed as already cancelled in counting all of the ^{3}P states):

	M_L 2	1	0
M_s 1		$\|p_1\alpha p_0\alpha\|$	$\|p_1\alpha p_{-1}\alpha\|$
0	$\|\mathrm{p}_1\alpha\mathrm{p}_1\beta\|$	$\|p_1\alpha p_0\beta\|$, $\|\mathrm{p}_0\alpha\mathrm{p}_1\beta\|$	$\|p_1\alpha p_{-1}\beta\|$, $\|\mathrm{p}_{-1}\alpha\mathrm{p}_1\beta\|$, $\|\mathrm{p}_0\alpha\mathrm{p}_0\beta\|$

It should be emphasized that the process of deleting or crossing off entries in various M_L, M_S boxes involves only counting how many states there are; by no means do we identify the particular L, S, M_L, M_S wave functions when we cross out any particular entry in a box. For example, when the $|\mathrm{p}_1\alpha\mathrm{p}_0\beta|$ product is deleted from the $M_L = 1$, $M_S = 0$ box in accounting for the states in the ^{3}P level, we do not claim that $|\mathrm{p}_1\alpha\mathrm{p}_0\beta|$ itself is a member of the ^{3}P level; the $|\mathrm{p}_0\alpha\mathrm{p}_1\beta|$ product state could just as well be eliminated when accounting for the ^{3}P states.

Returning to the p^2 example at hand, after the ^{3}P term symbol's states have been accounted for, the highest M_S value is 0 (hence there is an $S = 0$ state), and within this M_S value, the highest M_L value is 2 (hence there is an $L = 2$ state). This means there is a ^{1}D level with five states having $M_L = 2, 1, 0, -1, -2$. Deleting five appropriate entries from the above box (again denoting deletions by italics) leaves the following box:

	M_L 2	1	0
M_s 1		$\|p_1\alpha p_0\alpha\|$	$\|p_1\alpha p_{-1}\alpha\|$
0	$\|p_1\alpha p_1\beta\|$	$\|p_1\alpha p_0\beta\|$, $\|p_0\alpha p_1\beta\|$	$\|p_1\alpha p_{-1}\beta\|$, $\|p_{-1}\alpha p_1\beta\|$, $\|\mathrm{p}_0\alpha\mathrm{p}_0\beta\|$

The only remaining entry, which thus has the highest M_S and M_L values, has $M_S = 0$ and $M_L = 0$. Thus there is also a ^{1}S level in the p^2 configuration.

Thus, unlike the non-equivalent 2p^{1}3p^1 case, in which ^{3}P, ^{1}P, ^{3}D, ^{1}D, ^{3}S, and ^{1}S levels arise, only the ^{3}P, ^{1}D, and ^{1}S arise in the p^2 situation. It is necessary to carry out this "box method" whenever one is dealing with equivalent angular momenta.

If one has mixed equivalent and non-equivalent angular momenta, one can determine all possible couplings of the equivalent angular momenta using this method and then use the simpler vector coupling method to add the non-equivalent angular momenta to *each* of these coupled angular momenta. For example, the p^{2}d^1 configuration can be handled by vector coupling (using the straightforward non-equivalent procedure) $L = 2$ (the d orbital) and $S = 1/2$ (the third electron's spin) to *each* of ^{3}P, ^{1}D, and ^{1}S. The result is ^{4}F, ^{4}D, ^{4}P, ^{2}F, ^{2}D, ^{2}P, ^{2}G, ^{2}F, ^{2}D, ^{2}P, ^{2}S, and ^{2}D.

2.8 Rotations of molecules

2.8.1 Rotational motion for rigid diatomic and linear polyatomic molecules

This Schrödinger equation relates to the rotation of diatomic and linear polyatomic molecules. It also arises when treating the angular motions of electrons in any spherically symmetric potential.

A diatomic molecule with fixed bond length R rotating in the absence of any external potential is described by the following Schrödinger equation:

$$\frac{-\hbar^2}{2\mu}\left[\frac{1}{R^2 \sin\theta}\frac{\partial}{\partial\theta}\left(\sin\theta\frac{\partial}{\partial\theta}\right) + \frac{1}{R^2\sin^2\theta}\frac{\partial^2}{\partial\phi^2}\right]\psi = E\psi \tag{2.179}$$

or

$$\frac{L^2\psi}{2\mu R^2} = E\psi, \tag{2.180}$$

where L^2 is the square of the total angular momentum operator $L_x^2 + L_y^2 + L_z^2$ expressed in polar coordinates above. The angles θ and ϕ describe the orientation of the diatomic molecule's axis relative to a laboratory-fixed coordinate system, and μ is the reduced mass of the diatomic molecule $\mu = m_1 m_2/(m_1 + m_2)$. The differential operators can be seen to be exactly the same as those that arose in the hydrogen-like atom case as discussed above. Therefore, the same spherical harmonics that served as the angular parts of the wave function in the hydrogen-atom case now serve as the entire wave function for the so-called rigid rotor: $\psi = Y_{J,M}(\theta, \phi)$. These are exactly the same functions as we plotted earlier when we graphed the s ($L = 0$), p ($L = 1$), and d ($L = 2$) orbitals. The energy eigenvalues corresponding to each such eigenfunction are given as

$$E_J = \frac{\hbar^2 J(J+1)}{2\mu R^2} = BJ(J+1) \tag{2.181}$$

and are independent of M. Thus each energy level is labeled by J and is $(2J + 1)$-fold degenerate (because M ranges from $-J$ to J). Again, this is just like we saw when we looked at the hydrogen orbitals; the p orbitals are three-fold degenerate and the d orbitals are five-fold degenerate. The so-called rotational constant B (defined as $\hbar^2/2\mu R^2$) depends on the molecule's bond length and reduced mass. Spacings between successive rotational levels (which are of spectroscopic relevance because, as shown in Chapter 5, angular momentum selection rules often restrict the changes ΔJ in J that can occur upon photon absorption to 1,0, and -1) are given by

$$\Delta E = B(J+1)(J+2) - BJ(J+1) = 2B(J+1). \tag{2.182}$$

These energy spacings are of relevance to microwave spectroscopy which probes the rotational energy levels of molecules. In fact, microwave spectroscopy offers the most direct way to determine molecular rotational constants and hence molecular bond lengths.

The rigid rotor provides the most commonly employed approximation to the rotational energies and wave functions of linear molecules. As presented above, the model restricts the bond length to be fixed. Vibrational motion of the molecule gives rise to changes in R which are then reflected in changes in the rotational energy levels. The coupling between rotational and vibrational motion gives rise to rotational B constants that depend on vibrational state as well as dynamical couplings, called centrifugal distortions, that cause the total ro-vibrational energy of the molecule to depend on rotational and vibrational quantum numbers in a non-separable manner.

Within this "rigid rotor" model, the absorption spectrum of a rigid diatomic molecule should display a series of peaks, each of which corresponds to a specific $J \rightarrow J + 1$ transition. The energies at which these peaks occur should grow linearly with J. An example of such a progression of rotational lines is shown in Fig. 2.23. The energies at which the rotational transitions occur appear to fit the $\Delta E = 2B(J + 1)$ formula rather well. The intensities of transitions from level J to level $J + 1$ vary strongly with J primarily because the population of molecules in the absorbing level varies with J. These populations P_J are given, when the system is at equilibrium at temperature T, in terms of the degeneracy $(2J + 1)$ of the Jth level and the energy of this level $BJ(J + 1)$ by the Boltzmann formula:

$$P_J = Q^{-1}(2J + 1)\,\exp(-BJ(J + 1)/kT), \tag{2.183}$$

where Q is the rotational partition function:

$$Q = \sum_J (2J + 1)\,\exp(-BJ(J + 1)/kT). \tag{2.184}$$

For low values of J, the degeneracy is low and the $\exp(-BJ(J + 1)/kT)$ factor is near unity. As J increases, the degeracy grows linearly but the

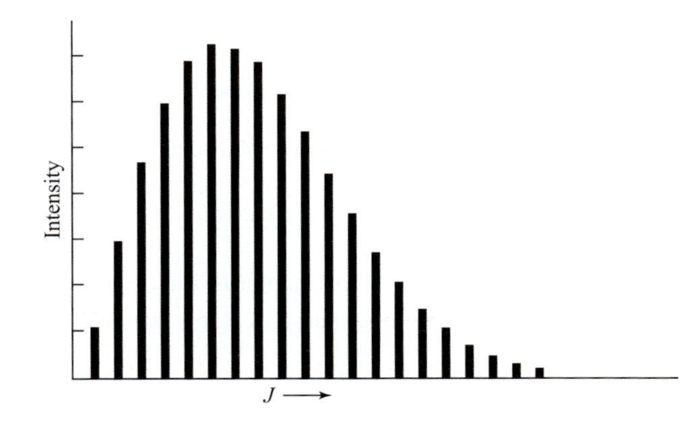

Figure 2.23 Typical rotational absorption profile showing intensity vs. J value of the absorbing level.

$\exp(-BJ(J+1)/kT)$ factor decreases more rapidly. As a result, there is a value of J, given by taking the derivative of $(2J+1)\exp(-BJ(J+1)/kT)$ with respect to J and setting it equal to zero,

$$2J_{\text{max}} + 1 = \sqrt{2kT/B}, \tag{2.185}$$

at which the intensity of the rotational transition is expected to reach its maximum. This behavior is clearly displayed in the above figure.

The eigenfunctions belonging to these energy levels are the spherical harmonics $Y_{L,M}(\theta,\phi)$ which are normalized according to

$$\int_0^{\pi}\int_0^{2\pi}(Y^*_{L,M}(\theta,\phi)\,Y_{L',M'}(\theta,\phi)\sin\theta d\theta d\phi) = \delta_{L,L'}\delta_{M,M'}. \tag{2.186}$$

As noted above, these functions are identical to those that appear in the solution of the angular part of hydrogen-like atoms. The above energy levels and eigenfunctions also apply to the rotation of rigid linear polyatomic molecules; the only difference is that the moment of inertia I entering into the rotational energy expression is given by

$$I = \sum_a m_a R_a^2, \tag{2.187}$$

where m_a is the mass of the ath atom and R_a is its distance from the center of mass of the molecule. This moment of inertia replaces μR^2 in the earlier rotational energy level expressions.

2.8.2 Rotational motions of rigid non-linear molecules

The rotational kinetic energy

The rotational kinetic energy operator for a rigid polyatomic molecule is

$$H_{\text{rot}} = J_a^2/2I_a + J_b^2/2I_b + J_c^2/2I_c, \tag{2.188}$$

where the $I_k(k = a, b, c)$ are the three principal moments of inertia of the molecule (the eigenvalues of the moment of inertia tensor). This tensor has elements in a Cartesian coordinate system $(K, K' = X, Y, Z)$, whose origin is located at the center of mass of the molecule, that can be computed as

$$I_{K,K} = \sum_j m_j \left(R_j^2 - R_{K,j}^2\right) \qquad (\text{for } K = K'), \tag{2.189}$$

$$I_{K,K'} = -\sum_j m_j R_{K,j} R_{K',j} \qquad (\text{for } K \neq K'). \tag{2.190}$$

As discussed in more detail in Chapter 6, the components of the quantum mechanical angular momentum operators along the three principal axes are

$$J_a = -i\hbar\cos\chi\left(\cot\theta\frac{\partial}{\partial\chi} - \frac{1}{\sin\theta}\frac{\partial}{\partial\phi}\right) + i\hbar\sin\chi\frac{\partial}{\partial\theta}, \tag{2.191}$$

$$J_b = i\hbar \sin\chi \left(\cot\theta \frac{\partial}{\partial\chi} - \frac{1}{\sin\theta}\frac{\partial}{\partial\phi}\right) + i\hbar\cos\chi \frac{\partial}{\partial\theta}, \tag{2.192}$$

$$J_c = -i\hbar\frac{\partial}{\partial\chi}. \tag{2.193}$$

The angles θ, ϕ, and χ are the Euler angles needed to specify the orientation of the rigid molecule relative to a laboratory-fixed coordinate system. The corresponding square of the total angular momentum operator J^2 can be obtained as

$$\begin{aligned} J^2 &= J_a^2 + J_b^2 + J_c^2 \\ &= -\frac{\partial^2}{\partial\theta^2} - \cot\theta\frac{\partial}{\partial\theta} + \frac{1}{\sin\theta}\left(\frac{\partial^2}{\partial\phi^2} + \frac{\partial^2}{\partial\chi^2} - 2\cos\theta\frac{\partial^2}{\partial\phi\partial\chi}\right), \end{aligned} \tag{2.194}$$

and the component along the lab-fixed Z-axis J_Z is $-i\hbar\partial/\partial\phi$ as we saw much earlier in this text.

The eigenfunctions and eigenvalues for special cases

(i) Spherical tops

When the three principal moment of inertia values are identical, the molecule is termed a spherical top. In this case, the total rotational energy can be expressed in terms of the total angular momentum operator J^2

$$H_{\text{rot}} = J^2/2I. \tag{2.195}$$

As a result, the eigenfunctions of H_{rot} are those of J^2 and J_a as well as J_Z, both of which commute with J^2 and with one another. J_Z is the component of **J** along the lab-fixed Z-axis and commutes with J_a because $J_Z = -i\hbar\partial/\partial\phi$ and $J_a = -i\hbar\partial/\partial\chi$ act on different angles. The energies associated with such eigenfunctions are

$$E(J, K, M) = \hbar^2 J(J+1)/2I^2, \tag{2.196}$$

for all K (i.e., J_a quantum numbers) ranging from $-J$ to J in unit steps and for all M (i.e., J_Z quantum numbers) ranging from $-J$ to J. Each energy level is therefore $(2J+1)^2$ degenerate because there are $2J+1$ possible K values and $2J+1$ possible M values for each J.

The eigenfunctions $|J, M, K\rangle$ of J^2, J_Z and J_a , are given in terms of the set of so-called rotation matrices $D_{J,M,K}$:

$$|J, M, K\rangle = \sqrt{\frac{2J+1}{8\pi^2}} D^*_{J,M,K}(\theta, \phi, \chi), \tag{2.197}$$

which obey

$$J^2|J, M, K\rangle = \hbar^2 J(J+1)|J, M, K\rangle, \tag{2.198}$$

$$J_a|J, M, K\rangle = \hbar K|J, M, K\rangle, \tag{2.199}$$

$$J_Z|J, M, K\rangle = \hbar M|J, M, K\rangle. \tag{2.200}$$

These $D_{J,M,K}$ functions are proportional to the spherical harmonics $Y_{J,M}(\theta, \phi)$ multiplied by $\exp(iK\chi)$, which reflects its χ-dependence.

(ii) Symmetric tops

Molecules for which two of the three principal moments of inertia are equal are called symmetric tops. Those for which the unique moment of inertia is smaller than the other two are termed prolate symmetric tops; if the unique moment of inertia is larger than the others, the molecule is an oblate symmetric top. An American football is prolate, and a frisbee is oblate.

Again, the rotational kinetic energy, which is the full rotational Hamiltonian, can be written in terms of the total rotational angular momentum operator J^2 and the component of angular momentum along the axis with the unique principal moment of inertia:

$$H_{\rm rot} = J^2/2I + J_a^2\{1/2I_a - 1/2I\} \quad \text{for prolate tops,} \tag{2.201}$$

$$H_{\rm rot} = J^2/2I + J_c^2\{1/2I_c - 1/2I\} \quad \text{for oblate tops.} \tag{2.202}$$

Here, the moment of inertia I denotes that moment that is common to two directions; that is, I is the non-unique moment of inertia. As a result, the eigenfunctions of $H_{\rm rot}$ are those of J^2 and J_a or J_c (and of J_Z), and the corresponding energy levels are

$$E(J, K, M) = \hbar^2 J(J+1)/2I^2 + \hbar^2 K^2\{1/2I_a - 1/2I\} \tag{2.203}$$

for prolate tops,

$$E(J, K, M) = \hbar^2 J(J+1)/2I^2 + \hbar^2 K^2\{1/2I_c - 1/2I\} \tag{2.204}$$

for oblate tops, again for K and M (i.e., J_a or J_c and J_Z quantum numbers, respectively) ranging from $-J$ to J in unit steps. Since the energy now depends on K, these levels are only $2J + 1$ degenerate due to the $2J + 1$ different M values that arise for each J value. Notice that for prolate tops, because I_a is smaller than I, the energies increase with increasing K for given J. In contrast, for oblate tops, since I_c is larger than I, the energies decrease with K for given J. The eigenfunctions $|J, M, K\rangle$ are the same rotation matrix functions as arise for the spherical-top case, so they do not require any further discussion at this time.

(iii) Asymmetric tops

The rotational eigenfunctions and energy levels of a molecule for which all three principal moments of inertia are distinct (a so-called asymmetric top) can not easily be expressed in terms of the angular momentum eigenstates and the J, M, and K quantum numbers. In fact, no one has ever solved the corresponding Schrödinger equation for this case. However, given the three principal moments of inertia I_a, I_b, and I_c, a matrix representation of each of the three contributions to the rotational Hamiltonian

$$H_{\text{rot}} = \frac{J_a^2}{2I_a} + \frac{J_b^2}{2I_b} + \frac{J_c^2}{2I_c} \tag{2.205}$$

can be formed within a basis set of the $\{|J, M, K\rangle\}$ rotation-matrix functions discussed earlier. This matrix will not be diagonal because the $|J, M, K\rangle$ functions are not eigenfunctions of the asymmetric top H_{rot}. However, the matrix can be formed in this basis and subsequently brought to diagonal form by finding its eigenvectors $\{C_{n,J,M,K}\}$ and its eigenvalues $\{E_n\}$. The vector coefficients express the asymmetric top eigenstates as

$$\Psi_n(\theta, \phi, \chi) = \sum_{J,M,K} C_{n,J,M,K}|J, M, K\rangle. \tag{2.206}$$

Because the total angular momentum J^2 still commutes with H_{rot}, each such eigenstate will contain only one J-value, and hence Ψ_n can also be labeled by a J quantum number:

$$\Psi_{n,J}(\theta, \phi, \chi) = \sum_{M,K} C_{n,J,M,K}|J, M, K\rangle. \tag{2.207}$$

To form the only non-zero matrix elements of $\mathrm{H_{rot}}$ within the $|\mathrm{J, M, K}\rangle$ basis, one can use the following properties of the rotation-matrix functions (see, for example, R. N. Zare, *Angular Momentum*, John Wiley, New York (1988)):

$$\begin{aligned}\langle J, M, K|J_a^2|J, M, K\rangle &= \langle J, M, K|J_b^2|J, M, K\rangle \\ &= 1/2\langle J, M, K|J^2 - J_c^2|J, M, K\rangle \\ &= \hbar^2[J(J+1) - K^2],\end{aligned} \tag{2.208}$$

$$\langle J, M, K|J_c^2|J, M, K\rangle = \hbar^2 K^2, \tag{2.209}$$

$$\begin{aligned}\langle J, M, K|J_a^2|J, M, K \pm 2\rangle &= -\langle J, M, K|J_b^2|J, M, K \pm 2\rangle \\ &= \hbar^2[J(J+1) - K(K \pm 1)]^{1/2}[J(J+1) \\ &\quad - (K \pm 1)(K \pm 2)]^{1/2},\end{aligned} \tag{2.210}$$

$$\langle J, M, K|J_c^2|J, M, K \pm 2\rangle = 0. \tag{2.211}$$

Each of the elements of J_c^2, J_a^2, and J_b^2 must, of course, be multiplied, respectively, by $1/2I_c$, $1/2I_a$, and $1/2I_b$ and summed together to form the matrix representation of H_{rot}. The diagonalization of this matrix then provides the asymmetric top energies and wave functions.

2.9 Vibrations of molecules

This Schrödinger equation forms the basis for our thinking about bond stretching and angle bending vibrations as well as collective vibrations called phonons in solids.

The radial motion of a diatomic molecule in its lowest ($J = 0$) rotational level can be described by the following Schrödinger equation:

$$-\frac{\hbar^2}{2\mu} r^{-2} \frac{\partial}{\partial r}\left(r^2 \frac{\partial}{\partial r}\right)\psi + V(r)\psi = E\psi, \tag{2.212}$$

where μ is the reduced mass $\mu = m_1 m_2/(m_1 + m_2)$ of the two atoms. If the molecule is rotating, then the above Schrödinger equation has an additional term $J(J+1)\hbar^2/2\mu r^{-2}\psi$ on its left-hand side. Thus, each rotational state (labeled by the rotational quantum number J) has its own vibrational Schrödinger equation and thus its own set of vibrational energy levels and wave functions. It is common to examine the $J = 0$ vibrational problem and then to use the vibrational levels of this state as approximations to the vibrational levels of states with non-zero J values (treating the vibration–rotation coupling via perturbation theory introduced in Section 4.1). Let us thus focus on the $J = 0$ situation.

By substituting $\psi = F(r)/r$ into this equation, one obtains an equation for $F(r)$ in which the differential operators appear to be less complicated:

$$-\frac{\hbar^2}{2\mu}\frac{d^2F}{dr^2} + V(r)F = EF. \tag{2.213}$$

This equation is exactly the same as the equation seen earlier in this text for the radial motion of the electron in the hydrogen-like atoms except that the reduced mass μ replaces the electron mass m and the potential $V(r)$ is not the Coulomb potential.

If the vibrational potential is approximated as a quadratic function of the bond displacement $x = r - r_e$ expanded about the equilibrium bond length r_e where V has its minimum,

$$V = 1/2k(r - r_e)^2, \tag{2.214}$$

the resulting harmonic-oscillator equation can be solved exactly. Because the potential V grows without bound as x approaches ∞ or $-\infty$, only bound-state solutions exist for this model problem. That is, the motion is confined by the nature of the potential, so no continuum states exist in which the two atoms bound together by the potential are dissociated into two separate atoms.

In solving the radial differential equation for this potential, the large-r behavior is first examined. For large r, the equation reads

$$\frac{d^2F}{dx^2} = \frac{1}{2}kx^2\left(\frac{2\mu}{\hbar^2}\right)F, \tag{2.215}$$

where $x = r - r_e$ is the bond displacement away from equilibrium. Defining $\xi = (\mu k/\hbar^2)^{1/4}x$ as a new scaled radial coordinate allows the solution of the large-r equation to be written as

$$F_{\text{large } r} = \exp(-\xi^2/2). \tag{2.216}$$

The general solution to the radial equation is then expressed as this large-r solution multiplied by a power series in the ζ variable:

$$F = \exp(-\xi^2/2) \sum_{n=0}^{\infty} \xi^n C_n, \tag{2.217}$$

where the C_n are coefficients to be determined. Substituting this expression into the full radial equation generates a set of recursion equations for the C_n amplitudes. As in the solution of the hydrogen-like radial equation, the series described by these coefficients is divergent unless the energy E happens to equal specific values. It is this requirement that the wave function not diverge so it can be normalized that yields energy quantization. The energies of the states that arise are given by

$$E_n = \hbar (k/\mu)^{1/2}(n + 1/2), \tag{2.218}$$

and the eigenfunctions are given in terms of the so-called Hermite polynomials $H_n(y)$ as follows:

$$\psi_n(x) = (n!2^n)^{-1/2}(\alpha/\pi)^{1/4} \exp(-\alpha x^2/2) H_n\left(\alpha^{1/2}x\right), \tag{2.219}$$

where $\alpha = (k\mu/\hbar^2)^{1/2}$. Within this harmonic approximation to the potential, the vibrational energy levels are evenly spaced:

$$\Delta E = E_{n+1} - E_n = \hbar (k/\mu)^{1/2}. \tag{2.220}$$

In experimental data such evenly spaced energy level patterns are seldom seen; most commonly, one finds spacings $E_{n+1} - E_n$ that decrease as the quantum number n increases. In such cases, one says that the progression of vibrational levels displays anharmonicity.

Because the Hermite functions H_n are odd or even functions of x (depending on whether n is odd or even), the wave functions $\psi_n(x)$ are odd or even. This splitting of the solutions into two distinct classes is an example of the effect of symmetry; in this case, the symmetry is caused by the symmetry of the harmonic potential with respect to reflection through the origin along the x-axis (i.e., changing x to $-x$). Throughout this text, many symmetries arise; in each case, symmetry properties of the potential cause the solutions of the Schrödinger equation to be decomposed into various symmetry groupings. Such symmetry decompositions are of great use because they provide additional quantum numbers (i.e., symmetry labels) by which the wave functions and energies can be labeled.

The basic idea underlying how such symmetries split the solutions of the Schrödinger equation into different classes relates to the fact that a symmetry operator (e.g., the reflection plane in the above example) commutes with the Hamiltonian. That is, the symmetry operator S obeys

$$SH\Psi = HS\Psi. \tag{2.221}$$

So S leaves H unchanged as it acts on H (this allows us to pass S through H in the above equation). Any operator that leaves the Hamiltonian (i.e., the energy) unchanged is called a symmetry operator.

If you have never learned about how point group symmetry can be used to help simplify the solution of the Schrödinger equation, this would be a good time to interrupt your reading and go to Chapter 4 and read the material there.

The harmonic oscillator energies and wave functions comprise the simplest reasonable model for vibrational motion. Vibrations of a polyatomic molecule are often characterized in terms of individual bond-stretching and angle-bending motions, each of which is, in turn, approximated harmonically. This results in a total vibrational wave function that is written as a product of functions, one for each of the vibrational coordinates.

Two of the most severe limitations of the harmonic oscillator model, the lack of anharmonicity (i.e., non-uniform energy level spacings) and lack of bond dissociation, result from the quadratic nature of its potential. By introducing model potentials that allow for proper bond dissociation (i.e., that do not increase without bound as $x \to \infty$), the major shortcomings of the harmonic oscillator picture can be overcome. The so-called Morse potential (see Fig. 2.24)

$$V(r) = D_e\{1 - \exp[-a(r - r_e)]\}^2, \tag{2.222}$$

is often used in this regard.

In the Morse potential function, D_e is the bond dissociation energy, r_e is the equilibrium bond length, and a is a constant that characterizes the "steepness" of the potential and thus affects the vibrational frequencies. The advantage of using the Morse potential to improve upon harmonic-oscillator-level predictions

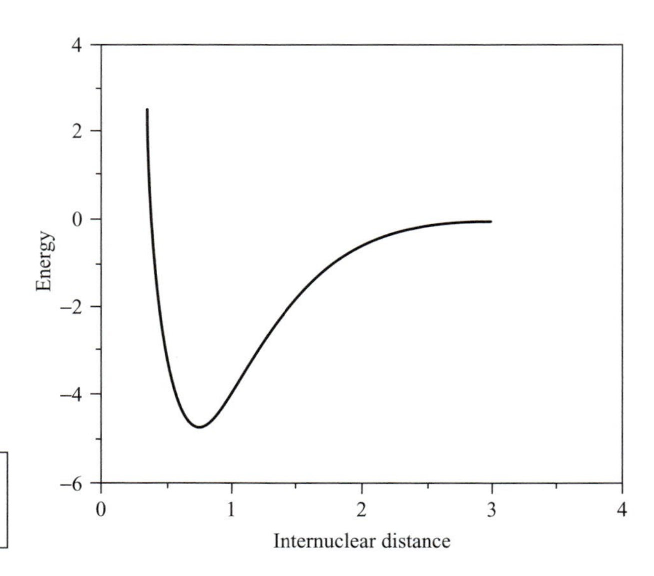

Figure 2.24 Morse potential energy as a function of bond length.

is that its energy levels and wave functions are also known exactly. The energies are given in terms of the parameters of the potential as follows:

$$E_n = \hbar \left(\frac{k}{\mu} \right)^{1/2} \left[\left(n + \frac{1}{2} \right) - \frac{\left(n + \frac{1}{2}\right)^2 \hbar \left(\frac{k}{\mu} \right)^{1/2}}{4D_e} \right], \tag{2.223}$$

where the force constant is given by $k = 2D_e\, a^2$. The Morse potential supports both bound states (those lying below the dissociation threshold for which vibration is confined by an outer turning point) and continuum states lying above the dissociation threshold. Its degree of anharmonicity is governed by the ratio of the harmonic energy $\hbar(k/\mu)^{1/2}$ to the dissociation energy D_e.

The eigenfunctions of the harmonic and Morse potentials display nodal character analogous to what we have seen earlier in the particle-in-a-box model problems. Namely, as the energy of the vibrational state increases, the number of nodes in the vibrational wave function also increases. The state having vibrational quantum number v has v nodes. I hope that by now the student is getting used to seeing the number of nodes increase as the quantum number and hence the energy grows.

Chapter 3

Characteristics of energy surfaces

3.1 Strategies for geometry optimization

The extension of the harmonic and Morse vibrational models to polyatomic molecules requires that the multidimensional energy surface be analyzed in a manner that allows one to approximate the molecule's motions in terms of many nearly independent vibrations. In this section, we will explore the tools that one uses to carry out such an analysis of the surface.

Many strategies that attempt to locate minima on molecular potential energy landscapes begin by approximating the potential energy V for geometries (collectively denoted in terms of $3N$ Cartesian coordinates $\{q_j\}$) in a Taylor series expansion about some "starting point" geometry (i.e., the current molecular geometry in an iterative process):

$$V(q_k) = V(0) + \sum_k (\partial V/\partial q_k) q_k + 1/2 \sum_{j,k} q_j H_{j,k} q_k + \cdots. \tag{3.1}$$

Here, $V(0)$ is the energy at the current geometry, $(\partial V/\partial q_k) = g_k$ is the gradient of the energy along the q_k coordinate, $H_{j,k} = (\partial^2 V/\partial q_j \partial q_k)$ is the second derivative or Hessian matrix, and q_k is the length of the "step" to be taken along this Cartesian direction. An example of an energy surface in only two dimensions is given in Fig. 3.1 where various special aspects are illustrated. For example, minima corresponding to stable molecular structures, transition states (first order saddle points) connecting such minima, and higher order saddle points are displayed.

If the only knowledge that is available is $V(0)$ and the gradient components (e.g., computation of the second derivatives is usually much more computationally taxing than is evaluation of the gradient), the linear approximation

$$V(q_k) = V(0) + \sum_k g_K q_k \tag{3.2}$$

suggests that one should choose "step" elements q_k that are opposite in sign from that of the corresponding gradient elements $g_k = (\partial V/\partial q_k)$. The magnitude of

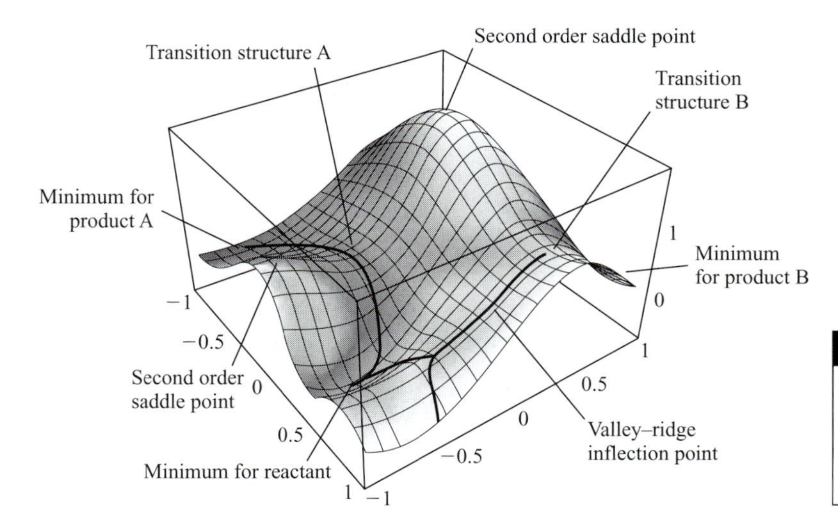

Figure 3.1 Two-dimensional potential surface showing minima, transition states, and paths connecting them.

the step elements is usually kept small in order to remain within the "trust radius" within which the linear approximation to V is valid to some predetermined desired precision.

When second derivative data is available, there are different approaches to predicting what step $\{q_k\}$ to take in search of a minimum. We first write the quadratic Taylor expansion

$$V(q_k) = V(0) + \sum_k g_k q_k + 1/2 \sum_{j,k} q_j H_{j,k} q_k \tag{3.3}$$

in matrix-vector notation

$$V(\mathbf{q}) = V(0) + \mathbf{q}^{\mathrm{T}} \cdot \mathbf{g} + 1/2\mathbf{q}^{\mathrm{T}} \cdot \mathbf{H} \cdot \mathbf{q} \tag{3.4}$$

with the elements$\{q_k\}$ collected into the column vector $\mathbf{q}$ whose transpose is denoted $\mathbf{q}^{\mathrm{T}}$. Introducing the unitary matrix $\mathbf{U}$ that diagonalizes the symmetric $\mathbf{H}$ matrix, the above equation becomes

$$V(\mathbf{q}) = V(0) + \mathbf{g}^{\mathrm{T}}\mathbf{U}\mathbf{U}^{\mathrm{T}}\mathbf{q} + 1/2\mathbf{q}^{\mathrm{T}}\mathbf{U}\mathbf{U}^{\mathrm{T}}\mathbf{H}\mathbf{U}\mathbf{U}^{\mathrm{T}}\mathbf{q}. \tag{3.5}$$

Because $\mathbf{U}^{\mathrm{T}}\mathbf{H}\mathbf{U}$ is diagonal,

$$(\mathbf{U}^{\mathrm{T}}\mathbf{H}\mathbf{U})_{k,l} = \delta_{k,l}\lambda_k \tag{3.6}$$

and has eigenvalues λ_k. For non-linear molecules, $3N - 6$ of these eigenvalues will be non-zero; for linear molecules, $3N - 5$ will be non-zero. The 5 or 6 zero eigenvalues of $\mathbf{H}$ have eigenvectors that describe translation and rotation of the entire molecule; they are zero because the energy surface V does not change if the molecule is rotated or translated. The eigenvectors of $\mathbf{H}$ form the columns of the array $\mathbf{U}$ that brings $\mathbf{H}$ to diagonal form:

$$\sum_{\lambda} H_{k,l} U_{l,m} = \lambda_m U_{k,m}. \tag{3.7}$$

Therefore, if we define

$$Q_m = \sum_k U^{\mathrm{T}}_{m,k} q_k \quad \text{and} \quad G_m = \sum_k U^T_{m,k} g_k \tag{3.8}$$

to be the component of the step $\{q_k\}$ and of the gradient along the mth eigenvector of $\mathbf{H}$, the quadratic expansion of V can be written in terms of steps along the $3N - 5$ or $3N - 6$ $\{Q_m\}$ directions that correspond to non-zero Hessian eigenvalues:

$$V(q_k) = V(0) + \sum_m G^{\mathrm{T}}_m Q_m + 1/2 \sum_m Q_m \lambda_m Q_m. \tag{3.9}$$

The advantage to transforming the gradient, step, and Hessian to the eigenmode basis is that each such mode (labeled m) appears in an independent uncoupled form in the expansion of V. This allows us to take steps along each of the Q_m directions in an independent manner with each step designed to lower the potential energy (as we search for minima).

For each eigenmode direction, one can ask for what step Q would the quantity $GQ + 1/2\,\lambda Q^2$ be a minimum. Differentiating this quadratic form with respect to Q and setting the result equal to zero gives

$$Q_m = -G_m/\lambda_m; \tag{3.10}$$

that is, one should take a step opposite the gradient but with a magnitude given by the gradient divided by the eigenvalue of the Hessian matrix. If the current molecular geometry is one that has all positive λ_m values, this indicates that one may be "close" to a minimum on the energy surface (because all λ_m are positive at minima). In such a case, the step $Q_m = -G_m/\lambda_m$ is opposed to the gradient along all $3N - 5$ or $3N - 6$ directions. The energy change that is expected to occur if the step $\{Q_m\}$ is taken can be computed by substituting $Q_m = -G_m/\lambda_m$ into the quadratic equation for V:

$$\begin{aligned} V(\text{after step}) &= V(0) + \sum_m G^{\mathrm{T}}_m(-G_m/\lambda_m) + 1/2 \sum_m \lambda_m(-G_m/\lambda_m)^2 \\ &= V(0) - 1/2 \sum_m \lambda_m(-G_m/\lambda_m)^2. \end{aligned} \tag{3.11}$$

This clearly suggests that the step will lead "downhill" in energy as long as all of the λ_m values are positive.

However, if one or more of the λ_m are negative at the current geometry, one is in a region of the energy surface that is not close to a minimum. In fact, if only one λ_m is negative, one anticipates being near a transition state (at which all gradient components vanish and all but one λ_m are positive with one λ_m negative). In such a case, the above analysis suggests taking a step $Q_m = -G_m/\lambda_m$ along all of the modes having positive λ_m, but taking a step of opposite direction $Q_n = +G_n/\lambda_n$ along the direction having negative λ_n.

In any event, once a step has been suggested within the eigenmode basis, one needs to express that step in terms of the original Cartesian coordinates q_k so that these Cartesian values can be altered within the software program to effect the predicted step. Given values for the $3N - 5$ or $3N - 6$ step components Q_m (n.b., the step components Q_m along the 5 or 6 modes having zero Hessian eigenvalues can be taken to be zero because they would simply translate or rotate the molecule), one must compute the $\{q_k\}$. To do so, we use the relationship

$$Q_m = \sum_k U^{\mathrm{T}}_{m,k} q_k \tag{3.12}$$

and write its inverse (using the unitary nature of the $\mathbf{U}$ matrix):

$$q_k = \sum_m U_{k,m} Q_m \tag{3.13}$$

to compute the desired Cartesian step components.

In using the Hessian-based approaches outlined above, one has to take special care when one or more of the Hessian eigenvalues is small. This often happens when:

(i) one has a molecule containing "soft modes" (i.e., degrees of freedom along which the energy varies little), or
(ii) one moves from a region of negative curvature into a region of positive curvature (or vice versa) – in such cases, the curvature must move through or near zero.

For these situations, the expression $Q_m = -G_m/\lambda_m$ can produce a very large step along the mode having small curvature. Care must be taken to not allow such incorrect artificially large steps to be taken.

Before closing this section, I should note that there are other important regions of potential energy surfaces that one must be able to locate and characterize. Above, we focused on local minima and transition states. In Chapter 8, we will discuss how to follow so-called reaction paths that connect these two kinds of stationary points using the type of gradient and Hessian information that we introduced earlier in this chapter.

Finally, it is sometimes important to find geometries at which two Born–Oppenheimer energy surfaces $V_1(\mathbf{q})$ and $V_2(\mathbf{q})$ intersect. First, let's spend a few minutes thinking about whether such surfaces can indeed intersect, because students often hear that surfaces do not intersect but, instead, undergo "avoided crossings". To understand the issue, let us assume that we have two wave functions Φ_1 and Φ_2 both of which depend on $3N - 6$ coordinates $\{\mathbf{q}\}$. These two functions are not assumed to be exact eigenfunctions of the Hamiltonian H, but likely are chosen to approximate such eigenfunctions. To find the improved functions Ψ_1 and Ψ_2 that more accurately represent the eigenstates, one usually forms linear combinations of Φ_1 and Φ_2,

$$\Psi_K = C_{K,1}\Phi_1 + C_{K,2}\Phi_2 \tag{3.14}$$

from which a 2×2 matrix eigenvalue problem arises:

$$\begin{vmatrix} H_{1,1} - E & H_{1,2} \\ H_{2,1} & H_{2,2} - E \end{vmatrix} = 0. \tag{3.15}$$

This quadratic equation has two solutions

$$2E_{\pm} = (H_{1,1} + H_{2,2}) \pm \sqrt{(H_{1,1} - H_{2,2})^2 + 4H_{1,2}^2}. \tag{3.16}$$

These two solutions can be equal (i.e., the two state energies can cross) only if the square root factor vanishes. Because this factor is a sum of two squares (each thus being positive quantities), this can only happen if two identities hold:

$$H_{1,1} = H_{2,2} \tag{3.17}$$

and

$$H_{1,2} = 0. \tag{3.18}$$

The main point then is that in the $3N - 6$ dimensional space, the two states will generally not have equal energy. However, in a space of two lower dimensions (because there are two conditions that must simultaneously be obeyed – $H_{1,1} = H_{2,2}$ and $H_{1,2} = 0$), their energies may be equal. They do not have to be equal, but it is possible that they are. It is based upon such an analysis that one usually says that potential energy surfaces in $3N - 6$ dimensions may undergo intersections in spaces of dimension $3N - 8$. If the two states are of different symmetry, the off-diagonal element $H_{1,2}$ vanishes automatically, so only one other condition is needed to realize crossing. So, we say that two states of different symmetry can cross in a space of dimension $3N - 7$.

To find the lower-dimensional space in which two surfaces cross, one must have available information about the gradients and Hessians of both functions V_1 and V_2. One then uses this information to locate a geometry at which the difference function $F = [V_1 - V_2]^2$ passes through zero by using conventional "root finding" methods designed to locate where $F = 0$. Once one such geometry ($\mathbf{q}_0$) has been located, one subsequently tries to follow the "seam" along which the function F remains zero. This is done by parameterizing steps away from ($\mathbf{q}_0$) in a manner that constrains such steps to have no component along the gradient of F (i.e., to lie in the tangent plane where F is constant). For a system with $3N - 6$ geometrical degrees of freedom, this seam will be a sub-surface of lower dimension ($3N - 8$ or $3N - 7$ as noted earlier). Such intersection seam location procedures are becoming more commonly employed, but are still under very active development. Locating these intersections is an important ingredient when one is interested in studying, for example, photochemical reactions in which the reactants and products may move from one electronic surface to another.

3.2 Normal modes of vibration

Having seen how one can use information about the gradients and Hessians on a Born–Oppenheimer surface to locate geometries corresponding to stable species, let us now move on to see how this same data are used to treat vibrations on this surface.

For a polyatomic molecule whose electronic energy depends on the $3N$ Cartesian coordinates of its N atoms, the potential energy V can be expressed (approximately) in terms of a Taylor series expansion about any of the local minima. Of course, different local minima (i.e., different isomers) will have different values for the equilibrium coordinates and for the derivatives of the energy with respect to these coordinates. The Taylor series expansion of the electronic energy is written as

$$V(q_k) = V(0) + \sum_k (\partial V/\partial q_k) q_k + 1/2 \sum_{j,k} q_j H_{j,k} q_k + \cdots, \tag{3.19}$$

where $V(0)$ is the value of the electronic energy at the stable geometry under study, q_k is the displacement of the kth Cartesian coordinate away from this starting position, $(\partial V/\partial q_k)$ is the gradient of the electronic energy along this direction, and the $H_{j,k}$ are the second derivative or Hessian matrix elements along these directions, $H_{j,k} = (\partial^2 V/\partial q_j \partial q_k)$. If the geometry corresponds to a stable species, the gradient terms will all vanish (meaning this geometry corresponds to a minimum, maximum, or saddle point), and the Hessian matrix will possess $3N - 5$ (for linear species) or $3N - 6$ (for non-linear molecules) positive eigenvalues and 5 or 6 zero eigenvalues (corresponding to 3 translational and 2 or 3 rotational motions of the molecule). If the Hessian has one negative eigenvalue, the geometry corresponds to a transition state. From now on, we assume that the geometry under study corresponds to that of a stable minimum about which vibrational motion occurs. The treatment of unstable geometries is of great importance to chemistry, but this material will be limited to vibrations of stable species.

3.2.1 The Newton equations of motion for vibration

The kinetic and potential energy matrices

Truncating the Taylor series at the quadratic terms (assuming these terms dominate because only small displacements from the equilibrium geometry are of interest), one has the so-called harmonic potential:

$$V(q_k) = V(0) + 1/2 \sum_{j,k} q_j H_{j,k} q_k. \tag{3.20}$$

The classical mechanical equations of motion for the $3N\{q_k\}$ coordinates can be written in terms of the above potential energy and the following kinetic energy function:

$$T = 1/2 \sum_j m_j \dot{q}_j^2, \tag{3.21}$$

where $\dot{q}_j$ denotes the time rate of change of the coordinate q_j and m_j is the mass of the atom on which the *j*th Cartesian coordinate resides. The Newton equations thus obtained are

$$m_j \ddot{q} = -\sum_k H_{j,k} q_k, \tag{3.22}$$

where the force along the *j*th coordinate is given by minus the derivative of the potential V along this coordinate $(\partial V/\partial q_j) = \sum_k H_{j,k} q_k$ within the harmonic approximation.

These classical equations can more compactly be expressed in terms of the time evolution of a set of so-called mass weighted Cartesian coordinates defined as

$$x_j = q_j (m_j)^{1/2}, \tag{3.23}$$

in terms of which the above Newton equations become

$$\ddot{x}_j = -\sum_k H'_{j,k} x_k \tag{3.24}$$

and the mass-weighted Hessian matrix elements are

$$H'_{j,k} = H_{j,k}(m_j m_k)^{-1/2}. \tag{3.25}$$

The harmonic vibrational energies and normal mode eigenvectors

Assuming that the x_j undergo some form of sinusoidal time evolution:

$$x_j(t) = x_j(0)\cos(\omega t), \tag{3.26}$$

and substituting this into the Newton equations produces a matrix eigenvalue equation:

$$\omega^2 x_j = \sum_k H'_{j,k} x_k \tag{3.27}$$

in which the eigenvalues are the squares of the so-called normal mode vibrational frequencies and the eigenvectors give the amplitudes of motion along each of the $3N$ mass-weighted Cartesian coordinates that belong to each mode. Hence, to perform a normal-mode analysis of a molecule, one forms the mass-weighted

Hessian matrix and then finds the $3N-5$ or $3N-6$ non-zero eigenvalues ω_j^2 as well as the corresponding eigenvectors $x_k^{(j)}$.

Within this harmonic treatment of vibrational motion, the total vibrational energy of the molecule is given as

$$E(v_1, v_2, \ldots, v_{3N-5 \text{ or } 6}) = \sum_{j=1}^{3N-5 \text{ or } 6} \hbar\omega_j(v_j + 1/2), \tag{3.28}$$

a sum of $3N-5$ or $3N-6$ independent contributions, one for each normal mode. The corresponding total vibrational wave function

$$\Psi = \prod_{j=1,3N-5 \text{ or } 6} \psi_{v_j}\left(x^{(j)}\right) \tag{3.29}$$

is a product of $3N-5$ or $3N-6$ harmonic oscillator functions $\psi_{v_j}(x^{(j)})$, one for each normal mode. The energy gap between one vibrational level and another in which one of the v_j quantum numbers is increased by unity (i.e., for fundamental vibrational transitions) is

$$\Delta E_{v_j \to v_j+1} = \hbar\omega_j. \tag{3.30}$$

The harmonic model thus predicts that the "fundamental" ($v=0 \to v=1$) and "hot band" ($v=1 \to v=2$) transitions should occur at the same energy, and the overtone ($v=0 \to v=2$) transitions should occur at exactly twice this energy.

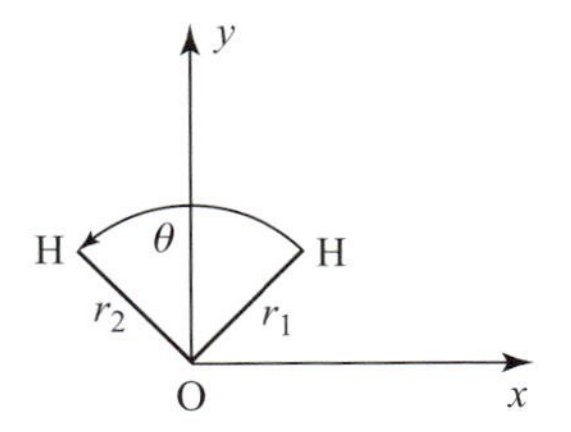

Figure 3.2 Water molecule showing its two bond lengths and bond angle.

3.2.2 The use of symmetry

Symmetry adapted modes

It is often possible to simplify the calculation of the normal mode frequencies and eigenvectors by exploiting molecular point group symmetry. For molecules that possess symmetry at a particular stable geometry, the electronic potential $V(q_j)$ displays symmetry with respect to displacements of symmetry equivalent Cartesian coordinates. For example, consider the water molecule at its C_{2v} equilibrium geometry as illustrated in Fig. 3.2. A very small movement of the H_2O molecule's left H atom in the positive x direction (Δx_L) produces the same change in the potential V as a correspondingly small displacement of the right H atom in the negative x direction ($-\Delta x_R$). Similarly, movement of the left H in the positive y direction (Δy_L) produces an energy change identical to movement of the right H in the positive y direction (Δy_R).

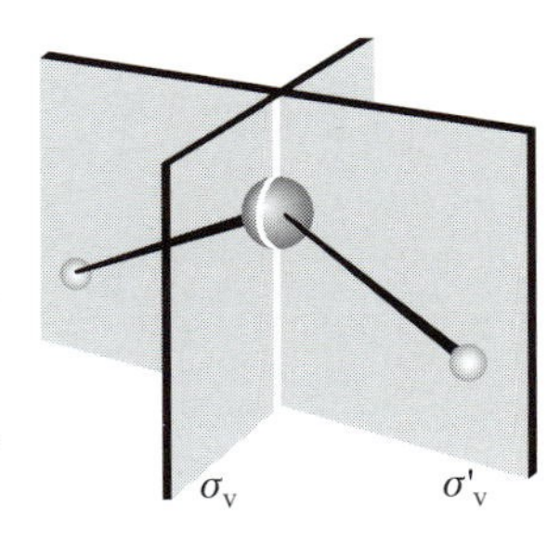

Figure 3.3 Two planes of symmetry of the water molecule.

The equivalence of the pairs of Cartesian coordinate displacements is a result of the fact that the displacement vectors are connected by the point group operations of the C_{2v} group. In particular, reflection of Δx_L through the yz plane (the two planes are depicted in Fig. 3.3) produces $-\Delta x_R$, and reflection of Δy_L through this same plane yields Δy_R.

More generally, it is possible to combine sets of Cartesian displacement coordinates $\{q_k\}$into so-called symmetry adapted coordinates $\{Q_{\Gamma,j}\}$, where the index Γ labels the irreducible representation in the appropriate point group and j labels the particular combination of that symmetry. These symmetry adapted coordinates can be formed by applying the point group projection operators (that are treated in detail in Chapter 4) to the individual Cartesian displacement coordinates.

To illustrate, again consider the H_2O molecule in the coordinate system described above. The $3N = 9$ mass weighted Cartesian displacement coordinates $(X_L, Y_L, Z_L, X_O, Y_O, Z_O, X_R, Y_R, Z_R)$ can be symmetry adapted by applying the following 4 projection operators:

$$P_{a_1} = 1 + \sigma_{yz} + \sigma_{xy} + C_2, \tag{3.31}$$

$$P_{b_1} = 1 + \sigma_{yz} - \sigma_{xy} - C_2, \tag{3.32}$$

$$P_{b_2} = 1 - \sigma_{yz} + \sigma_{xy} - C_2, \tag{3.33}$$

$$P_{a_2} = 1 - \sigma_{yz} - \sigma_{xy} + C_2, \tag{3.34}$$

to each of the 9 original coordinates (the symbol σ denotes reflection through a plane and C_2 means rotation about the molecule's C_2 axis). Of course, one will *not* obtain $9 \times 4 = 36$ independent symmetry adapted coordinates in this manner; many identical combinations will arise, and only 9 will be independent.

The independent combinations of a_1 *symmetry* (normalized to produce vectors of unit length) are

$$Q_{a_1,1} = 2^{-1/2}[X_L - X_R], \tag{3.35}$$

$$Q_{a_1,2} = 2^{-1/2}[Y_L + Y_R], \tag{3.36}$$

$$Q_{a_1,3} = [Y_O]. \tag{3.37}$$

Those of b_2 symmetry are

$$Q_{b_2,1} = 2^{-1/2}[X_L + X_R], \tag{3.38}$$

$$Q_{b_2,2} = 2^{-1/2}[Y_L - Y_R], \tag{3.39}$$

$$Q_{b_2,3} = [X_O], \tag{3.40}$$

and the combinations

$$Q_{b_1,1} = 2^{-1/2}[Z_L + Z_R], \tag{3.41}$$

$$Q_{b_1,2} = [Z_O] \tag{3.42}$$

are of b_1 symmetry, whereas

$$Q_{a_2,1} = 2^{-1/2}[Z_L - Z_R] \tag{3.43}$$

is of a_2 symmetry.

Point group symmetry of the harmonic potential

These nine $Q_{\Gamma,j}$ are expressed as unitary transformations of the original mass-weighted Cartesian coordinates:

$$Q_{\Gamma,j} = \sum_k C_{\Gamma,j,k} X_k. \tag{3.44}$$

These transformation coefficients $\{C_{\Gamma,j,k}\}$ can be used to carry out a unitary transformation of the 9×9 mass-weighted Hessian matrix. In so doing, we need only form blocks

$$H^{\Gamma}_{j,l} = \sum_{k,k'} C_{\Gamma,j,k} H_{k,k'} (m_k m_{k'})^{-1/2} C_{\Gamma,l,k'} \tag{3.45}$$

within which the symmetries of the two modes are identical. The off-diagonal elements

$$H^{\Gamma\Gamma'}_{j,1} = \sum_{k,k'} C_{\Gamma,j,k} H_{k,k'} (m_k m_{k'})^{-1/2} C_{\Gamma,l,k'} \tag{3.46}$$

vanish because the potential $V(q_j)$ (and the full vibrational Hamiltonian $H = T + V$) commutes with the C_{2v} point group symmetry operations.

As a result, the 9×9 mass-weighted Hessian eigenvalue problem can be subdivided into two 3×3 matrix problems (of a_1 and b_2 symmetry), one 2×2 matrix of b_1 symmetry and one 1×1 matrix of a_2 symmetry. The eigenvalues of each of these blocks provide the squares of the harmonic vibrational frequencies, the eigenvectors provide the normal mode displacements as linear combinations of the symmetry adapted $\{Q_{\Gamma,j}\}$.

Regardless of whether symmetry is used to block diagonalize the mass-weighted Hessian, six (for non-linear molecules) or five (for linear species) of the eigenvalues will equal zero. The eigenvectors belonging to these zero eigenvalues describe the three translations and two or three rotations of the molecule. For example,

$$\frac{1}{\sqrt{3}}[X_L + X_R + X_O], \tag{3.47}$$

$$\frac{1}{\sqrt{3}}[Y_L + Y_R + Y_O], \tag{3.48}$$

$$\frac{1}{\sqrt{3}}[Z_L + Z_R + Z_O] \tag{3.49}$$

are three translation eigenvectors of b_2, a_1 and b_1 symmetry, and

$$\frac{1}{\sqrt{2}}(Z_L - Z_R) \tag{3.50}$$

is a rotation (about the y-axis in Fig. 3.2) of a_2 symmetry. This rotation vector can be generated by applying the a_2 projection operator to Z_L or to Z_R. The other

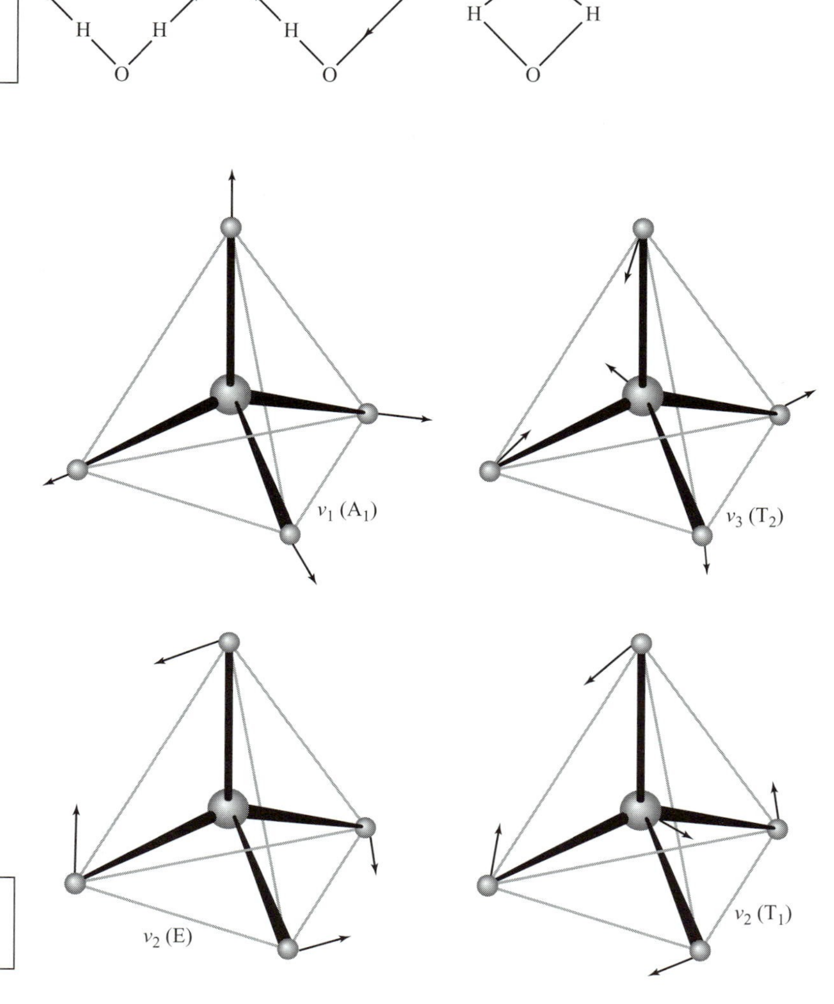

Figure 3.4 Symmetric and asymmetric stretch modes and bending mode of water.

Figure 3.5 Symmetries of vibrations of methane.

two rotations are of b_1 and b_2 symmetry and involve spinning of the molecule about the x- and z-axes of Fig. 3.2, respectively.

So, of the nine Cartesian displacements, three are of a_1 symmetry, three of b_2, two of b_1, and one of a_2. Of these, there are three translations (a_1, b_2, and b_1) and three rotations (b_2, b_1, and a_2). This leaves two vibrations of a_1 and one of b_2 symmetry. For the H_2O example treated here, the three non-zero eigenvalues of the mass-weighted Hessian are therefore of a_1, b_2, and a_1 symmetry. They describe the symmetric and asymmetric stretch vibrations and the bending mode, respectively, as illustrated in Fig. 3.4.

The method of vibrational analysis presented here can work for any polyatomic molecule. One knows the mass-weighted Hessian and then computes the non-zero eigenvalues, which then provide the squares of the normal mode vibrational frequencies. Point group symmetry can be used to block diagonalize this Hessian and to label the vibrational modes according to symmetry as we show in Fig. 3.5 for the CH_4 molecule in tetrahedral symmetry.

Chapter 4

Some important tools of theory

4.1 Perturbation theory and the variational method

In most practical applications of quantum mechanics to molecular problems, one is faced with the harsh reality that the Schrödinger equation pertinent to the problem at hand can not be solved exactly. To illustrate how desperate this situation is, I note that neither of the following two Schrödinger equations have ever been solved exactly (meaning analytically):

(i) The Schrödinger equation for the two electrons moving about the He nucleus:

$$\left[-\frac{\hbar^2}{2m_e}\nabla_1^2 - \frac{\hbar^2}{2m_e}\nabla_2^2 - \frac{2e^2}{r_1} - \frac{2e^2}{r_2} + \frac{e^2}{r_{1,2}}\right]\psi = E\psi. \tag{4.1}$$

(ii) The Schrödinger equation for the two electrons moving in an H_2 molecule even if the locations of the two nuclei (labeled A and B) are held clamped:

$$\left[-\frac{\hbar^2}{2m_e}\nabla_1^2 - \frac{\hbar^2}{2m_e}\nabla_2^2 - \frac{e^2}{r_{1,A}} - \frac{e^2}{r_{2,A}} - \frac{e^2}{r_{1,B}} - \frac{e^2}{r_{2,B}} + \frac{e^2}{r_{1,2}}\right]\psi = E\psi. \tag{4.2}$$

These two problems are examples of what is called the "three-body problem", meaning solving for the behavior of three bodies moving relative to one another. Motions of the sun, earth, and moon (even neglecting all the other planets and their moons) constitute another three-body problem. None of these problems, even the classical Newton equation for the sun, earth, and moon, have ever been solved exactly. So, what does one do when faced with trying to study real molecules using quantum mechanics?

There are two very powerful tools that one can use to "sneak up" on the solutions to the desired equations by first solving an easier "model" problem and then using the solutions to this problem to approximate the solutions to the real Schrödinger problem of interest. For example, to solve for the energies and wave functions of a boron atom, one could use hydrogenic 1s orbitals (but with $Z = 5$) and hydrogenic 2s and 2p orbitals with $Z = 3$ to account for the screening of the full nuclear charge by the two 1s electrons as a starting point. To solve for the vibrational energies of a diatomic molecule whose energy vs. bond length $E(R)$ is known, one could use the Morse oscillator wave functions as starting points.

But, once one has decided on a reasonable "starting point" model to use, how does one connect this model to the real system of interest? Perturbation theory and the variational method are the two tools that are most commonly used for this purpose.

4.1.1 Perturbation theory

In this method, one has available a set of equations for generating a sequence of approximations to the true energy E and true wave function ψ. I will now briefly outline the derivation of these working equations for you. First, one decomposes the true Hamiltonian H into a so-called zeroth order part H^0 (this is the Hamiltonian of the model problem one has chosen to use to represent the real system) and the difference $(H - H^0)$ which is called the perturbation and often denoted V:

$$H = H^0 + V. \tag{4.3}$$

The fundamental assumption of perturbation theory is that the wave functions and energies can be expanded in a Taylor series involving various powers of the perturbation. That is, one expands the energy E and the wave function ψ into zeroth, first, second, etc., order pieces which form the unknowns in this method:

$$E = E^0 + E^1 + E^2 + E^3 + \cdots, \tag{4.4}$$
$$\psi = \psi^0 + \psi^1 + \psi^2 + \psi^3 + \cdots. \tag{4.5}$$

Next, one substitutes these expansions for E of H and of ψ into $H\psi = E\psi$. This produces one equation whose right- and left-hand sides both contain terms of various "powers" in the perturbation. For example, terms of the form $E^1\psi^2$ and $V\psi^2$ and $E^0\psi^3$ are all of third power (also called third order). Next, one equates the terms on the left and right sides that are of the same order. This produces a set of equations, each containing all the terms of a given order. The zeroth, first, and second order such equations are given below:

$$H^0\psi^0 = E^0\psi^0, \tag{4.6}$$
$$H^0\psi^1 + V\psi^0 = E^0\psi^1 + E^1\psi^0, \tag{4.7}$$
$$H^0\psi^2 + V\psi^1 = E^0\psi^2 + E^1\psi^1 + E^2\psi^0. \tag{4.8}$$

The zeroth order equation simply instructs us to solve the zeroth order Schrödinger equation to obtain the zeroth order wave function ψ^0 and its zeroth order energy E^0. In the first order equation, the unknowns are ψ^1 and E^1 (recall that V is assumed to be known because it is the difference between the Hamiltonian one wants to solve and the model Hamiltonian H^0).

To solve the first order and higher order equations, one expands each of the corrections to the wave function ψ^K in terms of the complete set of wave functions of the zeroth order problem $\{\psi_J^0\}$. This means that one must be able to solve $H^0\psi_J^0 = E_J^0\psi_J^0$ not just for the zeroth order state one is interested in (denoted ψ^0 above) but for all of the other (e.g., excited states if ψ^0 is the ground state) zeroth order states $\{\psi_J^0\}$. For example, expanding ψ^1 in this manner gives

$$\psi^1 = \sum_J C_J^1 \psi_J^0. \tag{4.9}$$

Now, the unknowns in the first order equation become E^1 and the C_J^1 expansion coefficients. Substituting this expansion into $H^0\psi^1 + V\psi^0 = E^0\psi^1 + E^1\psi^0$ and solving for these unknowns produces the following final first order working equations:

$$E^1 = \langle\psi^0|V|\psi^0\rangle, \tag{4.10}$$

$$\psi^1 = \sum_J \psi_J^0 \left\{\langle\psi^0|V|\psi_J^0\rangle / \left(E^0 - E_J^0\right)\right\}, \tag{4.11}$$

where the index J is restricted such that ψ_J^0 not equal the state ψ^0 you are interested in. These are the fundamental working equations of first order perturbation theory. They instruct us to compute the average value of the perturbation taken over a probability distribution equal to $\psi^{0*}\psi^0$ to obtain the first order correction to the energy E^1. They also tell us how to compute the first order correction to the wave function in terms of coefficients multiplying various other zeroth order wave functions ψ_J^0.

An analogous approach is used to solve the second and higher order equations. Although modern quantum mechanics does indeed use high order perturbation theory in some cases, much of what the student needs to know is contained in the first and second order results to which I will therefore restrict our attention. The expression for the second order energy correction is found to be

$$E^2 = \sum_J \langle\psi^0|V|\psi_J^0\rangle^2 \Big/ \left(E^0 - E_J^0\right), \tag{4.12}$$

where again, the index J is restricted as noted above. Let's now consider an example problem that illustrates how perturbation theory is used.

Example problem for perturbation theory

As we discussed earlier, an electron moving in a conjugated bond framework can be modeled as a particle-in-a-box. An externally applied electric field of strength ε interacts with the electron in a fashion that can be described by adding the perturbation $V = e\varepsilon(x - \frac{L}{2})$ to the zeroth order Hamiltonian. Here, x is the position of the electron in the box, e is the electron's charge, and L is the length of the box.

First, we will compute the first order correction to the energy of the $n = 1$ state and the first order wave function for the $n = 1$ state. In the wave function calculation, we will only compute the contribution to ψ made by ψ_2^0 (this is just an approximation to keep things simple in this example). Let me now do all the steps needed to solve this part of the problem. Try to make sure you can do the algebra but also make sure you understand how we are using the first order perturbation equations.

$$V = e\varepsilon\left(x - \frac{L}{2}\right), \qquad \Psi_n^{(0)} = \left(\frac{2}{L}\right)^{\frac{1}{2}} \sin\left(\frac{n\pi x}{L}\right),$$

$$E_n^{(0)} = \frac{\hbar^2\pi^2 n^2}{2mL^2},$$

$$E_{n=1}^{(1)} = \left\langle \Psi_{n=1}^{(0)} \right| V \left| \Psi_{n=1}^{(0)} \right\rangle = \left\langle \Psi_{n=1}^{(0)} \right| e\varepsilon\left(x - \frac{L}{2}\right) \left| \Psi_{n=1}^{(0)} \right\rangle$$

$$= \left(\frac{2}{L}\right) \int_0^L \sin^2\left(\frac{\pi x}{L}\right) e\varepsilon\left(x - \frac{L}{2}\right) dx$$

$$= \left(\frac{2e\varepsilon}{L}\right) \int_0^L \sin^2\left(\frac{\pi x}{L}\right) x dx - \left(\frac{2e\varepsilon}{L}\right) \frac{L}{2} \int_0^L \sin^2\left(\frac{\pi x}{L}\right) dx.$$

The first integral can be evaluated using the following identity with $a = \frac{\pi}{L}$:

$$\int_0^L \sin^2(ax)\, x\, dx = \frac{x^2}{4} - \frac{x \sin(2ax)}{4a} - \frac{\cos(2ax)}{8a^2}\bigg|_0^L = \frac{L^2}{4}.$$

The second integral can be evaluated using the following identity with $\theta = \frac{\pi x}{L}$ and $d\theta = \frac{\pi}{L}\mathrm{d}x$:

$$\int_0^L \sin^2\left(\frac{\pi x}{L}\right) dx = \frac{L}{\pi} \int_0^\pi \sin^2\theta\, d\theta$$

$$\int_0^\pi \sin^2\theta\, d\theta = -\frac{1}{4}\sin(2\theta) + \frac{\theta}{2}\bigg|_0^\pi = \frac{\pi}{2}.$$

Making all of these appropriate substitutions we obtain:

$$E_{n=1}^{(1)} = \left(\frac{2e\varepsilon}{L}\right)\left(\frac{L^2}{4} - \frac{L}{2}\frac{L}{\pi}\frac{\pi}{2}\right) = 0,$$

$$\Psi_{n=1}^{(1)} = \frac{\left\langle \Psi_{n=2}^{(0)} \right| e\varepsilon\left(x - \frac{L}{2}\right) \left| \Psi_{n=1}^{(0)} \right\rangle \Psi_{n=2}^{(0)}}{E_{n=1}^{(0)} - E_{n=2}^{(0)}},$$

$$\Psi_{n=1}^{(1)} = \frac{\left(\frac{2}{L}\right) \int_0^L \sin\left(\frac{2\pi x}{L}\right) e\varepsilon\left(x - \frac{L}{2}\right) \sin\left(\frac{\pi x}{L}\right) dx}{\frac{\hbar^2\pi^2}{2mL^2(1^2-2^2)}} \left(\frac{2}{L}\right)^{\frac{1}{2}} \sin\left(\frac{2\pi x}{L}\right).$$

The two integrals in the numerator need to be evaluated:

$$\int_0^L x \sin\left(\frac{2\pi x}{L}\right) \sin\left(\frac{\pi x}{L}\right) dx, \qquad \text{and} \qquad \int_0^L \sin\left(\frac{2\pi x}{L}\right) \sin\left(\frac{\pi x}{L}\right) dx.$$

Using the integral $\int_x \cos(ax)dx = \frac{1}{a^2}\cos(ax) + \frac{x}{a}\sin(ax)$, and the integral $\int \cos(ax)dx = \frac{1}{a}\sin(ax)$, we obtain the following:

$$\int_0^L \sin\left(\frac{2\pi x}{L}\right) \sin\left(\frac{\pi x}{L}\right) dx = \frac{1}{2}\left[\int_0^L \cos\left(\frac{\pi x}{L}\right) dx - \int_0^L \cos\left(\frac{3\pi x}{L}\right) dx\right]$$

$$= \frac{1}{2}\left[\frac{L}{\pi}\sin\left(\frac{\pi x}{L}\right)\bigg|_0^L - \frac{L}{3\pi}\sin\left(\frac{3\pi x}{L}\right)\bigg|_0^L\right] = 0,$$

$$\int_0^L x \sin\left(\frac{2\pi x}{L}\right) \sin\left(\frac{\pi x}{L}\right) dx = \frac{1}{2}\left[\int_0^L x \cos\left(\frac{\pi x}{L}\right) dx - \int_0^L x \cos\left(\frac{3\pi x}{L}\right) dx\right]$$
$$= \frac{1}{2}\left[\left(\frac{L^2}{\pi^2}\cos\left(\frac{\pi x}{L}\right) + \frac{Lx}{\pi}\sin\left(\frac{\pi x}{L}\right)\right)\Bigg|_0^L\right.$$
$$\left. - \left(\frac{L^2}{9\pi^2}\cos\left(\frac{3\pi x}{L}\right) + \frac{Lx}{3\pi}\sin\left(\frac{3\pi x}{L}\right)\right)\Bigg|_0^L\right]$$
$$= \frac{L^2}{2\pi^2}(\cos\pi - \cos 0) + \frac{L^2}{2\pi}\sin\pi - 0$$
$$= \frac{-2L^2}{2\pi^2} - \frac{-2L^2}{18\pi^2} = \frac{L^2}{9\pi^2} - \frac{L^2}{\pi^2} = -\frac{8L^2}{9\pi^2}.$$

Making all of these appropriate substitutions we obtain

$$\Psi_{n=1}^{(1)} = \frac{\left(\frac{2}{L}\right)(e\varepsilon)\left(-\frac{8L^2}{9\pi^2} - \frac{L}{2}(0)\right)}{\frac{-3\hbar^2\pi^2}{2mL^2}}\left(\frac{2}{L}\right)^{\frac{1}{2}}\sin\left(\frac{2\pi x}{L}\right)$$
$$= \frac{32mL^3e\varepsilon}{27\hbar^2\pi^4}\left(\frac{2}{L}\right)^{\frac{1}{2}}\sin\left(\frac{2\pi x}{L}\right).$$

Now, let's compute the induced dipole moment caused by the polarization of the electron density due to the electric field effect using the equation $\mu_{\text{induced}} = -e\int \Psi^*(x - \frac{L}{2})\Psi\, dx$ with Ψ now being the sum of our zeroth and first order wave functions. In computing this integral, we neglect the term proportional to ε^2 because we are interested in only the term linear in ε because this is what gives the dipole moment. Again, allow me to do the algebra and see if you can follow.

$$\mu_{\text{induced}} = -e\int \Psi^*\left(x - \frac{L}{2}\right)\Psi dx, \qquad \text{where} \quad \Psi = \left(\Psi_1^{(0)} + \Psi_1^{(1)}\right).$$
$$\mu_{\text{induced}} = -e\int_0^L \left(\Psi_1^{(0)} + \Psi_1^{(1)}\right)^*\left(x - \frac{L}{2}\right)\left(\Psi_1^{(0)} + \Psi_1^{(1)}\right) dx$$
$$= -e\int_0^L \Psi_1^{(0)*}\left(x - \frac{L}{2}\right)\Psi_1^{(0)}dx - e\int_0^L \Psi_1^{(0)*}\left(x - \frac{L}{2}\right)\Psi_1^{(1)}dx$$
$$-e\int_0^L \Psi_1^{(1)*}\left(x - \frac{L}{2}\right)\Psi_1^{(0)}dx - e\int_0^L \Psi_1^{(1)*}\left(x - \frac{L}{2}\right)\Psi_1^{(1)}dx.$$

The first integral is zero (see the evaluation of this integral for $E_1^{(1)}$ above). The fourth integral is neglected since it is proportional to ε^2. The second and third integrals are the same and are combined to give

$$\mu_{\text{induced}} = -2e\int_0^L \Psi_1^{(0)*}\left(x - \frac{L}{2}\right)\Psi_1^{(1)}dx.$$

Substituting $\Psi_1^{(0)} = (\frac{2}{L})^{\frac{1}{2}} \sin(\frac{\pi x}{L})$ and $\Psi_1^{(1)} = \frac{32mL^3e\varepsilon}{27\hbar^2\pi^4}(\frac{2}{L})^{\frac{1}{2}} \sin(\frac{2\pi x}{L})$, we obtain

$$\mu_{\text{induced}} = -2e\frac{32mL^3e\varepsilon}{27\hbar^2\pi^4}\left(\frac{2}{L}\right)\int_0^L \sin\left(\frac{\pi x}{L}\right)\left(x - \frac{L}{2}\right)\sin\left(\frac{2\pi x}{L}\right)dx.$$

These integrals are familiar from what we did to compute Ψ^1; doing them we finally obtain

$$\mu_{\text{induced}} = -2e\frac{32mL^3e\varepsilon}{27\hbar^2\pi^4}\left(\frac{2}{L}\right)\left(-\frac{8L^2}{9\pi^2}\right)$$
$$= \frac{mL^4e^2\varepsilon}{\hbar^2\pi^6}\frac{2^{10}}{3^5}.$$

Now let's compute the polarizability, α, of the electron in the $n = 1$ state of the box, and try to understand physically why α should depend as it does upon the length of the box L. To compute the polarizability, we need to know that $\alpha = \frac{\partial\mu}{\partial\varepsilon}|_{\varepsilon=0}$. Using our induced moment result above, we then find

$$\alpha = \left(\frac{\partial\mu}{\partial\varepsilon}\right)_{\varepsilon=0} = \frac{mL^4e^2}{\hbar^2\pi^6}\frac{2^{10}}{3^5}.$$

Notice that this finding suggests that the larger the box (molecule), the more polarizable the electron density. This result also suggests that the polarizability of conjugated polyenes should vary non-linearly with the length of the conjugated chain.

4.1.2 The variational method

Let us now turn to the other method that is used to solve Schrödinger equations approximately, the variational method. In this approach, one must again have some reasonable wave function ψ^0 that is used to approximate the true wave function. Within this approximate wave function, one imbeds one or more variables $\{\alpha_J\}$ that one subsequently varies to achieve a minimum in the energy of ψ^0 computed as an expectation value of the true Hamiltonian H:

$$E(\{\alpha_J\}) = \langle\psi^0|H|\psi^0\rangle/\langle\psi^0|\psi^0\rangle. \tag{4.13}$$

The optimal values of the α_J parameters are determined by making

$$dE/d\alpha_J = 0, \tag{4.14}$$

to achieve the desired energy minimum (n.b., we also should verify that the second derivative matrix $(\partial^2E/\partial\alpha_J\,\partial\alpha_L)$ has all positive eigenvalues).

The theoretical basis underlying the variational method can be understood through the following derivation. Suppose that someone knew the exact eigenstates (i.e., true Ψ_K and true E_K) of the true Hamiltonian H. These states obey

$$H\Psi_K = E_K\Psi_K. \tag{4.15}$$

Because these true states form a complete set (it can be shown that the eigenfunctions of all the Hamiltonian operators we ever encounter have this property), our so-called "trial wave function" ψ^0 can, in principle, be expanded in terms of these Ψ_K:

$$\psi^0 = \sum_K C_K \Psi_K. \tag{4.16}$$

Before proceeding further, allow me to overcome one likely misconception. What I am going through now is only a derivation of the working formula of the variational method. The final formula will not require us to ever know the exact Ψ_K or the exact E_K, but we are allowed to use them as tools in our derivation because we know they exist even if we never know them.

With the above expansion of our trial function in terms of the exact eigenfunctions, let us now substitute this into the quantity $\langle\psi^0|H|\psi^0\rangle/\langle\psi^0|\psi^0\rangle$ that the variational method instructs us to compute:

$$E = \langle\psi^0|H|\psi^0\rangle/\langle\psi^0 \mid \psi^0\rangle = \left\langle \sum_K C_K \Psi_K \middle| H \middle| \sum_L C_L \Psi_L \right\rangle \Big/ \\ \times \left\langle \sum_K C_K \Psi_K \mid \sum_L C_L \Psi_L \right\rangle. \tag{4.17}$$

Using the fact that the Ψ_K obey $H\Psi_K = E_K \Psi_K$ and that the Ψ_K are orthonormal (I hope you remember this property of solutions to all Schrödinger equations that we discussed earlier)

$$\langle\Psi_K \mid \Psi_L\rangle = \delta_{K.L}, \tag{4.18}$$

the above expression reduces to

$$E = \sum_K \langle C_K \Psi_K|H|C_K \Psi_K\rangle \Big/ \left(\sum_K \langle C_K \Psi_K \mid C_K \Psi_K\rangle \right) \\ = \sum_K |C_K|^2 E_K \Big/ \sum_K |C_K|^2. \tag{4.19}$$

One of the basic properties of the kind of Hamiltonian we encounter is that they have a lowest-energy state. Sometimes we say they are bounded from below, which means their energy states do not continue all the way to minus infinity. There are systems for which this is not the case, but we will now assume that we are not dealing with such systems. This allows us to introduce the inequality $E_K \geq E_0$ which says that all of the energies are higher than or equal to the energy of the lowest state which we denote E_0. Introducing this inequality into the above expression gives

$$E \geq \sum_K |C_K|^2 E_0 \Big/ \sum_K |C_K|^2 = E_0. \tag{4.20}$$

This means that the variational energy, computed as $\langle\psi^0|H|\psi^0\rangle/\langle\psi^0 \mid \psi^0\rangle$ will lie above the true ground-state energy no matter what trial function ψ^0 we use.

The significance of the above result that $E \geq E_0$ is as follows. We are allowed to imbed into our trial wave function ψ^0 parameters that we can vary to make E, computed as $\langle\psi^0|H|\psi^0\rangle/\langle\psi^0 \mid \psi^0\rangle$, as low as possible because we know that we can never make $\langle\psi^0|H|\psi^0\rangle/\langle\psi^0 \mid \psi^0\rangle$ lower than the true ground-state energy. The philosophy then is to vary the parameters in ψ^0 to render E as low as possible, because the closer E is to E_0 the "better" is our variational wave function. Let me now demonstrate how the variational method is used in such a manner by solving an example problem.

Example variational problem

Suppose you are given a trial wave function of the form

$$\phi = \frac{Z_e^{\,3}}{\pi a_0^3} \exp\left(\frac{-Z_e r_1}{a_0}\right) \exp\left(\frac{-Z_e r_2}{a_0}\right)$$

to represent a two-electron ion of nuclear charge Z and suppose that you are lucky enough that I have already evaluated the $\langle\psi^0|H|\psi^0\rangle/\langle\psi^0 \mid \psi^0\rangle$ integral, which I'll call W, for you and found

$$W = \left(Z_e^{\,2} - 2ZZ_e + \frac{5}{8} Z_e\right) \frac{e^2}{a_0}.$$

Now, let's find the optimum value of the variational parameter Z_e for an arbitrary nuclear charge Z by setting $dW/dZ_e = 0$. After finding the optimal value of Z_e, we'll then find the optimal energy by plugging this Z_e into the above W expression. I'll do the algebra and see if you can follow.

$$\begin{aligned}
W &= \left(Z_e^{\,2} - 2ZZ_e + \frac{5}{8} Z_e\right) \frac{e^2}{a_0}, \\
\frac{dW}{dZ_e} &= \left(2Z_e - 2Z + \frac{5}{8}\right) \frac{e^2}{a_0} = 0, \\
& 2Z_e - 2Z + \frac{5}{8} = 0, \\
2Z_e &= 2Z - \frac{5}{8}, \\
Z_e &= Z - \frac{5}{16} = Z - 0.3125
\end{aligned}$$

(n.b., 0.3125 represents the shielding factor of one 1s electron to the other).

Now, using this optimal Z_e in our energy expression gives

$$\begin{aligned}
W &= Z_e \left(Z_e - 2Z + \frac{5}{8}\right) \frac{e^2}{a_0} \\
&= \left(Z - \frac{5}{16}\right) \left[\left(Z - \frac{5}{16}\right) - 2Z + \frac{5}{8}\right] \frac{e^2}{a_0}
\end{aligned}$$

$$
\begin{aligned}
&= \left(Z - \frac{5}{16}\right)\left(-Z + \frac{5}{16}\right)\frac{e^2}{a_0} \\
&= -\left(Z - \frac{5}{16}\right)\left(Z - \frac{5}{16}\right)\frac{e^2}{a_0} = -\left(Z - \frac{5}{16}\right)^2\frac{e^2}{a_0} \\
&= -(Z - 0.3125)^2(27.21)\text{ eV}
\end{aligned}
$$

(n.b., since a_0 is the Bohr radius 0.529 Å, $e^2/a_0 = 27.21$ eV).

Is this energy "any good"? The total energies of some two-electron atoms and ions have been experimentally determined to be:

Z	Atom	Energy (eV)
1	H^-	−14.35
2	He	−78.98
3	Li^+	−198.02
4	Be^{+2}	−371.5
5	B^{+3}	−599.3
6	C^{+4}	−881.6
7	N^{+5}	−1218.3
8	O^{+6}	−1609.5

Using our optimized expression for W, let's now calculate the estimated total energies of each of these atoms and ions as well as the percentage error in our estimate for each ion.

Z	Atom	Experimental (eV)	Calculated (eV)	% Error
1	H^-	−14.35	−12.86	10.38
2	He	−78.98	−77.46	1.92
3	Li^+	−198.02	−196.46	0.79
4	Be^{+2}	−371.5	−369.86	0.44
5	B^{+3}	−599.3	−597.66	0.27
6	C^{+4}	−881.6	−879.86	0.19
7	N^{+5}	−1218.3	−1216.48	0.15
8	O^{+6}	−1609.5	−1607.46	0.13

The energy errors are essentially constant over the range of Z, but produce a larger percentage error at small Z.

In 1928, when quantum mechanics was quite young, it was not known whether the isolated, gas-phase hydride ion, H^-, was stable with respect to dissociation into a hydrogen atom and an electron. Let's compare our estimated total energy for H^- to the ground-state energy of a hydrogen atom and an isolated electron (which is known to be -13.60 eV). When we use our expression for W and take $Z = 1$, we obtain $W = -12.86$ eV, which is greater than -13.6 eV ($H + e^-$), so this simple variational calculation erroneously predicts H^- to be unstable. More complicated variational treatments give a ground state energy of H^- of -14.35 eV, in agreement with experiment.

4.2 Point group symmetry

It is assumed that the reader has previously learned, in undergraduate inorganic or physical chemistry classes, how symmetry arises in molecular shapes and structures and what symmetry elements are (e.g., planes, axes of rotation, centers of inversion, etc.). For the reader who feels, after reading this section, that additional background is needed, the texts by Eyring, Walter, and Kimball or by Atkins and Friedman can be consulted. We review and teach here only that material that is of direct application to symmetry analysis of molecular orbitals and vibrations and rotations of molecules. We use a specific example, the ammonia molecule, to introduce and illustrate the important aspects of point group symmetry.

4.2.1 The C_{3v} symmetry group of ammonia – an example

The ammonia molecule NH_3 belongs, in its ground-state equilibrium geometry, to the C_{3v} point group. Its symmetry operations consist of two C_3 rotations, C_3, C_3^2 (rotations by 120° and 240°, respectively, about an axis passing through the nitrogen atom and lying perpendicular to the plane formed by the three hydrogen atoms), three vertical reflections, σ_v, σ_v', σ_v'', and the identity operation. Corresponding to these six operations are symmetry elements: the three-fold rotation axis, C_3 and the three symmetry planes σ_v, σ_v' and σ_v'' that contain the three NH bonds and the z-axis (see Fig. 4.1).

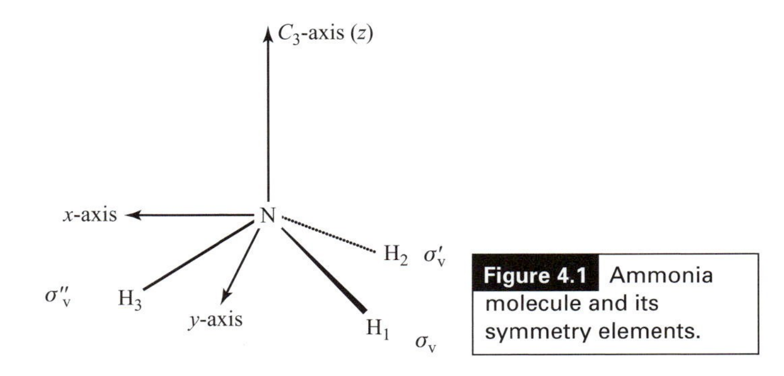

Figure 4.1 Ammonia molecule and its symmetry elements.

These six symmetry operations form a mathematical group. A group is defined as a set of objects satisfying four properties.

(i) A combination rule is defined through which two group elements are combined to give a result which we call the product. The product of two elements in the group must also be a member of the group (i.e., the group is closed under the combination rule).
(ii) One special member of the group, when combined with any other member of the group, must leave the group member unchanged (i.e., the group contains an identity element).
(iii) Every group member must have a reciprocal in the group. When any group member is combined with its reciprocal, the product is the identity element.
(iv) The associative law must hold when combining three group members (i.e., (AB)C must equal A(BC)).

The members of symmetry groups are symmetry operations; the combination rule is a successive operation. The identity element is the operation of doing nothing at all. The group properties can be demonstrated by forming a multiplication table. Let us label the rows of the table by the first operation and the columns by the second operation. Note that this order is important because most groups are *not commutative*. The C_{3v} group multiplication table is as follows:

First operation \ Second operation	E	C_3	C_3^2	σ_v	σ_v'	σ_v''
E	E	C_3	C_3^2	σ_v	σ_v'	σ_v''
C_3	C_3	C_3^2	E	σ_v'	σ_v''	σ_v
C_3^2	C_3^2	E	C_3	σ_v''	σ_v	σ_v'
σ_v	σ_v	σ_v''	σ_v'	E	C_3^2	C_3
σ_v'	σ_v'	σ_v	σ_v''	C_3	E	C_3^2
σ_v''	σ_v''	σ_v'	σ_v	C_3^2	C_3	E

Note the reflection plane labels do not move. That is, although we start with H_1 in the σ_v plane, H_2 in σ_v'', and H_3 in σ_v'', if H_1 moves due to the first symmetry operation, σ_v remains fixed and a different H atom lies in the σ_v plane.

4.2.2 Matrices as group representations

In using symmetry to help simplify molecular orbital (m.o.) or vibration/rotation energy level identifications, the following strategy is followed:

(i) A set of M objects belonging to the constituent atoms (or molecular fragments, in a more general case) is introduced. These objects are the orbitals of the individual atoms (or of the fragments) in the m.o. case; they are unit vectors along the x, y, and z directions located on each of the atoms, and represent displacements along each of these directions, in the vibration/rotation case.

(ii) Symmetry tools are used to combine these M objects into M new objects each of which belongs to a specific symmetry of the point group. Because the Hamiltonian (electronic in the m.o. case and vibration/rotation in the latter case) commutes with the symmetry operations of the point group, the matrix representation of **H** within the symmetry adapted basis will be "block diagonal". That is, objects of different symmetry will not interact; only interactions among those of the same symmetry need be considered.

To illustrate such symmetry adaptation, consider symmetry adapting the 2s orbital of N and the three 1s orbitals of the three H atoms. We begin by determining how these orbitals transform under the symmetry operations of the C_{3v} point group. The act of each of the six symmetry operations on the four atomic orbitals can be denoted as follows:

$$\begin{array}{ccc}
(S_N, S_1, S_2, S_3) & \xrightarrow{E} & (S_N, S_1, S_2, S_3) \\
 & \xrightarrow{C_3} & (S_N, S_3, S_1, S_2) \\
 & \xrightarrow{C_3^2} & (S_N, S_2, S_3, S_1) \\
 & \xrightarrow{\sigma_v} & (S_N, S_1, S_3, S_2) \\
 & \xrightarrow{\sigma_v''} & (S_N, S_3, S_2, S_1) \\
 & \xrightarrow{\sigma_v'} & (S_N, S_2, S_1, S_3)
\end{array} \tag{4.21}$$

Here we are using the active view that a C_3 rotation rotates the molecule by 120°. The equivalent passive view is that the 1s basis functions are rotated −120°. In the C_3 rotation, S_3 ends up where S_1 began, S_1 ends up where S_2 began, and S_2 ends up where S_3 began.

These transformations can be thought of in terms of a matrix multiplying a vector with elements (S_N, S_1, S_2, S_3). For example, if $D^{(4)}(C_3)$ is the representation matrix giving the C_3 transformation, then the above action of C_3 on the four basis orbitals can be expressed as

$$D^{(4)}(C_3)\begin{bmatrix} S_N \\ S_1 \\ S_2 \\ S_3 \end{bmatrix} = \begin{bmatrix} 1 & 0 & 0 & 0 \\ 0 & 0 & 0 & 1 \\ 0 & 1 & 0 & 0 \\ 0 & 0 & 1 & 0 \end{bmatrix}\begin{bmatrix} S_N \\ S_1 \\ S_2 \\ S_3 \end{bmatrix} = \begin{bmatrix} S_N \\ S_3 \\ S_1 \\ S_2 \end{bmatrix}. \tag{4.22}$$

We can likewise write matrix representations for each of the symmetry operations of the C_{3v} point group:

$$D^{(4)}\left(C_3^2\right) = \begin{bmatrix} 1 & 0 & 0 & 0 \\ 0 & 0 & 1 & 0 \\ 0 & 0 & 0 & 1 \\ 0 & 1 & 0 & 0 \end{bmatrix}, \quad D^{(4)}(E) = \begin{bmatrix} 1 & 0 & 0 & 0 \\ 0 & 1 & 0 & 0 \\ 0 & 0 & 1 & 0 \\ 0 & 0 & 0 & 1 \end{bmatrix},$$

$$D^{(4)}(\sigma_v) = \begin{bmatrix} 1 & 0 & 0 & 0 \\ 0 & 1 & 0 & 0 \\ 0 & 0 & 0 & 1 \\ 0 & 0 & 1 & 0 \end{bmatrix}, \quad D^{(4)}(\sigma_v') = \begin{bmatrix} 1 & 0 & 0 & 0 \\ 0 & 0 & 0 & 1 \\ 0 & 0 & 1 & 0 \\ 0 & 1 & 0 & 0 \end{bmatrix},$$

$$D^{(4)}(\sigma_v'') = \begin{bmatrix} 1 & 0 & 0 & 0 \\ 0 & 0 & 1 & 0 \\ 0 & 1 & 0 & 0 \\ 0 & 0 & 0 & 1 \end{bmatrix}. \tag{4.23}$$

It is easy to verify that a C_3 rotation followed by a σ_v reflection is equivalent to a σ_v' reflection alone. In other words

$$\sigma_v C_3 = \sigma_v', \qquad \text{or,} \qquad \begin{array}{ccc} & S_1 & \\ S_2 & & S_3 \end{array} \ \overset{C_3}{\rightarrow}\ \begin{array}{ccc} & S_3 & \\ S_1 & & S_2 \end{array} \ \overset{\sigma_v}{\rightarrow}\ \begin{array}{ccc} & S_3 & \\ S_2 & & S_1 \end{array} \tag{4.24}$$

Note that this same relationship is carried by the matrices:

$$\begin{aligned} D^{(4)}(\sigma_v)D^{(4)}(C_3) &= \begin{bmatrix} 1 & 0 & 0 & 0 \\ 0 & 1 & 0 & 0 \\ 0 & 0 & 0 & 1 \\ 0 & 0 & 1 & 0 \end{bmatrix} \begin{bmatrix} 1 & 0 & 0 & 0 \\ 0 & 0 & 0 & 1 \\ 0 & 1 & 0 & 0 \\ 0 & 0 & 1 & 0 \end{bmatrix} = \begin{bmatrix} 1 & 0 & 0 & 0 \\ 0 & 0 & 0 & 1 \\ 0 & 0 & 1 & 0 \\ 0 & 1 & 0 & 0 \end{bmatrix} \\ &= D^{(4)}(\sigma_v'). \end{aligned} \tag{4.25}$$

Likewise we can verify that $C_3\sigma_v = \sigma_v''$ directly and we can notice that the matrices also show the same identity:

$$\begin{aligned} D^{(4)}(C_3)D^{(4)}(\sigma_v) &= \begin{bmatrix} 1 & 0 & 0 & 0 \\ 0 & 0 & 0 & 1 \\ 0 & 1 & 0 & 0 \\ 0 & 0 & 1 & 0 \end{bmatrix} \begin{bmatrix} 1 & 0 & 0 & 0 \\ 0 & 1 & 0 & 0 \\ 0 & 0 & 0 & 1 \\ 0 & 0 & 1 & 0 \end{bmatrix} = \begin{bmatrix} 1 & 0 & 0 & 0 \\ 0 & 0 & 1 & 0 \\ 0 & 1 & 0 & 0 \\ 0 & 0 & 0 & 1 \end{bmatrix} \\ &= D^{(4)}(\sigma_v''). \end{aligned} \tag{4.26}$$

In fact, one finds that the six matrices, $D^{(4)}(R)$, when multiplied together in all 36 possible ways obey the same multiplication table as did the six symmetry operations. We say the matrices form a representation of the group because the matrices have all the properties of the group.

Characters of representations

One important property of a matrix is the sum of its diagonal elements which is called the trace of the matrix D and is denoted $\mathrm{Tr}(D)$:

$$\mathrm{Tr}(D) = \sum_i D_{ii} = \chi. \tag{4.27}$$

So, χ is called the trace or character of the matrix. In the above example

$$\chi(E) = 4, \tag{4.28}$$
$$\chi(C_3) = \chi\left(C_3^2\right) = 1, \tag{4.29}$$
$$\chi(\sigma_v) = \chi(\sigma_v') = \chi(\sigma_v'') = 2. \tag{4.30}$$

The importance of the characters of the symmetry operations lies in the fact that they do not depend on the specific basis used to form them. That is, they are invariant to a unitary or orthogonal transformation of the objects used to define the matrices. As a result, they contain information about the symmetry operation itself and about the *space* spanned by the set of objects. The significance of this observation for our symmetry adaptation process will become clear later.

Note that the characters of both rotations are the same as are those of all three reflections. Collections of operations having identical characters are called *classes*. Each operation in a *class* of operations has the same character as other members of the class. The character of a class depends on the space spanned by the basis of functions on which the symmetry operations act.

Another basis and another representation

Above we used (S_N, S_1, S_2, S_3) as a basis. If, alternatively, we use the one-dimensional basis consisting of the 1s orbital on the N atom, we obtain different characters, as we now demonstrate.

The act of the six symmetry operations on this S_N can be represented as follows:

$$\begin{array}{ccc} E & C_3 & C_3^2 \\ S_N \rightarrow S_N, & S_N \rightarrow S_N, & S_N \rightarrow S_N, \\ \sigma_v & \sigma_v' & \sigma_v'' \\ S_N \rightarrow S_N, & S_N \rightarrow S_N, & S_N \rightarrow S_N. \end{array} \tag{4.31}$$

We can represent this group of operations in this basis by the one-dimensional set of matrices:

$$\begin{array}{ccc} D^{(1)}(E) = 1, & D^{(1)}(C_3) = 1, & D^{(1)}\left(C_3^2\right) = 1, \\ D^{(1)}(\sigma_v) = 1, & D^{(1)}(\sigma_v'') = 1, & D^{(1)}(\sigma_v') = 1. \end{array} \tag{4.32}$$

Again we have

$$D^{(1)}(\sigma_v)D^{(1)}(C_3) = 1 \cdot 1 = D^{(1)}(\sigma_v''),$$
$$\text{and} \quad D^{(1)}(C_3)D^{(1)}(\sigma_v) = 1 \cdot 1 = D^{(1)}(\sigma_v'). \tag{4.33}$$

These six matrices form another representation of the group. In this basis, each character is equal to unity. The representation formed by allowing the six symmetry operations to act on the 1s N-atom orbital is clearly not the same as that formed when the same six operations acted on the (S_N, S_1, S_2, S_3) basis. We now need to learn how to further analyze the information content of a specific representation of the group formed when the symmetry operations act on any specific set of objects.

4.2.3 Reducible and irreducible representations

A reducible representation

Note that every matrix in the four-dimensional group representation labeled $D^{(4)}$ has the so-called block diagonal form

$$\begin{array}{c|ccc} 1 & 0 & 0 & 0 \\ \hline 0 & & & \\ 0 & & 3 \times 3 \text{ matrix} & \\ 0 & & & \end{array}$$

This means that these $D^{(4)}$ matrices are really a combination of two separate group representations (mathematically, it is called a *direct sum* representation). We say that $D^{(4)}$ is reducible into a one-dimensional representation $D^{(1)}$ and a three-dimensional representation formed by the 3×3 submatrices that we will call $D^{(3)}$.

$$D^{(3)}(E) = \begin{bmatrix} 1 & 0 & 0 \\ 0 & 1 & 0 \\ 0 & 0 & 1 \end{bmatrix}, \quad D^{(3)}(C_3) = \begin{bmatrix} 0 & 0 & 1 \\ 1 & 0 & 0 \\ 0 & 1 & 0 \end{bmatrix}, \quad D^{(3)}\left(C_3^2\right) = \begin{bmatrix} 0 & 1 & 0 \\ 0 & 0 & 1 \\ 1 & 0 & 0 \end{bmatrix},$$
$$D^{(3)}(\sigma_v) = \begin{bmatrix} 1 & 0 & 0 \\ 0 & 0 & 1 \\ 0 & 1 & 0 \end{bmatrix}, \quad D^{(3)}(\sigma_v') = \begin{bmatrix} 0 & 0 & 1 \\ 0 & 1 & 0 \\ 1 & 0 & 0 \end{bmatrix}, \quad D^{(3)}(\sigma_v'') = \begin{bmatrix} 0 & 1 & 0 \\ 1 & 0 & 0 \\ 0 & 0 & 1 \end{bmatrix}. \tag{4.34}$$

The characters of $D^{(3)}$ are $\chi(E) = 3$, $\chi(2C_3) = 0$, $\chi(3\sigma_v) = 1$. Note that we would have obtained this $D^{(3)}$ representation directly if we had originally chosen to examine the basis (S_1, S_2, S_3); also note that these characters are equal to those of $D^{(4)}$ minus those of $D^{(1)}$.

A change in basis

Now let us convert to a new basis that is a linear combination of the original (S_1, S_2, S_3) basis:

$$T_1 = S_1 + S_2 + S_3, \tag{4.35}$$

$$T_2 = 2S_1 - S_2 - S_3, \tag{4.36}$$

$$T_3 = S_2 - S_3. \tag{4.37}$$

(Don't worry about how we construct T_1, T_2, and T_3 yet. As will be demonstrated later, we form them by using symmetry projection operators defined below.) We determine how the "T" basis functions behave under the group operations by allowing the operations to act on the S_j and interpreting the results in terms of the T_i. In particular,

$$\begin{aligned}
&(T_1, T_2, T_3) \xrightarrow{\sigma_v} (T_1, T_2, -T_3), \qquad (T_1, T_2, T_3) \xrightarrow{E} (T_1, T_2, T_3),\\
&(T_1, T_2, T_3) \xrightarrow{\sigma'_v} (S_3 + S_2 + S_1,\ 2S_3 - S_2 - S_1,\ S_2 - S_1)\\
&\quad = \left(T_1, -\frac{1}{2}T_2 - \frac{3}{2}T_3, -\frac{1}{2}T_2 + \frac{1}{2}T_3\right);\\
&(T_1, T_2, T_3) \xrightarrow{\sigma''_v} (S_2 + S_1 + S_3,\ 2S_2 - S_1 - S_3,\ S_1 - S_3)\\
&\quad = \left(T_1, -\frac{1}{2}T_2 + \frac{3}{2}T_3, \frac{1}{2}T_2 + \frac{1}{2}T_3\right),\\
&(T_1, T_2, T_3) \xrightarrow{C_3} (S_3 + S_1 + S_2,\ 2S_3 - S_1 - S_2,\ S_1 - S_2)\\
&\quad = \left(T_1, -\frac{1}{2}T_2 - \frac{3}{2}T_3, \frac{1}{2}T_2 - \frac{1}{2}T_3\right),\\
&(T_1, T_2, T_3) \xrightarrow{C_3^2} (S_2 + S_3 + S_1,\ 2S_2 - S_3 - S_1,\ S_3 - S_1)\\
&\quad = \left(T_1, -\frac{1}{2}T_2 + \frac{3}{2}T_3, -\frac{1}{2}T_2 - \frac{1}{2}T_3\right).
\end{aligned} \tag{4.38}$$

So the matrix representations in the new T_i basis are

$$\begin{aligned}
&D^{(3)}(E) = \begin{bmatrix} 1 & 0 & 0 \\ 0 & 1 & 0 \\ 0 & 0 & 1 \end{bmatrix}, \qquad D^{(3)}(C_3) = \begin{bmatrix} 1 & 0 & 0 \\ 0 & -\frac{1}{2} & -\frac{3}{2} \\ 0 & +\frac{1}{2} & -\frac{1}{2} \end{bmatrix},\\
&D^{(3)}\left(C_3^2\right) = \begin{bmatrix} 1 & 0 & 0 \\ 0 & -\frac{1}{2} & +\frac{3}{2} \\ 0 & -\frac{1}{2} & -\frac{1}{2} \end{bmatrix}, \qquad D^{(3)}(\sigma_v) = \begin{bmatrix} 1 & 0 & 0 \\ 0 & 1 & 0 \\ 0 & 0 & -1 \end{bmatrix},\\
&D^{(3)}(\sigma'_v) = \begin{bmatrix} 1 & 0 & 0 \\ 0 & -\frac{1}{2} & -\frac{3}{2} \\ 0 & -\frac{1}{2} & +\frac{1}{2} \end{bmatrix}, \qquad D^{(3)}(\sigma''_v) = \begin{bmatrix} 1 & 0 & 0 \\ 0 & -\frac{1}{2} & +\frac{3}{2} \\ 0 & +\frac{1}{2} & +\frac{1}{2} \end{bmatrix}.
\end{aligned} \tag{4.39}$$

Reduction of the reducible representation

These six matrices can be verified to multiply just as the symmetry operations do; thus they form another three-dimensional representation of the group. We see that in the T_i basis the matrices are block diagonal. This means that the space spanned by the T_i functions, which is the same space as the S_j span, forms a reducible representation that can be decomposed into a one-dimensional space and a two-dimensional space (via formation of the T_i functions). Note that the characters (traces) of the matrices are not changed by the change in bases.

The one-dimensional part of the above reducible three-dimensional representation is seen to be the same as the totally symmetric representation we arrived at before, $D^{(1)}$. The two-dimensional representation that is left can be shown to be *irreducible*; it has the following matrix representations:

$$D^{(2)}(E) = \begin{bmatrix} 1 & 0 \\ 0 & 1 \end{bmatrix}, \quad D^{(2)}(C_3) = \begin{bmatrix} -\frac{1}{2} & -\frac{3}{2} \\ +\frac{1}{2} & -\frac{1}{2} \end{bmatrix}, \quad D^{(2)}\left(C_3^2\right) = \begin{bmatrix} -\frac{1}{2} & +\frac{3}{2} \\ -\frac{1}{2} & -\frac{1}{2} \end{bmatrix},$$
$$D^{(2)}(\sigma_v) = \begin{bmatrix} 1 & 0 \\ 0 & -1 \end{bmatrix}, \quad D^{(2)}(\sigma_v') = \begin{bmatrix} -\frac{1}{2} & -\frac{3}{2} \\ -\frac{1}{2} & +\frac{1}{2} \end{bmatrix}, \quad D^{(2)}(\sigma_v'') = \begin{bmatrix} -\frac{1}{2} & -\frac{3}{2} \\ -\frac{1}{2} & +\frac{1}{2} \end{bmatrix}. \tag{4.40}$$

The characters can be obtained by summing diagonal elements:

$$\chi(E) = 2, \qquad \chi(2C_3) = -1, \qquad \chi(3\sigma_v) = 0. \tag{4.41}$$

Rotations as a basis

Another one-dimensional representation of the group can be obtained by taking rotation about the z-axis (the C_3 axis) as the object on which the symmetry operations act:

$$R_z \xrightarrow{E} R_z, \qquad R_z \xrightarrow{C_3} R_z, \qquad R_z \xrightarrow{C_3^2} R_z,$$
$$R_z \xrightarrow{\sigma_v} -R_z, \qquad R_z \xrightarrow{\sigma_v''} -R_z, \qquad R_z \xrightarrow{\sigma_v'} -R_z. \tag{4.42}$$

In writing these relations, we use the fact that reflection reverses the sense of a rotation. The matrix representations corresponding to this one-dimensional basis are

$$D^{(1)}(E) = 1, \qquad D^{(1)}(C_3) = 1, \qquad D^{(1)}\left(C_3^2\right) = 1,$$
$$D^{(1)}(\sigma_v) = -1, \qquad D^{(1)}(\sigma_v'') = -1, \qquad D^{(1)}(\sigma_v') = -1. \tag{4.43}$$

These one-dimensional matrices can be shown to multiply together just like the symmetry operations of the C_{3v} group. They form an *irreducible* representation of the group (because it is one-dimensional, it can not be further reduced). Note that this one-dimensional representation is not identical to that found above for the 1s N-atom orbital, or the T_1 function.

Overview

We have found three distinct irreducible representations for the C_{3v} symmetry group; two different one-dimensional and one two-dimensional representations. Are there any more? An important theorem of group theory shows that the number of irreducible representations of a group is equal to the number of classes. Since there are three classes of operation (i.e., E, C_3 and σ_v), we have found all the irreducible representations of the C_{3v} point group. There are no more.

The irreducible representations have standard names; the first $D^{(1)}$ (that arising from the T_1 and $1s_N$ orbitals) is called A_1, the $D^{(1)}$ arising from R_z is called A_2 and $D^{(2)}$ is called E (not to be confused with the identity operation E). We will see shortly where to find and identify these names.

Thus, our original $D^{(4)}$ representation was a combination of two A_1 representations and one E representation. We say that $D^{(4)}$ is a direct sum representation: $D^{(4)} = 2A_1 \oplus E$. A consequence is that the characters of the combination representation $D^{(4)}$ can be obtained by adding the characters of its constituent irreducible representations.

	E	$2C_3$	$3\sigma_v$
A_1	1	1	1
A_1	1	1	1
E	2	−1	0
$2A_1 \oplus E$	4	1	2

How to decompose reducible representations in general

Suppose you were given only the characters (4,1, 2). How can you find out how many times A_1, E, and A_2 appear when you reduce $D^{(4)}$ to its irreducible parts? You want to find a linear combination of the characters of A_1, A_2 and E that add up (4,1, 2). You can treat the characters of matrices as vectors and take the dot product of A_1 with $D^{(4)}$

$$\begin{bmatrix} 1 & 1 & 1 & 1 & 1 & 1 \\ E & C_3 & & \sigma_v & & \end{bmatrix} \cdot \begin{bmatrix} 4 & E \\ 1 & C_3 \\ 1 & \\ 2 & \sigma_v \\ 2 & \\ 2 & \end{bmatrix} = 4 + 1 + 1 + 2 + 2 + 2 = 12. \quad (4.44)$$

The vector (1,1,1,1,1,1) is not normalized; hence to obtain the component of (4,1,1, 2, 2, 2) along a unit vector in the (1,1,1,1,1,1) direction, one must divide by the norm of (1,1,1,1,1,1); this norm is 6. The result is that the reducible representation contains $12/6 = 2$ A_1 components. Analogous projections in the

E and A_2 directions give components of 1 and 0, respectively. In general, to determine the number n_Γ of times irreducible representation Γ appears in the reducible representation with characters χ_{red}, one calculates

$$n_\Gamma = \frac{1}{g}\sum_R \chi_\Gamma(R)\chi_{red}(R), \tag{4.45}$$

where g is the order of the group and $\chi_\Gamma(R)$ are the characters of the Γth irreducible representation.

Commonly used bases

We could take *any* set of functions as a basis for a group representation. Commonly used sets include: coordinates (x, y, z) located on the atoms of a polyatomic molecule (their symmetry treatment is equivalent to that involved in treating a set of p orbitals on the same atoms), quadratic functions such as d orbitals $-xy,\ yz,\ xz,\ x^2 - y^2,\ z^2$, as well as rotations about the x, y and z axes. The transformation properties of these very commonly used bases are listed in the character tables shown in the Appendix.

Summary

The basic idea of symmetry analysis is that any basis of orbitals, displacements, rotations, etc. transforms either as one of the irreducible representations or as a direct sum (reducible) representation. Symmetry tools are used to first determine how the basis transforms under action of the symmetry operations. They are then used to decompose the resultant representations into their irreducible components.

4.2.4 Another example

The 2p orbitals of nitrogen

For a function to transform according to a specific irreducible representation means that the function, when operated upon by a point-group symmetry operator, yields a linear combination of the functions that transform according to that irreducible representation. For example, a $2p_z$ orbital (z is the C_3 axis of NH_3) on the nitrogen atom belongs to the A_1 representation because it yields unity times itself when C_3, C_3^2, σ_v, σ_v', σ_v'' or the identity operation act on it. The factor of 1 means that $2p_z$ has A_1 symmetry since the characters (the numbers listed opposite A_1 and below E, $2C_3$, and $3\sigma_v$ in the C_{3v} character table shown in the Appendix) of all six symmetry operations are 1 for the A_1 irreducible representation.

The $2p_x$ and $2p_y$ orbitals on the nitrogen atom transform as the E representation since C_3, C_3^2, σ_v, σ_v', σ_v'' and the identity operation map $2p_x$ and $2p_y$ among one another. Specifically,

$$C_3 \begin{bmatrix} 2p_x \\ 2p_y \end{bmatrix} = \begin{bmatrix} \cos 120^\circ & -\sin 120^\circ \\ \sin 120^\circ & \cos 120^\circ \end{bmatrix} \begin{bmatrix} 2p_x \\ 2p_y \end{bmatrix}, \quad (4.46)$$

$$C_3^2 \begin{bmatrix} 2p_x \\ 2p_y \end{bmatrix} = \begin{bmatrix} \cos 240^\circ & -\sin 240^\circ \\ \sin 240^\circ & \cos 240^\circ \end{bmatrix} \begin{bmatrix} 2p_x \\ 2p_y \end{bmatrix}, \quad (4.47)$$

$$E \begin{bmatrix} 2p_x \\ 2p_y \end{bmatrix} = \begin{bmatrix} 1 & 0 \\ 0 & 1 \end{bmatrix} \begin{bmatrix} 2p_x \\ 2p_y \end{bmatrix}, \quad (4.48)$$

$$\sigma_v \begin{bmatrix} 2p_x \\ 2p_y \end{bmatrix} = \begin{bmatrix} -1 & 0 \\ 0 & 1 \end{bmatrix} \begin{bmatrix} 2p_x \\ 2p_y \end{bmatrix}, \quad (4.49)$$

$$\sigma_v' \begin{bmatrix} 2p_x \\ 2p_y \end{bmatrix} = \begin{bmatrix} +\frac{1}{2} & +\frac{\sqrt{3}}{2} \\ +\frac{\sqrt{3}}{2} & -\frac{1}{2} \end{bmatrix} \begin{bmatrix} 2p_x \\ 2p_y \end{bmatrix}, \quad (4.50)$$

$$\sigma_v'' \begin{bmatrix} 2p_x \\ 2p_y \end{bmatrix} = \begin{bmatrix} +\frac{1}{2} & -\frac{\sqrt{3}}{2} \\ -\frac{\sqrt{3}}{2} & -\frac{1}{2} \end{bmatrix} \begin{bmatrix} 2p_x \\ 2p_y \end{bmatrix}. \quad (4.51)$$

The 2×2 matrices, which indicate how each symmetry operation maps $2p_x$ and $2p_y$ into some combinations of $2p_x$ and $2p_y$, are the representation matrices ($D^{(IR)}$) for that particular operation and for this particular irreducible representation (IR). For example,

$$\begin{bmatrix} +\frac{1}{2} & +\frac{\sqrt{3}}{2} \\ +\frac{\sqrt{3}}{2} & -\frac{1}{2} \end{bmatrix} = D^{(E)}(\sigma_v'). \quad (4.52)$$

This set of matrices have the same characters as the $D^{(2)}$ matrices obtained earlier when the T_i displacement vectors were analyzed, but the individual matrix elements are different because we used a different basis set (here $2p_x$ and $2p_y$; above it was T_2 and T_3). This illustrates the invariance of the trace to the specific representation; the trace only depends on the space spanned, not on the specific manner in which it is spanned.

A short-cut

A short-cut device exists for evaluating the trace of such representation matrices (that is, for computing the characters). The diagonal elements of the representation matrices are the projections along each orbital of the effect of the symmetry operation acting on that orbital. For example, a diagonal element of the C_3 matrix is the component of $C_3 2p_y$ along the $2p_y$ direction. More rigorously, it is $\int 2p_y^* C_3 2p_y \, d\tau$. Thus, the character of the C_3 matrix is the sum of $\int 2p_y^* C_3 2p_y \, d\tau$ and $\int 2p_x^* C_3 2p_x \, d\tau$. In general, the character χ of any symmetry operation S can be computed by allowing S to operate on each orbital ϕ_i, then projecting $S\phi_i$ along ϕ_i (i.e., forming $\int \phi_i^* S\phi_i \, d\tau$), and summing these terms,

$$\sum_i \int \phi_i^* S\phi_i \, d\tau = \chi(S). \quad (4.53)$$

If these rules are applied to the $2p_x$ and $2p_y$ orbitals of nitrogen within the C_{3v} point group, one obtains

$$\chi(E)=2, \qquad \chi(C_3)=\chi\left(C_3^2\right)=-1, \qquad \chi(\sigma_v)=\chi(\sigma_v'')=\chi(\sigma_v')=0. \tag{4.54}$$

This set of characters is the same as $D^{(2)}$ above and agrees with those of the E representation for the C_{3v} point group. Hence, $2p_x$ and $2p_y$ belong to or transform as the E representation. This is why (x, y) is to the right of the row of characters for the E representation in the C_{3v} character table shown in the Appendix. In similar fashion, the C_{3v} character table (please refer to this table now) states that $d_{x^2-y^2}$ and d_{xy} orbitals on nitrogen transform as E, as do d_{xy} and d_{yz}, but d_{z^2} transforms as A_1.

Earlier, we considered in some detail how the three $1s_H$ orbitals on the hydrogen atoms transform. Repeating this analysis using the short-cut rule just described, the traces (characters) of the 3×3 representation matrices are computed by allowing E, $2C_3$, and $3\sigma_v$ to operate on $1s_{H_1}$, $1s_{H_2}$, and $1s_{H_3}$ and then computing the component of the resulting function along the original function. The resulting characters are $\chi(E)=3$, $\chi(C_3)=\chi(C_3^2)=0$, and $\chi(\sigma_v)=\chi(\sigma_v')=\chi(\sigma_v'')=1$, in agreement with what we calculated before.

Using the orthogonality of characters taken as vectors we can reduce the above set of characters to $A_1 + E$. Hence, we say that our orbital set of three $1s_H$ orbitals forms a *reducible* representation consisting of the sum of A_1 and E IRs. This means that the three $1s_H$ orbitals can be combined to yield one orbital of A_1 symmetry and a *pair* that transform according to the E representation.

4.2.5 Projection operators: symmetry-adapted linear combinations of atomic orbitals

To generate the above A_1 and E symmetry-adapted orbitals, we make use of so-called symmetry projection operators P_E and P_{A_1}. These operators are given in terms of linear combinations of products of characters times elementary symmetry operations as follows:

$$P_{A_1} = \sum_S \chi_A(S)S, \tag{4.55}$$

$$P_E = \sum_S \chi_E(S)S, \tag{4.56}$$

where S ranges over C_3, C_3^2, σ_v, σ_v' and σ_v'' and the identity operation. The result of applying P_{A_1} to say $1s_{H_1}$ is

$$\begin{aligned} P_{A_1} 1s_{H_1} &= 1s_{H_1} + 1s_{H_2} + 1s_{H_3} + 1s_{H_2} + 1s_{H_3} + 1s_{H_1} \\ &= 2(1s_{H_1} + 1s_{H_2} + 1s_{H_3}) = \phi_{A1}, \end{aligned} \tag{4.57}$$

which is an (unnormalized) orbital having A_1 symmetry. Clearly, this same ϕ_{A_1} would be generated by P_{A_1} acting on $1s_{H_2}$ or $1s_{H_3}$. Hence, only one A_1 orbital exists. Likewise,

$$P_E\, 1s_{H_1} = 2 \times 1s_{H_1} - 1s_{H_2} - 1s_{H_3} \equiv \phi_{E,1}, \tag{4.58}$$

which is one of the symmetry-adapted orbitals having E symmetry. The other E orbital can be obtained by allowing P_E to act on $1s_{H_2}$ or $1s_{H_3}$:

$$P_E\, 1s_{H_2} = 2 \cdot 1s_{H_2} - 1s_{H_1} - 1s_{H_3} \equiv \phi_{E,2}, \tag{4.59}$$

$$P_E\, 1s_{H_3} = 2 \cdot 1s_{H_3} - 1s_{H_1} - 1s_{H_2} = \phi_{E,3}. \tag{4.60}$$

It might seem as though three orbitals having E symmetry were generated, but only two of these are really independent functions. For example, $\phi_{E,3}$ is related to $\phi_{E,1}$ and $\phi_{E,2}$ as follows:

$$\phi_{E,3} = -(\phi_{E,1} + \phi_{E,2}). \tag{4.61}$$

Thus, only $\phi_{E,1}$ and $\phi_{E,2}$ are needed to span the two-dimensional space of the E representation. If we include $\phi_{E,1}$ in our set of orbitals and require our orbitals to be orthogonal, then we must find numbers a and b such that $\phi'_E = a\phi_{E,2} + b\phi_{E,3}$ is orthogonal to $\phi_{E,1}$: $\int \phi'_E \phi_{E,1}\, d\tau = 0$. A straightforward calculation gives $a = -b$ or $\phi'_E = a(1s_{H_2} - 1s_{H_3})$ which agrees with what we used earlier to construct the T_i functions in terms of the S_j functions.

4.2.6 Summary

Let us now summarize what we have learned. Any given set of atomic orbitals $\{\phi_i\}$, atom-centered displacements or rotations can be used as a basis for the symmetry operations of the point group of the molecule. The characters $\chi(S)$ belonging to the operations S of this point group within any such space can be found by summing the integrals $\int \phi_i^* S\phi_i\, d\tau$ over all the atomic orbitals (or corresponding unit vector atomic displacements). The resultant characters will, in general, be reducible to a combination of the characters of the irreducible representations $\chi_i(S)$. To decompose the characters $\chi(S)$ of the reducible representation to a sum of characters $\chi_i(S)$ of the irreducible representation

$$\chi(S) = \sum_i n_i \chi_i(S), \tag{4.62}$$

it is necessary to determine how many times, n_i, the ith irreducible representation occurs in the reducible representation. The expression for n_i is

$$n_i = \frac{1}{g} \sum_S \chi(S)\chi_i(S), \tag{4.63}$$

in which g is the order of the point group – the total number of symmetry operations in the group (e.g., $g = 6$ for C_{3v}).

For example, the reducible representation $\chi(E) = 3$, $\chi(C_3) = 0$, and $\chi(\sigma_v) = 1$ formed by the three $1s_H$ orbitals discussed above can be decomposed as follows:

$$n_{A_1} = \frac{1}{6}(3 \cdot 1 + 2 \cdot 0 \cdot 1 + 3 \cdot 1 \cdot 1) = 1, \tag{4.64}$$

$$n_{A_2} = \frac{1}{6}(3 \cdot 1 + 2 \cdot 0 \cdot 1 + 3 \cdot 1 \cdot (-1)) = 0, \tag{4.65}$$

$$n_E = \frac{1}{6}(3 \cdot 2 + 2 \cdot 0 \cdot (-1) + 3 \cdot 1 \cdot 0) = 1. \tag{4.66}$$

These equations state that the three $1s_H$ orbitals can be combined to give one A_1 orbital and, since E is degenerate, one pair of E orbitals, as established above. With knowledge of the n_i, the symmetry-adapted orbitals can be formed by allowing the projectors

$$P_i = \sum_i \chi_i(S)S \tag{4.67}$$

to operate on each of the primitive atomic orbitals. How this is carried out was illustrated for the $1s_H$ orbitals in our earlier discussion. These tools allow a symmetry decomposition of any set of atomic orbitals into appropriate symmetry-adapted orbitals.

Before considering other concepts and group-theoretical machinery, it should once again be stressed that these same tools can be used in symmetry analysis of the translational, vibrational and rotational motions of a molecule. The twelve motions of NH_3 (three translations, three rotations, six vibrations) can be described in terms of combinations of displacements of each of the four atoms in each of three (x, y, z) directions. Hence, unit vectors placed on each atom directed in the x, y, and z directions form a basis for action by the operations $\{S\}$ of the point group. In the case of NH_3, the characters of the resultant 12×12 representation matrices form a reducible representation in the C_{2v} point group: $\chi(E) = 12$, $\chi(C_3) = \chi(C_3^2) = 0$, $\chi(\sigma_v) = \chi(\sigma_v') = \chi(\sigma_v'') = 2$. For example under σ_v, the H_2 and H_3 atoms are interchanged, so unit vectors on either one will not contribute to the trace. Unit z-vectors on N and H_1 remain unchanged as well as the corresponding y-vectors. However, the x-vectors on N and H_1 are reversed in sign. The total character for σ_v' of the H_2 and H_3 atoms are interchanged, so unit vectors on either one will not contribute to the trace. Unit z-vectors on N and H_1 remain unchanged as well as the corresponding y-vectors. However, the x-vectors on N and H_1 are reversed in sign. The total character for σ_v is thus $4 - 2 = 2$. This representation can be decomposed as follows:

$$n_{A_1} = \frac{1}{6}[1 \cdot 1 \cdot 12 + 2 \cdot 1 \cdot 0 + 3 \cdot 1 \cdot 2] = 3, \tag{4.68}$$

$$n_{A_2} = \frac{1}{6}[1 \cdot 1 \cdot 12 + 2 \cdot 1 \cdot 0 + 3 \cdot (-1) \cdot 2] = 1, \tag{4.69}$$

$$n_E = \frac{1}{6}[1 \cdot 2 \cdot 12 + 2 \cdot (-1) \cdot 0 + 3 \cdot 0 \cdot 2] = 4. \tag{4.70}$$

From the information on the right side of the C_{3v} character table, translations of all four atoms in the z, x and y directions transform as $A_1(z)$ and $E(x, y)$, respectively, whereas rotations about the $z(R_z)$, $x(R_x)$, and $y(R_y)$ axes transform as A_2 and E. Hence, of the twelve motions, three translations have A_1 and E symmetry and three rotations have A_2 and E symmetry. This leaves six vibrations, of which two have A_1 symmetry, none have A_2 symmetry, and two (pairs) have E symmetry. We could obtain symmetry-adapted vibrational and rotational bases by allowing symmetry projection operators of the irreducible representation symmetries to operate on various elementary Cartesian (x, y, z) atomic displacement vectors.

4.2.7 Direct product representations

Direct products in N-electron wave functions

We now turn to the symmetry analysis of orbital products. Such knowledge is important because one is routinely faced with constructing symmetry-adapted N-electron configurations that consist of products of N individual spin orbitals, one for each electron. A point-group symmetry operator S, when acting on such a product of orbitals, gives the product of S acting on each of the individual orbitals

$$S(\phi_1\phi_2\phi_3 \cdots \phi_N) = (S\phi_1)(S\phi_2)(S\phi_3) \cdots (S\phi_N). \tag{4.71}$$

For example, reflection of an N-orbital product through the σ_v plane in NH_3 applies the reflection operation to all N electrons.

Just as the individual orbitals formed a basis for action of the point-group operators, the configurations (N-orbital products) form a basis for the action of these same point-group operators. Hence, the various electronic configurations can be treated as functions on which S operates, and the machinery illustrated earlier for decomposing orbital symmetry can then be used to carry out a symmetry analysis of configurations.

Another short-cut makes this task easier. Since the symmetry-adapted individual orbitals $\{\phi_i, i = 1, \ldots, M\}$ transform according to irreducible representations, the representation matrices for the N-term products shown above consist of products of the matrices belonging to each ϕ_i. This matrix product is not a simple product but what is called a direct product. To compute the characters of the direct product matrices, one multiplies the characters of the individual matrices of the irreducible representations of the N orbitals that appear in the electron configuration. The direct-product representation formed by the orbital products can therefore be symmetry analyzed (reduced) using the same tools as we used earlier.

For example, if one is interested in knowing the symmetry of an orbital product of the form $a_1^2a_2^2e^2$ (note: lower case letters are used to denote the symmetry of electronic orbitals, whereas capital letters are reserved to label the overall configuration's symmetry) in C_{3v} symmetry, the following procedure is used. For each of the six symmetry operations in the C_{2v} point group, the *product* of the characters associated with each of the *six* spin orbitals (orbital multiplied by α or β spin) is formed:

$$\chi(S) = \prod_i \chi_i(S) = (\chi_{A_1}(S))^2 (\chi_{A_2}(S))^2 (\chi_E(S))^2. \tag{4.72}$$

In the specific case considered here, $\chi(E) = 4$, $\chi(2C_3) = 1$, and $\chi(3\sigma_v) = 0$. Notice that the contributions of any doubly occupied non-degenerate orbitals (e.g., a_1^2 and a_2^2) to these direct product characters $\chi(S)$ are unity because for *all* operators $(\chi_k(S))^2 = 1$ for any one-dimensional irreducible representation. As a result, only the singly occupied or degenerate orbitals need to be considered when forming the characters of the reducible direct-product representation $\chi(S)$. For this example this means that the direct-product characters can be determined from the characters $\chi_E(S)$ of the two active (i.e., non-closed-shell) orbitals – the e^2 orbitals. That is, $\chi(S) = \chi_E(S) \cdot \chi_E(S)$.

From the direct-product characters $\chi(S)$ belonging to a particular electronic configuration (e.g., $a_1^2a_2^2e^2$), one must still decompose this list of characters into a sum of irreducible characters. For the example at hand, the direct-product characters $\chi(S)$ decompose into one A_1, one A_2, and one E representation. This means that the e^2 configuration contains A_1, A_2, and E symmetry elements. Projection operators analogous to those introduced earlier for orbitals can be used to form symmetry-adapted orbital products from the individual basis orbital products of the form $a_1^2a_2^2e_x^m e_y^{m'}$, where m and m' denote the occupation (1 or 0) of the two degenerate orbitals e_x and e_y. When dealing with indistinguishable particles such as electrons, it is also necessary to further project the resulting orbital products to make them antisymmetric (for fermions) or symmetric (for bosons) with respect to interchange of any pair of particles. This step reduces the set of N-electron states that can arise. For example, in the above e^2 configuration case, only $^3A_2$, $^1A_1$, and ^1E states arise; the ^3E, $^3A_1$, and $^1A_2$ possibilities disappear when the antisymmetry projector is applied. In contrast, for an $e^1e'^1$ configuration, all states arise even after the wave function has been made antisymmetric. The steps involved in combining the point-group symmetry with permutational antisymmetry are illustrated in Chapter 10 of my *QMIC* text. In Appendix III of *Electronic Spectra and Electronic Structure of Polyatomic Molecules*, G. Herzberg, Van Nostrand Reinhold Co., New York, N.Y. (1966), the resolution of direct products among various representations within many point groups are tabulated.

Direct products in selection rules

Two states ψ_a and ψ_b that are eigenfunctions of a Hamiltonian $\mathbf{H}_0$ in the absence of some external perturbation (e.g., electromagnetic field or static electric field or potential due to surrounding ligands) can be "coupled" by the perturbation $\mathbf{V}$ only if the symmetries of **V** and of the two wave functions obey a so-called selection rule. In particular, only if the coupling integral

$$\int \psi_a^* \mathbf{V} \psi_b d\tau = V_{a,b} \tag{4.73}$$

is non-vanishing will the two states be coupled by **V**.

The role of symmetry in determining whether such integrals are non-zero can be demonstrated by noting that the integrand, considered as a whole, must contain a component that is invariant under all of the group operations (i.e., belongs to the totally symmetric representation of the group) if the integral is to not vanish. In terms of the projectors introduced above we must have

$$\sum_S \chi_A(S)\, S\psi_a^* \mathbf{V} \psi_b \tag{4.74}$$

not vanish. Here the subscript A denotes the totally symmetric representation of whatever point group applies. The symmetry of the product $\psi_a^* \mathbf{V} \psi_b$ is, according to what was covered earlier, given by the direct product of the symmetries of ψ_a^* of **V** and of ψ_b. So, the conclusion is that the integral will vanish unless this triple direct product contains, when it is reduced to its irreducible components, a component of the totally symmetric representation.

To see how this result is used, consider the integral that arises in formulating the interaction of electromagnetic radiation with a molecule within the electric-dipole approximation:

$$\int \psi_a^* \mathbf{r} \psi_b d\tau. \tag{4.75}$$

Here, **r** is the vector giving, together with e, the unit charge, the quantum mechanical dipole moment operator

$$\mathbf{r} = e\sum_n Z_n \mathbf{R}_n - e\sum_j \mathbf{r}_j, \tag{4.76}$$

where Z_n and $\mathbf{R}_n$ are the charge and position of the nth nucleus and $\mathbf{r}_j$ is the position of the jth electron. Now, consider evaluating this integral for the singlet $n \to \pi^*$ transition in formaldehyde. Here, the closed-shell ground state is of $^1A_1$ symmetry and the singlet excited state, which involves promoting an electron from the non-bonding b_2 lone pair orbital on the oxygen into the $\pi^* b_1$ orbital on the CO moiety, is of $^1A_2$ symmetry ($b_1 \times b_2 = a_2$). The direct product of the two wave function symmetries thus contains only a_2 symmetry. The three components (x, y, and z) of the dipole operator have, respectively, b_1, b_2, and a_1 symmetry. Thus, the triple direct products give rise to the following possibilities:

$$a_2 \times b_1 = b_2, \tag{4.77}$$

$$a_2 \times b_2 = b_1, \tag{4.78}$$

$$a_2 \times a_1 = a_2. \tag{4.79}$$

There is no component of a_1 symmetry in the triple direct product, so the integral vanishes. This allows us to conclude that the $n \rightarrow \pi^*$ excitation in formaldehyde is electric dipole forbidden.

4.2.8 Overview

We have shown how to make a symmetry decomposition of a basis of atomic orbitals (or Cartesian displacements or orbital products) into irreducible representation components. This tool is very helpful when studying spectroscopy and when constructing the orbital correlation diagrams that form the basis of the Woodward–Hoffmann rules. We also learned how to form the direct-product symmetries that arise when considering configurations consisting of products of symmetry-adapted spin orbitals. Finally, we learned how the direct product analysis allows one to determine whether or not integrals of products of wave functions with operators between them vanish. This tool is of utmost importance in determining selection rules in spectroscopy and for determining the effects of external perturbations on the states of the species under investigation.

Part II

Three primary areas of theoretical chemistry

Chapter 5
An overview of theoretical chemistry

In this chapter, many of the basic concepts and tools of theoretical chemistry are discussed only at an introductory level and without providing much of the background needed to fully comprehend them. Most of these topics are covered again in considerably more detail in Chapters 6–8, which focus on the three primary sub-disciplines of the field. The purpose of the present chapter is to give you an overview of the field that you will learn the details of in these later chapters.

What is theoretical chemistry about?

The science of chemistry deals with molecules including the radicals, cations, and anions they produce when fragmented or ionized. Chemists study isolated molecules (e.g., as occur in the atmosphere and in astronomical environments), solutions of molecules or ions dissolved in solvents, as well as solid, liquid, and plastic materials comprised of molecules. All such forms of molecular matter are what chemistry is about. Chemical science includes how to make molecules (synthesis), how to detect and quantitate them (analysis), how to probe their properties and how they undergo change as reactions occur (physical).

5.1 Molecular structure – bonding, shapes, electronic structures

One of the more fundamental issues chemistry addresses is molecular structure, which means how the molecule's atoms are linked together by bonds and what the interatomic distances and angles are. Another component of structure analysis relates to what the electrons are doing in the molecule; that is, how the molecule's orbitals are occupied and in which electronic state the molecule exists. For example, in the arginine molecule shown in Fig. 5.1, a HOOC– carboxylic acid group is linked to an adjacent carbon atom which itself is bonded to an $-NH_2$ amino group. Also connected to the α-carbon atom are a chain of three methylene $-CH_2-$ groups, an –NH– group, then a carbon atom attached both by a double bond to an imine –NH group and to an amino $-NH_2$ group.

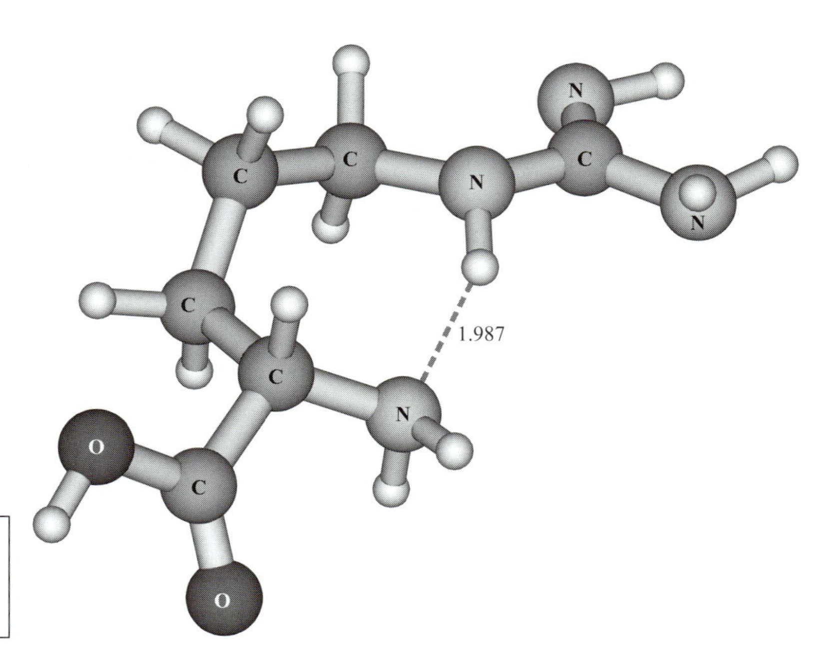

Figure 5.1 The arginine molecule in its non-zwitterion form with dotted hydrogen bond.

The connectivity among the atoms in arginine is dictated by the well-known valence preferences displayed by H, C, O, and N atoms. The internal bond angles are, to a large extent, also determined by the valences of the constituent atoms (i.e., the sp^3 or sp^2 nature of the bonding orbitals). However, there are other interactions among the several functional groups in arginine that also contribute to its ultimate structure. In particular, the hydrogen bond linking the α-amino group's nitrogen atom to the —NH— group's hydrogen atom causes this molecule to fold into a less extended structure than it otherwise might.

What does theory have to do with issues of molecular structure and why is knowledge of structure so important? It is important because the structure of a molecule has a very important role in determining the kinds of reactions that molecule will undergo, what kind of radiation it will absorb and emit, and to what "active sites" in neighboring molecules or nearby materials it will bind. A molecule's shape (e.g., rod-like, flat, globular, etc.) is one of the first things a chemist thinks of when trying to predict where, at another molecule or on a surface or a cell, the molecule will "fit" and be able to bind and perhaps react. The presence of lone pairs of electrons (which act as Lewis base sites), of π orbitals (which can act as electron donor and electron acceptor sites), and of highly polar or ionic groups guide the chemist further in determining where on the molecule's framework various reactant species (e.g., electrophylic or nucleophilic or radical) will be most strongly attracted. Clearly, molecular structure is a crucial aspect of the chemists' toolbox.

How does theory relate to molecular structure? As we discussed in the Background Material, the Born–Oppenheimer approximation leads us to use quantum mechanics to predict the energy E of a molecule for any positions ($\{R_a\}$) of its nuclei given the number of electrons N_e in the molecule (or ion). This means, for example, that the energy of the arginine molecule in its lowest electronic state (i.e., with the electrons occupying the lowest energy orbitals) can be determined for any location of the nuclei if the Schrödinger equation governing the movements of the electrons can be solved.

If you have not had a good class on how quantum mechanics is used within chemistry, I urge you to take the time needed to master the Background Material. In those pages, I introduce the central concepts of quantum mechanics and I show how they apply to several very important cases including:

1. *electrons moving in one, two, and three dimensions and how these models relate to electronic structures of polyenes and to electronic bands in solids,*
2. *the classical and quantum probability densities and how they differ,*
3. *time propagation of quantum wave functions,*
4. *the Hückel or tight-binding model of chemical bonding among atomic orbitals,*
5. *harmonic vibrations,*
6. *molecular rotations,*
7. *electron tunneling,*
8. *atomic orbitals' angular and radial characteristics,*
9. *and point group symmetry and how it is used to label orbitals and vibrations.*

You need to know this material if you wish to understand most of what this text offers, so I urge you to read the Background Material if your education to date has not yet adequately been exposed to it.

Let us now return to the discussion of how theory deals with molecular structure. We assume that we know the energy $E(\{R_a\})$ at various locations $\{R_a\}$ of the nuclei. In some cases, we denote this energy $V(R_a)$ and in others we use $E(R_a)$ because, within the Born–Oppenheimer approximation, the electronic energy E serves as the potential V for the molecule's vibrational motions. As discussed in the Background Material, one can then perform a search for the lowest energy structure (e.g., by finding where the gradient vector vanishes $\partial E/\partial R_a = 0$ and where the second derivative or Hessian matrix $(\partial^2 E/\partial R_a \partial R_b)$ has no negative eigenvalues). By finding such a local-minimum in the energy landscape, theory is able to determine a stable structure of such a molecule. The word stable is used to describe these structures not because they are lower in energy than all other possible arrangements of the atoms but because the curvatures, as given in terms of eigenvalues of the Hessian matrix $(\partial^2 E/\partial R_a \partial R_b)$, are positive at this particular geometry. The procedures by which minima on the energy landscape are found may involve simply testing whether the energy decreases or increases as each geometrical coordinate is varied by a small amount. Alternatively, if the gradients $\partial E/\partial R_a$ are known at a particular geometry, one can perform searches directed "downhill" along the negative of the gradient itself. By taking a small "step" along such a direction, one can move to a new geometry that is lower in energy. If not only the gradients $\partial E/\partial R_a$ but also the second derivatives $(\partial^2 E/\partial R_a \partial R_b)$ are known at some geometry, one can make a more "intelligent" step toward a geometry of lower energy. For additional details about how such geometry optimization searches are performed within modern computational chemistry software, see the Background Material where this subject was treated in greater detail.

It often turns out that a molecule has more than one stable structure (isomer) for a given electronic state. Moreover, the geometries that pertain to stable structures of excited electronic states are different than those obtained for the ground state (because the orbital occupancy and thus the nature of the bonding is different). Again using arginine as an example, its ground electronic state also has the structure shown in Fig. 5.2 as a stable isomer. Notice that this isomer and that shown earlier have the atoms linked together in identical manners, but in the second structure the α-amino group is involved in two hydrogen bonds while it is

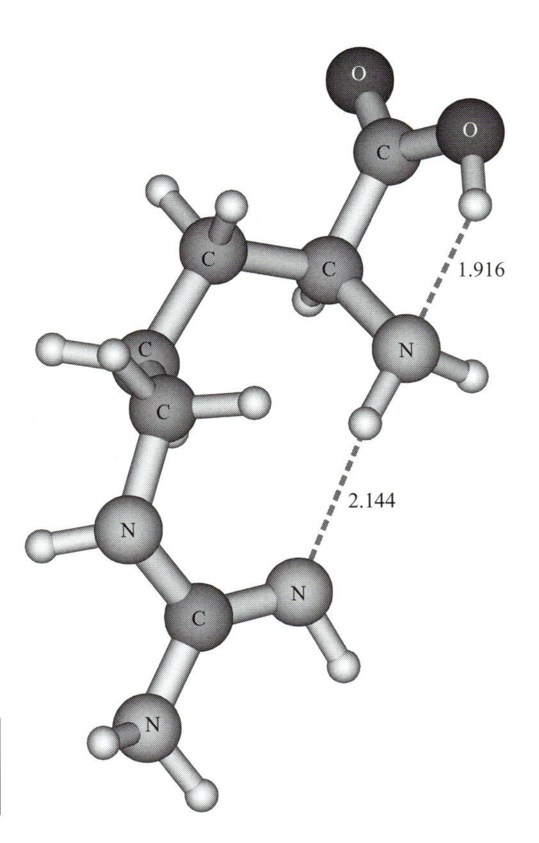

Figure 5.2 Another stable structure for the arginine molecule.

involved in only one in the former. In principle, the relative energies of these two geometrical isomers can be determined by solving the electronic Schrödinger equation while placing the constituent nuclei in the locations described in the two figures.

If the arginine molecule is excited to another electronic state, for example, by promoting a non-bonding electron on its C=O oxygen atom into the neighboring C—O π^* orbital, its stable structures will not be the same as in the ground electronic state. In particular, the corresponding C—O distance will be longer than in the ground state, but other internal geometrical parameters may also be modified (albeit probably less so than the C—O distance). Moreover, the chemical reactivity of this excited state of arginine will be different than that of the ground state because the two states have different orbitals available to react with attacking reagents.

In summary, by solving the electronic Schrödinger equation at a variety of geometries and searching for geometries where the gradient vanishes and the Hessian matrix has all positive eigenvalues, one can find stable structures of molecules (and ions). The Schrödinger equation is a necessary aspect of this process because the movement of the electrons is governed by this equation rather than by Newtonian classical equations. The information gained after carrying out such a geometry optimization process includes (1) all of the interatomic distances and internal angles needed to specify the equilibrium geometry $\{R_{a\text{eq}}\}$ and (2) the total electronic energy E at this particular geometry.

It is also possible to extract much more information from these calculations. For example, by multiplying elements of the Hessian matrix $(\partial^2 E/\partial R_a \partial R_b)$ by the inverse square roots of the atomic masses of the atoms labeled a and b, one forms the mass-weighted Hessian $(m_a m_b)^{-1/2}(\partial^2 E/\partial R_a \partial R_b)$ whose non-zero eigenvalues give the harmonic vibrational frequencies $\{\omega_k\}$ of the molecule. The eigenvectors $\{R_{k,a}\}$ of the mass-weighted Hessian matrix give the relative displacements in coordinates $R_{k,a}$ that accompany vibration in the kth normal mode (i.e., they describe the normal mode motions). Details about how these harmonic vibrational frequencies and normal modes are obtained were discussed earlier in the Background Material.

5.2 Molecular change – reactions, isomerization, interactions

5.2.1 Changes in bonding

Chemistry also deals with transformations of matter including changes that occur when molecules react, are excited (electronically, vibrationally, or rotationally), or undergo geometrical rearrangements. Again, theory forms the cornerstone that

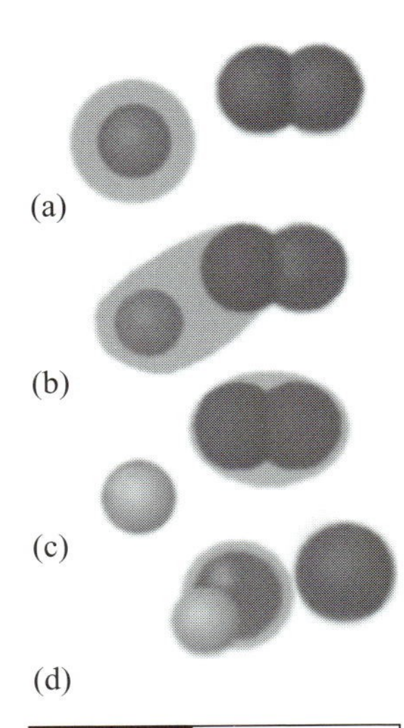

Figure 5.3 Two bimolecular reactions; (a) and (b) show an atom combining with a diatomic; (c) and (d) show an atom abstracting an atom from a diatomic.

allows experimental probes of chemical change to be connected to the molecular level and that allows simulations of such changes.

Molecular excitation may or may not involve altering the electronic structure of the molecule; vibrational and rotational excitation do not, but electronic excitation, ionization, and electron attachment do. As illustrated in Fig. 5.3 where a bimolecular reaction is displayed, chemical reactions involve breaking some bonds and forming others, and thus involve rearrangement of the electrons among various molecular orbitals. In this example, in part (a) an atom collides with the diatomic molecule and forms the bound triatomic (b). Alternatively, in (c) and (d), an atom collides with a diatomic to break one bond and form a new bond between two different atoms. Both such reactions are termed bimolecular because the basic step in which the reaction takes place requires a collision between two independent species (i.e., the atom and the diatomic).

A simple example of a unimolecular chemical reaction is offered by the arginine molecule considered above. In the first structure shown for arginine, the carboxylic acid group retains its HOOC— bonding. However, in the zwitterion structure of this same molecule, shown in Fig. 5.4, the HOOC— group has been deprotonated to produce a carboxylate anion group —COO^-, with the H^+ ion now bonded to the terminal imine group, thus converting it to an amino group and placing the net positive charge on the adjacent carbon atom. The unimolecular tautomerization reaction in which the two forms of arginine are interconverted involves breaking an O—H bond, forming a N—H bond, and changing a carbon–nitrogen double bond into a carbon–nitrogen single bond. In such a process, the electronic structure is significantly altered, and, as a result, the two isomers

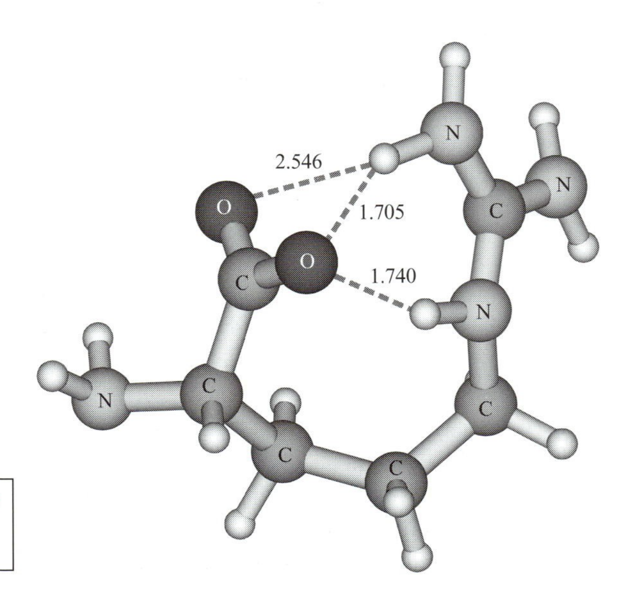

Figure 5.4 The arginine molecule in a zwitterion stable structure.

can display very different chemical reactivities toward other reagents. Notice that, once again, the ultimate structure of the zwitterion tautomer of arginine is determined by the valence preferences of its constitutent atoms as well as by hydrogen bonds formed among various functional groups (the carboxylate group and one amino group and one —NH— group).

5.2.2 Energy conservation

In any chemical reaction, as in all physical processes, total energy must be conserved. Reactions in which the summation of the strengths of all the chemical bonds in the reactants exceeds the sum of the bond strengths in the products are termed endothermic. For such reactions, energy must be provided to the reacting molecules to allow the reaction to occur. Exothermic reactions are those for which the bonds in the products exceed in strength those of the reactants. For exothermic reactions, no net energy input is needed to allow the reaction to take place. Instead, excess energy is generated and liberated when such reactions take place. In the former (endothermic) case, the energy needed by the reaction usually comes from the kinetic energy of the reacting molecules or molecules that surround them. That is, thermal energy from the environment provides the needed energy. Analogously, for exothermic reactions, the excess energy produced as the reaction proceeds is usually deposited into the kinetic energy of the product molecules and into that of surrounding molecules. For reactions that are very endothermic, it may be virtually impossible for thermal excitation to provide sufficient energy to effect reaction. In such cases, it may be possible to use a light source (i.e., photons whose energy can excite the reactant molecules) to induce reaction. When the light source causes electronic excitation of the reactants (e.g., one might excite one electron in the bound diatomic molecule discussed above from a bonding to an anti-bonding orbital), one speaks of inducing reaction by photochemical means.

5.2.3 Conservation of orbital symmetry – the Woodward–Hoffmann rules

As an example of how important it is to understand the changes in bonding that accompany a chemical reaction, let us consider a reaction in which 1,3-butadiene is converted, via ring-closure, to form cyclobutene. Specifically, focus on the four π orbitals of 1,3-butadiene as the molecule undergoes so-called disrotatory closing along which the plane of symmetry which bisects and is perpendicular to the C_2—C_3 bond is preserved. The orbitals of the reactant and product can be labeled as being even (e) or odd (o) under reflection through this symmetry plane. It is not appropriate to label the orbitals with respect to their symmetry under the plane containing the four C atoms because, although this plane is indeed a

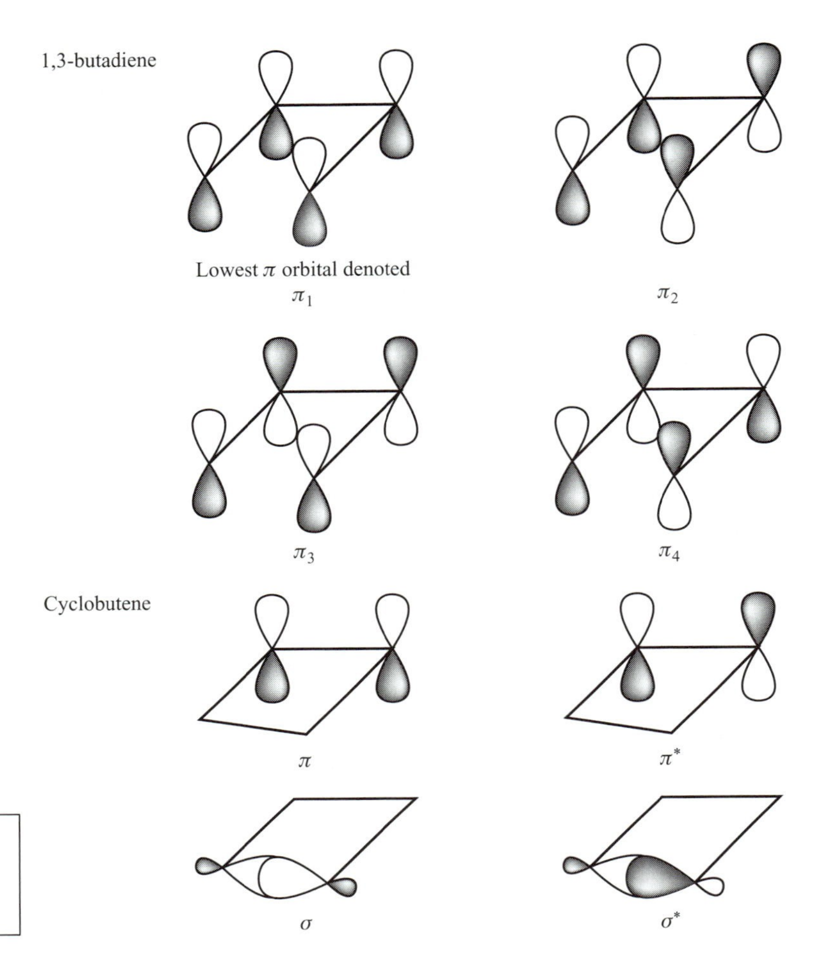

Figure 5.5 The active valence orbitals of 1,3-butadiene and of cyclobutene.

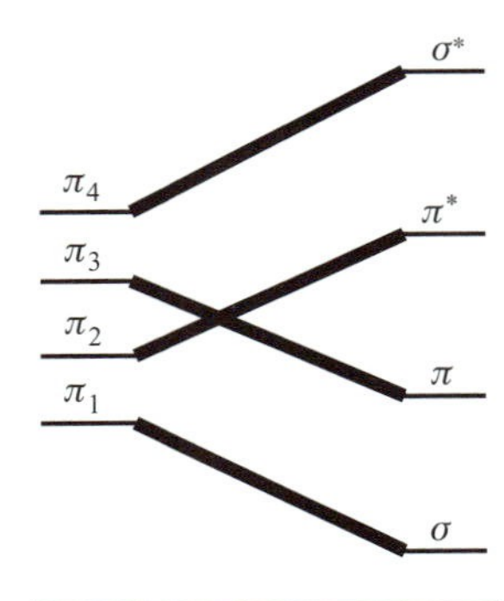

Figure 5.6 The orbital correlation diagram for the 1,3-butadiene to cyclobutene reaction.

symmetry operation for the reactants and products, it does not remain a valid symmetry throughout the reaction path. That is, we symmetry label the orbitals using only those symmetry elements that are preserved throughout the reaction path being examined.

The four π orbitals of 1,3-butadiene are of the following symmetries under the preserved symmetry plane (see the orbitals in Fig. 5.5): π_1 = e, π_2 = o, π_3 = e, π_4 = o. The π and π^* and σ and σ^* orbitals of the product cyclobutane, which evolve from the four orbitals of the 1,3-butadiene, are of the following symmetry and energy order: σ = e, π = e, π^* = o, σ^* = o. The Woodward–Hoffmann rules instruct us to arrange the reactant and product orbitals in order of increasing energy and to then connect these orbitals by symmetry, starting with the lowest energy orbital and going through the highest energy orbital. This process gives the so-called orbital correlation diagram shown in Fig. 5.6. We then need to consider how the electronic configurations in which the electrons are arranged as in the ground state of the reactants evolves as the reaction occurs.

We notice that the lowest two orbitals of the reactants, which are those occupied by the four π electrons of the reactant, do not connect to the lowest two orbitals of the products, which are the orbitals occupied by the two σ and two π electrons of the products. This causes the ground-state configuration of the reactants ($\pi_1^2\pi_2^2$) to evolve into an excited configuration ($\sigma^2\pi^{*2}$) of the products. This, in turn, produces an activation barrier for the thermal disrotatory rearrangement (in which the four active electrons occupy these lowest two orbitals) of 1,3-butadiene to produce cyclobutene.

If the reactants could be prepared, for example by photolysis, in an excited state having orbital occupancy $\pi_1^2\pi_2^1\pi_3^1$, then reaction along the path considered would not have any symmetry-imposed barrier because this singly excited configuration correlates to a singly excited configuration $\sigma^2\pi^1\pi^{*1}$ of the products. The fact that the reactant and product configurations are of equivalent excitation level causes there to be no symmetry constraints on the photochemically induced reaction of 1,3-butadiene to produce cyclobutene. In contrast, the thermal reaction considered first above has a symmetry-imposed barrier because the orbital occupancy is forced to rearrange (by the occupancy of two electrons from $\pi_2^2 = \pi^{*2}$ to $\pi^2 = \pi_3^2$) from the ground-state wave function of the reactant to smoothly evolve into that of the product. Of course, if the reactants could be generated in an excited state having $\pi_1^2\pi_3^2$ orbital occupancy, then products could also be produced directly in their ground electronic state. However, it is difficult, if not impossible, to generate such doubly excited electronic states, so it is rare that one encounters reactions being induced via such states.

It should be stressed that although these symmetry considerations may allow one to anticipate barriers on reaction potential energy surfaces, they have nothing to do with the thermodynamic energy differences of such reactions. What the above Woodward–Hoffmann symmetry treatment addresses is whether there will be symmetry-imposed barriers above and beyond any thermodynamic energy differences. The enthalpies of formation of reactants and products contain the information about the reaction's overall energy balance and need to be considered independently of the kind of orbital symmetry analysis just introduced.

As the above example illustrates, whether a chemical reaction occurs on the ground-state or an excited-state electronic surface is important to be aware of. This example shows that one might want to photo-excite the reactant molecules to cause the reaction to occur at an accelerated rate. With the electrons occupying the lowest-energy orbitals, the ring closure reaction can still occur, but it has to surmount a barrier to do so (it can employ thermal collisional energy to surmount this barrier), so its rate might be slow. If an electron is excited, there is no symmetry barrier to surmount, so the rate can be greater. Reactions that take place on excited states also have a chance to produce products in excited electronic states, and such excited-state products may emit light. Such reactions are called chemiluminescent because they produce light (luminescence) by way of a chemical reaction.

5.2.4 Rates of change

Rates of reactions play crucial roles in many aspects of our lives. Rates of various biological reactions determine how fast we metabolize food, and rates at which fuels burn in air determine whether an explosion or a calm flame will result. Chemists view the rate of any reaction among molecules (and perhaps photons or electrons if they are used to induce excitation in reactant molecules) to be related to (i) the frequency with which the reacting species encounter one another and (ii) the probability that a set of such species will react once they do encounter one another. The former aspects relate primarily to the concentrations of the reacting species and the speeds with which they are moving. The latter have more to do with whether the encountering species collide in a favorable orientation (e.g., do the enzyme and substrate "dock" properly, or does the Br^- ion collide with the H_3C— end of H_3C—Cl or with the Cl end in the S_N2 reaction that yields CH_3 Br + Cl^-?) and with sufficient energy to surmount any barrier that must be passed to effect breaking bonds in reactants to form new bonds in products.

The rates of reactions can be altered by changing the concentrations of the reacting species, by changing the temperature, or by adding a catalyst. Concentrations and temperature control the collision rates among molecules, and temperature also controls the energy available to surmount barriers. Catalysts are molecules that are not consumed during the reaction but which cause the rate of the reaction to be increased (species that slow the rate of a reaction are called inhibitors). Most catalysts act by providing orbitals of their own that interact with the reacting molecules' orbitals to cause the energies of the latter to be lowered as the reaction proceeds. In the ring-closure reaction cited earlier, the catalyst's orbitals would interact (i.e., overlap) with the 1,3-butadiene's π orbitals in a manner that lowers their energies and thus reduces the energy barrier that must be overcome for reaction to proceed.

In addition to being capable of determining the geometries (bond lengths and angles), energies, and vibrational frequencies of species such as the isomers of arginine discussed above, theory also addresses questions of how and how fast transitions among these isomers occur. The issue of how chemical reactions occur focuses on the mechanism of the reaction, meaning how the nuclei move and how the electronic orbital occupancies change as the system evolves from reactants to products. In a sense, understanding the mechanism of a reaction in detail amounts to having a mental moving picture of how the atoms and electrons move as the reaction is occurring.

The issue of how fast reactions occur relates to the rates of chemical reactions. In most cases, reaction rates are determined by the frequency with which the reacting molecules access a "critical geometry" (called the transition state or activated complex) near which bond breaking and bond forming takes place. The reacting molecules' potential energy along the path connecting reactants through a transition state to products is often represented as shown in Fig. 5.7.

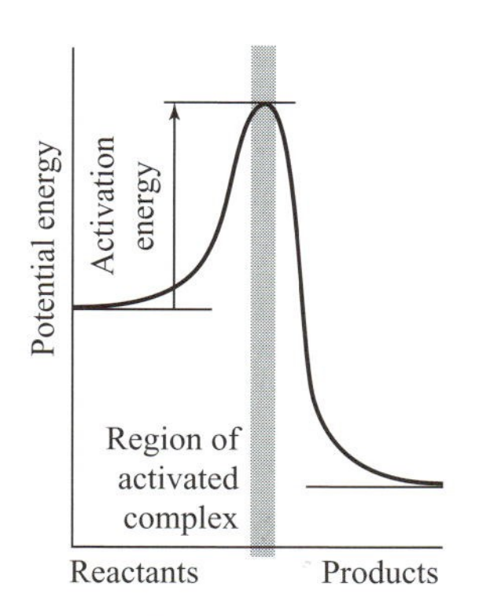

Figure 5.7 Energy vs. reaction progress plot showing the transition state or activated complex and the activation energy.

In this figure, the potential energy (i.e., the electronic energy without the nuclei's kinetic energy included) is plotted along a coordinate connecting reactants to products. The geometries and energies of the reactants, products, and of the activated complex can be determined using the potential energy surface searching methods discussed briefly above and detailed earlier in the Background Material. Chapter 8 provides more information about the theory of reaction rates and how such rates depend upon geometrical, energetic, and vibrational properties of the reacting molecules.

The frequencies with which the transition state is accessed are determined by the amount of energy (termed the activation energy E^*) needed to access this critical geometry. For systems at or near thermal equilbrium, the probability of the molecule gaining energy E^* is shown for three temperatures in Fig. 5.8. For such cases, chemical reaction rates usually display a temperature dependence characterized by linear plots of $\ln(k)$ vs. $1/T$. Of course, not all reactions involve molecules that have been prepared at or near thermal equilibrium. For example, in supersonic molecular beam experiments, the kinetic energy distribution of the colliding molecules is more likely to be of the type shown in Fig. 5.9. In this figure, the probability is plotted as a function of the relative speed with which reactant molecules collide. It is common in making such collision speed plots to include the v^2 "volume element" factor in the plot. That is, the normalized probability distribution for molecules having reduced mass μ to collide with relative velocity

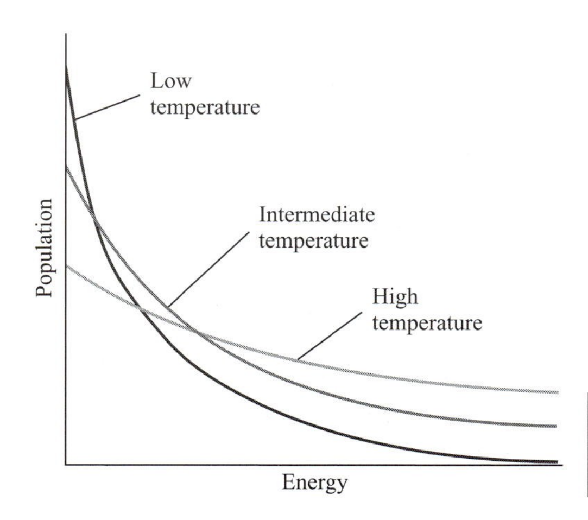

Figure 5.8 Distributions of energies at various temperatures.

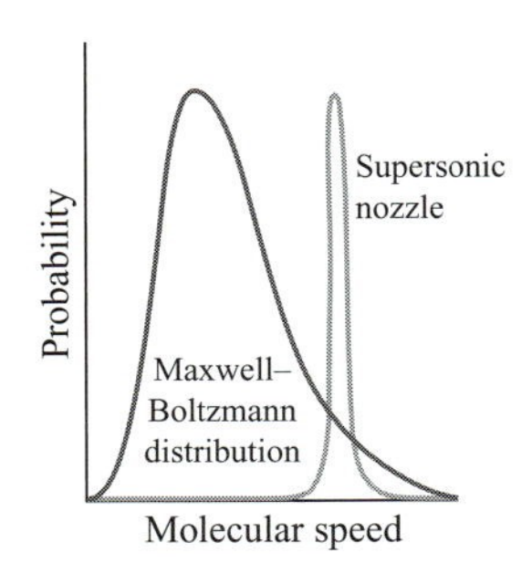

Figure 5.9 Molecular speed distributions in thermal and supersonic beam cases.

components v_x, v_y, v_z is

$$P(v_x, v_y, v_z)dv_x\, dv_y\, dv_z = (\mu/2\pi kT)^{3/2} \exp\left(-\mu\left(v_x^2 + v_y^2 + v_z^2\right)/2kT\right) \times dv_x\, dv_y\, dv_z. \quad (5.1)$$

Because only the total collisional kinetic energy is important in surmounting reaction barriers, we convert this Cartesian velocity component distribution to one in terms of $v = (v_x^2 + v_y^2 + v_z^2)^{1/2}$ the collision speed. This is done by changing from Cartesian to polar coordinates (in which the "radial" variable is v itself) and gives (after integrating over the two angular coordinates)

$$P(v)dv = 4\pi(\mu/2\pi kT)^{3/2} \exp(-\mu v^2/2kT)v^2 dv. \quad (5.2)$$

It is the v^2 factor in this speed distribution that causes the Maxwell–Boltzmann distribution to vanish at low speeds in Fig. 5.9.

Another kind of experiment in which non-thermal conditions are used to extract information about activation energies occurs within the realm of ion–molecule reactions where one uses collision-induced dissociation (CID) to break a molecule apart. For example, when a complex consisting of a Na^+ cation bound to a uracil molecule is accelerated by an external electric field to a kinetic energy E and subsequently allowed to impact into a gaseous sample of Xe atoms, the high-energy collision allows kinetic energy to be converted into internal energy. This collisional energy transfer may deposit into the Na^+(uracil) complex enough energy to fragment the Na^+ ... uracil attractive binding energy, thus producing Na^+ and neutral uracil fragments. If the signal for production of Na^+ is monitored as the collision energy E is increased, one generates a CID reaction rate profile such as I show in Fig. 5.10. On the vertical axis is plotted a quantity proportional to the rate at which Na^+ ions are formed. On the

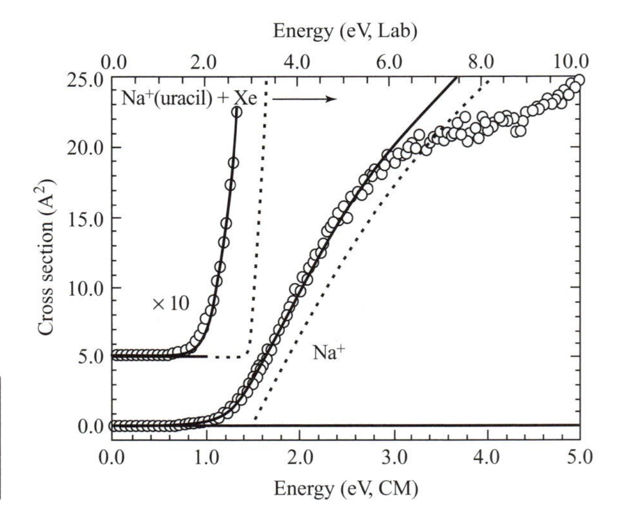

Figure 5.10 Reaction cross-section as a function of collision energy.

horizontal axis is plotted the collision energy E in two formats. The laboratory kinetic energy is simply 1/2 the mass of the Na^+(uracil) complex multiplied by the square of the speed of these ion complexes measured with respect to a laboratory-fixed coordinate frame. The center-of-mass (CM) kinetic energy is the amount of energy available between the Na^+(uracil) complex and the Xe atom, and is given by

$$E_{CM} = \frac{1}{2} \frac{m_{complex} m_{Xe}}{m_{complex} + m_{Xe}} v^2, \tag{5.3}$$

where v is the relative speed of the complex and the Xe atom, and m_{Xe} and $m_{complex}$ are the respective masses of the colliding partners.

The most essential lesson to learn from such a graph is that no dissociation occurs if E is below some critical "threshold" value, and the CID reaction

$$Na^+(\text{uracil}) \rightarrow Na^+ + \text{uracil} \tag{5.4}$$

occurs with higher and higher rate as the collision energy E increases beyond the threshold. For the example shown above, the threshold energy is *c*. 1.2–1.4 eV. These CID thresholds can provide us with estimates of reaction endothermicities and are especially useful when these energies are greatly in excess of what can be realized by simply heating the sample.

5.3 Statistical mechanics: treating large numbers of molecules in close contact

When one has a large number of molecules that undergo frequent collisions (thereby exchanging energy, momentum, and angular momentum), the behavior of this collection of molecules can often be described in a simple way. At first glance, it seems unlikely that the treatment of a large number of molecules could require far less effort than that required to describe one or a few such molecules.

To see the essence of what I am suggesting, consider a sample of 10 cm^3 of water at room temperature and atmospheric pressure. In this macroscopic sample, there are approximately 3.3×10^{23} water molecules. If one imagines having an "instrument" that could monitor the instantaneous speed of a selected molecule, one would expect the instrumental signal to display a very "jerky" irregular behavior if the signal were monitored on time scales of the order of the time between molecular collisions. On this time scale, the water molecule being monitored may be moving slowly at one instant, but, upon collision with a neighbor, may soon be moving very rapidly. In contrast, if one monitors the speed of this single water molecule over a very long time scale (i.e., much longer than the average time between collisions), one obtains an average square of the

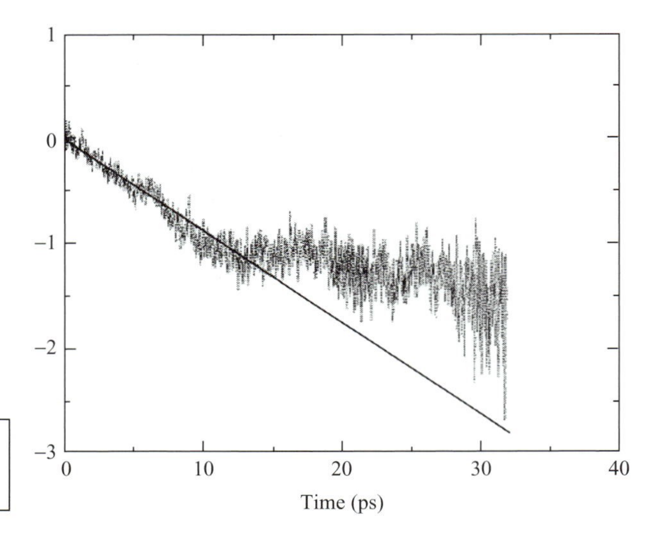

Figure 5.11 The energy possessed by a CN^- ion as a function of time.

speed that is related to the temperature T of the sample via $1/2\ mv^2 = 3/2kT$. This relationship holds because the sample is at equilibrium at temperature T.

An example of the kind of behavior I describe above is shown in Fig. 5.11. In this figure, on the vertical axis is plotted the log of the energy (kinetic plus potential) of a single CN^- anion in a solution with water as the solvent, as a function of time. The CN^- ion initially has excess vibrational energy in this simulation which was carried out in part to model the energy flow from this "hot" solute ion to the surrounding solvent molecules. One clearly sees the rapid jerks in energy that this ion experiences as it undergoes collisions with neighboring water molecules. These jerks occur approximately every 0.01 ps, and some of them correspond to collisions that take energy from the ion and others to collisions that give energy to the ion. On longer time scales (e.g., over 1–10 ps), we also see a gradual drop off in the energy content of the CN^- ion which illustrates the slow loss of its excess energy on the longer time scale.

Now, let's consider what happens if we monitor a large number of molecules rather than a single molecule within the 10 cm^3 sample of H_2O mentioned earlier. If we imagine drawing a sphere of radius R and monitoring the average speed of all water molecules within this sphere, we obtain a qualitatively different picture if the sphere is large enough to contain many water molecules. For large R, one finds that the average square of the speed of all the N water molecules residing inside the sphere (i.e., $\sum_{K=1,N} 1/2mv_K^2$) is independent of time (even when considered at a sequence of times separated by fractions of ps) and is related to the temperature T through $\sum_K 1/2mv_K^2 = (3N/2)kT$.

This example shows that, at equilibrium, the long-time average of a property of any single molecule is the same as the instantaneous average of this same property over a large number of molecules. For the single molecule, one achieves the average value of the property by averaging its behavior over time scales lasting for many, many collisions. For the collection of many molecules, the same average value is achieved (at any instant of time) because the number of molecules within the sphere (which is proportional to 4/3 πR^3) is so much larger than the number near the surface of the sphere (proportional to $4\pi R^2$) that the molecules interior to the sphere are essentially at equilibrium for all times.

Another way to say the same thing is to note that the fluctuations in the energy content of a single molecule are very large (i.e., the molecule undergoes frequent large jerks) but last a short time (i.e., the time between collisions). In contrast, for a collection of many molecules, the fluctuations in the energy for the whole collection are small at all times because fluctuations take place by exchange of energy with the molecules that are not inside the sphere (and thus relate to the surface area to volume ratio of the sphere).

So, if one has a large number of molecules that one has reason to believe are at thermal equilibrium, one can avoid trying to follow the instantaneous short-time detailed dynamics of any one molecule or of all the molecules. Instead, one can focus on the average properties of the entire collection of molecules. What this means for a person interested in theoretical simulations of such condensed-media problems is that there is no need to carry out a Newtonian molecular dynamics simulation of the system (or a quantum simulation) if it is at equilibrium because the long-time averages of whatever is calculated can be found another way. How one achieves this is through the "magic" of statistical mechanics and statistical thermodynamics. One of the most powerful of the devices of statistical mechanics is the so-called Monte-Carlo simulation algorithm. Such theoretical tools provide a direct way to compute equilibrium averages (and small fluctuations about such averages) for systems containing large numbers of molecules. In Chapter 7, I provide a brief introduction to the basics of this sub-discipline of theoretical chemistry where you will learn more about this exciting field.

Sometimes we speak of the equilibrium behavior or the dynamical behavior of a collection of molecules. Let me elaborate a little on what these phrases mean. Equilibrium properties of molecular collections include the radial and angular distribution functions among various atomic centers. For example, the O—O and O—H radial distribution functions in liquid water are shown in Fig. 5.12. Such properties represent averages, over long times or over a large collection of molecules, of some property that is not changing with time except on a very fast time scale corresponding to individual collisions.

In contrast, dynamical properties of molecular collections include the folding and unfolding processes that proteins and other polymers undergo; the migrations of protons from water molecule to water molecule in liquid water and along H_2O chains within ion channels; and the self-assembly of molecular monolayers on

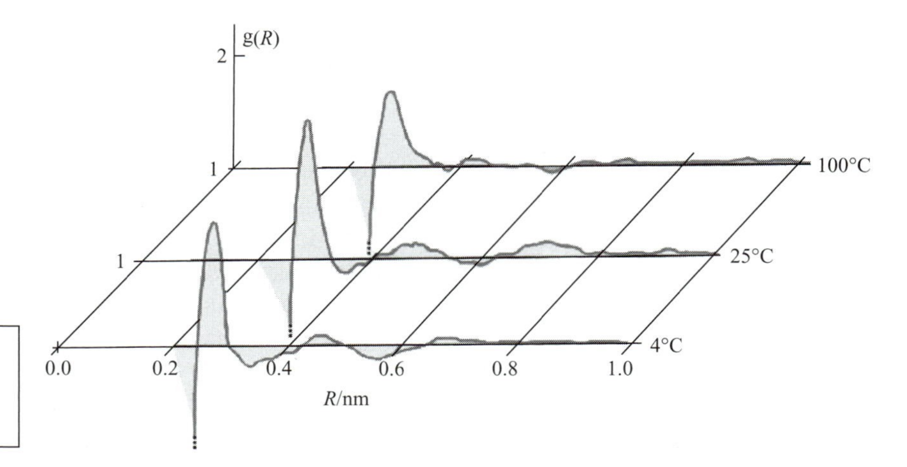

Figure 5.12 Radial O—O distribution functions at three temperatures.

solid surfaces as the concentration of the molecules in the liquid overlayer varies. These are properties that occur on time scales much longer than those between molecular collisions and on time scales that we wish to probe by some experiment or by simulation.

Having briefly introduced the primary areas of theoretical chemistry – structure, dynamics, and statistical mechanics, let us now examine each of them in somewhat greater detail, keeping in mind that Chapters 6–8 are where each is treated more fully.

Molecular structure: theory and experiment

5.4 Experimental probes of molecular shapes

I expect that you are wondering why I want to discuss how experiments measure molecular shapes in this text whose aim is to introduce you to the field of theoretical chemistry. In fact, theory and experimental measurement are very connected, and it is these connections that I wish to emphasize in the following discussion. In particular, I want to make it clear that experimental data can only be interpreted, and thus used to extract molecular properties, through the application of theory. So, theory does not replace experiment, but serves both as a complementary component of chemical research (via simulation of molecular properties) and as the means by which we connect laboratory data to molecular properties.

5.4.1 Rotational spectroscopy

Most of us use rotational excitation of molecules in our everyday life. In particular, when we cook in a microwave oven, the microwave radiation, which has a frequency in the 10^9–10^{11} s^{-1} range, inputs energy into the rotational motions of the (primarily) water molecules contained in the food. These rotationally "hot" water molecules then collide with neighboring molecules (i.e., other water as well as proteins and other molecules in the food and in the cooking vessel) to transfer some of their motional energy to them. Through this means, the translational kinetic energy of all the molecules inside the cooker increases. This process of rotation-to-translation energy transfer is how the microwave radiation ultimately heats the food, which cooks it. What happens when you put the food into the microwave oven in a metal container or with some other metal material? As shown in the Background Material, the electrons in metals exist in very delocalized orbitals called bands. These band orbitals are spread out throughout the entire piece of metal. The application of any external electric field (e.g., that belonging to the microwave radiation) causes these metal electrons to move throughout the metal. As these electrons accumulate more and more energy from the microwave radiation, they eventually have enough kinetic energy to be ejected into the surrounding air forming a discharge. This causes the sparking that we see when we make the mistake of putting anything metal into our microwave oven. Let's now learn more about how the microwave photons cause the molecules to become rotationally excited.

Using microwave radiation, molecules having dipole moment vectors ($\boldsymbol{\mu}$) can be made to undergo rotational excitation. In such processes, the time-varying electric field $\boldsymbol{E}\cos(\omega t)$ of the microwave electromagnetic radiation interacts with the molecules via a potential energy of the form $V = \mathbf{E} \cdot \boldsymbol{\mu} \cos(\omega t)$. This potential can cause energy to flow from the microwave energy source into the molecule's rotational motions when the energy of the former $\mathrm{h}\omega/2\pi$ matches the energy spacing between two rotational energy levels.

This idea of matching the energy of the photons to the energy spacings of the molecule illustrates the concept of resonance and is something that is ubiquitous in spectroscopy. Upon first hearing that the photon's energy must match an energy-level spacing in the molecule if photon absorption is to occur, it appears obvious and even trivial. However, upon further reflection, there is more to such resonance requirements than one might think. Allow me to illustrate using this microwave-induced rotational excitation example by asking you to consider why photons whose energies $\mathrm{h}\omega/2\pi$ considerably exceed the energy spacing ΔE will not be absorbed in this transition. That is, why is more than enough energy not good enough? The reason is that for two systems (in this case the photon's electric field and the molecule's rotation which causes its dipole moment to also rotate) to interact and thus exchange energy (this is what photon absorption is), they must have very nearly the same frequencies. If the photon's frequency (ω) exceeds the rotational frequency of the molecule by a significant amount, the molecule will experience an electric field that oscillates too quickly to induce a torque on the molecule's dipole that is always in the same direction and that lasts over a significant length of time. As a result, the rapidly oscillating electric field will not provide a coherent twisting of the dipole and hence will not induce rotational excitation.

One simple example from everyday life can further illustrate this issue. When you try to push your friend, spouse, or child on a swing, you move your arms in resonance with the swinging person's movement frequency. Each time the person returns to you, your arms are waiting to give a push in the direction that gives energy to the swinging individual. This happens over and over again; each time they return, your arms have returned to be ready to give another push in the same direction. In this case, we say that your arms move in resonance with the swing's motion and offer a coherent excitation of the swinger. If you were to increase greatly the rate at which your arms are moving in their up and down pattern, the swinging person would not always experience a push in the correct direction when they return to meet your arms. Sometimes they would feel a strong in-phase push, but other times they would feel an out-of-phase push in the opposite direction. The net result is that, over a long period of time, they would feel random "jerks" from your arms, and thus would not undergo smooth energy transfer from you. This is why too high a frequency (and hence too high an energy) does not induce excitation. Let us now return to the case of rotational excitation by microwave photons.

As we saw in the Background Material, for a rigid diatomic molecule, the rotational energy spacings are given by

$$E_{J+1} - E_J = 2(J+1)(\hbar^2/2I) = 2hcB(J+1), \tag{5.5}$$

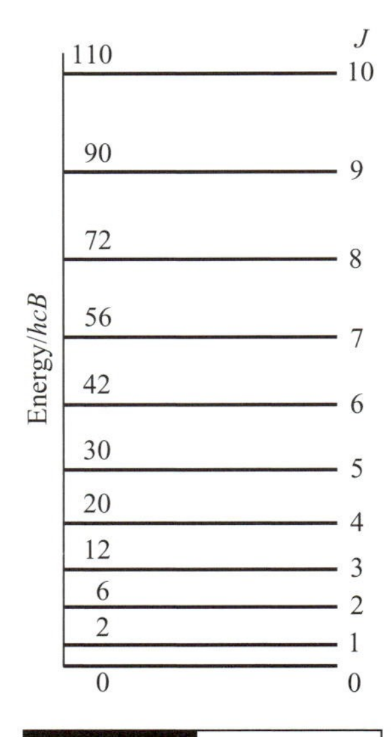

Figure 5.13 Rotational energy levels vs. rotational quantum number.

where I is the moment of inertia of the molecule given in terms of its equilibrium bond length r_e and its reduced mass $\mu = m_a m_b/(m_a + m_b)$ as $I = \mu r_e^2$. Thus, in principle, measuring the rotational energy level spacings via microwave spectroscopy allows one to determine r_e. The second identity above simply defines what is called the rotational constant B in terms of the moment of inertia. The rotational energy levels described above give rise to a manifold of levels of non-uniform spacing as shown in the Fig. 5.13. The non-uniformity in spacings is a result of the quadratic dependence of the rotational energy levels E_J on the rotational quantum number J:

$$E_J = J(J+1)(\hbar^2/2I). \tag{5.6}$$

Moreover, the level with quantum number J is $(2J+1)$-fold degenerate; that is, there are $2J+1$ distinct energy states and wave functions that have energy E_J and that are distinguished by a quantum number M. These $2J+1$ states have identical energy but differ among one another by the orientation of their angular momentum in space (i.e., the orientation of how they are spinning).

For polyatomic molecules, we know from the Background Material that things are a bit more complicated because the rotational energy levels depend on three so-called principal moments of inertia ($I_{a,}$, I_b, I_c) which, in turn, contain information about the molecule's geometry. These three principal moments are found by forming a 3×3 moment of inertia matrix having elements

$$I_{x,x} = \sum_a m_a[(R_a - R_{\mathrm{CofM}})^2 - (x_a - x_{\mathrm{CofM}})^2], \qquad \text{and}$$
$$I_{x,y} = \sum_a m_a[(x_a - x_{\mathrm{CofM}})(y_a - y_{\mathrm{CofM}})] \tag{5.7}$$

expressed in terms of the Cartesian coordinates of the nuclei (a) and of the center of mass in an arbitrary molecule-fixed coordinate system (analogous definitions hold for $I_{z,z}$, $I_{y,y}$, $I_{x,z}$ and $I_{y,z}$). The principal moments are then obtained as the eigenvalues of this 3×3 matrix.

For molecules with all three principal moments equal, the rotational energy levels are given by $E_{J,K} = \hbar^2 J(J+1)/2I$, and are independent of the K quantum number and of the M quantum number that again describes the orientation of how the molecule is spinning in space. Such molecules are called spherical tops. For molecules (called symmetric tops) with two principal moments equal (I_a) and one unique moment I_c, the energies depend on two quantum numbers J and K and are given by $E_{J,K} = \hbar^2 J(J+1)/2I_\mathrm{a} + \hbar^2 K^2(1/2I_\mathrm{c} - 1/2I_\mathrm{a})$. Species having all three principal moments of inertia unique, termed asymmetric tops, have rotational energy levels for which no analytic formula is yet known. The H_2O molecule, shown in Fig. 5.14, is such an asymmetric top molecule. More details about the rotational energies and wave functions were given in the Background Material.

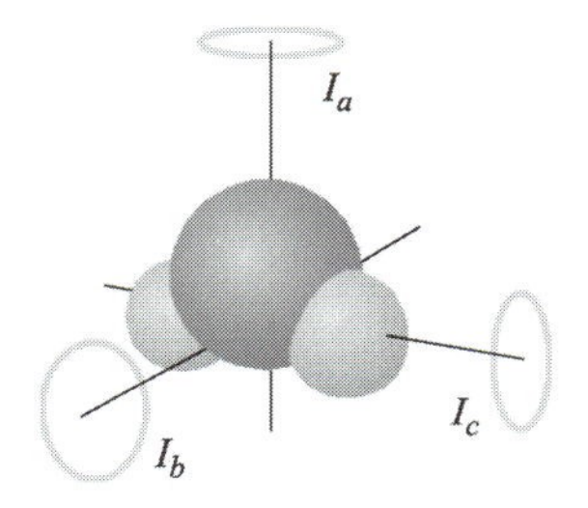

Figure 5.14 Water molecule showing its three distinct principal moments of inertia.

The moments of inertia that occur in the expressions for the rotational energy levels involve positions of atomic nuclei relative to the center of mass of the molecule. So, a microwave spectrum can, in principle, determine the moments of inertia and hence the geometry of a molecule. In the discussion given above, we treated these positions, and thus the moments of inertia as fixed (i.e., not varying with time). Of course, these distances are not unchanging with time in a real molecule because the molecule's atomic nuclei undergo vibrational motions. Because of this, it is the vibrationally averaged moment of inertia that must be incorporated into the rotational energy level formulas. Specifically, because the rotational energies depend on the inverses of moments of inertia, one must vibrationally average $(R_a - R_{\mathrm{CofM}})^{-2}$ over the vibrational motion that characterizes the molecule's movement. For species containing "stiff" bonds, the vibrational average $\langle v|(R_a - R_{\mathrm{CofM}})^{-2}|v\rangle$ of the inverse squares of atomic distances relative to the center of mass does not differ significantly from the equilibrium values $(R_{a,\mathrm{eq}} - R_{\mathrm{CofM}})^{-2}$ of the same distances. However, for molecules such as weak van der Waals complexes (e.g., $(H_2O)_2$ or Ar..HCl) that undergo "floppy" large amplitude vibrational motions, there may be large differences between the equilibrium and the vibrationally averaged values $\langle v|(R_a - R_{\mathrm{CofM}})^{-2}|v\rangle$. The proper treatment of the rotational energy level patterns in such floppy molecules is still very much under active study by theoretical chemists.

So, in the area of rotational spectroscopy theory plays several important roles:

(i) It provides the basic equations in terms of which the rotational line spacings relate to moments of inertia.
(ii) It allows one, given the distribution of geometrical bond lengths and angles characteristic of the vibrational state the molecule exists in, to compute the proper vibrationally averaged moment of inertia.
(iii) It can be used to treat large amplitude floppy motions (e.g., by simulating the nuclear motions on a Born–Oppenheimer energy surface), thereby allowing rotationally resolved spectra of such species to provide proper moment of inertia (and thus geometry) information.

5.4.2 Vibrational spectroscopy

The ability of molecules to absorb and emit infrared radiation as they undergo transitions among their vibrational energy levels is critical to our planet's health. It turns out that water and CO_2 molecules have bonds that vibrate in the 10^{13}–10^{14} s^{-1} frequency range which is within the infrared spectrum (10^{11}–10^{14} s^{-1}). As solar radiation (primarily visible and ultraviolet) impacts the earth's surface, it is absorbed by molecules with electronic transitions in this energy range (e.g, colored molecules such as those contained in plant leaves and other dark material). These molecules are thereby promoted to excited electronic states. Some such molecules re-emit the photons that excited them but most undergo so-called radiationless relaxation that allows them to return to their ground electronic state but with a substantial amount of internal vibrational energy. That is, these molecules become vibrationally very "hot". Subsequently, these hot molecules, as they undergo transitions from high-energy vibrational levels to lower-energy levels, emit infrared (IR) photons.

If our atmosphere were devoid of water vapor and CO_2, these IR photons would travel through the atmosphere and be lost into space. The result would be that much of the energy provided by the sun's visible and ultraviolet photons would be lost via IR emission. However, the water vapor and CO_2 do not allow so much IR radiation to escape. These greenhouse gases absorb the emitted IR photons to generate vibrationally hot water and CO_2 molecules in the atmosphere. These vibrationally excited molecules undergo collisions with other molecules in the atmosphere and at the earth's surface. In such collisions, some of their vibrational energy can be transferred to translational kinetic energy of the collision-partner molecules. In this manner, the temperature (which is a measure of the average translational energy) increases. Of course, the vibrationally hot molecules can also re-emit their IR photons, but there is a thick layer of such molecules forming a "blanket" around the earth, and all of these molecules are available to continually absorb and re-emit the IR energy. In this manner, the blanket keeps the IR radiation from escaping and thus keeps our atmosphere warm. Those of us who live in dry desert climates are keenly aware of such effects. Clear cloudless nights in the desert can become very cold, primarily because much of the day's IR energy production is lost to radiative emission through the atmosphere and into space. Let's now learn more about molecular vibrations, how IR radiation excites them, and what theory has to do with this.

When infrared (IR) radiation is used to excite a molecule, it is the vibrations of the molecule that are in resonance with the oscillating electric field $\mathbf{E}\cos(\omega t)$. Molecules that have dipole moments that vary as its vibrations occur interact with the IR electric field via a potential energy of the form $V = (\partial\mu/\partial Q)\cdot\mathbf{E}\cos(\omega t)$. Here $\partial\mu/\partial Q$ denotes the change in the molecule's dipole moment μ associated with motion along the vibrational normal mode labeled Q.

As the IR radiation is scanned, it comes into resonance with various vibrations of the molecule under study, and radiation can be absorbed. Knowing the frequencies at which radiation is absorbed provides knowledge of the vibrational energy level spacings in the molecule. Absorptions associated with transitions from the lowest vibrational level to the first excited level are called fundamental transitions. Those connecting the lowest level to the second excited state are called first overtone transitions. Excitations from excited levels to even higher levels are named hot-band absorptions.

Fundamental vibrational transitions occur at frequencies that characterize various functional groups in molecules (e.g., O—H stretching, H—N—H bending, N—H stretching, C—C stretching, etc.). As such, a vibrational spectrum offers an important "fingerprint" that allows the chemist to infer which functional groups are present in the molecule. However, when the molecule contains soft "floppy" vibrational modes, it is often more difficult to use information about the absorption frequency to extract quantitative information about the molecule's energy surface and its bonding structure. As was the case for rotational levels of such floppy molecules, the accurate treatment of large-amplitude vibrational motions of such species remains an area of intense research interest within the theory community.

In a polyatomic molecule with N atoms, there are many vibrational modes. The total vibrational energy of such a molecule can be approximated as a sum of terms, one for each of the $3N - 6$ (or $3N - 5$ for a linear molecule) vibrations:

$$E(v_1 \ldots v_{3N-5\,\mathrm{or}\,6}) = \sum_{J=1}^{3N-5\,\mathrm{or}\,6} \hbar\omega_j(v_j + 1/2). \tag{5.8}$$

Here, ω_j is the harmonic frequency of the jth mode and v_j is the vibrational quantum number associated with that mode. As we discussed in the Background Material, the vibrational wave functions are products of harmonic vibrational

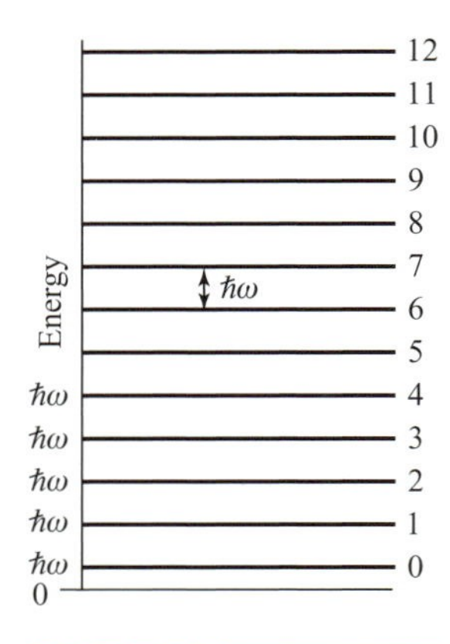

Figure 5.15 Harmonic vibrational energy levels vs. vibrational quantum number.

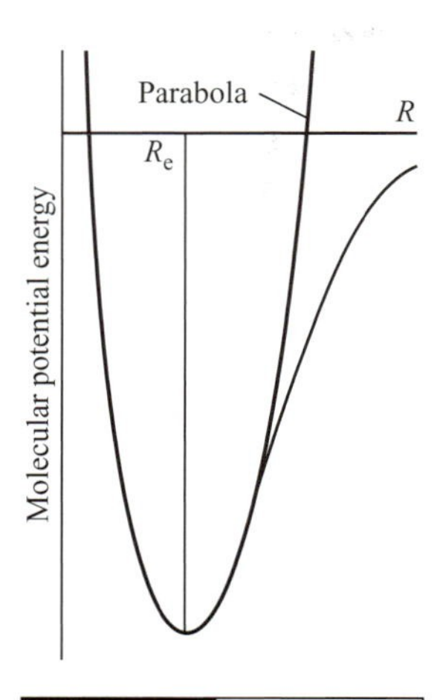

Figure 5.16 Harmonic (parabola) and anharmonic potentials.

functions for each mode:

$$\psi = \prod_{j=1,3N-5\,\text{or}\,6} \psi_{vj}\left(x^{(j)}\right), \tag{5.9}$$

and the spacings between energy levels in which one of the normal-mode quantum numbers increases by unity are expressed as

$$\Delta E_{v_j} = E(\cdots v_j + 1 \cdots) - E(\cdots v_j \cdots) = \hbar\omega_j. \tag{5.10}$$

That is, the spacings between successive vibrational levels of a given mode are predicted to be independent of the quantum number v within this harmonic model as shown in Fig. 5.15. In the Background Material, the details connecting the local curvature (i.e., Hessian matrix elements) in a polyatomic molecule's potential energy surface to its normal modes of vibration are presented.

Experimental evidence clearly indicates that significant deviations from the harmonic oscillator energy expression occur as the quantum number v_j grows. These deviations are explained in terms of the molecule's true potential $V(R)$ deviating strongly from the harmonic $1/2k(R - R_e)^2$ potential at higher energy as shown in the Fig. 5.16. At larger bond lengths, the true potential is "softer" than the harmonic potential, and eventually reaches its asymptote, which lies at the dissociation energy D_e above its minimum. This deviation of the true $V(R)$ from $1/2k(R - R_e)^2$ causes the true vibrational energy levels to lie below the harmonic predictions.

It is convention to express the experimentally observed vibrational energy levels along each of the $3N - 5$ or 6 independent modes in terms of an anharmonic formula similar to what we discussed for the Morse potential in the Background Material:

$$\begin{aligned} E(v_j) = \hbar[&\omega_j(v_j + 1/2) - (\omega x)_j(v_j + 1/2)^2 + (\omega y)_j(v_j + 1/2)^3 \\ &+ (\omega z)_j(v_j + 1/2)^4 + \cdots]. \end{aligned} \tag{5.11}$$

The first term is the harmonic expression. The next is termed the first anharmonicity; it (usually) produces a negative contribution to $E(v_j)$ that varies as $(v_j + 1/2)^2$. Subsequent terms are called higher anharmonicity corrections. The spacings between successive $v_j \to v_j + 1$ energy levels are then given by:

$$\Delta E_{v_j} = E(v_j + 1) - E(v_j) \tag{5.12}$$

$$= \hbar[\omega_j - 2(\omega x)_j(v_j + 1) + \cdots]. \tag{5.13}$$

A plot of the spacing between neighboring energy levels versus v_j should be linear for values of v_j where the harmonic and first anharmonicity terms dominate. The slope of such a plot is expected to be $-2\hbar(\omega x)_j$ and the small $-v_j$ intercept

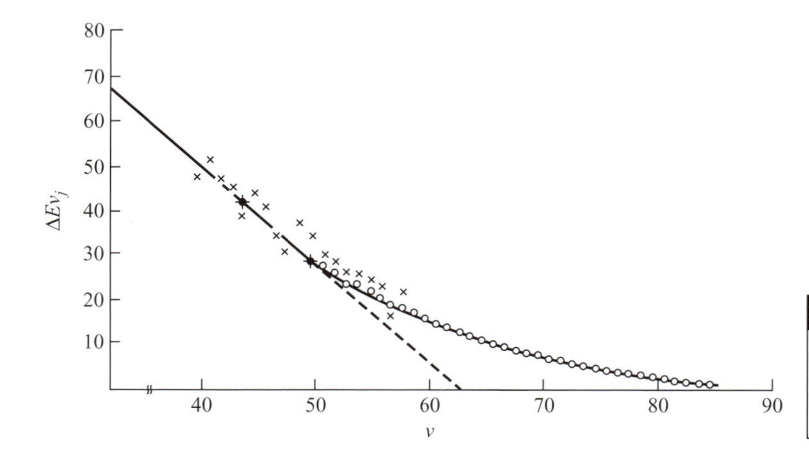

Figure 5.17 Birge–Sponer plot of vibrational energy spacings vs. quantum number.

should be $\hbar[\omega_j - 2(\omega x)_j]$. Such a plot of experimental data, which clearly can be used to determine the ω_j and $(\omega x)_j$ parameters of the vibrational mode of study, is shown in Fig. 5.17.

These so-called Birge–Sponer plots can also be used to determine dissociation energies of molecules if the vibration whose spacings are plotted corresponds to a bond-stretching mode. By linearly extrapolating such a plot of experimental ΔE_{vj} values to large v_j values, one can find the value of v_j at which the spacing between neighboring vibrational levels goes to zero. This value v_j, max specifies the quantum number of the last bound vibrational level for the particular bond-stretching mode of interest. The dissociation energy D_e can then be computed by adding to $1/2\hbar\omega_j$ (the zero point energy along this mode) the sum of the spacings between neighboring vibrational energy levels from $v_j = 0$ to $v_j = v_j$, max:

$$D_e = 1/2\hbar\omega_j + \sum_{v_j=0}^{v_j\,\max} \Delta E_{v_j}. \tag{5.14}$$

So, in the case of vibrational spectroscopy, theory allows us to:

(i) interpret observed infrared lines in terms of absorptions arising in localized functional groups;
(ii) extract dissociation energies if a long progression of lines is observed in a bond-stretching transition;
(iii) treat highly non-harmonic "floppy" vibrations by carrying out dynamical simulations on a Born–Oppenheimer energy surface.

5.4.3 X-ray crystallography

In x-ray crystallography experiments, one employs crystalline samples of the molecules of interest and makes use of the diffraction patterns produced by scattered x-rays to determine positions of the atoms in the molecule relative to one another using the famous Bragg formula:

$$n\lambda = 2d \sin \theta. \tag{5.15}$$

In this equation, λ is the wavelength of the x-rays, d is a spacing between layers (planes) of atoms in the crystal, θ is the angle through which the x-ray beam is scattered, and n is an integer $(1, 2, \ldots)$ that labels the order of the scattered beam.

Because the x-rays scatter most strongly from the inner-shell electrons of each atom, the interatomic distances obtained from such diffraction experiments are, more precisely, measures of distances between high electron densities in the neighborhoods of various atoms. X-rays interact most strongly with the inner-shell electrons because it is these electrons whose characteristic Bohr frequencies of motion are (nearly) in resonance with the high frequency of such radiation. For this reason, x-rays can be viewed as being scattered from the core electrons that reside near the nuclear centers within a molecule. Hence, x-ray diffraction data offers a very precise and reliable way to probe inter-atomic distances in molecules.

The primary difficulties with x-ray measurements are:

(i) One needs to have crystalline samples (often, materials simply can not be grown as crystals).
(ii) One learns about inter-atomic spacings as they occur in the crystalline state, not as they exist, for example, in solution or in gas-phase samples. This is especially problematic for biological systems where one would like to know the structure of the bio-molecule as it exists within the living organism.

Nevertheless, x-ray diffraction data and its interpretation through the Bragg formula provide one of the most widely used and reliable ways for probing molecular structure.

5.4.4 NMR spectroscopy

NMR spectroscopy probes the absorption of radio-frequency (RF) radiation by the nuclear spins of the molecule. The most commonly occurring spins in natural samples are ^{1}H (protons), ^{2}H (deuterons), ^{13}C and ^{15}N nuclei. In the presence of an external magnetic field $B_0\mathbf{z}$ along the z-axis, each such nucleus has its spin states split in energy by an amount given by $B_0(1 - \sigma_k)\gamma_k M_I$, where M_I is the component of the kth nucleus' spin angular momentum along the z-axis, B_0 is the strength of the external magnetic field, and γ_k is a so-called gyromagnetic factor (i.e., a constant) that is characteristic of the kth nucleus. This splitting of magnetic spin levels by a magnetic field is called the Zeeman effect, and it is illustrated in Fig. 5.18. The factor $(1 - \sigma_k)$ is introduced to describe the screening of the external B-field at the kth nucleus caused by the electron cloud that surrounds this nucleus. In effect, $B_0(1 - \sigma_k)$ is the magnetic field experienced local to the kth nucleus. It is this $(1 - \sigma_k)$ screening that gives rise to the phenomenon of chemical shifts in NMR spectroscopy, and it is this factor that allows NMR measurements of shielding factors (σ_κ) to be related, by theory, to the electronic environment of a nucleus. In Fig. 5.19 we display the chemical shifts of proton and ^{13}C nuclei in a variety of chemical bonding environments. Because the M_I quantum number changes in steps of unity and because each photon possesses one unit of angular momentum, the RF energy $\hbar\omega$ that will be in resonance with the kth nucleus' Zeeman-split levels is given by $\hbar\omega = B_0(1 - \sigma_k)\gamma_k$.

In most NMR experiments, a fixed RF frequency is employed and the external magnetic field is scanned until the above resonance condition is met. Determining at what B_0 value a given nucleus absorbs RF radiation allows one to determine the local shielding $(1 - \sigma_\kappa)$ for that nucleus. This, in turn, provides information about the electronic environment local to that nucleus as illustrated in Fig. 5.19. This data tells the chemist a great deal about the molecule's structure because it suggests what kinds of functional groups occur within the molecule.

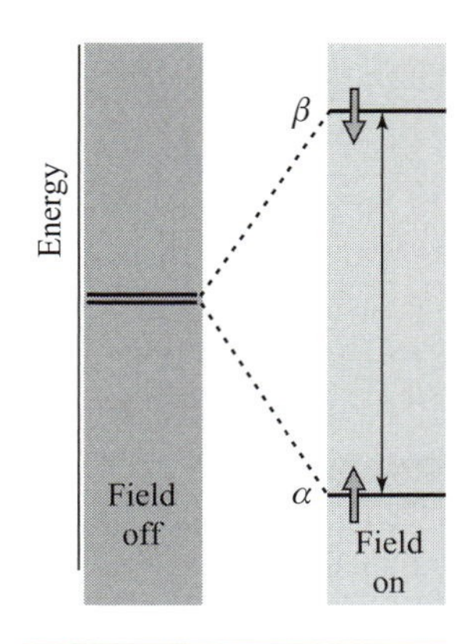

Figure 5.18 Splitting of magnetic nucleus' two levels caused by magnetic field.

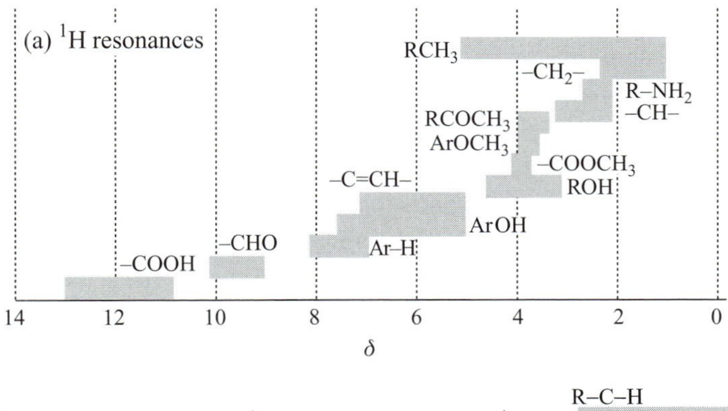

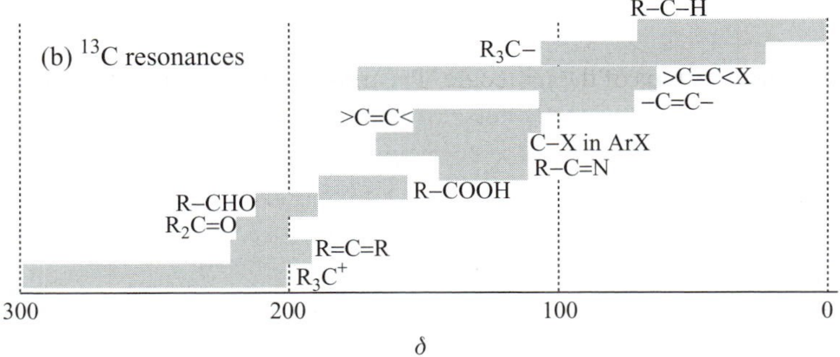

Figure 5.19 Chemical shifts characterizing various electronic environments for protons and for carbon-13 nuclei.

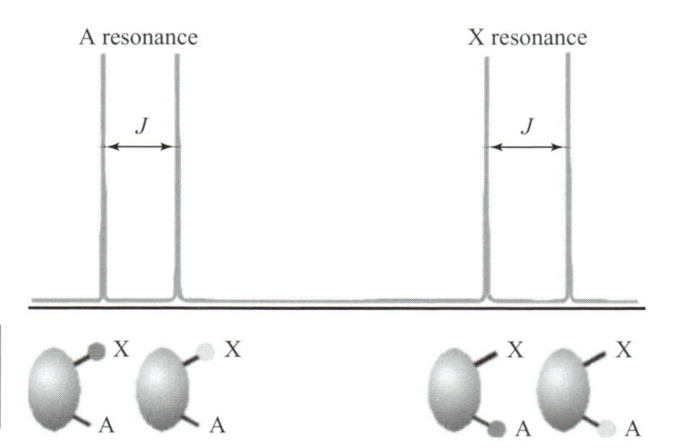

Figure 5.20 Splitting pattern characteristic of AX case.

To extract even more geometrical information from NMR experiments, one makes use of another feature of nuclear spin states. In particular, it is known that the energy levels of a given nucleus (e.g., the kth one) are altered by the presence of other nearby nuclear spins. These spin–spin coupling interactions give rise to splittings in the energy levels of the kth nucleus that alter the above energy expression as follows:

$$E_M = B_0(1 - \sigma_k)\gamma_k M + JMM', \tag{5.16}$$

where M is the z-component of the kth nuclear spin angular momentum, M' is the corresponding component of a nearby nucleus causing the splitting, and J is called the spin–spin coupling constant between the two nuclei.

Examples of how spins on neighboring centers split the NMR absorption lines of a given nucleus are shown in Figs. 5.20–5.22 for three common cases. The first involves a nucleus (labeled A) that is close enough to one other magnetically active nucleus (labeled X); the second involves a nucleus (A) that is close to two equivalent nuclei (X2); and the third describes a nucleus (A) close to three equivalent nuclei (X3). In Fig. 5.20 are illustrated the splitting in the X nucleus' absorption due to the presence of a single A neighbor nucleus (right) and the splitting in the A nucleus' absorption (left) caused by the X nucleus. In both of these examples, the X and A nuclei have only two M_I values, so they must be spin-1/2 nuclei. This kind of splitting pattern would, for example, arise for a $^{13}C{-}H$ group in the benzene molecule where A = ^{13}C and X = ^{1}H. The (AX2) splitting pattern shown in Fig. 5.21 would, for example, arise in the ^{13}C spectrum of a $-CH_2-$ group, and illustrates the splitting of the A nucleus' absorption line by the four spin states that the two equivalent X spins can occupy. Again, the lines shown would be consistent with X and A both having spin 1/2 because they each assume only two M_I values. In Fig. 5.22 is the kind of splitting pattern (AX3) that would apply to the ^{13}C NMR absorptions for a $-CH_3$ group. In this case, the spin-1/2 A line is split by the eight spin states that the three equivalent spin-1/2 H nuclei can occupy.

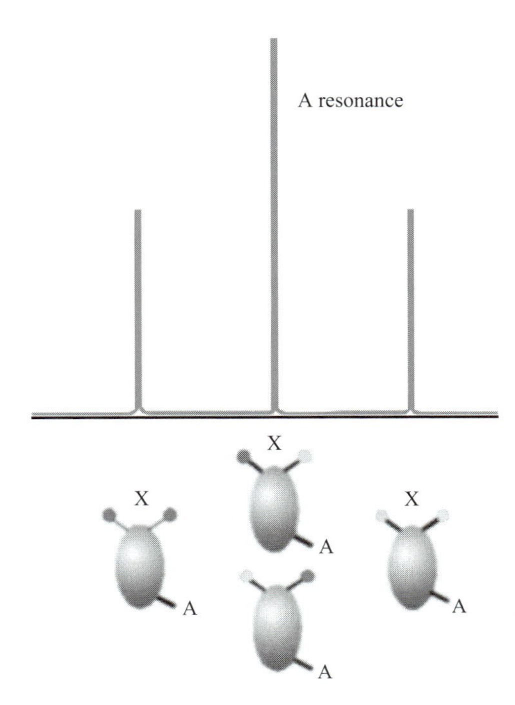

Figure 5.21 Splitting pattern characteristic of AX2 case.

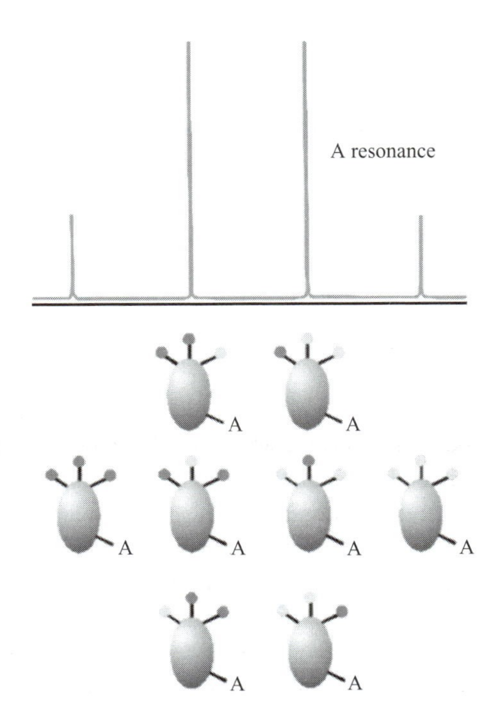

Figure 5.22 Splitting pattern characteristic of AX3 case.

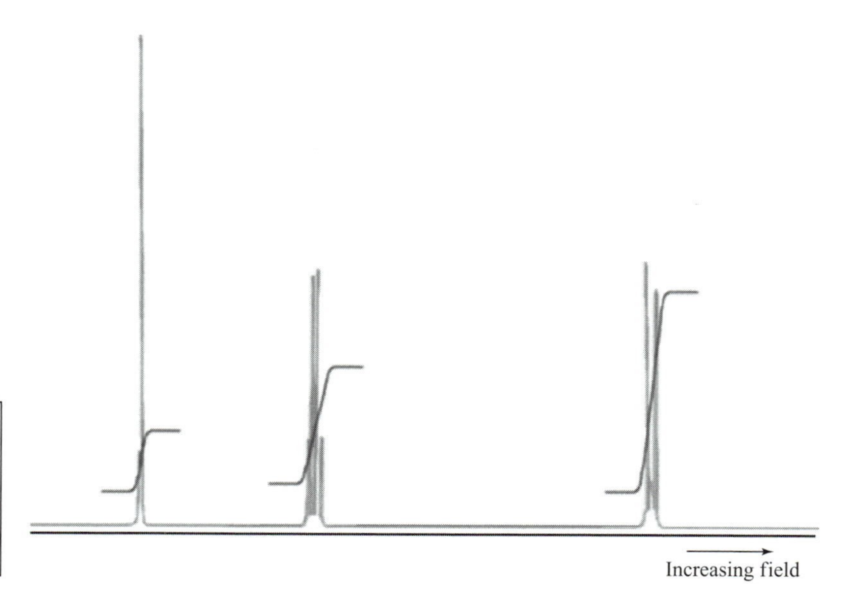

Figure 5.23 Proton NMR of ethanol showing splitting of three distinct protons as well as integrated intensities of the three sets of peaks.

The magnitudes of these J coupling constants depend on the distances R between the two nuclei to the inverse sixth power (i.e., as R^{-6}). They also depend on the γ values of the two interacting nuclei. In the presence of splitting caused by nearby (usually covalently bonded) nuclei, the NMR spectrum of a molecule consists of sets of absorptions (each belonging to a specific nuclear type in a particular chemical environment and thus having a specific chemical shift) that are split by their couplings to the other nuclei. Because of the spin–spin coupling's strong decay with internuclear distance, the magnitude and pattern of the splitting induced on one nucleus by its neighbors provides a clear signature of what the neighboring nuclei are (i.e., through the number of M' values associated with the peak pattern) and how far these nuclei are (through the magnitude of the J constant, knowing it is proportional to R^{-6}). This near-neighbor data, combined with the chemical shift functional group data, offer powerful information about molecular structure.

An example of a full NMR spectrum is given in Fig. 5.23 where the ^{1}H spectrum (i.e., only the proton absorptions are shown) of $H_3C–H_2C–OH$ appears along with plots of the integrated intensities under each set of peaks. The latter data suggests the total number of nuclei corresponding to that group of peaks. Notice how the OH proton's absorption, the absorption of the two equivalent protons on the $–CH_2–$ group, and that of the three equivalent protons in the $–CH_3$ group occur at different field strengths (i.e., have different chemical shifts). Also note how the OH peak is split only slightly because this proton is distant from any others, but the CH_3 protons' peak is split by the neighboring $–CH_2–$ group's protons in an AX2 pattern. Finally, the $–CH_2–$ protons' peak is split by the neighboring $–CH_3$ group's three protons (in an AX3 pattern).

In summary, NMR spectroscopy is a very powerful tool that:

(i) Allows us to extract inter-nuclear distances (or at least tell how many near-neighbor nuclei there are) and thus geometrical information by measuring coupling constants J and subsequently using the theoretical expressions that relate J values to R^{-6} values.
(ii) Allows us to probe the local electronic environment of nuclei inside molecules by measuring chemical shifts or shielding σ_I and then using the theoretical equations relating the two quantities. Knowledge about the electronic environment tells one about the degree of polarity in bonds connected to that nuclear center.
(iii) Tells us, through the splitting patterns associated with various nuclei, the number and nature of the neighbor nuclei, again providing a wealth of molecular structure information.

5.5 Theoretical simulation of structures

We have seen how microwave, infrared, and NMR spectroscopy as well as x-ray diffraction data, when subjected to proper interpretation using the appropriate theoretical equations, can be used to obtain a great deal of structural information about a molecule. As discussed in the Background Material, theory is also used to probe molecular structure in another manner. That is, not only does theory offer the equations that connect the experimental data to the molecular properties, but it also allows one to "simulate" a molecule. This simulation is done by solving the Schrödinger equation for the motions of the electrons to generate a potential energy surface (PES), $E(R)$, after which this energy landscape can be searched for points where the gradients along all directions vanish. An example of such a PES is shown in Fig. 5.24 for a simple case in which

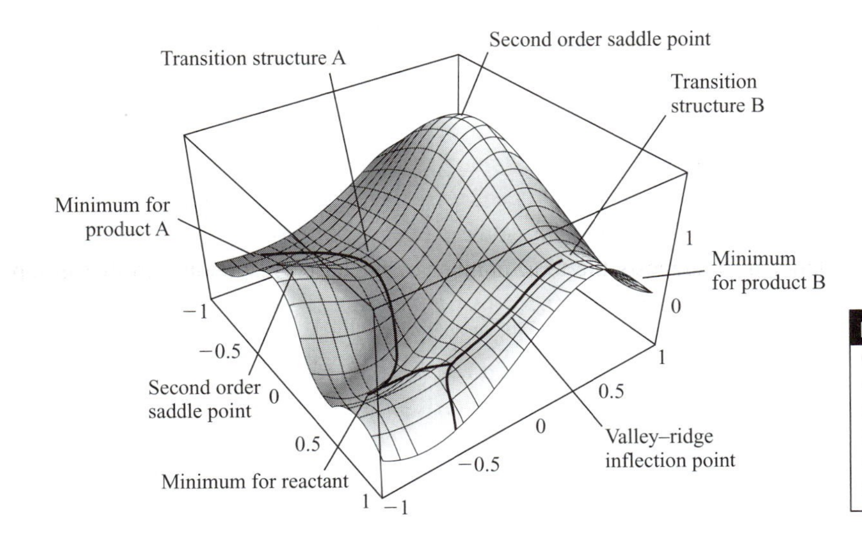

Figure 5.24 Potential energy surface in two dimensions showing reactant and product minima, transition states, and paths connecting them.

the energy depends on only two geometrical parameters. Even in such a case, one can find several local minima and transition state structures connecting them.

Among the "stationary points" on the PES, those at which all eigenvalues of the second derivative (Hessian) matrix are positive represent geometrically stable isomers of the molecule. Those stationary points on the PES at which all but one Hessian eigenvalue are positive and one is negative represent transition state structures that connect pairs of stable isomers.

Once the stable isomers of a molecule lying within some energy interval above the lowest such isomer have been identified, the vibrational motions of the molecule within the neighborhood of each such isomer can be described either by solving the Schrödinger equation for the vibrational wave functions $\chi_v(Q)$ belonging to each normal mode or by solving the classical Newton equations of motion using the gradient $\partial E/\partial Q$ of the PES to compute the forces along each molecular distortion direction Q:

$$F_Q = -\partial E/\partial Q. \tag{5.17}$$

The decision about whether to use the Schrödinger or Newtonian equations to treat the vibrational motion depends on whether one wishes (needs) to properly include quantum effects (e.g., zero-point motion and wave function nodal patterns) in the simulation.

Once the vibrational motions have been described for a particular isomer, and given knowledge of the geometry of that isomer, one can evaluate the moments of inertia, one can properly vibrationally average all of the R^{-2} quantities that enter into these moments, and, hence, one can simulate the microwave spectrum of the molecule. Also, given the Hessian matrix for this isomer, one can form its mass-weighted variant whose non-zero eigenvalues give the normal-mode harmonic frequencies of vibration of that isomer and whose eigenvectors describe the atomic motions that correspond to these vibrations. Moreover, the solution of the electronic Schrödinger equation allows one to compute the NMR shielding σ_I values at each nucleus as well as the spin–spin coupling constants J between pairs of nuclei (the treatment of these subjects is beyond the level of this text; you can find it in *Molecular Electronic Structure Theory* by Helgaker, *et al.*). Again, using the vibrational motion knowledge, one can average the σ and J values over this motion to gain vibrationally averaged σ_I and $J_{I,I'}$ values that best simulate the experimental parameters.

One carries out such a theoretical simulation of a molecule for various reasons. Especially in the "early days" of developing theoretical tools to solve the electronic Schrödinger equation or the vibrational motion problem, one would do so for molecules whose structures and IR and NMR spectra were well known. The purpose in such cases was to calibrate the accuracy of the theoretical methods against well-known experimental data. Now that theoretical tools have been reasonably well tested and can be trusted (within known limits of accuracy), one often uses theoretically simulated structural and spectroscopic properties to identify spectral features whose molecular origin is not known. That is, one compares the theoretical spectra of a variety of test molecules to the observed spectral features to attempt to identify the molecule that produced the spectra.

It is also common to use simulations to examine species that are especially difficult to generate in reasonable quantities in the laboratory and species that do not persist for long times. Reactive radicals, cations and anions are often difficult to generate in the laboratory and may be impossible to retain in sufficient concentrations and for a sufficient duration to permit experimental characterization. In such cases, theoretical simulation of the properties of these molecules may be the most reliable way to access such data.

Chemical change

5.6 Experimental probes of chemical change

Many of the same tools that are used to determine the structures of molecules can be used to follow the changes that the molecule undergoes as it is involved in a chemical reaction. Specifically, for any reaction in which one kind of molecule A is converted into another kind B, one needs to have:

(i) the ability to identify, via some physical measurement, the experimental signatures of both A and B,
(ii) the ability to relate the magnitude of these experimental signals to the concentrations [A] and [B] of these molecules, and
(iii) the ability to monitor these signals as functions of time so that these concentrations can be followed as time evolves.

The third requirement is what allows one to determine the rates at which the A and B molecules are reacting.

Many of the experimental tools used to identify molecules (e.g., NMR allows one to identify functional groups and near-neighbor functional groups, IR also allows functional groups to be seen) and to determine their concentrations have restricted time scales over which they can be used. For example, NMR spectra require that the sample be studied for *c.* 1 second or more to obtain a useable signal. Likewise, a mass spectroscopic analysis of a mixture of reacting species requires many seconds or minutes to carry out. These restrictions, in turn, limit the rates of reactions that can be followed using these experimental tools (e.g., one can not use NMR or mass spectroscopy to follow a reaction that occurs on a time scale of 10^{-12} s).

Especially for very fast reactions and for reactions involving unstable species that can not easily be handled, so-called pump–probe experimental approaches are often used. For example, suppose one were interested in studying the reaction of Cl radicals (e.g., as formed in the decomposition of chlorofluorocarbons (CFCs) by ultraviolet light) with ozone to generate ClO and O_2:

$$Cl + O_3 \rightarrow ClO + O_2. \quad (5.18)$$

One can not simply deposit a known amount of Cl radicals from a vessel into a container in which gaseous O_3 of a known concentration has been prepared; the Cl radicals will recombine and react with other species, making their concentrations difficult to determine. So, alternatively, one places known concentrations of some Cl radical precursor (e.g., a CFC or some other X-Cl species) and ozone into a reaction vessel. One then uses, for example, a very short light pulse whose photon's frequencies are tuned to a transition that will cause the X-Cl precursor to undergo rapid photodissociation:

$$h\nu + \text{X-Cl} \rightarrow X + Cl. \quad (5.19)$$

Because the "pump" light source used to prepare the Cl radicals is of very short duration (δt) and because the X-Cl dissociation is prompt, one knows, to within δt, the time at which the Cl radicals begin to react with the ozone. The initial concentration of the Cl radicals can be known if the quantum yield for the $h\nu$ + X-Cl $\rightarrow$ X + Cl reaction is known. This means that the intensity of photons, the probability of photon absorption by X-Cl, and the fraction of excited X-Cl molecules that dissociate to produce X + Cl must be known. Such information is available (albeit, from rather tedious earlier studies) for a variety of X-Cl precursors.

So, knowing the time at which the Cl radicals are formed and their initial concentrations, one then allows the Cl + O_3 + $h\nu \rightarrow$ ClO + O_2 reaction to proceed for some time duration Δt. One then, at $t = \Delta t$, uses a second light source to "probe" either the concentration of the ClO, the O_2 or the O_3, to determine the extent of progress of the reaction. Which species is so monitored depends on the availability of light sources whose frequencies these species absorb. Such probe experiments are carried out at a series of time delays Δt, the result of which is the determination of the concentrations of some product or reactant species at various times after the initial pump event created the reactive Cl radicals. In this way, one can monitor, for example, the ClO concentration as a function of time after the Cl begins to react with the O_3. If one has reason to believe that the reaction occurs in a single bimolecular event as

$$Cl + O_3 \rightarrow ClO + O_2 \quad (5.20)$$

one can then extract the rate constant k for the reaction by using the following kinetic scheme:

$$d[ClO]/dt = k[Cl][O_3]. \quad (5.21)$$

If the initial concentration of O_3 is large compared to the amount of Cl that is formed in the pump event, $[O_3]$ can be taken as constant and known. If the initial concentration of Cl is denoted $[Cl]_0$, and the concentration of ClO is called x, this kinetic equation reduces to

$$dx/dt = k([Cl]_0 - x)[O_3], \tag{5.22}$$

the solution of which is

$$[ClO] = x = [Cl]_0\{1 - \exp(-k[O_3]t)\}. \tag{5.23}$$

So, knowing the [ClO] concentration as a function of time delay t, and knowing the initial ozone concentration $[O_3]$ as well as the initial Cl radical concentration, one can find the rate constant k.

Such pump–probe experiments are necessary when one wants to study species that must be generated and allowed to react immediately. This is essentially always the case when one or more of the reactants is a highly reactive species such as a radical. There is another kind of experiment that can be used to probe very fast reactions if the reaction and its reverse reaction can be brought into equilibrium to the extent that reactants and products both exist in measurable concentrations. For example, consider the reaction of an enzyme E and a substrate S to form the enzyme-substrate complex ES:

$$E + S \Leftrightarrow ES. \tag{5.24}$$

At equilibrium, the forward rate

$$k_f[E]_{eq}[S]_{eq} \tag{5.25}$$

and the reverse rate

$$k_r[ES]_{eq} \tag{5.26}$$

are equal:

$$k_f[E]_{eq}[S]_{eq} = k_r[ES]_{eq}. \tag{5.27}$$

The idea behind so called "perturbation techniques" is to begin with a reaction that is in such an equilibrium condition and to then use some external means to slightly perturb the equilibrium. Because both the forward and reverse rates are assumed to be very fast, it is essential to use a perturbation that can alter the concentrations very quickly. This usually precludes simply adding a small amount of one or more of the reacting species to the reaction vessel. Instead, one usually employs a fast light source or electric field pulse to perturb the equilibrium to one side or the other. For example, if the reaction thermochemistry is known, the equilibrium constant K_{eq} can be changed by rapidly heating the sample (e.g, with a fast laser pulse that is absorbed and rapidly heats the sample) and using

$$d \ln K_{eq}/dT = \Delta H/(RT^2) \tag{5.28}$$

to calculate the change in K_{eq} and thus the changes in concentrations caused by the sudden heating. Alternatively, if the polarity of the reactants and products is substantially different, one may use a rapidly applied electric field to quickly change the concentrations of the reactant and product species.

In such experiments, the concentrations of the species are "shifted" by a small amount δ as a result of the application of the perturbation, so that

$$[\mathrm{ES}] = [\mathrm{ES}]_{\mathrm{eq}} - \delta, \tag{5.29}$$

$$[\mathrm{E}] = [\mathrm{E}]_{\mathrm{eq}} + \delta, \tag{5.30}$$

$$[\mathrm{S}] = [\mathrm{S}]_{\mathrm{eq}} + \delta, \tag{5.31}$$

once the perturbation has been applied and then turned off. Subsequently, the following rate law will govern the time evolution of the concentration change δ:

$$-d\delta/dt = -k_{\mathrm{r}}([\mathrm{ES}]_{\mathrm{eq}} - \delta) + k_{\mathrm{f}}\,([\mathrm{E}]_{\mathrm{eq}} + \delta)([\mathrm{S}]_{\mathrm{eq}} + \delta). \tag{5.32}$$

Assuming that δ is very small (so that the term involving δ^2 can be neglected) and using the fact that the forward and reverse rates balance at equilibrium, this equation for the time evolution of δ can be reduced to

$$-d\delta/dt = (k_{\mathrm{r}} + k_{\mathrm{f}}\,[\mathrm{S}]_{\mathrm{eq}} + k_{\mathrm{f}}\,[\mathrm{E}_{\mathrm{eq}}])\delta. \tag{5.33}$$

So, the concentration deviations from equilibrium will return to equilibrium (i.e., δ will decay to zero) exponentially with an effective rate coefficient that is equal to a sum of terms:

$$k_{\mathrm{eff}} = k_{\mathrm{r}} + k_{\mathrm{f}}\,[\mathrm{S}]_{\mathrm{eq}} + k_{\mathrm{f}}\,[\mathrm{E}_{\mathrm{eq}}] \tag{5.34}$$

involving both the forward and reverse rate constants.

So, by quickly perturbing an equilibrium reaction mixture for a short period of time and subsequently following the concentrations of the reactants or products as they return to their equilibrium values, one can extract the effective rate coefficient k_{eff}. Doing this at a variety of different initial equilibrium concentrations (e.g., $[\mathrm{S}]_{\mathrm{eq}}$ and $[\mathrm{E}]_{\mathrm{eq}}$), and seeing how k_{eff} changes, one can then determine both the forward and reverse rate constants.

Both the pump-probe and the perturbation methods require that one be able to quickly create (or perturb) concentrations of reactive species and that one have available an experimental probe that allows one to follow the concentrations of at least some of the species as time evolves. Clearly, for very fast reactions, this means that one must use experimental tools that can respond on a very short time scale. Modern laser technology and molecular beam methods have provided the most widely used of such tools. These experimental approaches are discussed in some detail in Chapter 8.

5.7 Theoretical simulation of chemical change

The most common theoretical approach to simulating a chemical reaction is to use Newtonian dynamics to follow the motion on a Born–Oppenheimer electronic energy surface. If the molecule of interest contains few (N) atoms, such a surface could be computed (using the methods discussed in Chapter 6) at a large number of molecular geometries $\{Q_K\}$ and then fit to an analytical function $E(\{q_J\})$ of the $3N - 6$ or $3N - 5$ variables denoted $\{q_J\}$. Knowing E as a function of these variables, one can then compute the forces

$$F_J = -\partial E/\partial q_J \tag{5.35}$$

along each coordinate, and then use the Newton equations

$$m_J d^2 q_J/dt^2 = F_J \tag{5.36}$$

to follow the time evolution of these coordinates and hence the progress of the reaction. The values of the coordinates $\{q_J(t_L)\}$ at a series of discrete times t_L constitute what is called a classical trajectory. To simulate a chemical reaction, one begins the trajectory with initial coordinates characteristic of the reactant species (i.e., within one of the "valleys" on the reactant side of the potential surface) and one follows the trajectory long enough to determine whether the collision results in:

(i) a non-reactive outcome characterized by final coordinates describing reactant not product molecules, or
(ii) a reactive outcome that is recognized by the final coordinates describing product molecules rather than reactants.

However, if the molecule contains more than three or four atoms, it is more common to not compute the Born–Oppenheimer energy at a set of geometries and then fit this data to an analytical form. Instead, one begins a trajectory at some initial coordinates $\{q_J(0)\}$ and with some initial momenta $\{p_J(0)\}$ and then uses the Newton equations, usually in the finite-difference form:

$$q_J = q_J(0) + (p_J(0)/m_J)\delta t, \tag{5.37}$$

$$p_J = p_J(0) - (\partial E/\partial q_J)(t = 0)\delta t, \tag{5.38}$$

to propagate the coordinates and momenta forward in time by a small amount δt. Here, $(\partial E/\partial q_J)(t = 0)$ denotes the gradient of the BO energy computed at the $\{q_J(0)\}$ values of the coordinates. The above propagation procedure is then used again, but with the values of q_J and p_J appropriate to time $t = \delta t$ as "new" $t = 0$ coordinates and momenta, to generate yet another set of $\{q_J\}$ and $\{p_J\}$ values. In such "direct dynamics" approaches, the energy gradients, which produce the forces, are computed only at geometries that the classical trajectory encounters along its time propagation. In the earlier procedure, in which the BO energy is fit to an analytical form, one often computes E at geometries that the trajectory never accesses.

In carrying out such a classical trajectory simulation of a chemical reaction, there are other issues that must be addressed. In particular, one can essentially never use any single trajectory to simulate a reaction carried out in a laboratory setting. One must perform a series of such trajectory calculations with a variety of different initial coordinates and momenta chosen in a manner to represent the experimental conditions of interest. For example, suppose one were to wish to model a molecular beam experiment in which a beam of species A having a well-defined kinetic energy E_A collides with a beam of species B having kinetic energy E_B as shown in Fig. 5.25. Even though the A and B molecules all collide at right angles and with specified kinetic energies (and thus specified initial momenta), not all of these collisions occur "head on". Fig 5.26 illustrates this point. Here, we show two collisions between an A and a B molecule, both of which have identical A and B velocities V_A and V_B, respectively. What differs in the two events is their distance of closest approach. In the collision shown on

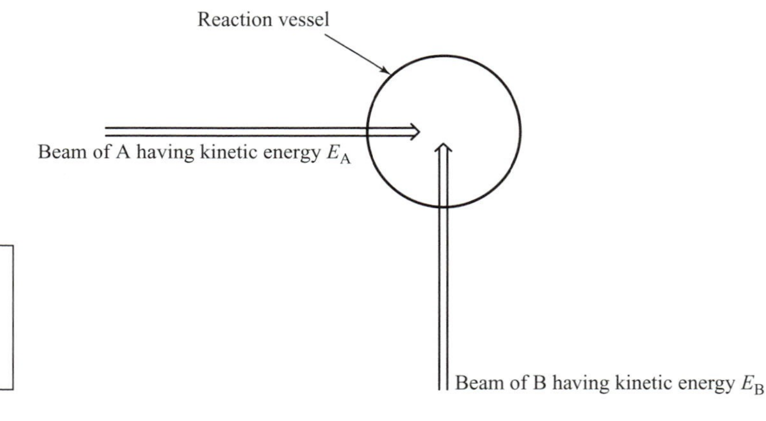

Figure 5.25 Crossed beam experiment in which A and B molecules collide in a reaction vessel.

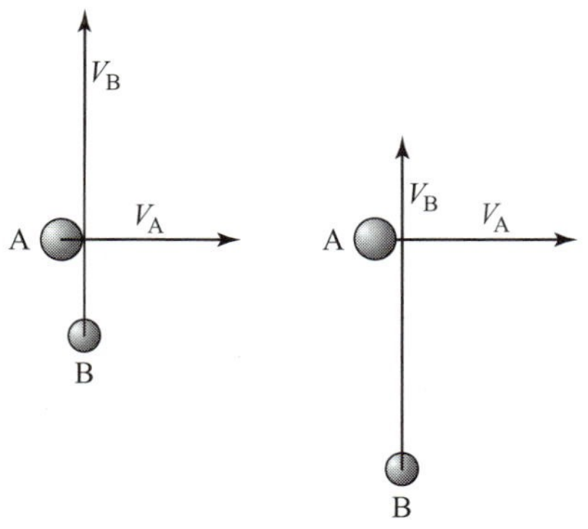

V_B V_A A B V_B V_A A B

Figure 5.26 Two A + B collisions. In the first, the A and B have a small distance of closest approach; in the second this distance is larger.

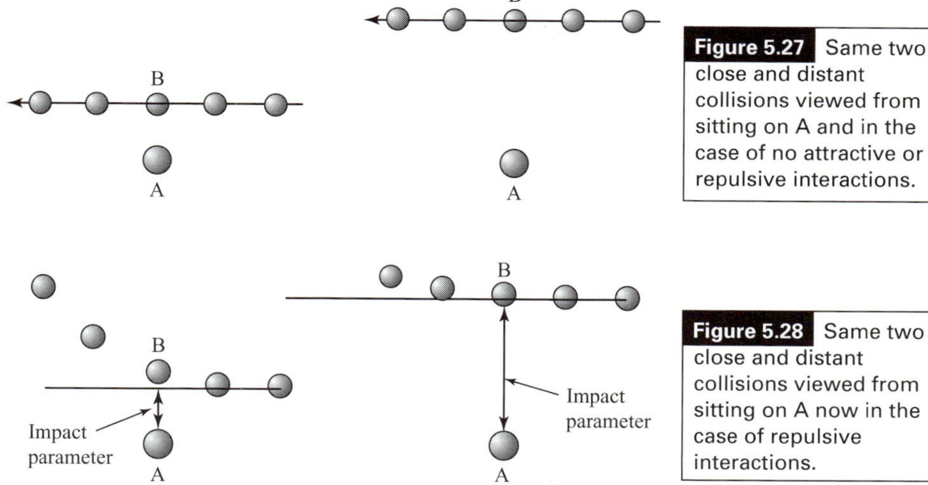

Figure 5.27 Same two close and distant collisions viewed from sitting on A and in the case of no attractive or repulsive interactions.

Figure 5.28 Same two close and distant collisions viewed from sitting on A now in the case of repulsive interactions.

the left, the A and B come together closely. However, in the left collision, the A molecule is moving away from the region where B would strike it before B has reached it. These two cases can be viewed from a different perspective that helps to clarify their differences. In Fig. 5.27, we illustrate these two collisions viewed from a frame of reference located on the A molecule. In this figure, we show the location of the B molecule relative to A at a series of times, showing B moving from right to left. In the figure on the left, the B molecule clearly undergoes a closer collision than is the case on the right. The distance of closest approach in each case is called the impact parameter and it represents the distance of closest approach if the colliding partners did not experience any attractive or repulsive interactions (as the above figures would be consistent with). Of course, when A and B have forces acting between them, the trajectories shown above would be modified to look more like those shown in Fig. 5.28. In both of these trajectories, repulsive intermolecular forces cause the trajectory to move away from its initial path which defines the respective impact parameters.

So, even in this molecular beam example in which both colliding molecules have well-specified velocities, one must carry out a number of classical trajectories, each with a different impact parameter b to simulate the laboratory event. In practice, the impact parameters can be chosen to range from $b = 0$ (i.e., a "head on" collision) to some maximum value b_{max} beyond which the A and B molecules no longer interact (and thus can no longer undergo reaction). Each trajectory is followed long enough to determine whether it leads to geometries characteristic of the product molecules. The fraction of such trajectories, weighted by the volume element $2\pi\ b\ db$ for trajectories with impact parameters in the range between b and $b + db$, then gives the averaged fraction of trajectories that react.

In most simulations of chemical reactions, there are more initial conditions that also must be sampled (i.e., trajectories with a variety of initial variables must be followed) and properly weighted. For example:

(i) if there is a range of velocities for the reactants A and/or B, one must follow trajectories with velocities in this range and weigh the outcomes (i.e., reaction or not) of such trajectories appropriately (e.g., with a Maxwell–Boltzmann weighting factor), and
(ii) if the reactant molecules have internal bond lengths, angles, and orientations, one must follow trajectories with different initial values of these variables and properly weigh each such trajectory (e.g., using the vibrational state's coordinate probability distribution as a weighting factor for the initial values of that coordinate).

As a result, to properly simulate a laboratory experiment of a chemical reaction, it usually requires one to follow a very large number of classical trajectories. Fortunately, such a task is well suited to distributed parallel computing, so it is currently feasible to do so even for rather complex reactions.

There is a situation in which the above classical trajectory approach can be foolish to pursue, even if there is reason to believe that a classical Newton description of the nuclear motions is adequate. This occurs when one has a rather high barrier to surmount to evolve from reactants to products and when the fraction of trajectories whose initial conditions permit this barrier to be accessed is very small. In such cases, one is faced with the reactive trajectories being very "rare" among the full ensemble of trajectories needed to properly simulate the laboratory experiment. Certainly, one can apply the trajectory following technique outlined above, but if one observes, for example, that only one trajectory in 10^6 produces a reaction, one does not have adequate statistics to determine the reaction probability. One could subsequently run 10^8 trajectories (chosen again to represent the same experiment), and see whether 100 or 53 or 212 of these trajectories react, thereby increasing the precision of your reaction probability. However, it may be computationally impractical to perform 100 times as many trajectories to achieve better accuracy in the reaction probability.

When faced with such rare-event situations, one is usually better off using an approach that breaks the problem of determining what fraction of the (properly weighted) initial conditions produce reaction into two parts:

(i) among all of the (properly weighted) initial conditions, what fraction can access the high-energy barrier? and
(ii) of those that do access the high barrier, how many react?

This way of formulating the reaction probability question leads to the transition state theory (TST) method that is treated in detail in Chapter 8, along with some of its more common variants.

Briefly, the answer to the first question posed above involves computing the quasi-equilibrium fraction of reacting species that reach the barrier region in terms of the partition functions of statistical mechanics. This step becomes practical if the chemical reactants can be assumed to be in some form of thermal equilibrium (which is where these kinds of models are useful). In the simplest form of TST, the answer to the second question posed above is taken to be "all trajectories that reach the barrier react". In more sophisticated variants, other models are introduced to take into consideration that not all trajectories that cross over the barrier indeed proceed onward to products and that some trajectories may tunnel through the barrier near its top. I will leave further discussion of the TST to Chapter 8.

In addition to the classical trajectory and TST approaches to simulating chemical reactions, there are more quantum approaches. These techniques should be used when the nuclei involved in the reaction include hydrogen or deuterium nuclei. A discussion of the details involved in quantum propagation is beyond the level of this chapter, so I will delay it until Chapter 8.

Chapter 6
Electronic structures

Electrons are the "glue" that holds the nuclei together in the chemical bonds of molecules and ions. Of course, it is the nuclei's positive charges that bind the electrons to the nuclei. The competitions among Coulomb repulsions and attractions as well as the existence of non-zero electronic and nuclear kinetic energies make the treatment of the full electronic–nuclear Schrödinger equation an extremely difficult problem. Electronic structure theory deals with the quantum states of the electrons, usually within the Born–Oppenheimer approximation (i.e., with the nuclei held fixed). It also addresses the forces that the electrons' presence creates on the nuclei; it is these forces that determine the geometries and energies of various stable structures of the molecule as well as transition states connecting these stable structures. Because there are ground and excited electronic states, each of which has different electronic properties, there are different stable-structure and transition-state geometries for each such electronic state. Electronic structure theory deals with all of these states, their nuclear structures, and the spectroscopies (e.g., electronic, vibrational, rotational) connecting them.

Theoretical treatment of electronic structure: atomic and molecular orbital theory

In Chapter 5's discussion of molecular structure, I introduced you to the strategies that theory uses to interpret experimental data relating to such matters, and how and why theory can also be used to simulate the behavior of molecules. In carrying out simulations, the Born–Oppenheimer electronic energy $E(R)$ as a function of the $3N$ coordinates of the N atoms in the molecule plays a central role. It is on this landscape that one searches for stable isomers and transition states, and it is the second derivative (Hessian) matrix of this function that provides the harmonic vibrational frequencies of such isomers. In the present chapter, I want to provide you with an introduction to the tools that we use to solve the electronic Schrödinger equation to generate $E(R)$ and the electronic wave function $\Psi(r \mid R)$. In essence, this treatment will focus on orbitals of atoms and molecules and how we obtain and interpret them.

For an atom, one can approximate the orbitals by using the solutions of the hydrogenic Schrödinger equation discussed in the Background Material. Although such functions are not proper solutions to the actual N-electron Schrödinger equation (believe it or not, no one has ever solved exactly any such equation for $N > 1$) of any atom, they can be used as perturbation or variational starting-point approximations when one may be satisfied with qualitatively accurate answers. In particular, the solutions of this one-electron hydrogenic problem form the qualitative basis for much of atomic and molecular orbital theory. As discussed in detail in the Background Material, these orbitals are labeled by n, l, and m quantum numbers for the bound states and by l and m quantum numbers and the energy E for the continuum states.

Much as the particle-in-a-box orbitals are used to qualitatively describe π-electrons in conjugated polyenes or electronic bands in solids, these so-called hydrogen-like orbitals provide qualitative descriptions of orbitals of atoms with more than a single electron. By introducing the concept of screening as a way to represent the repulsive interactions among the electrons of an atom, an effective nuclear charge $Z_{\rm eff}$ can be used in place of Z in the hydrogenic $\psi_{n,l,m}$ and $E_{n,l}$ formulas of the Background Material to generate approximate atomic orbitals to be filled by electrons in a many-electron atom. For example, in the crudest approximation of a carbon atom, the two 1s electrons experience the full nuclear attraction, so $Z_{\rm eff} = 6$ for them, whereas the 2s and 2p electrons are screened by the two 1s electrons, so $Z_{\rm eff} = 4$ for them. Within this approximation, one then occupies two 1s orbitals with $Z = 6$, two 2s orbitals with $Z = 4$ and two 2p orbitals with $Z = 4$ in forming the full six-electron product wave function of the lowest-energy state of carbon:

$$\Psi(1, 2, \ldots, 6) = \psi_{1s}\alpha(1)\psi_{1s}\beta(2)\psi_{2s}\alpha(3)\ldots\psi_{2p(0)}\beta(6). \tag{6.1}$$

However, such approximate orbitals are not sufficiently accurate to be of use in quantitative simulations of atomic and molecular structure. In particular, their energies do not properly follow the trends in atomic orbital (AO) energies that are taught in introductory chemistry classes and that are shown pictorially in Fig. 6.1. For example, the relative energies of the 3d and 4s orbitals are not adequately described in a model that treats electron repulsion effects in terms of a simple screening factor. So, now it is time to examine how we can move beyond the screening model and take the electron repulsion effects, which cause the inter-electronic couplings that render the Schrödinger equation insoluble, into account in a more reliable manner.

6.1 Orbitals

6.1.1 The Hartree description

The energies and wave functions within the most commonly used theories of atomic structure are assumed to arise as solutions of a Schrödinger equation

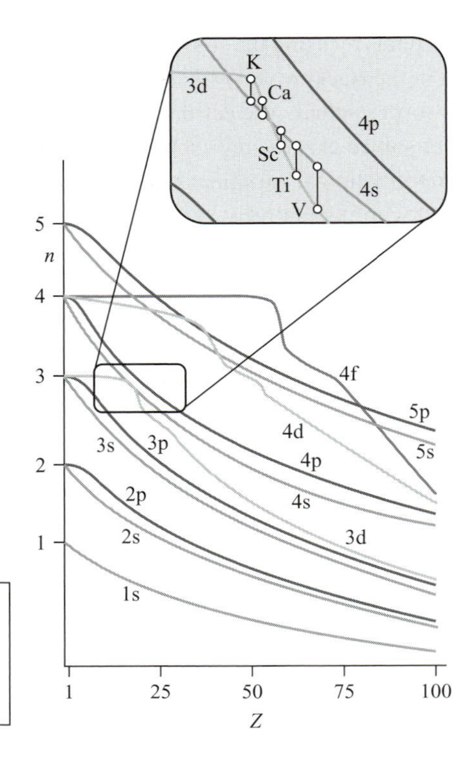

Figure 6.1 Energies of atomic orbitals as functions of nuclear charge for neutral atoms.

whose Hamiltonian $h_e(r)$ possess three kinds of energies:

(i) Kinetic energy, whose average value is computed by taking the expectation value of the kinetic energy operator $-\hbar^2/2m\nabla^2$ with respect to any particular solution $\phi_J(r)$ to the Schrödinger equation: $\text{KE} = \langle\phi_J| - \hbar^2/2m\nabla^2|\phi_J\rangle$;

(ii) Coulomb attraction energy with the nucleus of charge Z: $\langle\phi_J| - Ze^2/r|\phi_J\rangle$;

(iii) Coulomb repulsion energies with all of the $n - 1$ other electrons, which are assumed to occupy other atomic orbitals (AOs) denoted ϕ_K, with this energy computed as $\sum_K \langle\phi_J(r)\phi_K(r')|(e^2/|r - r'|)|\phi_J(r)\phi_K(r')\rangle$.

The so-called Dirac notation $\langle\phi_J(r)\phi_K(r')|(e^2/|r - r'|)|\phi_J(r)\phi_K(r')\rangle$ is used to represent the six-dimensional Coulomb integral $J_{J,K} = \int |\phi_J(r)|^2|\phi_K(r')|^2 (e^2/|r - r'|)drdr'$ that describes the Coulomb repulsion between the charge density $|\phi_J(r)|^2$ for the electron in ϕ_J and the charge density $|\phi_K(r')|^2$ for the electron in ϕ_K. Of course, the sum over K must be limited to exclude $K = J$ to avoid counting a "self-interaction" of the electron in orbital ϕ_J with itself.

The total energy ε_J of the orbital ϕ_J is the sum of the above three contributions:

$$\varepsilon_J = \langle\phi_J| - \frac{\hbar^2}{2m}\nabla^2|\phi_J\rangle + \langle\phi_J| - \frac{Ze^2}{|r|}|\phi_J\rangle + \sum_K \langle\phi_J(r)\phi_K(r')|\frac{e^2}{|r - r'|}|\phi_J(r)\phi_K(r')\rangle. \quad (6.2)$$

This treatment of the electrons and their orbitals is referred to as the Hartree-level of theory. As stated above, when screened hydrogenic AOs are used to approximate the ϕ_J and ϕ_K orbitals, the resultant ε_J values do not produce accurate predictions. For example, the negative of ε_J should approximate the ionization energy for removal of an electron from the AO ϕ_J. Such ionization potentials (IPs) can be measured, and the measured values do not agree well with the theoretical values when a crude screening approximation is made for the AOs.

6.1.2 The LCAO expansion

To improve upon the use of screened hydrogenic AOs, it is most common to approximate each of the Hartree AOs $\{\phi_K\}$ as a linear combination of so-called basis AOs $\{\chi_\mu\}$:

$$\phi_J = \sum_\mu C_{J,\mu}\chi_\mu, \tag{6.3}$$

using what is termed the linear-combination-of-atomic-orbitals (LCAO) expansion. In this equation, the expansion coefficients $\{C_{J,\mu}\}$ are the variables that are to be determined by solving the Schrödinger equation

$$h_e\phi_J = \varepsilon_J\phi_J. \tag{6.4}$$

After substituting the LCAO expansion for ϕ_J into this Schrödinger equation, multiplying on the left by one of the basis AOs χ_ν, and then integrating over the coordinates of the electron in ϕ_J, one obtains

$$\sum_\mu \langle\chi_\nu|h_e|\chi_\mu\rangle C_{J,\mu} = \varepsilon_J \sum_\mu \langle\chi_\nu \mid \chi_\mu\rangle C_{J,\mu}. \tag{6.5}$$

This is a matrix eigenvalue equation in which the ε_J and $\{C_{J,\mu}\}$ appear as eigenvalues and eigenvectors. The matrices $\langle\chi_\nu|h_e|\chi_\mu\rangle$ and $\langle\chi_\nu \mid \chi_\mu\rangle$ are called the Hamiltonian and overlap matrices, respectively. An explicit expression for the former is obtained by introducing the earlier definition of h_e:

$$\begin{aligned}\langle\chi_\nu|h_e|\chi_\mu\rangle &= \langle\chi_\nu| - \frac{\hbar^2}{2m}\nabla^2|\chi_\mu\rangle + \langle\chi_\nu| - \frac{Ze^2}{|r|}|\chi_\mu\rangle \\ &\quad + \sum_{K,\eta,\gamma} C_{K,\eta}C_{K,\gamma}\langle\chi_\nu(r)\chi_\eta(r')|\left(\frac{e^2}{|r-r'|}\right)|\chi_\mu(r)\chi_\gamma(r')\rangle.\end{aligned} \tag{6.6}$$

An important thing to notice about the form of the matrix Hartree equations is that to compute the Hamiltonian matrix, one must know the LCAO coefficients $\{C_{K,\gamma}\}$ of the orbitals which the electrons occupy. On the other hand, these LCAO coefficients are supposed to be found by solving the Hartree matrix eigenvalue equations. This paradox leads to the need to solve these equations iteratively in a so-called self-consistent field (SCF) technique. In the SCF process, one inputs an initial approximation to the $\{C_{K,\gamma}\}$ coefficients. This then allows one to form the Hamiltonian matrix defined above. The Hartree matrix equations $\sum_\mu \langle \chi_\nu | h_e | \chi_\mu \rangle C_{J,\mu} = \varepsilon_J \sum_\mu \langle \chi_\nu \mid \chi_\mu \rangle C_{J,\mu}$ are then solved for "new" $\{C_{K,\gamma}\}$ coefficients and for the orbital energies $\{\varepsilon_K\}$. The new LCAO coefficients of those orbitals that are occupied are then used to form a "new" Hamiltonian matrix, after which the Hartree equations are again solved for another generation of LCAO coefficients and orbital energies. This process is continued until the orbital energies and LCAO coefficients obtained in successive iterations do not differ appreciably. Upon such convergence, one says that a self-consistent field has been realized because the $\{C_{K,\gamma}\}$ coefficients are used to form a Coulomb field potential that details the electron–electron interactions.

6.1.3 AO basis sets

STOs and GTOs

As noted above, it is possible to use the screened hydrogenic orbitals as the $\{\chi_\mu\}$. However, much effort has been expended at developing alternative sets of functions to use as basis orbitals. The result of this effort has been to produce two kinds of functions that currently are widely used.

The basis orbitals commonly used in the LCAO process fall into two primary classes:

(i) Slater-type orbitals (STOs) $\chi_{n,l,m}(r, \theta, \phi) = N_{n,l,m,\zeta} Y_{l,m}(\theta, \phi) r^{n-1} \exp^{(-\zeta r)}$ are characterized by quantum numbers n, l, and m and exponents (which characterize the orbital's radial "size") ζ. The symbol $N_{n,l,m,\zeta}$ denotes the normalization constant.

(ii) Cartesian Gaussian-type orbitals (GTOs) $\chi_{a,b,c}(r, \theta, \phi) = N'_{a,b,c,\alpha} x^a y^b z^c \exp(-\alpha r^2)$ are characterized by quantum numbers a, b, and c, which detail the angular shape and direction of the orbital, and exponents α which govern the radial "size".

For both types of AOs, the coordinates r, θ, and ϕ refer to the position of the electron relative to a set of axes attached to the nucleus on which the basis orbital is located. Note that Slater-type orbitals (STOs) are similar to hydrogenic orbitals in the region close to the nucleus. Specifically, they have a non-zero slope near the nucleus (i.e., $d/dr(\exp(-\zeta r))_{r=0} = -\zeta$). In contrast, GTOs have zero slope near $r = 0$ because $d/dr(\exp(-\alpha r^2))_{r=0} = 0$. We say that STOs display a "cusp" at $r = 0$ that is characteristic of the hydrogenic solutions, whereas GTOs do not.

Although STOs have the proper "cusp" behavior near nuclei, they are used primarily for atomic and linear-molecule calculations because the multi-center integrals $\langle \chi_\mu(1)\chi_\kappa(2)|e^2/|r_1 - r_2\|\chi_\nu(1)\chi_\gamma(2)\rangle$ which arise in polyatomic-molecule calculations (we will discuss these integrals later in this chapter) can not efficiently be evaluated when STOs are employed. In contrast, such integrals can routinely be computed when GTOs are used. This fundamental advantage of GTOs has led to the dominance of these functions in molecular quantum chemistry.

To overcome the primary weakness of GTO functions (i.e., their radial derivatives vanish at the nucleus), it is common to combine two, three, or more GTOs, with combination coefficients which are fixed and not treated as LCAO parameters, into new functions called contracted GTOs or CGTOs. Typically, a series of radially tight, medium, and loose GTOs are multiplied by contraction coefficients and summed to produce a CGTO which approximates the proper "cusp" at the nuclear center (although no such combination of GTOs can exactly produce such a cusp because each GTO has zero slope at $r = 0$).

Although most calculations on molecules are now performed using Gaussian orbitals, it should be noted that other basis sets can be used as long as they span enough of the regions of space (radial and angular) where significant electron density resides. In fact, it is possible to use plane wave orbitals of the form $\chi(r, \theta, \phi) = N \exp[\mathrm{i}(k_x r \sin\theta \cos\phi + k_y r \sin\theta \sin\phi + k_z r \cos\theta)]$, where N is a normalization constant and k_x, k_y, and k_z are quantum numbers detailing the momenta of the orbital along the x, y, and z Cartesian directions. The advantage to using such "simple" orbitals is that the integrals one must perform are much easier to handle with such functions. The disadvantage is that one must use many such functions to accurately describe sharply peaked charge distributions of, for example, inner-shell core orbitals.

Much effort has been devoted to developing and tabulating in widely available locations sets of STO or GTO basis orbitals for main-group elements and transition metals. This ongoing effort is aimed at providing standard basis set libraries which:

(i) Yield predictable chemical accuracy in the resultant energies.
(ii) Are cost effective to use in practical calculations.
(iii) Are relatively transferable so that a given atom's basis is flexible enough to be used for that atom in various bonding environments (e.g., hybridization and degree of ionization).

The fundamental core and valence basis

In constructing an atomic orbital basis, one can choose from among several classes of functions. First, the size and nature of the primary core and valence basis must be specified. Within this category, the following choices are common:

(i) A minimal basis in which the number of CGTO orbitals is equal to the number of core and valence atomic orbitals in the atom.
(ii) A double-zeta (DZ) basis in which twice as many CGTOs are used as there are core and valence atomic orbitals. The use of more basis functions is motivated by a desire to provide additional variational flexibility so the LCAO process can

generate molecular orbitals of variable diffuseness as the local electronegativity of the atom varies.

(iii) A triple-zeta (TZ) basis in which three times as many CGTOs are used as the number of core and valence atomic orbitals (of course, there are quadruple-zeta and higher-zeta bases also).

Optimization of the orbital exponents (ζs or αs) and the GTO-to-CGTO contraction coefficients for the kind of bases described above have undergone explosive growth in recent years. The theory group at the Pacific Northwest National Labs (PNNL) offer a World Wide Web site from which one can find (and even download in a form prepared for input to any of several commonly used electronic structure codes) a wide variety of Gaussian atomic basis sets. This site can be accessed at http://www.emsl.pnl.gov:2080/forms/basisform.html.

Polarization functions

One usually enhances any core and valence basis set with a set of so-called polarization functions. They are functions of one higher angular momentum than appears in the atom's valence orbital space (e.g, d-functions for C, N, and O and p-functions for H), and they have exponents (ζ or α) which cause their radial sizes to be similar to the sizes of the valence orbitals (i.e., the polarization p orbitals of the H atom are similar in size to the 1s orbital). Thus, they are not orbitals which describe the atom's valence orbital with one higher l-value; such higher-l valence orbitals would be radially more diffuse.

The primary purpose of polarization functions is to give additional angular flexibility to the LCAO process in forming bonding orbitals between pairs of valence atomic orbitals. This is illustrated in Fig. 6.2 where polarization d_π orbitals on C and O are seen to contribute to formation of the bonding π orbital of a carbonyl group by allowing polarization of the carbon atom's p_π orbital toward the right and of the oxygen atom's p_π orbital toward the left. Polarization functions are essential in strained ring compounds because they provide the angular flexibility needed to direct the electron density into regions between bonded atoms, but they are also important in unstrained compounds when high accuracy is required.

Diffuse functions

When dealing with anions or Rydberg states, one must further augment the AO basis set by adding so-called diffuse basis orbitals. The valence and polarization functions described above do not provide enough radial flexibility to adequately describe either of these cases. The PNNL web site data base cited above offers a good source for obtaining diffuse functions appropriate to a variety of atoms.

Once one has specified an atomic orbital basis for each atom in the molecule, the LCAO-MO procedure can be used to determine the $C_{\mu,i}$ coefficients that describe the occupied and virtual (i.e., unoccupied) orbitals. It is important to

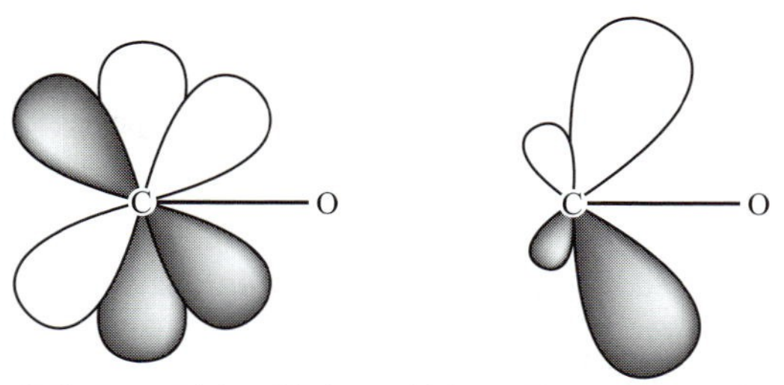

Carbon p_π and d_π orbitals combining to form a bent π orbital

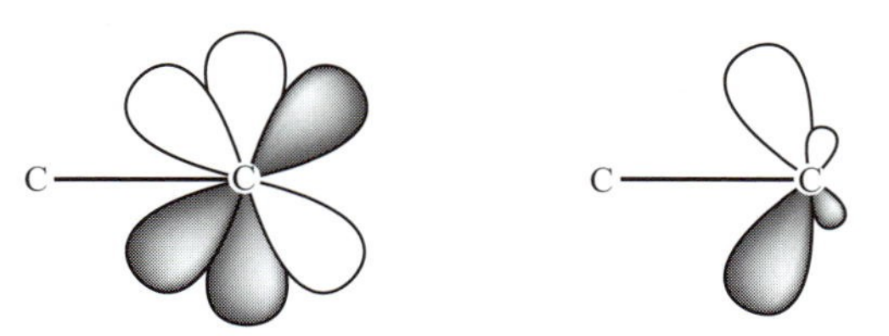

Oxygen p_π and d_π orbitals combining to form a bent π orbital

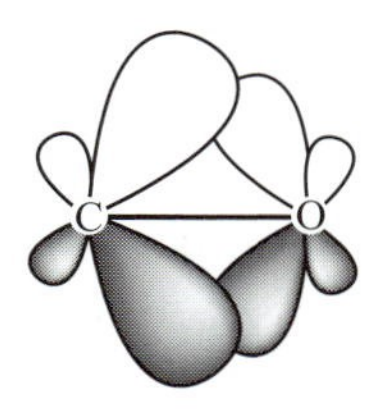

π bond formed from C and O bent (polarized) AOs

Figure 6.2 Oxygen and carbon form a π bond that uses the polarization functions on each atom.

keep in mind that the basis orbitals are not themselves the SCF orbitals of the isolated atoms; even the proper atomic orbitals are combinations (with atomic values for the $C_{\mu,i}$ coefficients) of the basis functions. The LCAO-MO-SCF process itself determines the magnitudes and signs of the $C_{\mu,i}$. In particular, it is alternations in the signs of these coefficients that allow radial nodes to form.

6.1.4 The Hartree–Fock approximation

Unfortunately, the Hartree approximation discussed above ignores an important property of electronic wave functions – their permutational antisymmetry. The full Hamiltonian

$$H = \sum_j \left(-\frac{\hbar^2}{2m} \nabla_j^2 - \frac{Ze^2}{r_j} \right) + \frac{1}{2} \sum_{j,k} \frac{e^2}{|r_j - r_k|} \tag{6.7}$$

is invariant (i.e., is left unchanged) under the operation $P_{i,j}$ in which a pair of electrons have their labels (i, j) permuted. We say that H commutes with the permutation operator $P_{i,j}$. This fact implies that any solution Ψ to $H\Psi = E\Psi$ must also be an eigenfunction of $P_{i,j}$. Because permutation operators are idempotent, which means that if one applies P twice, one obtains the identity $PP = 1$, it can be seen that the eigenvalues of P must be either $+1$ or -1. That is, if $P\Psi = c\Psi$, then $PP\Psi = cc\Psi$, but $PP = 1$ means that $cc = 1$, so $c = +1$ or -1.

As a result of H commuting with electron permutation operators and of the idempotency of P, the eigenfunctions Ψ must either be odd or even under the application of any such permutation. Particles whose wave functions are even under P are called Bose particles or bosons, those for which Ψ is odd are called fermions. Electrons belong to the latter class of particles.

The simple spin-orbital product function used in Hartree theory

$$\Psi = \prod_{k=1,N} \phi_k \tag{6.8}$$

does not have the proper permutational symmetry. For example, the Be atom function $\Psi = 1s\alpha(1)\, 1s\beta(2)\, 2s\alpha(3)\, 2s\beta(4)$ is not odd under the interchange of the labels of electrons 3 and 4; instead one obtains $1s\alpha(1)\, 1s\beta(2)\, 2s\alpha(4)\, 2s\beta(3)$. However, such products of spin-orbitals (i.e., orbitals multiplied by α or β spin functions) can be made into properly antisymmetric functions by forming the determinant of an $N \times N$ matrix whose row index labels the spin-orbital and whose column index labels the electrons. For example, the Be atom function $1s\alpha(1)\, 1s\beta(2)\, 2s\alpha(3)\, 2s\beta(4)$ produces the 4×4 matrix whose determinant is shown below

$$\begin{vmatrix} 1s\alpha(1) & 1s\alpha(2) & 1s\alpha(3) & 1s\alpha(4) \\ 1s\beta(1) & 1s\beta(2) & 1s\beta(3) & 1s\beta(4) \\ 2s\alpha(1) & 2s\alpha(2) & 2s\alpha(3) & 2s\alpha(4) \\ 2s\beta(1) & 2s\beta(2) & 2s\beta(3) & 2s\beta(4) \end{vmatrix} \tag{6.9}$$

Clearly, if one interchanges any columns of this determinant, one changes the sign of the function. Moreover, if a determinant contains two or more rows that are identical (i.e., if one attempts to form such a function having two or more spin-orbitals equal), it vanishes. This is how such antisymmetric wave functions embody the Pauli exclusion principle.

A convenient way to write such a determinant is as follows:

$$\sum_P (-1)^p \phi_{P1}(1)\phi_{P2}(2)\ldots\phi_{PN}(N), \tag{6.10}$$

where the sum is over all $N!$ permutations of the N spin-orbitals and the notation $(-1)^p$ means that a -1 is affixed to any permutation that involves an odd number of pairwise interchanges of spin-orbitals and a $+1$ sign is given to any that involves an even number. To properly normalize such a determinental wave function, one must multiply it by $(N!)^{-1/2}$. So, the final result is that wave functions of the form

$$\Psi = (N!)^{-1/2} \sum_P (-1)^p \phi_{P1}(1)\phi_{P2}(2)\ldots\phi_{PN}(N) \tag{6.11}$$

have the proper permutational antisymmetry. Note that such functions consist of a sum of $N!$ factors, all of which have exactly the same number of electrons occupying the same number of spin orbitals; the only difference among the $N!$ terms involves which electron occupies which spin-orbital. For example, in the $1s\alpha\ 2s\alpha$ function appropriate to the excited state of He, one has

$$\Psi = (2)^{-1/2}\{1s\alpha(1)\,2s\alpha(2) - 2s\alpha(1)\,1s\alpha(2)\}. \tag{6.12}$$

This function is clearly odd under the interchange of the labels of the two electrons, yet each of its two components has one electron in a $1s\alpha$ spin-orbital and another electron in a $2s\alpha$ spin-orbital.

Although having to make Ψ antisymmetric appears to complicate matters significantly, it turns out that the Schrödinger equation appropriate to the spin-orbitals in such an antisymmetrized product wave function is nearly the same as the Hartree Schrödinger equation treated earlier. In fact, the resultant equation is

$$\begin{aligned} h_e\phi_J &= \left[-\frac{\hbar^2}{2m}\nabla^2 - \frac{Ze^2}{r} + \sum_K \langle\phi_K(r')|\frac{e^2}{|r-r'|}|\phi_K(r')\rangle\right]\phi_J(r) \\ &\quad - \sum_K \langle\phi_K(r')|\frac{e^2}{|r-r'|}|\phi_J(r')\rangle\phi_K(r) \\ &= \varepsilon_J\phi_J(r). \end{aligned} \tag{6.13}$$

In this expression, which is known as the Hartree–Fock equation, the same kinetic and nuclear attraction potentials occur as in the Hartree equation. Moreover, the same Coulomb potential

$$\begin{aligned} \sum_K \int \phi_K(r')e^2/|r-r'|\phi_K(r')dr' &= \sum_K \langle\phi_K(r')|e^2/|r-r'|\,|\,\phi_K(r')\rangle \\ &= \sum_K J_K(r) \end{aligned} \tag{6.14}$$

appears. However, one also finds a so-called exchange contribution to the Hartree–Fock potential that is equal to $\sum_L \langle\phi_L(r')|(e^2/|r-r'|)|\phi_J(r')\rangle\phi_L(r)$ and is often written in short-hand notation as $\sum_L K_L\phi_J(r)$. Notice that the Coulomb and exchange terms cancel for the $L = J$ case; this causes the artificial self-interaction term $J_L\phi_L(r)$ that can appear in the Hartree equations (unless one explicitly eliminates it) to automatically cancel with the exchange term $K_L\phi_L(r)$ in the Hartree–Fock equations.

When the LCAO expansion of each Hartree–Fock (HF) spin-orbital is substituted into the above HF Schrödinger equation, a matrix equation is again obtained:

$$\sum_\mu \langle\chi_\nu|h_e|\chi_\mu\rangle C_{J,\mu} = \varepsilon_J \sum_\mu \langle\chi_\nu\,|\,\chi_\mu\rangle C_{J,\mu}, \tag{6.15}$$

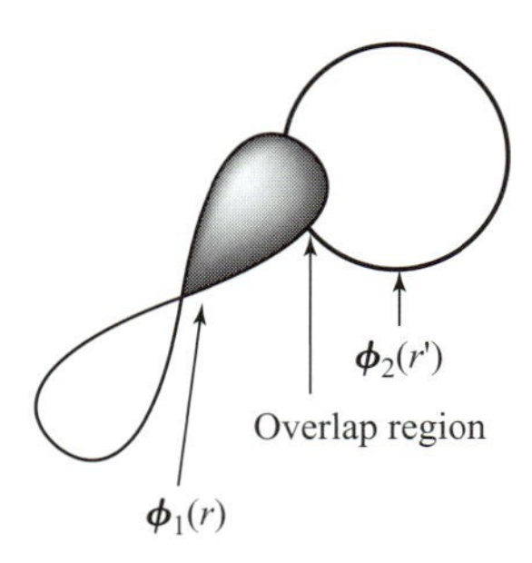

Figure 6.3 An s and a p orbital and their overlap region.

where the overlap integral $\langle \chi_\nu \mid \chi_\mu \rangle$ is as defined earlier, and the h_e matrix element is

$$\langle \chi_\nu | h_e | \chi_\mu \rangle = \langle \chi_\nu | -\frac{\hbar^2}{2m} \nabla^2 | \chi_\mu \rangle + \left\langle \chi_\nu \mid -\frac{Ze^2}{|r|} \chi_\mu \right\rangle$$
$$+ \sum_{K,\eta,\gamma} C_{K,\eta} C_{K,\gamma} \left[\langle \chi_\nu(r) \chi_\eta(r') | \left(\frac{e^2}{|r-r'|} \right) | \chi_\mu(r) \chi_\gamma(r') \rangle \right.$$
$$\left. - \langle \chi_\nu(r) \chi_\eta(r') | \left(\frac{e^2}{|r-r'|} \right) | \chi_\gamma(r) \chi_\mu(r') \rangle \right]. \tag{6.16}$$

Clearly, the only difference between this expression and the corresponding result of Hartree theory is the presence of the last term, the exchange integral. The SCF interative procedure used to solve the Hartree equations is again used to solve the HF equations.

Next, I think it is useful to reflect on the physical meaning of the Coulomb and exchange interactions between pairs of orbitals. For example, the Coulomb integral $J_{1,2} = \int |\phi_1(r)|^2 e^2/|r - r'| \phi_2(r')|^2 dr dr'$ appropriate to the two orbitals shown in Fig. 6.3 represents the Coulomb repulsion energy $e^2/|r - r'|$ of two charge densities, $|\phi_1|^2$ and $|\phi_2|^2$, integrated over all locations r and r' of the two electrons.

In contrast, the exchange integral $K_{1,2} = \int \phi_1(r)\phi_2(r') e^2/|r - r'| \phi_2(r)\phi_1(r') dr dr'$ can be thought of as the Coulomb repulsion between two electrons whose coordinates r and r' are both distributed throughout the "overlap region" $\phi_1\phi_2$. This overlap region is where both ϕ_1 and ϕ_2 have appreciable magnitude, so exchange integrals tend to be significant in magnitude only when the two orbitals involved have substantial regions of overlap.

Finally, a few words are in order about one of the most computer time-consuming parts of any Hartree–Fock calculation (or those discussed later) – the task of evaluating and transforming the two-electron integrals $\langle \chi_\nu(r)\chi_\eta(r')| (e^2/|r - r'|) | \chi_\mu(r)\chi_\gamma(r') \rangle$. Even when M GTOs are used as basis functions, the evaluation of $M^4/8$ of these integrals poses a major hurdle. For example, with 500 basis orbitals, there will be of the order of 7.8×10^9 such integrals. With each integral requiring 2 words of disk storage, this would require at least 1.5×10^4 Mwords of disk storage. Even in the era of modern computers that possess 100 Gby disks, this is a significant requirement. One of the more important technical advances that is under much current development is the efficient calculation of such integrals when the product functions $\chi_\nu(r)\chi_\mu(r)$ and $\chi_\gamma(r')\chi_\eta(r')$ that display the dependence on the two electrons' coordinates r and r' are spatially distant. In particular, multipolar expansions of these product functions are used to obtain more efficient approximations to their integrals when these functions are far apart. Moreover, such expansions offer a reliable way to "ignore" (i.e., approximate as zero) many integrals whose product functions are sufficiently distant. Such approaches show considerable promise for reducing the $M^4/8$ two-electron integral list to one whose size scales much less strongly with the size of the AO basis.

Koopmans' theorem

The HF-SCF equations $h_e\phi_i = \varepsilon_i\phi_i$ imply that the orbital energies ε_i can be written as

$$\begin{aligned}\varepsilon_i &= \langle\phi_i|h_e|\phi_i\rangle = \langle\phi_i|T+V|\phi_i\rangle + \sum_{j(\text{occupied})} \langle\phi_i|J_j - K_j|\phi_i\rangle \\ &= \langle\phi_i|T+V|\phi_i\rangle + \sum_{j(\text{occupied})} [J_{i,j} - K_{i,j}],\end{aligned} \tag{6.17}$$

where $T + V$ represents the kinetic (T) and nuclear attraction (V) energies, respectively. Thus, ε_i is the average value of the kinetic energy plus Coulomb attraction to the nuclei for an electron in ϕ_i plus the sum over all of the spin-orbitals occupied in Ψ of Coulomb minus exchange interactions.

If ϕ_i is an occupied spin-orbital, the $j = i$ term $[J_{i,i} - K_{i,i}]$ disappears in the above sum and the remaining terms in the sum represent the Coulomb minus exchange interaction of ϕ_i with all of the $N - 1$ other occupied spin-orbitals. If ϕ_i is a virtual spin-orbital, this cancellation does not occur because the sum over j does not include $j = i$. So, one obtains the Coulomb minus exchange interaction of ϕ_i with all N of the occupied spin-orbitals in Ψ. Hence the energies of occupied orbitals pertain to interactions appropriate to a total of N electrons, while the energies of virtual orbitals pertain to a system with $N + 1$ electrons.

Let us consider the following model of the detachment or attachment of an electron in an N-electron system.

(i) In this model, both the parent molecule and the species generated by adding or removing an electron are treated at the single-determinant level.
(ii) The Hartree–Fock orbitals of the parent molecule are used to describe both species. It is said that such a model neglects "orbital relaxation" (i.e., the reoptimization of the spin-orbitals to allow them to become appropriate to the daughter species).

Within this model, the energy difference between the daughter and the parent can be written as follows (ϕ_k represents the particular spin-orbital that is added or removed): for electron detachment:

$$E^{N-1} - E^N = -\varepsilon_k; \tag{6.18}$$

and for electron attachment:

$$E^N - E^{N+1} = -\varepsilon_k. \tag{6.19}$$

So, within the limitations of the HF, frozen-orbital model, the ionization potentials (IPs) and electron affinities (EAs) are given as the negative of the occupied and virtual spin-orbital energies, respectively. This statement is referred to as

Koopmans' theorem; it is used extensively in quantum chemical calculations as a means of estimating IPs and EAs and often yields results that are qualitatively correct (i.e., ± 0.5 eV).

Orbital energies and the total energy

The total HF-SCF electronic energy can be written as

$$E = \sum_{i(\text{occupied})} \langle \phi_i | T + V | \phi_i \rangle + \sum_{i>j(\text{occupied})} [J_{i,j} - K_{i,j}], \tag{6.20}$$

and the sum of the orbital energies of the occupied spin-orbitals is given by

$$\sum_{i(\text{occupied})} \varepsilon_i = \sum_{i(\text{occupied})} \langle \phi_i | T + V | \phi_i \rangle + \sum_{i,j(\text{occupied})} [J_{i,j} - K_{i,j}]. \tag{6.21}$$

These two expressions differ in a very important way; the sum of occupied orbital energies double counts the Coulomb minus exchange interaction energies. Thus, within the Hartree–Fock approximation, the sum of the occupied orbital energies is not equal to the total energy. This finding teaches us that we can not think of the total electronic energy of a given orbital occupation in terms of the orbital energies alone. We need to also keep track of the inter-electron Coulomb and exchange energies.

6.1.5 Molecular orbitals

Before moving on to discuss methods that go beyond the HF model, it is appropriate to examine some of the computational effort that goes into carrying out an SCF calculation on molecules. The primary differences that appear when molecules rather than atoms are considered are:

(i) The electronic Hamiltonian h_e contains not only one nuclear-attraction Coulomb potential $\sum_j Ze^2/r_j$ but a sum of such terms, one for each nucleus in the molecule $\sum_a \sum_j Z_a e^2/|r_j - R_a|$, whose locations are denoted R_a.

(ii) One has AO basis functions of the type discussed above located on each nucleus of the molecule. These functions are still denoted $\chi_\mu(r - R_a)$, but their radial and angular dependences involve the distance and orientation of the electron relative to the particular nucleus on which the AO is located.

Other than these two changes, performing a SCF calculation on a molecule (or molecular ion) proceeds just as in the atomic case detailed earlier. Let us briefly review how this iterative process occurs.

Once atomic basis sets have been chosen for each atom, the one- and two-electron integrals appearing in the h and overlap matrices must be evaluated. There are numerous highly efficient computer codes that allow such integrals to be computed for s, p, d, f, and even g, h, and i basis functions. After executing one of these "integral packages" for a basis with a total of M functions, one has available (usually on the computer's hard disk) of the order of $M^2/2$ one-electron ($\langle\chi_\mu|h_e|\chi_\nu\rangle$ and $\langle\chi_\mu \mid \chi_\nu\rangle$) and $M^4/8$ two-electron ($\langle\chi_\mu\chi_\delta \mid \chi_\nu\chi_\kappa\rangle$) integrals. When treating extremely large atomic orbital basis sets (e.g., 500 or more basis functions), modern computer programs calculate the requisite integrals but never store them on the disk. Instead, their contributions to the $\langle\chi_\mu|h_e|\chi_\nu\rangle$ matrix elements are accumulated "on the fly" after which the integrals are discarded.

Shapes, sizes, and energies of orbitals

Each molecular spin-orbital (MO) that results from solving the HF-SCF equations for a molecule or molecular ion consists of a sum of components involving all of the basis AOs:

$$\phi_j = \sum_\mu C_{j,\mu}\chi_\mu. \tag{6.22}$$

In this expression, the $C_{j,\mu}$ are referred to as LCAO-MO coefficients because they tell us how to linearly combine AOs to form the MOs. Because the AOs have various angular shapes (e.g., s, p, or d shapes) and radial extents (i.e., different orbital exponents), the MOs constructed from them can be of different shapes and radial sizes. Let's look at a few examples to see what I mean.

The first example arises when two H atoms combine to form the H_2 molecule. The valence AOs on each H atom are the 1s AOs; they combine to form the two valence MOs (σ and σ^*) depicted in Fig. 6.4. The bonding MO labeled σ has LCAO-MO coefficients of equal sign for the two 1s AOs, as a result of which this MO has the same sign near the left H nucleus (A) as near the right H nucleus (B). In contrast, the antibonding MO labeled σ^* has LCAO-MO coefficients of different sign for the A and B 1s AOs. As was the case in the Hückel or

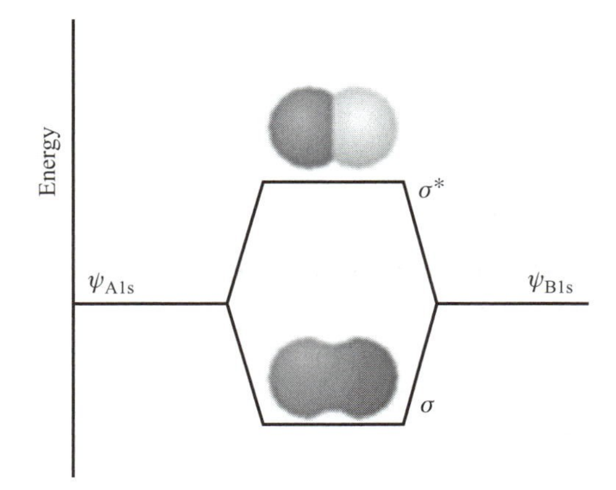

Figure 6.4 Two 1s hydrogen atomic orbitals combine to form a bonding and antibonding molecular orbital.

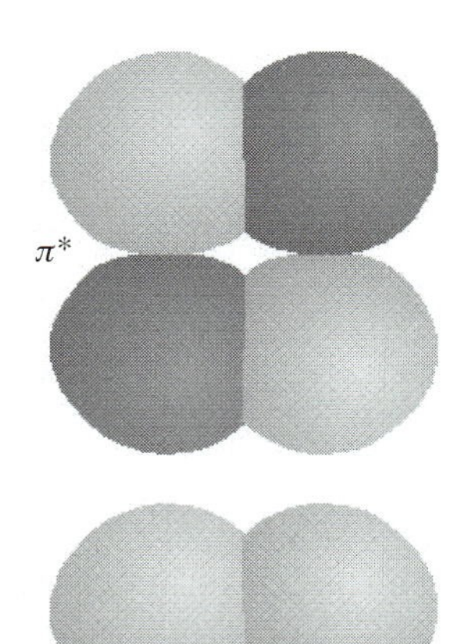

Figure 6.5 Two p_π atomic orbitals on carbon atoms combine to form a bonding and antibonding molecular orbital.

tight-binding model outlined in the Background Material, the energy splitting between the two MOs depends on the overlap $\langle \chi_{1sA} \mid \chi_{1sB} \rangle$ between the two AOs.

An analogous pair of bonding and antibonding MOs arises when two p orbitals overlap "sideways" as in ethylene to form π and π^* MOs which are illustrated in Fig. 6.5. The shapes of these MOs clearly are dictated by the shapes of the AOs that comprise them and the relative signs of the LCAO-MO coefficients that relate the MOs to AOs. For the π MO, these coefficients have the same sign on the left and right atoms; for the π^* MO, they have opposite signs.

I should stress that the signs and magnitudes of the LCAO-MO coefficients arise as eigenvectors of the HF-SCF matrix eigenvalue equation:

$$\sum_{\mu} \langle \chi_\nu | h_e | \chi_\mu \rangle C_{j,\mu} = \varepsilon_j \sum_{\mu} \langle \chi_\nu \mid \chi_\mu \rangle C_{j,\mu}. \tag{6.23}$$

It is a characteristic of such eigenvalue problems for the lower energy eigenfunctions to have fewer nodes than the higher energy solutions as we learned from several examples that we solved in the Background Material.

Another thing to note about the MOs shown above is that they will differ in their quantitative details, but not in their overall shapes, when various functional groups are attached to the ethylene molecule's C atoms. For example, if electron withdrawing groups such as Cl, OH or Br are attached to one of the C atoms, the attractive potential experienced by a π electron near that C atom will be enhanced. As a result, the bonding MO will have larger LCAO-MO coefficients $C_{k,\mu}$ belonging to the "tighter" basis AOs χ_μ on this C atom. This will make the bonding π MO more radially compact in this region of space, although its nodal character and gross shape will not change. Alternatively, an electron donating group such as H_3C— or t-butyl attached to one of the C centers will cause the π MO to be more diffuse (by making its LCAO-MO coefficients for more diffuse basis AOs larger).

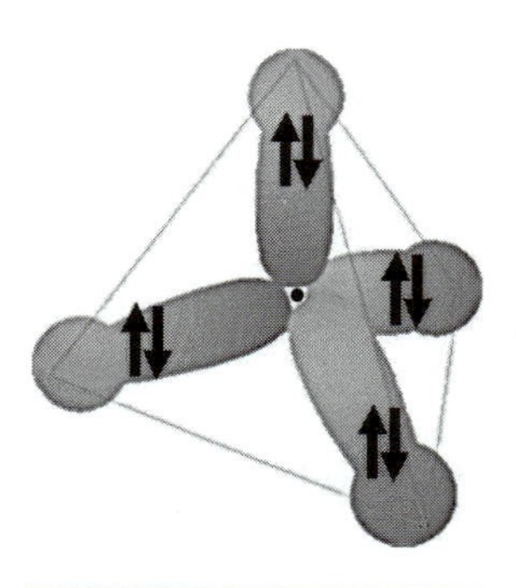

Figure 6.6 The four C—H bonds in methane.

In addition to MOs formed primarily of AOs of one type (i.e., for H_2 it is primarily s-type orbitals that form the σ and σ^* MOs; for ethylene's π bond, it is primarily the C 2p AOs that contribute), there are bonding and antibonding MOs formed by combining several AOs. For example, the four equivalent C—H bonding MOs in CH_4 shown in Fig. 6.6 each involve C 2s and 2p as well as H 1s basis AOs.

The energies of the MOs depend on two primary factors: the energies of the AOs from which the MOs are constructed and the overlap between these AOs. The pattern in energies for valence MOs formed by combining pairs of first-row atoms to form homonuclear diatomic molecules is shown in Fig. 6.7. In this figure, the core MOs formed from the 1s AOs are not shown, but only those MOs formed from 2s and 2p AOs appear. The clear trend toward lower orbital

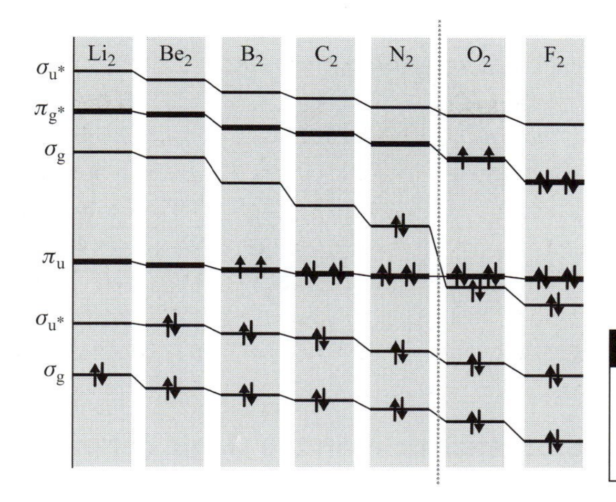

Figure 6.7 Energies of the valence molecular orbitals in homonuclear diatomics involving first-row atoms.

energies as one moves from left to right is due primarily to the trends in orbital energies of the constituent AOs. That is, F being more electronegative than N has a lower-energy 2p orbital than does N.

Bonding, antibonding, non-bonding, and Rydberg orbitals

As noted above, when valence AOs combine to form MOs, the relative signs of the combination coefficients determine, along with the AO overlap magnitudes, the MO's energy and nodal properties. In addition to the bonding and antibonding MOs discussed and illustrated earlier, two other kinds of MOs are important to know about.

Non-bonding MOs arise, for example, when an orbital on one atom is not directed toward and overlapping with an orbital on a neighboring atom. For example, the lone pair orbitals on H_2O or on the oxygen atom of $H_2C{=}O$ are non-bonding orbitals. They still are described in the LCAO-MO manner, but their $C_{\mu,i}$ coefficients do not contain dominant contributions from more than one atomic center.

Finally, there is a type of orbital that all molecules possess but that is ignored in most elementary discussions of electronic structure. All molecules have so-called Rydberg orbitals. These orbitals can be thought of as large diffuse orbitals that describe the regions of space an electron would occupy if it were in the presence of the corresponding molecular cation. Two examples of such Rydberg orbitals are shown in Fig. 6.8. On the left, we see the Rydberg orbital of NH_4 and on the right, that of $H_3N{-}CH_3$. The former species can be thought of as a closed-shell ammonium cation NH_4^+ around which a Rydberg orbital resides. The latter is protonated methyl amine with its Rydberg orbital.

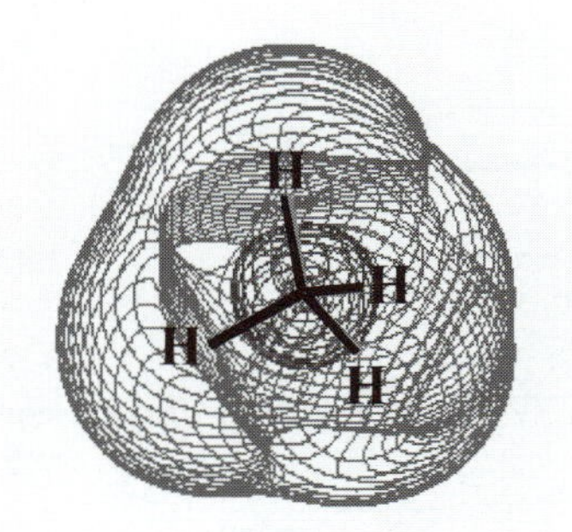

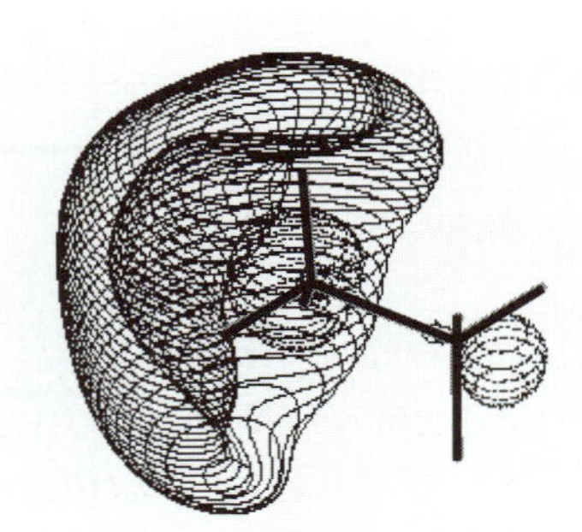

Figure 6.8 Rydberg orbitals of NH_4^+ and of protonated methyl amine.

6.2 Deficiencies in the single determinant model

To achieve reasonable chemical accuracy (e.g., ± 5 kcal mol^{-1}) in electronic structure calculations, one can not describe the wave function Ψ in terms of a single determinant. The reason such a wave function is inadequate is because the spatial probability density functions are not correlated. This means the probability of finding one electron at position **r** is independent of where the other electrons are, which is absurd because the electrons' mutual Coulomb repulsion causes them to "avoid" one another. This mutual avoidance is what we call electron correlation because the electrons' motions, as reflected in their spatial probability densities, are correlated (i.e., inter-related). Let us consider a simple example to illustrate this problem with single determinant functions. The $|1s\alpha(r)\ 1s\beta(r')|$ determinant, when written as

$$|1s\alpha(r)\ 1s\beta(r')| = 2^{-1/2}\{1s\alpha(r)\ 1s\beta(r') - 1s\alpha(r')1s\beta(r)\} \tag{6.24}$$

can be multiplied by itself to produce the two-electron spin- and spatial-probability density:

$$\begin{aligned} P(r, r') = 1/2\{&[1s\alpha(r)\ 1s\beta(r')]^2 + [1s\alpha(r')\ 1s\beta(r)]^2 \\ &- 1s\alpha(r)\ 1s\beta(r')\ 1s\alpha(r')\ 1s\beta(r) - 1s\alpha(r')\ 1s\beta(r)\ 1s\alpha(r)\ 1s\beta(r')\}. \end{aligned} \tag{6.25}$$

If we now integrate over the spins of the two electrons and make use of

$$\langle\alpha \mid \alpha\rangle = \langle\beta \mid \beta\rangle = 1, \quad \text{and} \quad \langle\alpha \mid \beta\rangle = \langle\beta \mid \alpha\rangle = 0, \tag{6.26}$$

we obtain the following spatial (i.e., with spin absent) probability density:

$$P(r, r') = |1s(r)|^2\ |1s(r')|^2. \tag{6.27}$$

This probability, being a product of the probability density for finding one electron at r times the density of finding another electron at r', clearly has no correlation in it. That is, the probability of finding one electron at r does not depend on where (r') the other electron is. This product form for $P(r, r')$ is a direct result of the single-determinant form for Ψ, so this form must be wrong if electron correlation is to be accounted for.

6.2.1 Electron correlation

Now, we need to ask how Ψ should be written if electron correlation effects are to be taken into account. As we now demonstrate, it turns out that one can account for electron avoidance by taking Ψ to be a combination of two or more determinants that differ by the promotion of two electrons from one orbital to another orbital. For example, in describing the π^2 bonding electron pair of an olefin or the ns^2 electron pair in alkaline earth atoms, one mixes in doubly excited determinants of the form $(\pi^*)^2$ or np^2, respectively.

Briefly, the physical importance of such doubly excited determinants can be made clear by using the following identity involving determinants:

$$C_1|\ldots\phi\alpha\phi\beta\ldots| - C_2|\ldots\phi'\alpha\phi'\beta\ldots| = C_1/2\{|\ldots(\phi - x\phi')\alpha(\phi + x\phi')\beta\ldots| \\ - |\ldots(\phi - x\phi')\beta(\phi + x\phi')\alpha\ldots|\}, \quad (6.28)$$

where

$$x = (C_2/C_1)^{1/2}. \quad (6.29)$$

This allows one to interpret the combination of two determinants that differ from one another by a double promotion from one orbital (ϕ) to another (ϕ') as equivalent to a singlet coupling (i.e., having $\alpha\beta$-$\beta\alpha$ spin function) of two different orbitals $(\phi - x\phi')$ and $(\phi + x\phi')$ that comprise what are called polarized orbital pairs. In the simplest embodiment of such a configuration interaction (CI) description of electron correlation, each electron pair in the atom or molecule is correlated by mixing in a configuration state function (CSF) in which that electron pair is "doubly excited" to a correlating orbital.

In the olefin example mentioned above, the two non-orthogonal polarized orbital pairs involve mixing the π and π^* orbitals to produce two left-right polarized orbitals as depicted in Fig. 6.9. In this case, one says that the π^2 electron pair

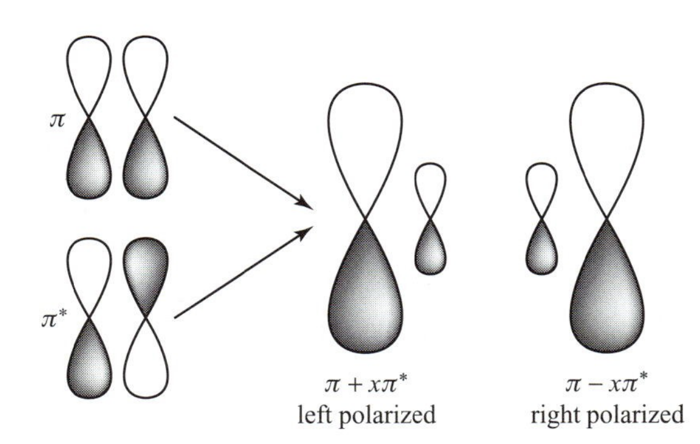

Figure 6.9 Left and right polarized orbitals of an olefin.

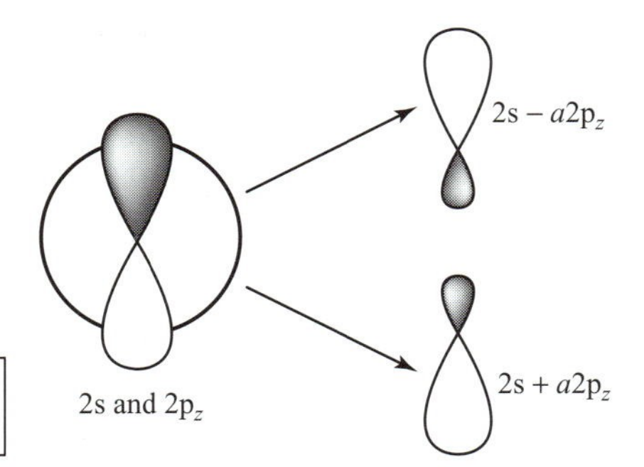

Figure 6.10 Angularly polarized orbital pairs.

undergoes left-right correlation when the $(\pi^*)^2$ determinant is mixed into the CI wave function.

In the alkaline earth atom case, the polarized orbital pairs are formed by mixing the ns and np orbitals (actually, one must mix in equal amounts of p_x, p_y, and p_z orbitals to preserve overall ^{1}S symmetry in this case), and give rise to angular correlation of the electron pair. Such a pair of polarized orbitals is shown in Fig. 6.10. More specifically, the following four determinants are found to have the largest amplitudes in Ψ:

$$\Psi \cong C_1|1s^2 2s^2| - C_2\left[|1s^2 2p_x^2| + |1s^2 2p_y^2| + |1s^2 2p_z^2|\right]. \tag{6.30}$$

The fact that the latter three terms possess the same amplitude C_2 is a result of the requirement that a state of ^{1}S symmetry is desired. It can be shown that this function is equivalent to:

$$\begin{aligned}\Psi \cong 1/6\, C_1|1s\alpha\, 1s\beta\{&[(2s - a2p_x)\alpha(2s + a2p_x)\beta - (2s - a2p_x)\beta(2s + a2p_x)\alpha]\\ &+[(2s - a2p_y)\alpha(2s + a2p_y)\beta - (2s - a2p_y)\beta(2s + a2p_y)\alpha]\\ &+[(2s - a2p_z)\alpha(2s + a2p_z)\beta - (2s - a2p_z)\beta(2s + a2p_z)\alpha]\}|,\end{aligned} \tag{6.31}$$

where $a = \sqrt{3C_2/C_1}$.

Here two electrons occupy the 1s orbital (with opposite, α and β spins), and are thus not being treated in a correlated manner, while the other pair resides in 2s/2p polarized orbitals in a manner that instantaneously correlates their motions. These polarized orbital pairs ($2s \pm a2p_{x,y,\text{or } z}$) are formed by combining the 2s orbital with the $2p_{x,y,\text{or } z}$ orbital in a ratio determined by C_2/C_1.

This ratio C_2/C_1 can be shown using perturbation theory to be proportional to the magnitude of the coupling $\langle 1s^2 2s^2|H|1s^2 2p^2\rangle$ between the two configurations involved and inversely proportional to the energy difference $[\langle 1s^2 2s^2|H|1s^2 2s^2\rangle - \langle 1s^2 2p^2|H|1s^2 2p^2\rangle]$ between these configurations. In general, configurations that have similar Hamiltonian expectation values and that are coupled strongly give rise to strongly mixed (i.e., with large $|C_2/C_1|$ ratios) polarized orbital pairs.

In each of the three equivalent terms in the alkaline earth wave function, one of the valence electrons moves in a $2s + a2p$ orbital polarized in one direction while the other valence electron moves in the $2s - a2p$ orbital polarized in the opposite direction. For example, the first term $[(2s - a2p_x)\alpha\ (2s + a2p_x)\beta - (2s - a2p_x)\beta\ (2s + a2p_x)\alpha]$ describes one electron occupying a $2s - a2p_x$ polarized orbital while the other electron occupies the $2s + a2p_x$ orbital. The electrons thus reduce their Coulomb repulsion by occupying different regions of space; in the SCF picture $1s^2 2s^2$, both electrons reside in the same 2s region of space. In this particular example, the electrons undergo angular correlation to avoid one another.

The use of doubly excited determinants is thus seen as a mechanism by which Ψ can place electron pairs, which in the single-configuration picture occupy the same orbital, into different regions of space (i.e., each one into a different member of the polarized orbital pair) thereby lowering their mutual Coulombic repulsion. Such electron correlation effects are extremely important to include if one expects to achieve chemically meaningful accuracy (i.e., ± 5 kcal mol^{-1}).

6.2.2 Essential configuration interaction

There are occasions in which the inclusion of two or more determinants in Ψ is essential to obtaining even a qualitatively correct description of the molecule's electronic structure. In such cases, we say that we are including essential correlation effects. To illustrate, let us consider the description of the two electrons in a single covalent bond between two atoms or fragments that we label X and Y. The fragment orbitals from which the bonding σ and antibonding σ^* MOs are formed we will label s_X and s_Y, respectively.

Several spin- and spatial-symmetry adapted two-electron determinants can be formed by placing two electrons into the σ and σ^* orbitals. For example, to describe the singlet determinant corresponding to the closed-shell σ^2 orbital occupancy, a single Slater determinant

$$^1\sum(0) = |\sigma\alpha\sigma\beta| = (2)^{-1/2}\{\sigma\alpha(1)\sigma\beta(2) - \sigma\beta(1)\sigma\alpha(2)\} \tag{6.32}$$

suffices. An analogous expression for the $(\sigma^*)^2$ determinant is given by

$$^1\overset{**}{\sum}(0) = |\sigma^*\alpha\ \sigma^*\beta| = (2)^{-1/2}\{\sigma^*\alpha(1)\ \sigma^*\beta(2) - \sigma^*\alpha(2)\ \sigma^*\beta(1)\}. \tag{6.33}$$

Also, the $M_S = 1$ component of the triplet state having $\sigma\sigma^*$ orbital occupancy can be written as a single Slater determinant:

$$^3\overset{*}{\sum}(1) = |\sigma\alpha\ \sigma^*\alpha| = (2)^{-1/2}\{\sigma\alpha(1)\ \sigma^*\alpha(2) - \sigma^*\alpha(1)\ \sigma\alpha(2)\}, \tag{6.34}$$

as can the $M_S = -1$ component of the triplet state

$$^3\overset{*}{\sum}(-1) = |\sigma\beta\ \sigma^*\beta| = (2)^{-1/2}\{\sigma\beta(1)\ \sigma^*\beta(2) - \sigma^*\beta(1)\ \sigma\beta(2)\}. \tag{6.35}$$

However, to describe the singlet and $M_S = 0$ triplet states belonging to the $\sigma\sigma^*$ occupancy, two determinants are needed:

$$^1\sum^*(0) = \frac{1}{\sqrt{2}}[|\sigma\alpha\,\sigma^*\beta| - |\sigma\beta\,\sigma^*\alpha|] \quad (6.36)$$

is the singlet and

$$^3\sum^*(0) = \frac{1}{\sqrt{2}}[|\sigma\alpha\,\sigma^*\beta| + |\sigma\beta\,\sigma^*\alpha|] \quad (6.37)$$

is the triplet. In each case, the spin quantum number S, its z-axis projection M_S, and the Λ quantum number are given in the conventional $^{2S+1}\Lambda(M_S)$ term symbol notation.

As the distance R between the X and Y fragments is changed from near its equilibrium value of R_e and approaches infinity, the energies of the σ and σ^* orbitals vary in a manner well known to chemists as depicted in Fig. 6.11 if X and Y are identical.

If X and Y are not identical, the s_X and s_Y orbitals still combine to form a bonding σ and an antibonding σ^* orbital. The energies of these orbitals, for R values ranging from near R_e to $R \to \infty$, are depicted in Fig. 6.12 for the case in which X is more electronegative than Y.

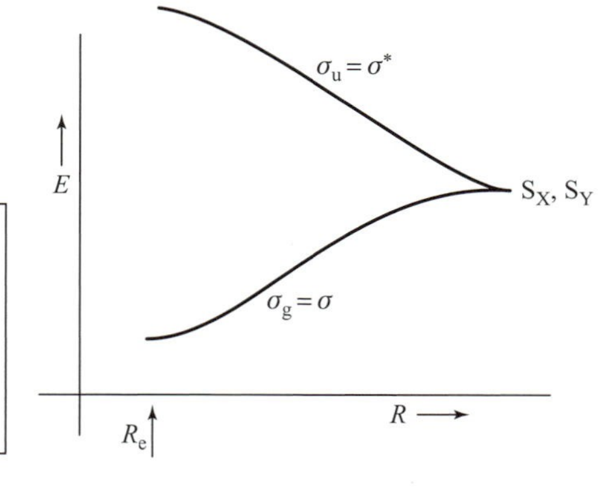

Figure 6.11 Orbital correlation diagram showing two σ-type orbitals combining to form a bonding and an antibonding molecular orbital.

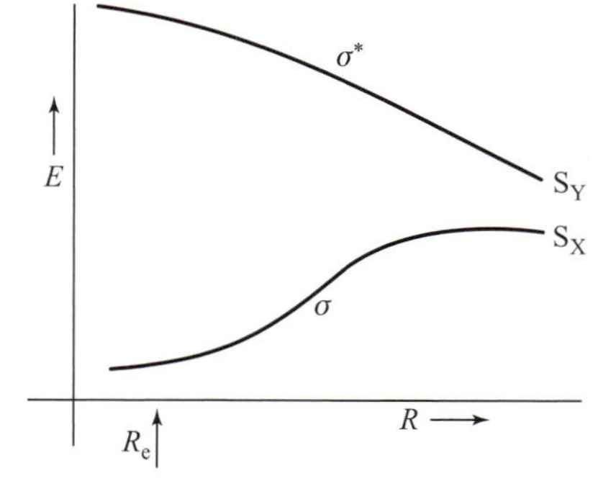

Figure 6.12 Orbital correlation diagram for σ-type orbitals in the heteronuclear case.

The energy variation in these orbital energies gives rise to variations in the energies of the six determinants listed above. As $R \rightarrow \infty$, the determinants' energies are difficult to "intuit" because the σ and σ^* orbitals become degenerate (in the homonuclear case) or nearly so (in the $X \neq Y$ case). To pursue this point and arrive at an energy ordering for the determinants that is appropriate to the $R \rightarrow \infty$ region, it is useful to express each such function in terms of the fragment orbitals s_X and s_Y that comprise σ and σ^*. To do so, the LCAO-MO expressions for σ and σ^*,

$$\sigma = C[s_X + z s_Y] \tag{6.38}$$

and

$$\sigma * = C^*[z s_X - s_Y], \tag{6.39}$$

are substituted into the Slater determinant definitions given above. Here C and C^* are the normalization constants. The parameter z is 1.0 in the homonuclear case and deviates from 1.0 in relation to the s_X and s_Y orbital energy difference (if s_X lies below s_Y, then $z < 1.0$; if s_X lies above s_Y, $z > 1.0$).

Let us examine the X = Y case to keep the analysis as simple as possible. The process of substituting the above expressions for σ and σ^* into the Slater determinants that define the singlet and triplet functions can be illustrated as follows for the ${}^1\sum(0)$ case:

$$\begin{aligned}{}^1\sum(0) &= |\sigma\alpha\ \sigma\beta| = C^2|(s_X + s_Y)\alpha\ (s_X + s_Y)\beta| \\ &= C^2[|s_X\alpha\ s_X\beta| + |s_Y\alpha\ s_Y\beta| + |s_X\alpha\ s_Y\beta| + |s_Y\alpha\ s_X\beta|]. \end{aligned} \tag{6.40}$$

The first two of these atomic-orbital-based Slater determinants ($|s_X\alpha\ s_X\beta|$ and $|s_Y\alpha\ s_Y\beta|$) are called "ionic" because they describe atomic orbital occupancies, which are appropriate to the $R \rightarrow \infty$ region that correspond to X: + X and X + X: valence bond structures, while $|s_X\alpha\ s_Y\beta|$ and $|s_Y\alpha\ s_X\beta|$ are called "covalent" because they correspond to X· + X· structures.

In similar fashion, the remaining five determinant functions may be expressed in terms of fragment-orbital-based Slater determinants. In so doing, use is made of the antisymmetry of the Slater determinants $|\phi_1\phi_2\phi_3| = -|\phi_1\phi_3\phi_2|$, which implies that any determinant in which two or more spin-orbitals are identical vanishes, $|\phi_1\phi_2\phi_2| = -|\phi_1\phi_2\phi_3| = 0$. The result of decomposing the MO-based determinants into their fragment-orbital components is as follows:

$$\begin{aligned}{}^1\overset{**}{\sum}(0) &= |\sigma^*\alpha\ \sigma^*\beta| \\ &= C^{*2}[|s_X\alpha\ s_X\beta| + |s_Y\alpha\ s_Y\beta| - |s_X\alpha\ s_Y\beta| - |s_Y\alpha\ s_X\beta|], \end{aligned} \tag{6.41}$$

$$\begin{aligned}{}^1\overset{*}{\sum}(0) &= \frac{1}{\sqrt{2}}[|\sigma\alpha\ \sigma^*\beta| - |\sigma\beta\ \sigma^*\alpha|] \\ &= CC^*\sqrt{2}[|s_X\alpha\ s_X\beta| - |s_Y\alpha\ s_Y\beta|], \end{aligned} \tag{6.42}$$

$$^3\overset{*}{\sum}(1) = |\sigma\alpha\ \sigma^*\alpha| \\ = CC^*2|s_Y\alpha\ s_X\alpha|, \quad (6.43)$$

$$^3\overset{*}{\sum}(0) = \frac{1}{\sqrt{2}}[|\sigma\alpha\ \sigma^*\beta| + |\sigma\beta\ \sigma^*\alpha|] \\ = CC^*\sqrt{2}[|s_Y\alpha\ s_X\beta| - |s_X\alpha\ s_Y\beta|], \quad (6.44)$$

$$^3\overset{*}{\sum}(-1) = |\sigma\alpha\ \sigma^*\alpha| \\ = CC^*2|s_Y\beta\ s_X\beta|. \quad (6.45)$$

These decompositions of the six valence determinants into fragment-orbital or valence bond components allow the $R = \infty$ energies of these states to be specified. For example, the fact that both $^1\sum$ and $^1\sum^{**}$ contain 50% ionic and 50% covalent structures implies that, as $R \to \infty$, both of their energies will approach the average of the covalent and ionic atomic energies $1/2[E(\mathrm{X}\cdot) + E(\mathrm{X}\cdot) + E(\mathrm{X}) + E(\mathrm{X}\!:)]$. The $^1\sum^*$ energy approaches the purely ionic value $E(\mathrm{X}) + E(\mathrm{X}\!:)$ as $R \to \infty$. The energies of $^3\sum^*(0)$, $^3\sum^*(1)$ and $^3\sum^*(-1)$ all approach the purely covalent value $E(\mathrm{X}\cdot) + E(\mathrm{X}\cdot)$ as $R \to \infty$.

The behaviors of the energies of the six valence determinants as R varies are depicted in Fig. 6.13 for situations in which the homolytic bond cleavage is energetically favored (i.e., for which $E(\mathrm{X}\cdot) + E(\mathrm{X}\cdot) < E(\mathrm{X}) + \mathrm{E}(\mathrm{X}\!:)$).

It is essential to realize that the energies of the determinants do not represent the energies of the true electronic states. For R-values at which the determinant energies are separated widely, the true state energies are rather well approximated by individual $^1\sum$ determinant energies; such is the case near R_e.

However, at large R, the situation is very different, and it is in such cases that what we term essential configuration interaction occurs. Specifically, for the X =Y example, the $^1\sum$ and $^1\sum^{**}$ determinants undergo essential CI coupling to form a pair of states of $^1\sum$ symmetry (the $^1\sum^*$ CSF cannot partake in this CI mixing because it is of ungerade symmetry; the $^3\sum^*$ states can not mix

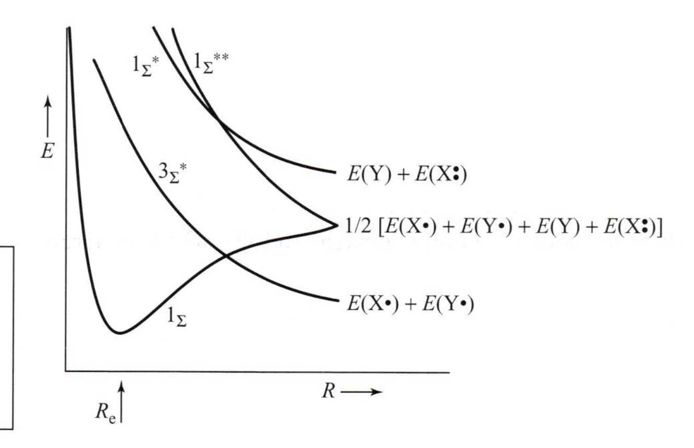

Figure 6.13 Configuration correlation diagram showing how the determinants' energies vary with R.

because they are of triplet spin symmetry). The CI mixing of the $^1\sum$ and $^1\sum^{**}$ determinants is described in terms of a 2×2 secular problem

$$\begin{bmatrix} \langle^1\sum|H|^1\sum\rangle & \langle^1\sum|H|^1\sum^{**}\rangle \\ \langle^1\sum^{**}|H|^1\sum\rangle & \langle^1\sum^{**}|H|^1\sum^{**}\rangle \end{bmatrix} \begin{bmatrix} \mathrm{A} \\ \mathrm{B} \end{bmatrix} = E \begin{bmatrix} \mathrm{A} \\ \mathrm{B} \end{bmatrix}. \tag{6.46}$$

The diagonal entries are the determinants' energies depicted in Fig. 6.13. The off-diagonal coupling matrix elements can be expressed in terms of an exchange integral between the σ and σ^* orbitals:

$$\begin{aligned} \left\langle^1\sum\right|H\left|^1\sum^{**}\right\rangle &= \langle|\sigma\alpha\ \sigma\beta|H\ |\ |\sigma^*\alpha\ \sigma^*\beta|\rangle \\ &= \langle\sigma\sigma|\frac{1}{r_{12}}|\sigma^*\sigma^*\rangle = K_{\sigma\sigma^*}. \end{aligned} \tag{6.47}$$

At $R \to \infty$, where the $^1\sum$ and $^1\sum^{**}$ determinants are degenerate, the two solutions to the above CI matrix eigenvalue problem are

$$E_{\mp} = 1/2[E(\mathrm{X}\cdot) + E(\mathrm{X}\cdot) + E(\mathrm{X}) + E(\mathrm{X}\cdot)]_{\mp}\langle\sigma\sigma|\frac{1}{r_{12}}|\sigma^*\sigma^*\rangle, \tag{6.48}$$

with respective amplitudes for the $^1\sum$ and $^1\sum^{**}$ CSFs given by

$$A_{\mp} = \pm\frac{1}{\sqrt{2}}, \qquad B_{\mp} = \mp\frac{1}{\sqrt{2}}. \tag{6.49}$$

The first solution thus has

$$\Psi_{-} = \frac{1}{\sqrt{2}}[|\sigma\alpha\ \sigma\beta| - |\sigma^*\alpha\ \sigma^*\beta|], \tag{6.50}$$

which, when decomposed into atomic orbital components, yields

$$\Psi_{-} = \frac{1}{\sqrt{2}}[|\mathrm{s_X}\alpha\ \mathrm{s_Y}\beta| - |\mathrm{s_X}\beta\ \mathrm{s_Y}\alpha|]. \tag{6.51}$$

The other root has

$$\begin{aligned} \Psi_{+} &= \frac{1}{\sqrt{2}}[|\sigma\alpha\ \sigma\beta| + |\sigma^*\alpha\ \sigma^*\beta|] \\ &= \frac{1}{\sqrt{2}}[|\mathrm{s_X}\alpha\ \mathrm{s_X}\beta| + |\mathrm{s_Y}\alpha\ \mathrm{s_Y}\beta|]. \end{aligned} \tag{6.52}$$

So, we see that $^1\sum$ and $^1\sum^{**}$, which both contain 50% ionic and 50% covalent parts, combine to produce Ψ_{-} which is purely covalent and Ψ_{+} which is purely ionic.

The above essential CI mixing of $^1\sum$ and $^1\sum^{**}$ as $R \to \infty$ qualitatively alters the energy diagrams shown above. Descriptions of the resulting valence singlet and triplet $\sum$ states are given in Fig. 6.14 for homonuclear situations in which covalent products lie below the ionic fragments.

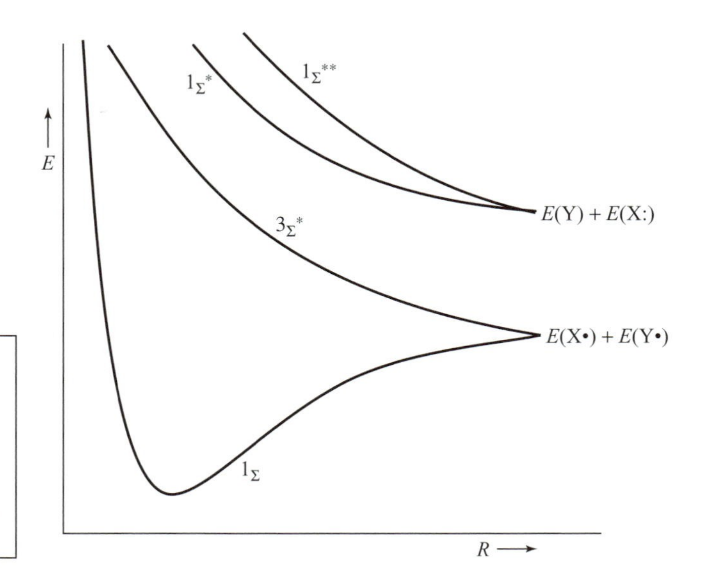

Figure 6.14 State correlation diagram showing how the energies of the states, comprised of combinations of determinants, vary with *R*.

6.2.3 Various approaches to electron correlation

There are numerous procedures currently in use for determining the "best" wave function that is usually expressed in the form

$$\Psi = \sum_I C_I \Phi_I, \tag{6.53}$$

where Φ_I is a spin- and space-symmetry-adapted configuration state function (CSF) that consists of one or more determinants $|\phi_{I1}\phi_{I2}\phi_{I3}\ldots\phi_{IN}|$ combined to produce the desired symmetry. In all such wave functions, there are two kinds of parameters that need to be determined – the C_I coefficients and the LCAO-MO coefficients describing the ϕ_{Ik} in terms of the AO basis functions. The most commonly employed methods used to determine these parameters include the following methods.

The CI method

In this approach, the LCAO-MO coefficients are determined first usually via a single-configuration SCF calculation. The C_I coefficients are subsequently determined by making the expectation value $\langle\Psi|H|\Psi\rangle\langle\Psi \mid \Psi\rangle$ variationally stationary.

The CI wave function is most commonly constructed from spin- and spatial-symmetry adapted combinations of determinants called configuration state functions (CSFs) Φ_J that include:

(i) The so-called reference CSF that is the SCF wave function used to generate the molecular orbitals ϕ_i.

(ii) CSFs generated by carrying out single, double, triple, etc. level "excitations" (i.e., orbital replacements) relative to the reference CSF. CI wave functions limited to include contributions through various levels of excitation are denoted S (singly), D (doubly), SD (singly and doubly), SDT (singly, doubly, and triply) excited.

The orbitals from which electrons are removed can be restricted to focus attention on correlations among certain orbitals. For example, if excitations out of core orbitals are excluded, one computes a total energy that contains no core correlation energy. The number of CSFs included in the CI calculation can be large. CI wave functions including 5000 to 50 000 CSFs are routine, and functions with one to several billion CSFs are within the realm of practicality.

The need for such large CSF expansions can be appreciated by considering (i) that each electron pair requires at least two CSFs to form polarized orbital pairs, (ii) there are of the order of $N(N-1)/2 = X$ electron pairs for a molecule containing N electrons, hence (iii) the number of terms in the CI wave function scales as 2^X. For a molecule containing ten electrons, there could be $2^{45} = 3.5 \times 10^{13}$ terms in the CI expansion. This may be an overestimate of the number of CSFs needed, but it demonstrates how rapidly the number of CSFs can grow with the number of electrons.

The Hamiltonian matrix elements $H_{I,J}$ between pairs of CSFs are, in practice, evaluated in terms of one- and two-electron integrals over the molecular orbitals. Prior to forming the $H_{I,J}$ matrix elements, the one- and two-electron integrals, which can be computed only for the atomic (e.g., STO or GTO) basis, must be transformed to the molecular orbital basis. This transformation step requires computer resources proportional to the fifth power of the number of basis functions, and thus is one of the more troublesome steps in most configuration interaction calculations. Further details of such calculations are beyond the scope of this text, but are treated in my *QMIC* text.

Perturbation theory

This method uses the single-configuration SCF process to determine a set of orbitals $\{\phi_i\}$. Then, with a zeroth order Hamiltonian equal to the sum of the N electrons' Fock operators $H^0 = \sum_{i=1,N} h_e(i)$, perturbation theory is used to determine the C_I amplitudes for the other CSFs. The Møller–Plesset perturbation (MPPT) procedure is a special case in which the above sum of Fock operators is used to define H^0. The amplitude for the reference CSF is taken as unity and the other CSFs' amplitudes are determined by using $H - H^0$ as the perturbation.

In the MPPT method, once the reference CSF is chosen and the SCF orbitals belonging to this CSF are determined, the wave function Ψ and energy E are determined in an order-by-order manner. The perturbation equations determine what CSFs to include through any particular order. This is one of the primary strengths of this technique; it does not require one to make further choices, in contrast to the CI treatment where one needs to choose which CSFs to include.

For example, the first order wave function correction Ψ^1 is

$$\Psi^1 = -\sum_{i<j,m<n} [\langle i,j|1/r_{12}|m,n\rangle - \langle i,j|1/r_{12}|n,m\rangle] \times [\varepsilon_m - \varepsilon_i + \varepsilon_n - \varepsilon_j]^{-1} \left|\Phi_{i,j}^{m,n}\right\rangle, \tag{6.54}$$

where the SCF orbital energies are denoted ε_k and $\Phi_{i,j}^{m,n}$ represents a CSF that is doubly excited (ϕ_i and ϕ_j are replaced by ϕ_m and ϕ_n) relative to the SCF wave function Φ. Only doubly excited CSFs contribute to the first order wave function; the fact that the contributions from singly excited configurations vanish in Ψ^1 is known as the Brillouin theorem.

The energy E is given through second order as

$$E = E_{\text{SCF}} - \sum_{i<j,\,m<n} |\langle i,j|1/r_{12}|m,n\rangle - \langle i,j|1/r_{12}|n,m\rangle|^2/[\varepsilon_m - \varepsilon_i + \varepsilon_n - \varepsilon_j]. \tag{6.55}$$

Both Ψ and E are expressed in terms of two-electron integrals $\langle i,j|1/r_{12}|m,n\rangle$ (that are sometimes denoted $\langle i,j \mid k,1\rangle$) coupling the virtual spin-orbitals ϕ_m and ϕ_n to the spin-orbitals from which electrons were excited ϕ_i and ϕ_j as well as the orbital energy differences $[\varepsilon_m - \varepsilon_i + \varepsilon_n - \varepsilon_j]$ accompanying such excitations. Clearly, major contributions to the correlation energy are made by double excitations into virtual orbitals $\phi_m\phi_n$ with large $\langle i,j|1/r_{12}|m,n\rangle$ integrals and small orbital energy gaps $[\varepsilon_m - \varepsilon_i + \varepsilon_n - \varepsilon_j]$. In higher order corrections, contributions from CSFs that are singly, triply, etc. excited relative to Φ appear, and additional contributions from the doubly excited CSFs also enter. The various orders of MPPT are usually denoted MPn (e.g., MP2 means second order MPPT).

The coupled-cluster method

As noted above, when the Hartree–Fock wave function Ψ^0 is used as the zeroth order starting point in a perturbation expansion, the first (and presumably most important) corrections to this function are the doubly excited determinants. In early studies of CI treatments of electron correlation, it was also observed that double excitations had the largest C_J coefficients after the SCF wave function, which has the very largest C_J. Moreover, in CI studies that included single, double, triple, and quadruple level excitations relative to the dominant SCF determinant, it was observed that quadruple excitations had the next largest C_J amplitudes after the double excitations. And, very importantly, it was observed that the amplitudes C_{abcd}^{mnpq} of the quadruply excited CSFs Φ_{abcd}^{mnpq} could be very closely approximated as products of the amplitudes $C_{ab}^{mn}C_{cd}^{pq}$ of the doubly excited CSFs Φ_{ab}^{mn} and Φ_{cd}^{pq}. This observation prompted workers to suggest that a more compact and efficient expansion of the correlated wave function might be realized by writing Ψ as

$$\Psi = \exp(T)\Phi, \tag{6.56}$$

where Φ is the SCF determinant and the operator T appearing in the exponential is taken to be a sum of operators

$$T = T_1 + T_2 + T_3 + \cdots + T_N \tag{6.57}$$

that create single (T_1), double (T_2), etc. level excited CSFs when acting on Φ. This way of writing Ψ is called the coupled-cluster (CC) form for Ψ.

In any practical calculation, this sum of T_n operators would be truncated to keep the calculation practical. For example, if excitation operators higher than T_3 were neglected, then one would use $T \approx T_1 + T_2 + T_3$. However, even when T is so truncated, the resultant Ψ would contain excitations of higher order. For example, using the truncation just introduced, we would have

$$\begin{aligned}\Psi = [&1 + T_1 + T_2 + T_3 + 1/2(T_1 + T_2 + T_3)(T_1 + T_2 + T_3)\\ &+ 1/6(T_1 + T_2 + T_3)(T_1 + T_2 + T_3)(T_1 + T_2 + T_3) + \cdots]\Phi.\end{aligned} \tag{6.58}$$

This function contains single excitations (in $T_1\Phi$), double excitations (in $T_2\Phi$ and in $T_1T_1\Phi$), triple excitations (in $T_3\Phi$, $T_2T_1\Phi$, $T_1T_2\Phi$, and $T_1T_1T_1\Phi$), and quadruple excitations in a variety of terms including $T_3T_1\Phi$ and $T_2T_2\Phi$, as well as even higher level excitations. By the design of this wave function, the quadruple excitations $T_2T_2\Phi$ will have amplitudes given as products of the amplitudes of the double excitations $T_2\Phi$ just as were found by earlier CI workers to be most important. Hence, in CC theory, we say that quadruple excitations include "unlinked" products of double excitations arising from the T_2T_2 product; the quadruple excitations arising from $T_4\Phi$ would involve linked terms and would have amplitudes that are not products of double-excitation amplitudes.

After writing Ψ in terms of an exponential operator, one is faced with determining the amplitudes of the various single, double, etc. excitations generated by the T operator acting on Φ. This is done by writing the Schrödinger equation as

$$H\exp(T)\Phi = E\exp(T)\Phi, \tag{6.59}$$

and then multiplying on the left by $\exp(-T)$ to obtain

$$\exp(-T)H\exp(T)\Phi = E\Phi. \tag{6.60}$$

The CC energy is then calculated by multiplying this equation on the left by Φ^* and integrating over the coordinates of all the electrons:

$$\langle\Phi \mid \exp(-T)H\exp(T)\Phi\rangle = E. \tag{6.61}$$

In practice, the combination of operators appearing in this expression is rewritten and dealt with as follows:

$$\begin{aligned}E = \langle\Phi|&T + [H, T] + 1/2[[H, T], T] + 1/6[[[H, T], T], T]\\ &+ 1/24[[[[H, T], T], T], T]|\Phi\rangle.\end{aligned} \tag{6.62}$$

This so-called Baker–Campbell–Hausdorf expansion of the exponential operators can be shown to truncate exactly after the fourth power term shown here. So, once the various operators and their amplitudes that comprise T are known, E is computed using the above expression that involves various powers of the T operators.

The equations used to find the amplitudes (e.g., those of the T_2 operator $\sum_{a,b,m,n} t_{ab}^{mn} T_{ab}^{mn}$, where the t_{ab}^{mn} are the amplitudes and T_{ab}^{mn} are the excitation operators) of the various excitation levels are obtained by multiplying the above Schrödinger equation on the left by an excited determinant of that level and integrating. For example, the equation for the double excitations is

$$0 = \langle \Phi_{ab}^{mn} \mid T + [H,T] + 1/2[[H,T],T] + 1/6[[[H,T],T],T] + 1/24[[[[H,T],T],T],T] | \Phi \rangle. \tag{6.63}$$

The zero arises from the fact that $\langle \Phi_{ab}^{mn} \mid \Phi \rangle = 0$; that is, the determinants are orthonormal. The number of such equations is equal to the number of doubly excited determinants Φ_{ab}^{mn}, which is equal to the number of unknown t_{ab}^{mn} amplitudes. So, the above quartic equations must be solved to determine the amplitudes appearing in the various T_J operators. Then, as noted above, once these amplitudes are known, the energy E can be computed using the earlier quartic equation.

Clearly, the CC method contains additional complexity as a result of the exponential expansion form of the wave function Ψ. However, it is this way of writing Ψ that allows us to automatically build in the fact that products of double excitations are the dominant contributors to quadruple excitations (and $T_2T_2T_2$ is the dominant component of six-fold excitations, not T_6). In fact, the CC method is today the most accurate tool that we have for calculating molecular electronic energies and wave functions.

The density functional method

These approaches provide alternatives to the conventional tools of quantum chemistry which move beyond the single-configuration picture by adding to the wave function more configurations whose amplitudes they each determine in their own way. As noted earlier, these conventional approaches can lead to a very large number of CSFs in the correlated wave function, and, as a result, a need for extraordinary computer resources.

The density functional approaches are different. Here one solves a set of orbital-level equations

$$\left[-\frac{\hbar^2}{2m_e}\nabla^2 - \sum_a \frac{Z_a e^2}{|\mathbf{r} - \mathbf{R}_a|} + \int \rho(\mathbf{r}') \frac{e^2}{|\mathbf{r} - \mathbf{r}'|} d\mathbf{r}' + U(\mathbf{r}) \right] \phi_i = \varepsilon_i \phi_i \tag{6.64}$$

in which the orbitals $\{\phi_i\}$ "feel" potentials due to the nuclear centers (having charges Z_a), Coulombic interaction with the total electron density $\rho(\mathbf{r}')$, and a so-called exchange-correlation potential denoted $U(\mathbf{r}')$. The particular electronic state for which the calculation is being performed is specified by forming a corresponding density $\rho(\mathbf{r}')$. Before going further in describing how DFT calculations are carried out, let us examine the origins underlying this theory.

The so-called Hohenberg–Kohn theorem states that the ground-state electron density $\rho(\mathbf{r})$ describing an N-electron system uniquely determines the potential $V(\mathbf{r})$ in the molecule's electronic Hamiltonian

$$H = \sum_j \left\{ -\frac{\hbar^2}{2m_e}\nabla_j^2 + V(r_j) + \frac{e^2}{2}\sum_{k\neq j}\frac{1}{r_{j,k}} \right\}, \tag{6.65}$$

and, because H determines the ground-state energy and wave function of the system, the ground-state density $\rho(\mathbf{r})$ therefore determines the ground-state properties of the system. The fact that $\rho(r)$ determines $V(r)$ is important because it is $V(r)$ that specifies where the nuclei are located.

The proof of this theorem proceeds as follows:

(i) $\rho(\mathbf{r})$ determines the number of electrons N because $\int \rho(\mathbf{r})d^3r = N$.

(ii) Assume that there are two distinct potentials (aside from an additive constant that simply shifts the zero of total energy) $V(\mathbf{r})$ and $V'(\mathbf{r})$ which, when used in H and H', respectively, to solve for a ground state, produce E_0, $\Psi(r)$ and E_0', $\Psi'(r)$ that have the same one-electron density: $\int |\Psi|^2 dr_2\, dr_3 \ldots dr_N = \rho(\mathbf{r}) = \int |\Psi'|^2 dr_2\, dr_3 \ldots dr_N$.

(iii) If we think of Ψ' as a trial variational wave function for the Hamiltonian H, we know that

$$\begin{aligned} E_0 < \langle\Psi'|H|\Psi'\rangle &= \langle\Psi'|H'|\Psi'\rangle + \int \rho(\mathbf{r})[V(\mathbf{r}) - V'(\mathbf{r})]d^3r \\ &= E_0' + \int \rho(\mathbf{r})[V(\mathbf{r}) - V'(\mathbf{r})]d^3r. \end{aligned} \tag{6.66}$$

(iv) Similarly, taking Ψ as a trial function for the H' Hamiltonian, one finds that

$$E_0' < E_0 + \int \rho(\mathbf{r})[V'(\mathbf{r}) - V(\mathbf{r})]d^3r. \tag{6.67}$$

(v) Adding the equations in (iii) and (iv) gives

$$E_0 + E_0' < E_0 + E_0', \tag{6.68}$$

a clear contradiction unless the electronic state of interest is degenerate.

Hence, there can not be two distinct potentials V and V' that give the same non-degenerate ground state $\rho(\mathbf{r})$. So, the ground-state density $\rho(\mathbf{r})$ uniquely determines N and V, and thus H, and therefore Ψ and E_0. Furthermore, because Ψ determines all properties of the ground state, then $\rho(\mathbf{r})$, in principle, determines all such properties. This means that even the kinetic energy and the electron–electron interaction energy of the ground state are determined by $\rho(\mathbf{r})$. It is easy to see that $\int \rho(\mathbf{r})V(\mathbf{r})\,d^3r = V[\rho]$ gives the average value of the electron–nuclear (plus any additional one-electron additive potential) interaction in terms of the ground-state density $\rho(\mathbf{r})$. However, how are the kinetic energy $T[\rho]$ and the electron–electron interaction $V_{ee}[\rho]$ energy expressed in terms of ρ?

The main difficulty with DFT is that the Hohenberg–Kohn theorem shows that the ground-state values of T, V_{ee}, V, etc. are all unique functionals of the ground state ρ (i.e., that they can, in principle, be determined once ρ is given), but it does not tell us what these functional relations are.

To see how it might make sense that a property such as the kinetic energy, whose operator $-\hbar^2/2m_e\nabla^2$ involves derivatives, can be related to the electron density, consider a simple system of N non-interacting electrons moving in a three-dimensional cubic "box" potential. The energy states of such electrons are known to be

$$E = (h^2/8m_eL^2)\left(n_x^2 + n_y^2 + n_z^2\right), \tag{6.69}$$

where L is the length of the box along the three axes, and n_x, n_y, and n_z are the quantum numbers describing the state. We can view $n_x^2 + n_y^2 + n_z^2 = R^2$ as defining the squared radius of a sphere in three dimensions, and we realize that the density of quantum states in this space is one state per unit volume in the n_x, n_y, n_z space. Because n_x, n_y, and n_z must be positive integers, the volume covering all states with energy less than or equal to a specified energy $E = (h^2/8m_eL^2)R^2$ is 1/8 the volume of the sphere of radius R:

$$\Phi(E) = \frac{1}{8}\left(\frac{4\pi}{3}\right)R^3 = \left(\frac{\pi}{6}\right)\left(\frac{8m_eL^2E}{h^2}\right)^{3/2}. \tag{6.70}$$

Since there is one state per unit of such volume, $\Phi(E)$ is also the number of states with energy less than or equal to E, and is called the integrated density of states. The number of states $g(E)dE$ with energy between E and $E + dE$, the density of states, is the derivative of Φ:

$$g(E) = \frac{d\Phi}{dE} = \left(\frac{\pi}{4}\right)\left(\frac{8m_eL^2}{h^2}\right)^{3/2} E^{1/2}. \tag{6.71}$$

If we calculate the total energy for N electrons that doubly occupy all states having energies up to the so-called Fermi energy (i.e., the energy of the highest occupied molecular orbital HOMO), we obtain the ground-state energy:

$$E_0 = 2\int_0^{E_F} g(E)E\,dE = \left(\frac{8\pi}{5}\right)\left(\frac{2m_e}{h^2}\right)^{3/2} L^3E_F^{5/2}. \tag{6.72}$$

The total number of electrons N can be expressed as

$$N = 2\int_0^{E_F} g(E)\,dE = \left(\frac{8\pi}{3}\right)\left(\frac{2m_e}{h^2}\right)^{3/2} L^3E_F^{3/2}, \tag{6.73}$$

which can be solved for E_F in terms of N to then express E_0 in terms of N instead of in terms of E_F:

$$E_0 = \left(\frac{3h^2}{10m_e}\right)\left(\frac{3}{8\pi}\right)^{2/3} L^3 \left(\frac{N}{L^3}\right)^{5/3}. \tag{6.74}$$

This gives the total energy, which is also the kinetic energy in this case because the potential energy is zero within the "box", in terms of the electron density $\rho(x, y, z) = (N/L^3)$. It therefore may be plausible to express kinetic energies in terms of electron densities $\rho(\mathbf{r})$, but it is by no means clear how to do so for "real" atoms and molecules with electron–nuclear and electron–electron interactions operative.

In one of the earliest DFT models, the Thomas–Fermi theory, the kinetic energy of an atom or molecule is approximated using the above kind of treatment on a "local" level. That is, for each volume element in $\mathbf{r}$ space, one assumes the expression given above to be valid, and then one integrates over all $\mathbf{r}$ to compute the total kinetic energy:

$$T_{TF}[\rho] = \int \left(\frac{3h^2}{10m_e}\right)\left(\frac{3}{8\pi}\right)^{2/3} [\rho(\mathbf{r})]^{5/3} d^3r = C_F \int [\rho(\mathbf{r})]^{5/3} d^3r, \tag{6.75}$$

where the last equality simply defines the C_F constant. Ignoring the correlation and exchange contributions to the total energy, this T is combined with the electron–nuclear V and Coulombic electron–electron potential energies to give the Thomas–Fermi total energy:

$$\begin{aligned} E_{0,TF}[\rho] = C_F \int [\rho(\mathbf{r})]^{5/3} d^3r + \int V(\mathbf{r})\rho(\mathbf{r}) d^3r \\ + e^2/2 \int \rho(\mathbf{r})\rho(\mathbf{r}')/|\mathbf{r} - \mathbf{r}'| \, d^3r d^3r'. \end{aligned} \tag{6.76}$$

This expression is an example of how E_0 is given as a local density functional approximation (LDA). The term local means that the energy is given as a functional (i.e., a function of ρ) which depends only on $\rho(\mathbf{r})$ at points in space but not on $\rho(\mathbf{r})$ at more than one point in space or on spatial derivatives of $\rho(\mathbf{r})$.

Unfortunately, the Thomas–Fermi energy functional does not produce results that are of sufficiently high accuracy to be of great use in chemistry. What is missing in this theory are (i) the exchange energy and (ii) the electronic correlation energy. Moreover, the kinetic energy is treated only in the approximate manner described.

Dirac was able to address the exchange energy for the "uniform electron gas" (N Coulomb interacting electrons moving in a uniform positive background charge whose magnitude balances the charge of the N electrons). If the exact expression for the exchange energy of the uniform electron gas is applied on a local level, one obtains the commonly used Dirac local density approximation to the exchange energy:

$$E_{\mathrm{ex,Dirac}}[\rho] = -C_x \int [\rho(\mathbf{r})]^{4/3}\, d^3r, \tag{6.77}$$

with $C_x = (3/4)(3/\pi)^{1/3}$. Adding this exchange energy to the Thomas–Fermi total energy $E_{0,\mathrm{TF}}[\rho]$ gives the so-called Thomas–Fermi–Dirac (TFD) energy functional.

Because electron densities vary rather strongly spatially near the nuclei, corrections to the above approximations to $T[\rho]$ and $E_{\mathrm{ex.Dirac}}$ are needed. One of the more commonly used so-called gradient-corrected approximations is that invented by Becke, and referred to as the Becke88 exchange functional:

$$E_{\mathrm{ex}}(\mathrm{Becke88}) = E_{\mathrm{ex,Dirac}}[\rho] - \gamma \int x^2 \rho^{4/3} (1 + 6\gamma x \sin h^{-1} x)^{-1} d\mathbf{r}, \tag{6.78}$$

where $x = \rho^{-4/3}|\nabla\rho|$, and γ is a parameter chosen so that the above exchange energy can best reproduce the known exchange energies of specific electronic states of the inert gas atoms (Becke finds γ to equal 0.0042). A common gradient correction to the earlier $T[\rho]$ is called the Weizsacker correction and is given by

$$\delta T_{\mathrm{Weizsacker}} = \frac{1}{72} \frac{\hbar}{m_e} \int \frac{|\nabla\rho(\mathbf{r})|^2}{\rho(\mathbf{r})} d\mathbf{r}. \tag{6.79}$$

Although the above discussion suggests how one might compute the ground-state energy once the ground-state density $\rho(\mathbf{r})$ is given, one still needs to know how to obtain ρ. Kohn and Sham (KS) introduced a set of so-called KS orbitals obeying the following equation:

$$\left\{ -\frac{\hbar^2}{2m}\nabla^2 + V(\mathbf{r}) + \frac{e^2}{2} \int \frac{\rho(\mathbf{r}')}{|\mathbf{r} - \mathbf{r}'|} d\mathbf{r}' + U_{\mathrm{xc}}(\mathbf{r}) \right\} \phi_j = \varepsilon_j \phi_j, \tag{6.80}$$

where the so-called exchange-correlation potential $U_{\mathrm{xc}}(\mathbf{r}) = \delta E_{\mathrm{xc}}[\rho]/\delta\rho(\mathbf{r})$ could be obtained by functional differentiation if the exchange-correlation energy functional $E_{\mathrm{xc}}[\rho]$ were known. KS also showed that the KS orbitals $\{\phi_j\}$ could be used to compute the density ρ by simply adding up the orbital densities multiplied by orbital occupancies n_j:

$$\rho(\mathbf{r}) = \sum_j n_j |\phi_j(\mathbf{r})|^2 \tag{6.81}$$

(here $n_j = 0$, 1, or 2 is the occupation number of the orbital ϕ_j in the state being studied) and that the kinetic energy should be calculated as

$$T = \sum_j n_j \langle \phi_j(\mathbf{r})| - \hbar^2/2m\nabla^2 |\phi_j(\mathbf{r})\rangle. \tag{6.82}$$

The same investigations of the idealized "uniform electron gas" that identified the Dirac exchange functional found that the correlation energy (per electron) could also be written exactly as a function of the electron density ρ of the system, but only in two limiting cases – the high-density limit (large ρ) and the low-density limit. There still exists no exact expression for the correlation energy even for the uniform electron gas that is valid at arbitrary values of ρ. Therefore, much work has been devoted to creating efficient and accurate interpolation formulas connecting the low- and high-density uniform electron gas. One such expression is

$$E_{\mathrm{C}}[\rho] = \int \rho(\mathbf{r})\varepsilon_{\mathrm{c}}(\rho)d\mathbf{r}, \tag{6.83}$$

where

$$\varepsilon_{\mathrm{c}}(\rho) = \frac{A}{2}\left\{\ln\frac{x}{X} + \frac{2b}{Q}\tan^{-1}\left(\frac{Q}{2x+b}\right) - \frac{bx_0}{X_0}\right. \times \left.\left[\ln\frac{(x-x_0)^2}{X} + \frac{2(b+2x_0)}{Q}\tan^{-1}\left(\frac{Q}{2x+b}\right)\right]\right\} \tag{6.84}$$

is the correlation energy per electron. Here $x = r_{\mathrm{s}}^{1/2}$, $X = x^2 + bx + c$, $X_0 = x_0^2 + bx_0 + c$ and $Q = (4c - b^2)^{1/2}$, $A = 0.0621814$, $x_0 = -0.409286$, $b = 13.0720$, and $c = 42.7198$. The parameter r_{s} is how the density ρ enters since $4/3\pi r_s^3$ is equal to $1/\rho$; that is, r_{s} is the radius of a sphere whose volume is the effective volume occupied by one electron. A reasonable approximation to the full $E_{\mathrm{xc}}[\rho]$ would contain the Dirac (and perhaps gradient corrected) exchange functional plus the above $E_{\mathrm{C}}[\rho]$, but there are many alternative approximations to the exchange-correlation energy functional. Currently, many workers are doing their best to "cook up" functionals for the correlation and exchange energies, but no one has yet invented functionals that are so reliable that most workers agree to use them.

To summarize, in implementing any DFT, one usually proceeds as follows:

(i) An atomic orbital basis is chosen in terms of which the KS orbitals are to be expanded.

(ii) Some initial guess is made for the LCAO-KS expansion coefficients $C_{j,a}: \phi_j = \sum_a C_{j,a}\chi_a$.

(iii) The density is computed as $\rho(\mathbf{r}) = \sum_j n_j|\phi_j(\mathbf{r})|^2$. Often, $\rho(\mathrm{r})$ itself is expanded in an atomic orbital basis, which need not be the same as the basis used for the ϕ_j, and the expansion coefficients of ρ are computed in terms of those of the ϕ_j. It is also common to use an atomic orbital basis to expand $\rho^{1/3}(\mathbf{r})$ which, together with ρ, is needed to evaluate the exchange-correlation functional's contribution to E_0.

(iv) The current iteration's density is used in the KS equations to determine the Hamiltonian $\{-\hbar^2/2m\nabla^2 + V(\mathbf{r}) + e^2/2\int \rho(\mathbf{r}')/|\mathbf{r}-\mathbf{r}'|\,d\mathbf{r}' + U_{\mathrm{xc}}(\mathbf{r})\}$ whose "new" eigenfunctions $\{\phi_j\}$ and eigenvalues $\{\varepsilon_j\}$ are found by solving the KS equations.

(v) These new ϕ_j are used to compute a new density, which, in turn, is used to solve a new set of KS equations. This process is continued until convergence is reached (i.e., until the ϕ_j used to determine the current iteration's ρ are the same ϕ_j that arise as solutions on the next iteration.

(vi) Once the converged $\rho(\mathbf{r})$ is determined, the energy can be computed using the earlier expression

$$E[\rho] = \sum_j n_j \langle \phi_j(\mathbf{r})| - \hbar^2/2m\nabla^2|\phi_j(\mathbf{r})\rangle + \int V(\mathbf{r})\rho(\mathbf{r})d\mathbf{r}$$
$$+ e^2/2 \int \rho(\mathbf{r})\rho(\mathbf{r}')/|\mathbf{r} - \mathbf{r}'|\, d\mathbf{r}\, d\mathbf{r}' + E_{xc}[\rho]. \quad (6.85)$$

Energy difference methods

In addition to the methods discussed above for treating the energies and wave functions as solutions to the electronic Schrödinger equation, there exists a family of tools that allow one to compute energy differences "directly" rather than by finding the energies of pairs of states and subsequently subtracting them. Various energy differences can be so computed: differences between two electronic states of the same molecule (i.e., electronic excitation energies ΔE), differences between energy states of a molecule and the cation or anion formed by removing or adding an electron (i.e., ionization potentials (IPs) and electron affinities (EAs)).

Because of space limitations, we will not be able to elaborate much further on these methods. However, it is important to stress that:

(i) These so-called Green function or propagator methods utilize essentially the same input information (e.g., atomic orbital basis sets) and perform many of the same computational steps (e.g., evaluation of one- and two-electron integrals, formation of a set of mean-field molecular orbitals, transformation of integrals to the MO basis, etc.) as do the other techniques discussed earlier.
(ii) These methods are now rather routinely used when ΔE, IP, or EA information is sought.

The basic ideas underlying most if not all of the energy difference methods are:

(i) One forms a reference wave function Ψ (this can be of the SCF, MPn, CI, CC, DFT, etc. variety); the energy differences are computed relative to the energy of this function.
(ii) One expresses the final-state wave function Ψ' (i.e., that describing the excited, cation, or anion state) in terms of an operator Ω acting on the reference Ψ: $\Psi' = \Omega\Psi$. Clearly, the Ω operator must be one that removes or adds an electron when one is attempting to compute IPs or EAs, respectively.
(iii) One writes equations which Ψ and Ψ' are expected to obey. For example, in the early development of these methods, the Schrödinger equation itself was assumed to be obeyed, so $H\Psi = E\Psi$ and $H\Psi' = E'\Psi'$ are the two equations.
(iv) One combines $\Omega\Psi = \Psi'$ with the equations that Ψ and Ψ' obey to obtain an equation that Ω must obey. In the above example, one (a) uses $\Omega\Psi = \Psi'$ in the Schrödinger equation for Ψ', (b) allows Ω to act from the left on the Schrödinger

equation for Ψ, and (c) subtracts the resulting two equations to achieve $(H\Omega - \Omega H)\Psi = (E' - E)\Omega\Psi$, or, in commutator form $[H, \Omega]\Psi = \Delta E\Omega\Psi$.

(v) One can, for example, express Ψ in terms of a superposition of configurations $\Psi = \sum_J C_J \Phi_J$ whose amplitudes C_J have been determined from a CI or MPn calculation and express Ω in terms of operators $\{O_K\}$ that cause single, double, etc. level excitations (for the IP (EA) cases, Ω is given in terms of operators that remove (add), remove and singly excite (add and singly excite, etc.) electrons): $\Omega = \sum_K D_K O_K$.

(vi) Substituting the expansions for Ψ and for Ω into the equation of motion (EOM) $[H, \Omega]\Psi = \Delta E\Omega\Psi$, and then projecting the resulting equation on the left against a set of functions (e.g., $\{O_{K'}|\Psi\rangle\}$) gives a matrix eigenvalue–eigenvector equation

$$\sum_K \langle O_{K'}\Psi \mid [H, O_K]\Psi\rangle D_K = \Delta E \sum_K \langle O_{K'}\Psi \mid O_K\Psi\rangle D_K \tag{6.86}$$

to be solved for the D_K operator coefficients and the excitation (or IP or EA) energies ΔE. Such are the working equations of the EOM (or Green function or propagator) methods.

In recent years, these methods have been greatly expanded and have reached a degree of reliability where they now offer some of the most accurate tools for studying excited and ionized states. In particular, the use of time-dependent variational principles has allowed a much more rigorous development of equations for energy differences and non-linear response properties. In addition, the extension of the EOM theory to include coupled-cluster reference functions now allows one to compute excitation and ionization energies using some of the most accurate *ab initio* tools.

The Slater–Condon rules

To form Hamiltonian matrix elements $H_{K,L}$ between any pair of Slater determinants, one uses the so-called Slater–Condon rules. These rules express all non-vanishing matrix elements involving either one- or two-electron operators. One-electron operators are additive and appear as

$$F = \sum_i f(i); \tag{6.87}$$

two-electron operators are pairwise additive and appear as

$$G = \sum_{ij} g(i, j). \tag{6.88}$$

The Slater–Condon rules give the matrix elements between two determinants

$$|\rangle = |\phi_1\phi_2\phi_3\cdots\phi N| \tag{6.89}$$

and

$$|'\rangle = |\phi'_1\phi'_2\phi'_3 \cdots \phi' N| \tag{6.90}$$

for any quantum mechanical operator that is a sum of one- and two-electron operators $(F + G)$. It expresses these matrix elements in terms of one- and two-electron integrals involving the spin-orbitals that appear in $|\rangle$ and $|'\rangle$ and the operators f and g.

As a first step in applying these rules, one must examine $|\rangle$ and $|'\rangle$ and determine by how many (if any) spin-orbitals $|\rangle$ and $|'\rangle$ differ. In so doing, one may have to reorder the spin-orbitals in one of the determinants to achieve maximal coincidence with those in the other determinant; it is essential to keep track of the number of permutations (N_p) that one makes in achieving maximal coincidence. The results of the Slater–Condon rules given below are then multiplied by $(-1)^{N_\mathrm{p}}$ to obtain the matrix elements between the original $|\rangle$ and $|'\rangle$. The final result does not depend on whether one chooses to permute $|\rangle$ or $|'\rangle$.

The Hamiltonian is, of course, a specific example of such an operator; the electric dipole operator $\sum_i e\mathbf{r}_i$ and the electronic kinetic energy $-\hbar^2/2m_\mathrm{e} \sum_i \nabla_i^2$ are examples of one-electron operators (for which one takes $g = 0$); the electron–electron Coulomb interaction $\sum_{i>j} e^2/r_{ij}$ is a two-electron operator (for which one takes $f = 0$).

Once maximal coincidence has been achieved, the Slater–Condon (SC) rules provide the following prescriptions for evaluating the matrix elements of any operator $F + G$ containing a one-electron part $F = \sum_i f(i)$ and a two-electron part $G = \sum_{ij} g(i, j)$:

(i) If $|\rangle$ and $|'\rangle$ are identical, then

$$\langle|F + G|\rangle = \sum_i \langle\phi_i|f|\phi_i\rangle + \sum_{i>j} [\langle\phi_i\phi_j|g|\phi_i\phi_j\rangle - \langle\phi_i\phi_j|g|\phi_j\phi_i\rangle], \tag{6.91}$$

where the sums over i and j run over all spin-orbitals in $|\rangle$;

(ii) If $|\rangle$ and $|'\rangle$ differ by a single spin-orbital mismatch ($\phi_\mathrm{p} \neq \phi'_p$),

$$\langle|F + G|'\rangle = \langle\phi_p|f|\phi'_p\rangle + \sum_j [\langle\phi_p\phi_j|g|\phi'_p\phi_j\rangle - \langle\phi_p\phi_j|g|\phi_j\phi'_p\rangle], \tag{6.92}$$

where the sum over j runs over all spin-orbitals in $|\rangle$ except ϕ_p;

(iii) If $|\rangle$ and $|'\rangle$ differ by two spin-orbitals ($\phi_p \neq \phi'_p$ and $\phi_q \neq \phi'_q$),

$$\langle|F + G|'\rangle = \langle\phi_p\phi_q|g|\phi'_p\phi'_q\rangle - \langle\phi_p\phi_q|g|\phi'_q\phi'_p\rangle \tag{6.93}$$

(note that the F contribution vanishes in this case);

(iv) If $|\rangle$ and $|'\rangle$ differ by three or more spin orbitals, then

$$\langle|F + G|'\rangle = 0; \tag{6.94}$$

(v) For the identity operator I, the matrix elements $\langle|I|'\rangle = 0$ if $|\rangle$ and $|'\rangle$ differ by one or more spin-orbitals (i.e., the Slater determinants are orthonormal if their spin-orbitals are).

Recall that each of these results is subject to multiplication by a factor of $(-1)^{N_p}$ to account for possible ordering differences in the spin-orbitals in $|\rangle$ and $|'\rangle$.

In these expressions,

$$\langle \phi_i | f | \phi_j \rangle \tag{6.95}$$

is used to denote the one-electron integral

$$\int \phi_i^*(r) f(r) \phi_j(r) dr \tag{6.96}$$

and

$$\langle \phi_i \phi_j | g | \phi_k \phi_l \rangle \tag{6.97}$$

(or in short hand notation $\langle ij \mid kl \rangle$) represents the two-electron integral

$$\int \phi_i^*(r) \phi_j^*(r') g(r, r') \phi_k(r)\, \phi_l(r') dr\, dr'. \tag{6.98}$$

The notation $\langle ij \mid kl \rangle$ introduced above gives the two-electron integrals for the $g(r, r')$ operator in the so-called Dirac notation, in which the i and k indices label the spin-orbitals that refer to the coordinates r and the j and l indices label the spin-orbitals referring to coordinates r'. The r and r' denote r, θ, ϕ, σ and $r', \theta', \phi', \sigma'$ (with σ and σ' being the α or β spin functions).

If the operators f and g do not contain any electron spin operators, then the spin integrations implicit in these integrals (all of the ϕ_i are spin-orbitals, so each ϕ is accompanied by an α or β spin function and each ϕ^* involves the adjoint of one of the α or β spin functions) can be carried out as $\langle \alpha \mid \alpha \rangle = 1$, $\langle \alpha \mid \beta \rangle = 0$, $\langle \beta \mid \alpha \rangle = 0$, $\langle \beta \mid \beta \rangle = 1$, thereby yielding integrals over spatial orbitals.

Atomic units

The electronic Hamiltonian that appears throughout this text is commonly expressed in the literature and in other texts in so-called atomic units (au). In that form, it is written as follows:

$$H_e = \sum_j \left\{ -\frac{1}{2}\nabla_j^2 - \sum_a \frac{Z_a}{r_{j,a}} \right\} + \sum_{j<k} \frac{1}{r_{j,k}}. \tag{6.99}$$

Atomic units are introduced to remove all of the $\hbar$, e, and m_e factors from the Schrödinger equation.

To effect the unit transformation that results in the Hamiltonian appearing as above, one notes that the kinetic energy operator scales as r_j^{-2} whereas the Coulomb potentials scale as r_j^{-1} and as $r_{j,k}^{-1}$. So, if each of the Cartesian coordinates of the electrons and nuclei were expressed as a unit of length a_0 multiplied by a dimensionless length factor, the kinetic energy operator would involve terms of the form $(-\hbar^2/2(a_0)^2 m_e)\nabla_j^2$, and the Coulomb potentials would appear as $Z_a e^2/(a_0) r_{j,a}$ and $e^2/(a_0) r_{j,k}$, with the $r_{j,a}$ and $r_{j,k}$ factors now referring to the dimensionless coordinates. A factor of e^2/a_0 (which has units of energy since

a_0 has units of length) can then be removed from the Coulomb and kinetic energies, after which the kinetic energy terms appear as $(-\hbar^2/2(e^2 a_0)m_e)\nabla_j^2$ and the potential energies appear as $Z_a/r_{j,a}$ and $1/r_{j,k}$. Then, choosing $a_0 = \hbar^2/e^2 m_e$ changes the kinetic energy terms into $-1/2\nabla_j^2$; as a result, the entire electronic Hamiltonian takes the form given above in which no e^2, m_e, or $\hbar^2$ factors appear. The value of the so-called Bohr radius $a_0 = \hbar^2/e^2 m_e$ turns out to be 0.529 Å, and the so-called Hartree energy unit e^2/a_0, which factors out of H_e, is 27.21 eV or 627.51 kcal mol^{-1}.

6.3 Molecules embedded in condensed media

Often one wants to model the behavior of a molecule or ion that is not isolated as it might be in a gas-phase experiment. When one attempts to describe a system that is embedded, for example, in a crystal lattice, in a liquid or in a glass, one has to have some way to treat both the effects of the surrounding "medium" on the molecule of interest and the motions of the medium's constituents. In so-called quantum mechanics–molecular mechanics (QM–MM) approaches to this problem, one treats the molecule or ion of interest using the electronic structure methods outlined earlier in this chapter, but with one modification. The one-electron component of the Hamiltonian, which contains the electron–nuclei Coulomb potential $\sum_{a,i}(-Z_a e^2/|r_i - R_a|)$, is modified to also contain a term that describes the potential energy of interaction of the electrons and nuclei with the surrounding medium. In the simplest such models, this solvation potential depends only on the dielectric constant of the surroundings. In more sophisticated models, the surroundings are represented by a collection of (fractional) point charges that may also be attributed with local dipole moments and polarizabilities that allow them to respond to changes in the internal charge distribution of the molecule or ion. The locations of such partial charges and the magnitudes of their dipoles and polarizabilities are determined to make the resultant solvation potential reproduce known (from experiment or other simulations) solvation characteristics (e.g., solvation energy, radial distribution functions) in a variety of calibration cases.

In addition to describing how the surroundings affect the Hamiltonian of the molecule or ion of interest, one needs to describe the motions or spatial distributions of the medium's constituent atoms or molecules. This is usually done within a purely classical treatment of these degrees of freedom. That is, if equilibrium properties of the solvated system are to be simulated, then Monte-Carlo (MC) sampling (this subject is treated in Chapter 7) of the surrounding medium's coordinates is used. Within such a MC sampling, the potential energy of the entire system is calculated as a sum of two parts:

(i) the electronic energy of the solute molecule or ion, which contains the interaction energy of the molecule's electrons and nuclei with the surrounding medium, plus

(ii) the intra-medium potential energy, which is taken to be of a simple molecular mechanics (MM) force field character (i.e., to depend on inter-atomic distances and internal angles in an analytical and easily computed manner).

If, alternatively, dynamical characteristics of the solvated species are to be simulated, a classical molecular dynamics (MD) treatment is used. In this approach, the solute-medium and internal-medium potential energies are handled in the same way as in the MC case but where the time evolution of the medium's coordinates are computed using the MD techniques discussed in Chapter 7.

6.4 High-end methods for treating electron correlation

Although their detailed treatment is beyond the scope of this text, it is important to appreciate that new approaches are always under development in all areas of theoretical chemistry. In this section, I want to introduce you to two tools that are proving to offer the highest precision in the treatment of electron correlation energies. These are the so-called quantum Monte-Carlo and $r_{1,2}$ approaches to this problem.

6.4.1 Quantum Monte-Carlo

In this method, one first re-writes the time-dependent Schrödinger equation

$$i\hbar\frac{d\Psi}{dt} = -\frac{\hbar^2}{2m_e}\sum_j \nabla_j^2\Psi + V\Psi \tag{6.100}$$

for negative imaginary values of the time variable t (i.e., one simply replaces t by $-i\tau$). This gives

$$\frac{d\Psi}{d\tau} = \frac{\hbar}{2m_e}\sum_j \nabla_j^2\Psi - \frac{V}{\hbar}\Psi, \tag{6.101}$$

which is analogous to the well-known diffusion equation

$$\frac{dC}{dt} = D\nabla^2 C + SC. \tag{6.102}$$

The re-written Schrödinger equation can be viewed as a diffusion equation in the $3N$ spatial coordinates of the N electrons with a diffusion coefficient D that is related to the electrons' mass m_e by

$$D = \hbar/2m_e. \tag{6.103}$$

The so-called source and sink term S in the diffusion equation is related to the electron–nuclear and electron–electron Coulomb potential energies denoted V:

$$S = -V/\hbar. \tag{6.104}$$

In regions of space where V is large and negative (i.e., where the potential is highly attractive), V is large and negative, so S is large and positive. This causes the concentration C of the diffusing material to accumulate in such regions. Likewise, where V is positive, C will decrease. Clearly by recognizing Ψ as the "concentration" variable in this analogy, one understands that Ψ will accumulate where V is negative and will decay where V is positive, as one expects.

So far, we see that the "trick" of taking t to be negative and imaginary causes the electronic Schrödinger equation to look like a $3N$-dimensional diffusion equation. Why is this useful and why does this trick "work"? It is useful because, as we see in Chapter 7, Monte-Carlo methods are highly efficient tools for solving certain equations; it turns out that the diffusion equation is one such case. So, the Monte-Carlo approach can be used to solve the imaginary-time-dependent Schrödinger equation even for systems containing many electrons. But, what does this imaginary time mean?

To understand the imaginary time trick, let us recall that any wave function (e.g., the trial wave function with which one begins to use Monte-Carlo methods to propagate the diffusing Ψ function) Φ can be written in terms of the exact eigenfunctions $\{\Psi_K\}$ of the Hamiltonian

$$H = -\hbar^2/2m_e \sum_j \nabla_j^2 + V \tag{6.105}$$

as follows:

$$\Phi = \sum_K C_K \Psi_K. \tag{6.106}$$

If the Monte-Carlo method can, in fact, be used to propagate forward in time such a function but with $t = -i\tau$, then it will, in principle, generate the following function at such an imaginary time:

$$\Phi = \sum_K C_K \Psi_K \exp(-iE_K t/\hbar) = \sum_K C_K \Psi_K \exp(-E_K \tau/\hbar). \tag{6.107}$$

As τ increases, the relative amplitudes $\{C_K \exp(-E_K \tau/\hbar)\}$ of all states but the lowest state (i.e., that with smallest E_K) will decay compared to the amplitude $C_0 \exp(-E_0 \tau/\hbar)$ of the lowest state. So, the time-propagated wave function will, at long enough τ, be dominated by its lowest-energy component. In this way, the quantum Monte-Carlo propagation method can generate a wave function in $3N$ dimensions that approaches the ground-state wave function.

It has turned out that this approach, which avoids tackling the N-electron correlation problem "head-on", has proven to yield highly accurate energies and wave functions that display the proper cusps near nuclei as well as the negative cusps (i.e., the wave function vanishes) whenever two electrons' coordinates approach one another. Finally, it turns out that by using a "starting function" Φ

of a given symmetry and radial nodal structure, this method can be extended to converge to the lowest-energy state of the chosen symmetry and nodal structure. So, the method can be used on excited states also. In the next chapter, you will learn how the Monte-Carlo tools can be used to simulate the behavior of many-body systems (e.g., the N-electron system we just discussed) in a highly efficient and easily parallelized manner.

6.4.2 The $r_{1,2}$ method

In this approach to electron correlation, one employs a trial variational wave function that contains components that depend on the inter-electron distances $r_{i,j}$ explicitly. By so doing, one does not rely on the polarized orbital pair approach introduced earlier in this chapter to represent all of the correlations among the electrons. An example of such an explicitly correlated wave function is

$$\psi = |\phi_1\phi_2\phi_3\cdots\phi_N|\left(1 + a\sum_{i<j} r_{i,j}\right), \quad (6.108)$$

which consists of an antisymmetrized product of N spin-orbitals multiplied by a factor that is symmetric under interchange of any pair of electrons and contains the electron–electron distances in addition to a single variational parameter a. Such a trial function is said to contain linear $-r_{1,2}$ correlation factors. Of course, it is possible to write many other forms for such an explicitly correlated trial function. For example, one could use

$$\psi = |\phi_1\phi_2\phi_3\cdots\phi_N| \exp\left(-a\sum_{i<j} r_{i,j}\right) \quad (6.109)$$

as a trial function. Both the linear and the exponential forms have been used in developing this tool of quantum chemistry. Because the integrals that must be evaluated when one computes the Hamiltonian expectation value $\langle\psi|H|\psi\rangle$ are most computationally feasible (albeit still very taxing) when the linear form is used, this particular parameterization is currently the most widely used.

Both the $r_{1,2}$ and quantum Monte-Carlo methods currently are used when one wishes to obtain the absolute highest precision in an electronic structure calculation. The computational requirements of both of these methods are very high, so, at present, they can only be used on species containing fewer than c. 100 electrons. However, with the power and speed of computers growing as fast as they are, it is likely that these high-end methods will be more and more widely used as time goes by.

Experimental probes of electronic structure

6.5 Visible and ultraviolet spectroscopy

Visible and ultraviolet spectroscopies are used to study transitions between states of a molecule or ion in which the electrons' orbital occupancy changes. We call these electronic transitions, and they usually require light in the 5000 cm^{-1} to 100 000 cm^{-1} regime. When such transitions occur, the initial and final states generally differ in their electronic, vibrational, and rotational energies because any change to the electrons' orbital occupancy will induce changes in the vibrational and rotational character. Excitations of inner-shell and core orbital electrons may require even higher energy photons, as would excitations that eject an electron. The interpretation of all such spectroscopic data relies heavily on theory as this section is designed to illustrate.

6.5.1 Electronic transition dipole and use of point group symmetry

The interaction of electomagnetic radiation with a molecule's electrons and nuclei can be treated using perturbation theory. Because this is not a text specializing in spectroscopy, we will not go into this derivation here. If you are interested in seeing this treatment, my *QMIC* text covers it in some detail as do most books on molecular spectroscopy. The result is a standard expression

$$R_{\mathrm{i,f}} = (2\pi/\hbar^2) g(\omega_{\mathrm{f,i}}) \mid \mathbf{E}_0 \cdot \langle \Phi_{\mathrm{f}} | \boldsymbol{\mu} | \Phi_{\mathrm{i}} \rangle |^2 \tag{6.110}$$

for the rate of photon absorption between initial Φ_{i} and final Φ_{f} states. In this equation, $g(\omega)$ is the intensity of the photon source at the frequency ω, $\omega_{\mathrm{f,i}}$ is the frequency corresponding to the transition under study, and $\mathbf{E}_0$ is the electric field vector of the photon field. The vector $\boldsymbol{\mu}$ is the electric dipole moment of the electrons and nuclei in the molecule.

Because each of these wave functions is a product of an electronic ψ_{e}, a vibrational and a rotational function, we realize that the electronic integral appearing in this rate expression involves

$$\langle \psi_{\mathrm{ef}} | \boldsymbol{\mu} | \psi_{\mathrm{ei}} \rangle = \mu_{\mathrm{f,i}}(\mathbf{R}), \tag{6.111}$$

a transition dipole matrix element between the initial ψ_{ei} and final ψ_{ef} electronic wave functions. This element is a function of the internal vibrational coordinates of the molecule, and is a vector locked to the molecule's internal axis frame.

Molecular point group symmetry can often be used to determine whether a particular transition's dipole matrix element will vanish and, as a result, the electronic transition will be "forbidden" and thus predicted to have zero intensity. If the direct product of the symmetries of the initial and final electronic states ψ_{ei} and ψ_{ef} does not match the symmetry of the electric dipole operator (which has the symmetry of its x, y, and z components; these symmetries can be read off the right-most column of the character tables), the matrix element will vanish.

For example, the formaldehyde molecule H_2CO has a ground electronic state that has $^1A_1$ symmetry in the C_{2v} point group. Its $\pi \rightarrow \pi^*$ singlet excited state also has $^1A_1$ symmetry because both the π and π^* orbitals are of b_1 symmetry. In contrast, the lowest $n \rightarrow \pi^*$ (these orbitals are shown in Fig. 6.15) singlet excited state is of $^1A_2$ symmetry because the highest energy oxygen centered non-bonding orbital is of b_2 symmetry and the π^* orbital is of b_1 symmetry, so the Slater determinant in which both the n and π^* orbitals are singly occupied has its symmetry dictated by the $b_2 \times b_1$ direct product, which is A_2.

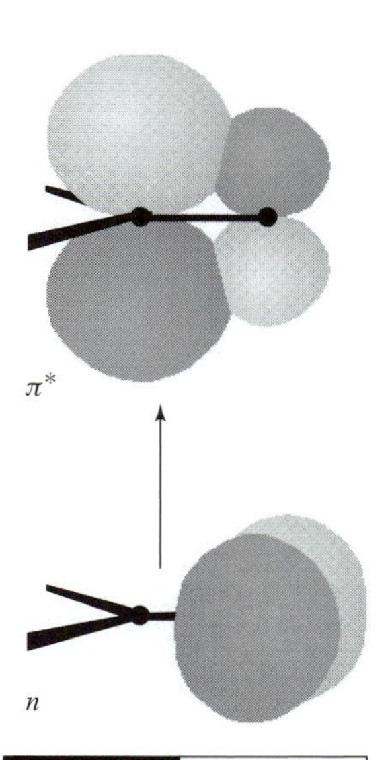

Figure 6.15 Electronic transition from the non-bonding n orbital to the antibonding π^* orbital of formaldehyde.

The $\pi \rightarrow \pi^*$ transition thus involves ground ($^1A_1$) and excited ($^1A_1$) states whose direct product ($A_1 \times A_1$) is of A_1 symmetry. This transition thus requires that the electric dipole operator possess a component of A_1 symmetry. A glance at the C_{2v} point group's character table shows that the molecular z-axis is of A_1 symmetry. Thus, if the light's electric field has a non-zero component along the C_2 symmetry axis (the molecule's z-axis), the $\pi \rightarrow \pi^*$ transition is predicted to be allowed. Light polarized along either of the molecule's other two axes cannot induce this transition.

In contrast, the $n \rightarrow \pi^*$ transition has a ground–excited state direct product of A_2 symmetry. The C_{2v}'s point group character table shows that the electric dipole operator (i.e., its x, y, and z components in the molecule-fixed frame) has no component of A_2 symmetry; thus, light of no electric field orientation can induce this $n \rightarrow \pi^*$ transition. We thus say that the $n \rightarrow \pi^*$ transition is forbidden.

The above examples illustrate one of the most important applications of visible–UV spectroscopy. The information gained in such experiments can be used to infer the symmetries of the electronic states and hence of the orbitals occupied in these states. It is in this manner that this kind of experiment probes electronic structures.

6.5.2 The Franck–Condon factors

Beyond such electronic symmetry analysis, it is also possible to derive vibrational selection rules for electronic transitions that are allowed. It is conventional to expand $\mu_{\mathrm{f,i}}(\mathbf{R})$ in a power series about the equilibrium geometry of the initial electronic state (since this geometry is characteristic of the molecular structure prior to photon absorption):

$$\mu_{\mathrm{f,i}}(\mathbf{R}) = \mu_{\mathrm{f,i}}(\mathbf{R}_\mathrm{e}) + \sum_a \frac{\partial \mu_{\mathrm{f,i}}}{\partial R_a}(R_a - R_{a,\mathrm{e}}) + \cdots . \tag{6.112}$$

The first term in this expansion, when substituted into the integral over the vibrational coordinates, gives $\mu_{\mathrm{f,i}}(\mathbf{R}_\mathrm{e})\langle \chi_{\mathrm{vf}} \mid \chi_{\mathrm{vi}} \rangle$, which has the form of the electronic transition dipole multiplied by the "overlap integral" between the initial and final

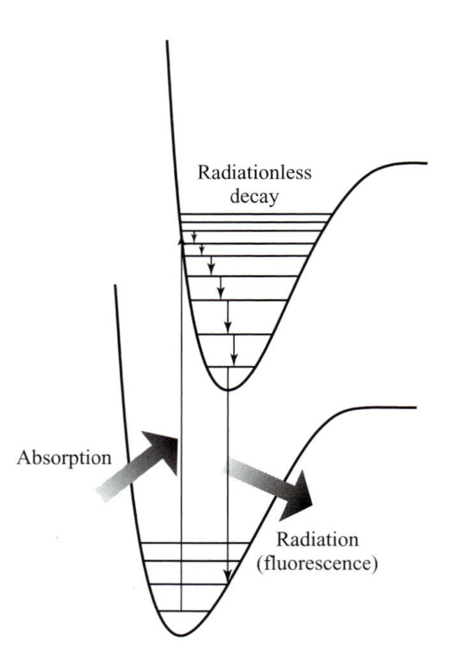

Figure 6.16 Absorption from one initial state to one final state followed by relaxation and then emission from the lowest state of the upper surface.

vibrational wave functions. The $\mu_{f,i}(\mathbf{R}_e)$ factor was discussed above; it is the electronic transition integral evaluated at the equilibrium geometry of the absorbing state. Symmetry can often be used to determine whether this integral vanishes, as a result of which the transition will be "forbidden".

The vibrational overlap integrals $\langle \chi_{vf} \mid \chi_{vi} \rangle$ do not necessarily vanish because χ_{vf} and χ_{vi} are eigenfunctions of different vibrational Hamiltonians. χ_{vf} is an eigenfunction whose potential energy is the final electronic state's energy surface; χ_{vi} has the initial electronic state's energy surface as its potential. The squares of these $\langle \chi_{vf} \mid \chi_{vi} \rangle$ integrals, which are what eventually enter into the transition rate expression $R_{i,f} = (2\pi/\hbar^2)\, g(\omega_{f,i}) \mid \mathbf{E}_0 \cdot \langle \Phi_f|\mu|\Phi_i\rangle|^2$, are called "Franck–Condon factors". Their relative magnitudes play strong roles in determining the relative intensities of various vibrational "bands" (i.e., peaks) within a particular electronic transition's spectrum. In Fig. 6.16, I show two potential energy curves and illustrate the kinds of absorption (and emission) transitions that can occur when the two electronic states have significantly different geometries.

Whenever an electronic transition causes a large change in the geometry (bond lengths or angles) of the molecule, the Franck–Condon factors tend to display the characteristic "broad progression" shown in Fig. 6.17 when considered for one initial-state vibrational level v_i and various final-state vibrational levels v_f. Notice that as one moves to higher v_f values, the energy spacing between the states $(E_{vf} - E_{vf-1})$ decreases; this, of course, reflects the anharmonicity in the excited-state vibrational potential. For this example, the transition to the $v_f = 2$ state has the largest Franck–Condon factor. This means that the overlap of the initial

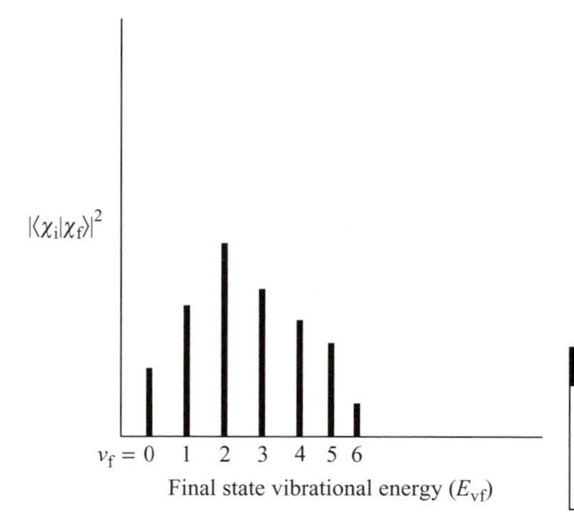

Figure 6.17 Broad Franck–Condon progression characteristic of large geometry change.

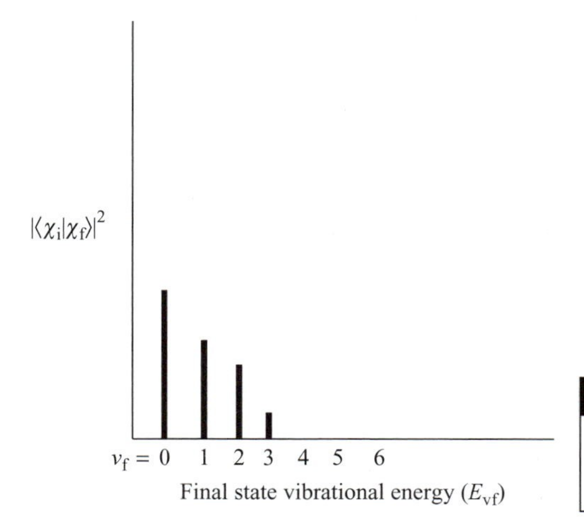

Figure 6.18 Franck–Condon profile characteristic of small geometry change.

state's vibrational wave function χ_{vi} is largest for the final state's χ_{vf} function with $v_f = 2$.

As a qualitative rule of thumb, the larger the geometry difference between the initial- and final-state potentials, the broader will be the Franck–Condon profile (as shown in Fig. 6.17) and the larger the v_f value for which this profile peaks. Differences in harmonic frequencies between the two states can also broaden the Franck–Condon profile.

If the initial and final states have very similar geometries and frequencies along the mode that is excited when the particular electronic excitation is realized, the type of Franck–Condon profile shown in Fig. 6.18 may result.

Another feature that is important to emphasize is the relation between absorption and emission when the two states' energy surfaces have different equilibrium geometries or frequencies. Subsequent to photon absorption to form an excited

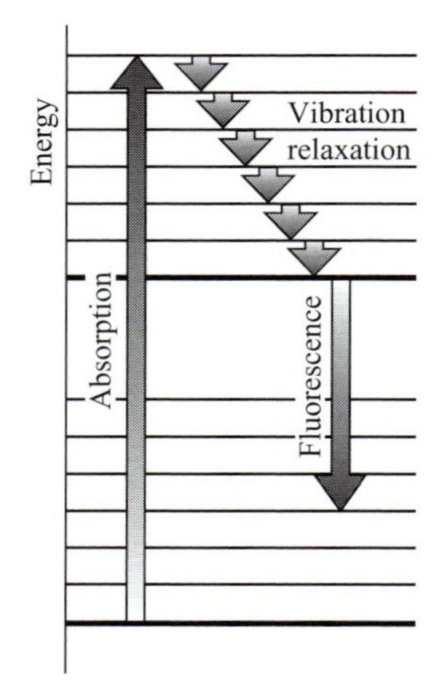

Figure 6.19 Absorption followed by relaxation to lower vibrational levels of the upper state.

electronic state but prior to photon emission, the molecule usually undergoes collisions with other nearby molecules. This, of course, is especially true in condensed-phase experiments. These collisions cause the excited molecule to lose much of its vibrational and rotational energy, thereby "relaxing" it to lower levels on the excited electronic surface. This relaxation process is illustrated in Fig. 6.19. Subsequently, the electronically excited molecule can undergo photon emission (also called fluorescence) to return to its ground electronic state as shown in Fig. 6.20. The Franck–Condon principle discussed earlier also governs the relative intensities of the various vibrational transitions arising in such emission processes. Thus, one again observes a set of peaks in the emission spectrum as shown in Fig. 6.21.

There are two differences between the lines that occur in emission and in absorption. First, the emission lines are shifted to the red (i.e., to lower energy or longer wavelength) because they occur at transition energies connecting the lowest vibrational level of the upper electronic state to various levels of the lower state. In contrast, the absorption lines connect the lowest vibrational level of the ground state to various levels of the upper state. These relationships are shown in Figure 6.22. The second difference relates to the spacings among the vibrational lines. In emission, these spacings reflect the energy spacings between vibrational levels of the ground state, whereas in absorption they reflect spacings between vibrational levels of the upper state.

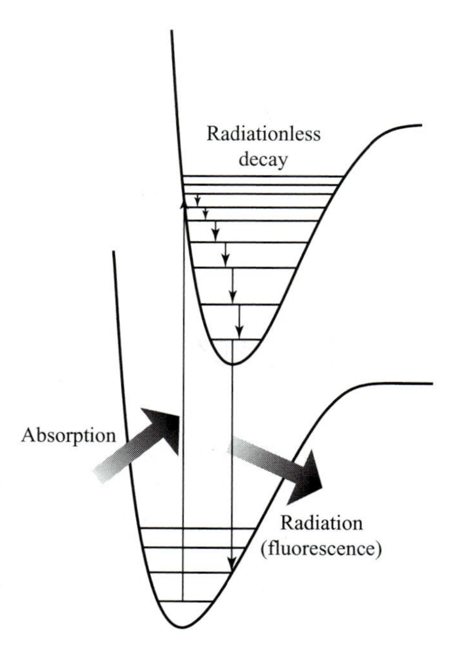

Figure 6.20 Fluorescence from lower levels of the upper surface.

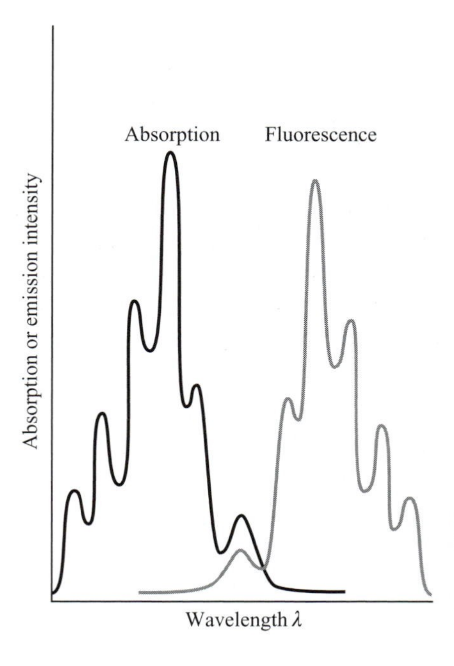

Figure 6.21 Absorption and emission spectra with the latter red shifted.

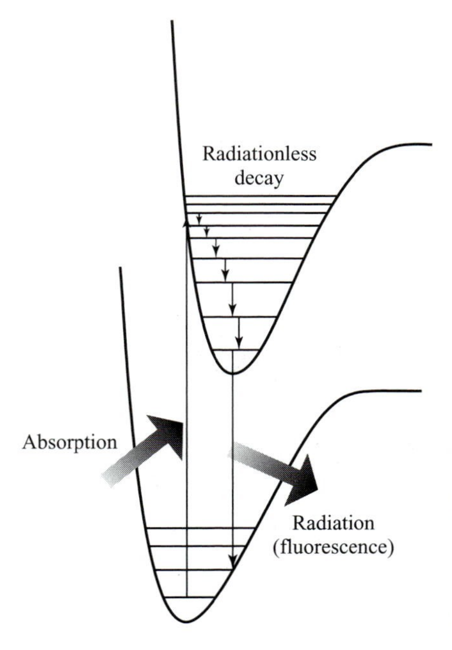

Figure 6.22 Absorption to high states on the upper surface, relaxation, and emission from lower states of the upper surface.

The above examples illustrate how vibrationally resolved visible–UV absorption and emission spectra can be used to gain valuable information about:

(i) the vibrational energy level spacings of the upper and ground electronic states (these spacings, in turn, reflect the strengths of the bonds existing in these states),
(ii) the change in geometry accompanying the ground-to-excited state electronic transition as reflected in the breadth of the Franck–Condon profiles (these changes also tell us about the bonding changes that occur as the electronic transition occurs).

So, again we see how visible–UV spectroscopy can be used to learn about the electronic structure of molecules in various electronic states.

6.5.3 Time correlation function expressions for transition rates

The above so-called "golden-rule" expressions for the rates of photon-induced transitions are written in terms of the initial and final electronic/vibrational/rotational states of the molecule. There are situations in which these states simply can not be reliably known. For example, the higher vibrational states of a large polyatomic molecule or the states of a molecule that strongly interacts with surrounding solvent molecules are such cases. In such circumstances, it is possible to recast the golden-rule formula into a form that is more amenable to introducing specific physical models that lead to additional insights.

Specifically, by using so-called equilibrium averaged time correlation functions, it is possible to obtain rate expressions appropriate to a large number of molecules that exist in a distribution of initial states (e.g., for molecules that occupy many possible rotational and perhaps several vibrational levels at room temperature). As we will soon see, taking this route to expressing spectroscopic transition rates also allows us to avoid having to know each vibrational–rotational wave function of the two electronic states involved; this is especially useful for large molecules or molecules in condensed media where such knowledge is likely not available.

To begin re-expressing the spectroscopic transition rates, the expression obtained earlier,

$$R_{\mathrm{i,f}} = (2\pi/\hbar^2) g(\omega_{\mathrm{f,i}}) \mid \mathbf{E}_0 \cdot \langle \Phi_{\mathrm{f}} | \boldsymbol{\mu} | \Phi_{\mathrm{i}} \rangle |^2, \tag{6.113}$$

appropriate to transitions between a particular initial state Φ_{i} and a specific final state Φ_{f}, is rewritten as

$$R_{\mathrm{i,f}} = (2\pi/\hbar^2) \int g(\omega) \mid \mathbf{E}_0 \cdot \langle \Phi_{\mathrm{f}} | \mu | \Phi_{\mathrm{i}} \rangle |^2 \delta(\omega_{\mathrm{f,i}} - \omega)\, d\omega. \tag{6.114}$$

Here, the $\delta(\omega_{\mathrm{f,i}} - \omega)$ function is used to specifically enforce the "resonance condition" which states that the photons' frequency ω must be resonant with the transition frequency $\omega_{\mathrm{f,i}}$. The following integral identity can be used to replace the δ-function:

$$\delta(\omega_{\text{f,i}} - \omega) = \frac{1}{2\pi} \int_{-\infty}^{\infty} \exp[i(\omega_{\text{f,i}} - \omega)t]\, dt \tag{6.115}$$

by a form that is more amenable to further development. Then, the state-to-state rate of transition becomes

$$R_{\text{i,f}} = (1/\hbar^2) \int g(\omega) \mid \mathbf{E}_0 \cdot \langle \Phi_\text{f} | \boldsymbol{\mu} | \Phi_\text{i} \rangle |^2 \int_{-\infty}^{\infty} \exp[i(\omega_{\text{f,i}} - \omega)t]\, dt d\omega. \tag{6.116}$$

If this expression is then multiplied by the equilibrium probability ρ_i that the molecule is found in the state Φ_i and summed over all such initial states and summed over all final states Φ_f that can be reached from Φ_i with photons of energy $\hbar\omega$, the equilibrium averaged rate of photon absorption by the molecular sample is obtained:

$$R_{\text{eq ave}} = (1/\hbar^2) \sum_{\text{f,i}} \rho_\text{i} \int g(\omega) \mid \mathbf{E}_0 \cdot \langle \Phi_\text{f} | \boldsymbol{\mu} | \Phi_\text{i} \rangle |^2 \int_{-\infty}^{\infty} \exp[i(\omega_{\text{f,i}} - \omega)t]\, dt d\omega. \tag{6.117}$$

This expression is appropriate for an ensemble of molecules that can be in various initial states Φ_i with probabilities ρ_i. The corresponding result for transitions that originate in a particular state (Φ_i) but end up in any of the "allowed" (by energy and selection rules) final states reads

$$R_\text{i} = (1/\hbar^2) \sum_{\text{f}} \rho_\text{i} \int g(\omega) \mid \mathbf{E}_0 \cdot \langle \Phi_\text{f} | \boldsymbol{\mu} | \Phi_\text{i} \rangle |^2 \int_{-\infty}^{\infty} \exp[i(\omega_{\text{f,i}} - \omega)t]\, dt d\omega. \tag{6.118}$$

As we discuss in Chapter 7, for an ensemble in which the number of molecules, the temperature, and the system volume are specified, ρ_i takes the form

$$\rho_\text{i} = g_\text{i} \exp(-E_\text{i}^0/kT)/Q \tag{6.119}$$

where Q is the partition function of the molecules and g_i is the degeneracy of the state Φ_i whose energy is E_i^0. If you are unfamiliar with partition functions and do not want to simply "trust me" in the analysis of time correlation functions that we are about to undertake, I suggest you interrupt your study of Chapter 6 and read up through Section 7.1.3 at this time.

In the above expression for $R_{\text{eq ave}}$, a double sum occurs. Writing out the elements that appear in this sum in detail, one finds

$$\sum_{\text{i,f}} \rho_\text{i} \mathbf{E}_0 \cdot \langle \Phi_\text{i} | \boldsymbol{\mu} | \Phi_\text{f} \rangle \mathbf{E}_0 \cdot \langle \Phi_\text{f} | \boldsymbol{\mu} | \Phi_\text{i} \rangle \exp[i(\omega_{\text{f,i}})t]. \tag{6.120}$$

In situations in which one is interested in developing an expression for the intensity arising from transitions to all allowed final states, the sum over these final states can be carried out explicitly by first writing

$$\langle \Phi_\text{f} | \boldsymbol{\mu} | \Phi_\text{i} \rangle \exp[i(\omega_{\text{f,i}})t] = \langle \Phi_\text{f} | \exp(iHt/\hbar) \mu \exp(-iHt/\hbar) | \Phi_\text{i} \rangle \tag{6.121}$$

and then using the fact that the set of states $\{\Phi_k\}$ are complete and hence obey

$$\sum_k |\Phi_k\rangle\langle\Phi_k| = 1. \tag{6.122}$$

The result of using these identities as well as the Heisenberg definition of the time-dependence of the dipole operator

$$\mu(t) = \exp(iHt/\hbar)\mu\exp(-iHt/\hbar), \tag{6.123}$$

is

$$\sum_{\mathrm{i}} \rho_{\mathrm{i}}\langle\Phi_{\mathrm{i}}|\mathbf{E}_0\cdot\boldsymbol{\mu}\,\mathbf{E}_0\cdot\mu(t)|\Phi_i\rangle. \tag{6.124}$$

In this form, one says that the time dependence has been reduced to that of an equilibrium averaged (i.e., as reflected in the $\sum_{\mathrm{i}} \rho_{\mathrm{i}}\langle\Phi_{\mathrm{i}}||\Phi_{\mathrm{i}}\rangle$ expression) time correlation function involving the component of the dipole operator along the external electric field at $t = 0$, $(\mathbf{E}_0 \cdot \boldsymbol{\mu})$ and this component at a different time t, $(\mathbf{E}_0 \cdot \mu(t))$.

If $\omega_{\mathrm{f,i}}$ is positive (i.e., in the photon absorption case), the above expression will yield a non-zero contribution when multiplied by $\exp(-i\omega t)$ and integrated over positive ω-values. If $\omega_{\mathrm{f,i}}$ is negative (as for stimulated photon emission), this expression will contribute, when multiplied by $\exp(-i\omega t)$, for negative ω-values. In the latter situation, ρ_{i} is the equilibrium probability of finding the molecule in the (excited) state from which emission will occur; this probability can be related to that of the lower state ρ_{f} by

$$\begin{aligned}\rho_{\mathrm{excited}} &= \rho_{\mathrm{lower}}\exp\left[-\left(\mathrm{E}^0_{\mathrm{excited}} - \mathrm{E}^0_{\mathrm{lower}}\right)/kT\right]\\ &= \rho_{\mathrm{lower}}\exp[-\hbar\omega/kT].\end{aligned} \tag{6.125}$$

The absorption and emission cases can be combined into a single expression for the net rate of photon absorption by recognizing that the latter process leads to photon production, and thus must be entered with a negative sign. The resultant expression for the net rate of decrease of photons is

$$\begin{aligned}R_{\mathrm{eq\ ave\ net}} &= (1/\hbar^2)\sum_{\mathrm{i}}\rho_{\mathrm{i}}\\ &\times\iint g(\omega)\langle\Phi_{\mathrm{i}}|(\mathbf{E}_0\cdot\boldsymbol{\mu})\mathbf{E}_0\cdot\mu(t)|\Phi_{\mathrm{i}}\rangle[1-\exp(-\hbar\omega/kT)]\exp(-\mathrm{i}\omega t)\,d\omega\,dt.\end{aligned} \tag{6.126}$$

It is convention to introduce the so-called "line shape" function $I(\omega)$:

$$I(\omega) = \sum_{\mathrm{i}}\rho_{\mathrm{i}}\int\langle\Phi_{\mathrm{i}}|(\mathbf{E}_0\cdot\boldsymbol{\mu})\mathbf{E}_0\cdot\mu(t)|\Phi_{\mathrm{i}}\rangle\exp(-i\omega t)\,dt, \tag{6.127}$$

in terms of which the net photon absorption rate is

$$R_{\mathrm{eq\ ave\ net}} = (1/\hbar^2)[1-\exp(-\hbar\omega/kT)]\int g(\omega)I(\omega)\,d\omega. \tag{6.128}$$

The function

$$C(t) = \sum_{\mathrm{i}} \rho_{\mathrm{i}} \langle \Phi_{\mathrm{i}} | (\mathbf{E}_0 \cdot \boldsymbol{\mu}) \mathbf{E}_0 \cdot \mu(t) | \Phi_i \rangle \tag{6.129}$$

is called the equilibrium averaged time correlation function of the component of the electric dipole operator along the direction of the external electric field $\mathbf{E}_0$. Its Fourier transform is $I(\omega)$, the spectral line shape function. The convolution of $I(\omega)$ with the light source's $g(\omega)$ function, multiplied by $[1 - \exp(\hbar\omega/kT)]$, the correction for stimulated photon emission, gives the net rate of photon absorption.

Although the correlation function expression for the photon absorption rate is equivalent to the state-to-state expression from which it was derived, we notice that

(i) $C(t)$ does not contain explicit reference to the final-state wave functions Φ_{f}; instead,
(ii) $C(t)$ requires us to describe how the dipole operator changes with time.

That is, in the time correlation framework, one is allowed to use models of the time evolution of the system to describe the spectra. This is especially appealing for large complex molecules and molecules in condensed media because, for such systems, it would be hopeless to attempt to find the final-state wave functions, but it is reasonable (albeit challenging) to model the system's time evolution. It turns out that a very wide variety of spectroscopic and thermodynamic properties (e.g., light scattering intensities, diffusion coefficients, and thermal conductivity) can be expressed in terms of molecular time correlation functions. The *Statistical Mechanics* text by McQuarrie has a good treatment of many of these cases. Let's now examine how such time evolution issues are used within the correlation function framework for the specific photon absorption case.

6.5.4 Line broadening mechanisms

If the rotational motion of the system's molecules is assumed to be entirely unhindered (e.g., by any environment or by collisions with other molecules), it is appropriate to express the time dependence of each of the dipole time correlation functions listed above in terms of a "free rotation" model. For example, when dealing with diatomic molecules, the electronic–vibrational–rotational $C(t)$ appropriate to a specific electronic–vibrational transition becomes

$$\begin{aligned} C(t) = {} & (q_{\mathrm{r}}\, q_{\mathrm{v}} q_{\mathrm{e}} q_{\mathrm{t}})^{-1} \sum_J (2J+1) \exp\left(-\frac{h^2 J(J+1)}{8\pi^2 IkT}\right) \exp\left(-\frac{h\nu_{\mathrm{vib}} v_{\mathrm{i}}}{kT}\right) \\ & \times g_{\mathrm{ie}} \langle \phi_J | \mathbf{E}_0 \cdot \boldsymbol{\mu}_{\mathrm{i,f}}(\mathbf{R}_{\mathrm{e}}) \mathbf{E}_0 \cdot \mu_{\mathrm{i,f}}(\mathbf{R}_{\mathrm{e}}, t) | \phi_J \rangle \mid \langle \chi_{\mathrm{iv}} | \chi_{\mathrm{fv}} \rangle |^2 \\ & \times \exp\left(\frac{i h \nu_{\mathrm{vib}} t + i \Delta E_{\mathrm{i,f}}\, t}{\hbar}\right). \end{aligned} \tag{6.130}$$

Here,

$$q_r = (8\pi^2 IkT/h^2) \tag{6.131}$$

is the rotational partition function (I being the molecule's moment of inertia $I = \mu R_e^2$, and $h^2J(J+1)/(8\pi^2 I)$ the molecule's rotational energy for the state with quantum number J and degeneracy $2J+1$),

$$q_v = \frac{\exp\left(-\frac{h\nu_{vib}}{2kT}\right)}{1-\exp\left(-\frac{h\nu_{vib}}{kT}\right)} \tag{6.132}$$

is the vibrational partition function (ν_{vib} being the vibrational frequency), g_{ie} is the degeneracy of the initial electronic state,

$$q_t = (2\pi mkT/h^2)^{3/2} V \tag{6.133}$$

is the translational partition function for the molecules of mass m moving in volume V, and $\Delta E_{i,f}$ is the adiabatic electronic energy spacing. The origins of such partition functions are treated in Chapter 7.

The functions $\langle \phi_J | \mathbf{E}_0 \cdot \mu_{i,f}(\mathbf{R}_e)\mathbf{E}_0 \cdot \mu_{i,f}(\mathbf{R}_e, t)|\phi_J\rangle$ describe the time evolution of the electronic transition dipole vector for the rotational state J. In a "free-rotation" model, this function is taken to be of the form:

$$\langle \phi_J | \mathbf{E}_0 \cdot \mu_{i,f}(\mathbf{R}_e)\mathbf{E}_0 \cdot \mu_{i,f}(\mathbf{R}_e, t)|\phi_J\rangle = \langle \phi_J | \mathbf{E}_0 \cdot \mu_{i,f}(\mathbf{R}_e)\mathbf{E}_0 \cdot \mu_{i,f}(\mathbf{R}_e)|\phi_J\rangle \cos(\omega_J t) \tag{6.134}$$

where ω_J is the rotational frequency (in cycles per second) for rotation of the molecule in the state labeled by J. This oscillatory time dependence, combined with the $\exp[(ih\nu_{vib}t + i\Delta E_{i,f}t)/\hbar]$ time dependence arising from the electronic and vibrational factors, produce, when this $C(t)$ function is Fourier transformed to generate $I(\omega)$, a series of δ-function "peaks". The intensities of these peaks are governed by the

$$(q_r q_v q_e q_t)^{-1} \sum_J (2J+1)\exp(-h^2 J(J+1)/(8\pi^2 IkT))\exp(-h\nu_{vib}vi/kT)g_{ie} \tag{6.135}$$

Boltzmann population factors as well as by the $|\langle \chi_{iv}|\chi_{fv}\rangle|^2$ Franck–Condon factors and the $\langle \phi_J | \mathbf{E}_0 \cdot \mu_{i,f}(\mathbf{R}_e)\mathbf{E}_0 \cdot \mu_{i,f}(\mathbf{R}_e,0)|\phi_J\rangle$ terms.

This same analysis can be applied to the pure rotation and vibration–rotation $C(t)$ time dependences with analogous results. In the former, δ-function peaks are predicted to occur at

$$\omega = \pm\omega_J, \tag{6.136}$$

and in the latter at

$$\omega = \omega_{fv,iv} \pm \omega_J, \tag{6.137}$$

with the intensities governed by the time-independent factors in the corresponding expressions for $C(t)$.

In experimental measurements, such sharp δ-function peaks are, of course, not observed. Even when very narrow bandwidth laser light sources are used (i.e., for which $g(\omega)$ is an extremely narrowly peaked function), spectral lines are found to possess finite widths. Let us now discuss several sources of line broadening, some of which will relate to deviations from the "unhindered" rotational motion model introduced above.

Doppler broadening

In the above expressions for $C(t)$, the averaging over initial rotational, vibrational, and electronic states is explicitly shown. There is also an average over the translational motion implicit in all of these expressions. Its role has not (yet) been emphasized because the molecular energy levels, whose spacings yield the characteristic frequencies at which light can be absorbed or emitted, do not depend on translational motion. However, the frequency of the electromagnetic field experienced by moving molecules does depend on the velocities of the molecules, so this issue must now be addressed.

Elementary physics classes express the so-called Doppler shift of a wave's frequency induced by relative movement of the light source and the molecule as follows:

$$\omega_{\text{observed}} = \omega_{\text{nominal}}(1 + v_z/c)^{-1} \approx \omega_{\text{nominal}}(1 - v_z/c + \cdots). \tag{6.138}$$

Here, ω_{nominal} is the frequency of the unmoving light source seen by unmoving molecules, v_z is the velocity of relative motion of the light source and molecules, c is the speed of light, and ω_{observed} is the Doppler shifted frequency (i.e., the frequency seen by the molecules). The second identity is obtained by expanding, in a power series, the $(1 + v_z/c)^{-1}$ factor, and is valid in truncated form when the molecules are moving with speeds significantly below the speed of light.

For all of the cases considered earlier, a $C(t)$ function is subjected to Fourier transformation to obtain a spectral line shape function $I(\omega)$, which then provides the essential ingredient for computing the net rate of photon absorption. In this Fourier transform process, the variable ω is assumed to be the frequency of the electromagnetic field experienced by the molecules. The above considerations of Doppler shifting then lead one to realize that the correct functional form to use in converting $C(t)$ to $I(\omega)$ is

$$I(\omega) = \int C(t) \exp[-it\omega(1 - v_z/c)]\, dt, \tag{6.139}$$

where ω is the nominal frequency of the light source.

As stated earlier, within $C(t)$ there is also an equilibrium average over translational motion of the molecules. For a gas-phase sample undergoing random collisions and at thermal equilibrium, this average is characterized by the well-known

Maxwell–Boltzmann velocity distribution:

$$\left(\frac{m}{2\pi kT}\right)^{3/2} \exp\left(-\frac{m\left(v_x^2+v_y^2+v_z^2\right)}{2kT}\right) dv_x dv_y dv_z. \quad (6.140)$$

Here m is the mass of the molecules and v_x, v_y, and v_z label the velocities along the lab-fixed Cartesian coordinates.

Defining the z-axis as the direction of propagation of the light's photons and carrying out the averaging of the Doppler factor over such a velocity distribution, one obtains

$$\begin{aligned}
&\int_{-\infty}^{\infty} \exp\left[-it\omega\left(1-\frac{v_z}{c}\right)\right]\left(\frac{m}{2\pi kT}\right)^{3/2} \exp\left(-\frac{m\left(v_x^2+v_y^2+v_z^2\right)}{2kT}\right) dv_x dv_y dv_z \\
&= \exp(-i\omega t)\int_{-\infty}^{\infty}\left(\frac{m}{2\pi kT}\right)^{1/2} \exp\left(\frac{i\omega t v_z}{c}\right)\exp\left(-\frac{mv_z^2}{2kT}\right) dv_z \\
&= \exp(-i\omega t)\exp\left(-\frac{\omega^2 t^2 kT}{2mc^2}\right). \quad (6.141)
\end{aligned}$$

This result, when substituted into the expressions for $C(t)$, yields expressions identical to those given for the three cases treated above but with one modification. The translational motion average need no longer be considered in each $C(t)$; instead, the earlier expressions for $C(t)$ must each be multiplied by a factor $\exp[-\omega^2 t^2 kT/(2mc^2)]$ that embodies the translationally averaged Doppler shift. The spectral line shape function $I(\omega)$ can then be obtained for each $C(t)$ by simply Fourier transforming:

$$I(\omega) = \int_{-\infty}^{\infty} \exp(-i\omega t) C(t)\, dt. \quad (6.142)$$

When applied to the rotation, vibration–rotation, or electronic–vibration–rotation cases within the "unhindered" rotation model treated earlier, the Fourier transform involves integrals of the form

$$\int_{-\infty}^{\infty} \exp(-i\omega t)\exp\left[-\frac{\omega^2 t^2 kT}{2mc^2}\right]\exp\left[i\left(\omega_{\text{fv,iv}}+\frac{\Delta E_{\text{i,f}}}{\hbar}\pm\omega_J\right)t\right] dt. \quad (6.143)$$

This integral would arise in the electronic–vibration–rotation case; the other two cases would involve integrals of the same form but with the $\Delta E_{\text{i,f}}/\hbar$ absent in the vibration–rotation situation and with $\omega_{\text{fv,iv}} + \Delta E_{\text{i,f}}/\hbar$ missing for pure rotation transitions. All such integrals can be carried out analytically and yield

$$\sqrt{\frac{2mc^2\pi}{\omega^2 kT}}\exp\left[-(\omega-\omega_{\text{fv,iv}}-\Delta E_{\text{i,f}}/\hbar\pm\omega_J)^2\frac{mc^2}{2\omega^2 kT}\right]. \quad (6.144)$$

The result is a series of Gaussian "peaks" in ω-space, centered at

$$\omega = \omega_{\text{fv,iv}} + \Delta E_{\text{i,f}}/\hbar \pm \omega_J, \quad (6.145)$$

with widths (σ) determined by

$$\sigma^2 = \frac{\omega^2 kT}{mc^2}, \tag{6.146}$$

given the temperature T and the mass of the molecules m. The hotter the sample, the faster the molecules are moving on average, and the broader is the distribution of Doppler shifted frequencies experienced by these molecules. The net result then of the Doppler effect is to produce a line shape function that is similar to the "unhindered" rotation model's series of δ-functions but with each δ-function peak broadened into a Gaussian shape.

If spectra can be obtained to accuracy sufficient to determine the Doppler width of the spectral lines, such knowledge can be used to estimate the temperature of the system. This can be useful when dealing with systems that can not be subjected to alternative temperature measurements. For example, the temperatures of stars can be estimated (if their velocity relative to the earth is known) by determining the Doppler shifts of emission lines from them. Alternatively, the relative speed of a star from the earth may be determined if its temperature is known. As another example, the temperature of hot gases produced in an explosion can be probed by measuring Doppler widths of absorption or emission lines arising from molecules in these gases.

Pressure broadening

To include the effects of collisions on the rotational motion part of any of the above $C(t)$ functions, one must introduce a model for how such collisions change the dipole-related vectors that enter into $C(t)$. The most elementary model used to address collisions applies to gaseous samples which are assumed to undergo unhindered rotational motion until struck by another molecule at which time a "kick" is applied to the dipole vector and after which the molecule returns to its unhindered rotational movement.

The effects of such infrequent collision-induced kicks are treated within the so-called pressure broadening (sometimes called collisional broadening) model by modifying the free-rotation correlation function through the introduction of an exponential damping factor $\exp(-|t|/\tau)$:

$$\begin{aligned}&\langle \phi_J | \mathbf{E}_0 \cdot \mu_{\mathrm{i,f}}(\mathbf{R}_\mathrm{e}) \mathbf{E}_0 \cdot \mu_{\mathrm{i,f}}(\mathbf{R}_\mathrm{e}, 0) | \phi_J \rangle \cos \frac{hJ(J+1)t}{4\pi I} \\ &\quad \Rightarrow \langle \phi_J | \mathbf{E}_0 \cdot \mu_{\mathrm{i,f}}(\mathbf{R}_\mathrm{e}) \mathbf{E}_0 \cdot \mu_{\mathrm{i,f}}(\mathbf{R}_\mathrm{e}, 0) | \phi_J \rangle \cos \frac{hJ(J+1)t}{4\pi I} \exp\left(-\frac{|t|}{\tau}\right).\end{aligned} \tag{6.147}$$

This damping function's time scale parameter τ is assumed to characterize the average time between collisions and thus should be inversely proportional to the collision frequency. Its magnitude is also related to the effectiveness with which collisions cause the dipole function to deviate from its unhindered rotational motion (i.e., related to the collision strength). In effect, the exponential damping causes the time correlation function $\langle\phi_J|\mathbf{E}_0\cdot\mu_{\rm i,f}(\mathbf{R}_e)\mathbf{E}_0\cdot\mu_{\rm i,f}(\mathbf{R}_e,t)|\phi_J\rangle$ to "lose its memory" and to decay to zero. This "memory" point of view is based on viewing $\langle\phi_J|\mathbf{E}_0\cdot\mu_{\rm i,f}(\mathbf{R}_e)\mathbf{E}_0\cdot\mu_{\rm i,f}(\mathbf{R}_e,t)|\phi_J\rangle$ as the projection of $\mathbf{E}_0\cdot\mu_{\rm i,f}(\mathbf{R}_e,t)$ along its $t=0$ value $\mathbf{E}_0\cdot\mu_{\rm i,f}(\mathbf{R}_e,0)$ as a function of time t.

Introducing this additional $\exp(-|t|/\tau)$ time dependence into $C(t)$ produces, when $C(t)$ is Fourier transformed to generate $I(\omega)$, integrals of the form

$$\int_{-\infty}^{\infty}\exp(-i\omega t)\exp\left(-\frac{|t|}{\tau}\right)\exp\left[-\frac{\omega^2t^2kT}{2mc^2}\right]\exp\left[i\left(\omega_{\rm fv,iv}+\frac{\Delta E_{\rm i,f}}{\hbar}\pm\omega_J\right)t\right]dt. \tag{6.148}$$

In the limit of very small Doppler broadening, the $[\omega^2t^2kT/(2mc^2)]$ factor can be ignored (i.e., $\exp[-\omega^2t^2kT/(2mc^2)]$ set equal to unity), and

$$\int_{-\infty}^{\infty}\exp(-i\omega t)\exp\left(-\frac{|t|}{\tau}\right)\exp\left[i\left(\omega_{\rm fv,iv}+\frac{\Delta E_{\rm i,f}}{\hbar}\pm\omega_J\right)t\right]dt \tag{6.149}$$

results. This integral can be performed analytically and generates

$$\frac{1}{4\pi}\left\{\frac{1/\tau}{(1+\tau)^2+(\omega-\omega_{\rm fv,iv}-\Delta E_{\rm i,f}/\hbar\pm\omega_J)^2}\right.\\ \left.+\frac{1/\tau}{(1+\tau)^2+(\omega+\omega_{\rm fv,iv}+\Delta E_{\rm i,f}/\hbar\mp\omega_J)^2}\right\}, \tag{6.150}$$

a pair of Lorentzian peaks in ω-space centered again at

$$\omega=\pm[\omega_{\rm fv,iv}+\Delta E_{\rm i,f}/\hbar\pm\omega_J]. \tag{6.151}$$

The full width at half height of these Lorentzian peaks is $2/\tau$. One says that the individual peaks have been pressure or collisionally broadened.

When the Doppler broadening can not be neglected relative to the collisional broadening, the above integral

$$\int_{-\infty}^{\infty}\exp(-i\omega t)\exp\left(-\frac{|t|}{\tau}\right)\exp\left[-\frac{\omega^2t^2kT}{2mc^2}\right]\exp\left[i\left(\omega_{\rm fv,iv}+\frac{\Delta E_{\rm i.f}}{\hbar}\pm\omega_J\right)t\right]dt \tag{6.152}$$

is more difficult to perform. Nevertheless, it can be carried out and again produces a series of peaks centered at

$$\omega=\omega_{\rm fv,iv}+\Delta E_{\rm i,f}/\hbar\pm\omega_J, \tag{6.153}$$

but whose widths are determined both by Doppler and pressure broadening effects. The resultant line shapes are thus no longer purely Lorentzian or Gaussian (which are compared in Fig. 6.23 for both functions having the same full width at half height and the same integrated area), but have a shape that is called a Voigt shape.

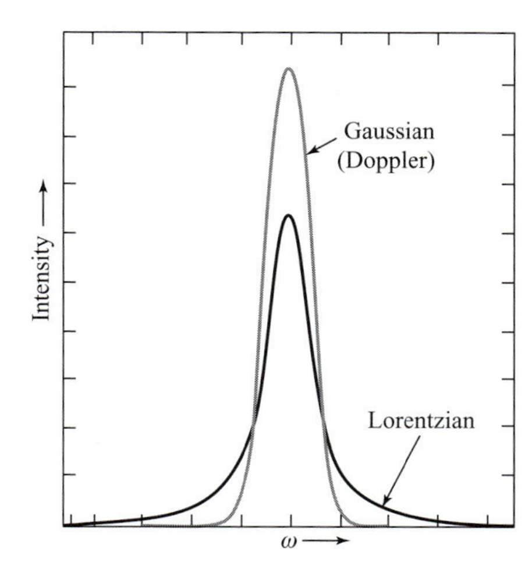

Figure 6.23 Typical forms of Gaussian and Lorentzian peaks.

Experimental measurements of line widths that allow one to extract widths originating from collisional broadening provide information (through τ) on the frequency of collisions and the "strength" of these collisions. By determining τ at a series of gas densities, one can separate the collision-frequency dependence and determine the strength of the individual collisions (meaning how effective each collision is in reorienting the molecule's dipole vector).

Rotational diffusion broadening

Molecules in liquids and very dense gases undergo such frequent collisions with the other molecules that the mean time between collisions is short compared to the rotational period for their unhindered rotation. As a result, the time dependence of the dipole-related correlation functions can no longer be modeled in terms of free rotation that is interrupted by (infrequent) collisions and Doppler shifted. Instead, a model that describes the incessant buffeting of the molecule's dipole by surrounding molecules becomes appropriate. For liquid samples in which these frequent collisions cause the dipole to undergo angular motions that cover all angles (i.e., in contrast to a frozen glass or solid in which the molecule's dipole would undergo strongly perturbed pendular motion about some favored orientation), the so-called rotational diffusion model is often used.

In this picture, the rotation-dependent part of $C(t)$ is expressed as

$$\begin{aligned}&\langle\phi_J|\mathbf{E}_0\cdot\mu_{\text{i,f}}(\mathbf{R}_\text{e})\mathbf{E}_0\cdot\mu_{\text{i,f}}(\mathbf{R}_\text{e},t)|\phi_J\rangle\\&\quad=\langle\phi_J|\mathbf{E}_0\cdot\mu_{\text{i,f}}(\mathbf{R}_\text{e})\mathbf{E}_0\cdot\mu_{\text{i,f}}(\mathbf{R}_\text{e},0)|\phi_J\rangle\exp(-2D_{\text{rot}}|t|),\end{aligned}\tag{6.154}$$

where D_{rot} is the rotational diffusion constant whose magnitude details the time decay in the averaged value of $\mathbf{E}_0\cdot\mu_{\text{i,f}}(\mathbf{R}_\text{e},t)$ at time t with respect to its value at time $t=0$; the larger D_{rot}, the faster is this decay.

As with pressure broadening, this exponential time dependence, when subjected to Fourier transformation, yields

$$\int_{-\infty}^{\infty} \exp(-i\omega t)\exp(-2D_{\text{rot}}|t|)\exp[-\omega^2 t^2 kT/(2mc^2)] \times \exp[i(\omega_{\text{fv,iv}} + \Delta E_{\text{i,f}}/\hbar \pm \omega_J)t]\,dt. \tag{6.155}$$

Again, in the limit of very small Doppler broadening, the $[\omega^2 t^2 kT/(2mc^2)]$ factor can be ignored (i.e., $\exp[-\omega^2 t^2 kT/(2mc^2)]$ set equal to unity), and

$$\int_{-\infty}^{\infty} \exp(-i\omega t)\exp(-2D_{\text{rot}}|t|)\exp[i(\omega_{\text{fv,iv}} + \Delta E_{\text{i,f}}/\hbar \pm \omega_J)t]\,dt \tag{6.156}$$

results. This integral can be evaluated analytically and generates

$$\frac{1}{4\pi}\left\{\frac{2D_{\text{rot}}}{(2D_{\text{rot}})^2 + (\omega - \omega_{\text{fv,iv}} - \Delta E_{\text{i,f}}/\hbar \pm \omega_J)^2} + \frac{2D_{\text{rot}}}{(2D_{\text{rot}})^2 + (\omega + \omega_{\text{fv,iv}} + \Delta E_{\text{i,f}}/\hbar \mp \omega_J)^2}\right\}, \tag{6.157}$$

a pair of Lorentzian peaks in ω-space centered again at

$$\omega = \pm[\omega_{\text{fv,iv}} + \Delta E_{\text{i,f}}/\hbar \pm \omega_J]. \tag{6.158}$$

The full width at half height of these Lorentzian peaks is $4D_{\text{rot}}$. In this case, one says that the individual peaks have been broadened via rotational diffusion. In such cases, experimental measurement of line widths yields valuable information about how fast the molecule is rotationally diffusing in its condensed environment.

Lifetime or Heisenberg homogeneous broadening

Whenever the absorbing species undergoes one or more processes that depletes its numbers, we say that it has a finite lifetime. For example, a species that undergoes unimolecular dissociation has a finite lifetime as does an excited state of a molecule that decays by spontaneous emission of a photon. Any process that depletes the absorbing species contributes another source of time dependence for the dipole time correlation functions $C(t)$ discussed above. This time dependence is usually modeled by appending, in a multiplicative manner, a factor $\exp(-|t|/\tau)$. This, in turn, modifies the line shape function $I(\omega)$ in a manner much like that discussed when treating the rotational diffusion case:

$$\int_{-\infty}^{\infty} \exp(-i\omega t)\exp(-|t|)\exp[-\omega^2 t^2 kT/(2mc^2)]\exp[i(\omega_{\text{fv,iv}} + \Delta E_{\text{i,f}}/\hbar \pm \omega_J)t]\,dt. \tag{6.159}$$

Not surprisingly, when the Doppler contribution is small, one obtains

$$\frac{1}{4\pi}\left\{\frac{1/\tau}{(1/\tau)^2 + (\omega - \omega_{\text{fv,iv}} - \Delta E_{\text{i,f}}/\hbar \pm \omega_J)^2} + \frac{1/\tau}{(1/\tau)^2 + (\omega + \omega_{\text{fv,iv}} + \Delta E_{\text{i,f}}/\hbar \mp \omega_J)^2}\right\}. \tag{6.160}$$

In these Lorentzian lines, the parameter τ describes the kinetic decay lifetime of the molecule. One says that the spectral lines have been lifetime or Heisenberg broadened by an amount proportional to $1/\tau$. The latter terminology arises because the finite lifetime of the molecular states can be viewed as producing, via the Heisenberg uncertainty relation $\Delta E \Delta t > \hbar$, states whose energy is "uncertain" to within an amount ΔE.

Site inhomogeneous broadening

Among the above line broadening mechanisms, the pressure, rotational diffusion, and lifetime broadenings are all of the homogeneous variety. This means that each and every molecule in the sample is affected in exactly the same manner by the broadening process. For example, one does not find some molecules with short lifetimes and others with long lifetimes in the Heisenberg case; the entire ensemble of molecules is characterized by a single lifetime.

In contrast, Doppler broadening is inhomogeneous in nature because each molecule experiences a broadening that is characteristic of its particular velocity v_z. That is, the fast molecules have their lines broadened more than do the slower molecules. Another important example of inhomogeneous broadening is provided by so-called site broadening. Molecules imbedded in a liquid, solid, or glass do not, at the instant of their photon absorption, all experience exactly the same interactions with their surroundings. The distribution of instantaneous "solvation" environments may be rather "narrow" (e.g., in a highly ordered solid matrix) or quite "broad" (e.g., in a liquid at high temperature or in a super-critical liquid). Different environments produce different energy level splittings $\omega = \omega_{\mathrm{fv,iv}} + \Delta E_{\mathrm{i,f}}/\hbar \pm \omega_J$ (because the initial and final states are "solvated" differently by the surroundings) and thus different frequencies at which photon absorption can occur. The distribution of energy level splittings causes the sample to absorb at a range of frequencies as illustrated in Fig. 6.24 where homogeneous and inhomogeneous line shapes are compared.

The spectral line shape function $I(\omega)$ is therefore further broadened when site inhomogeneity is present and significant. These effects can be modeled by

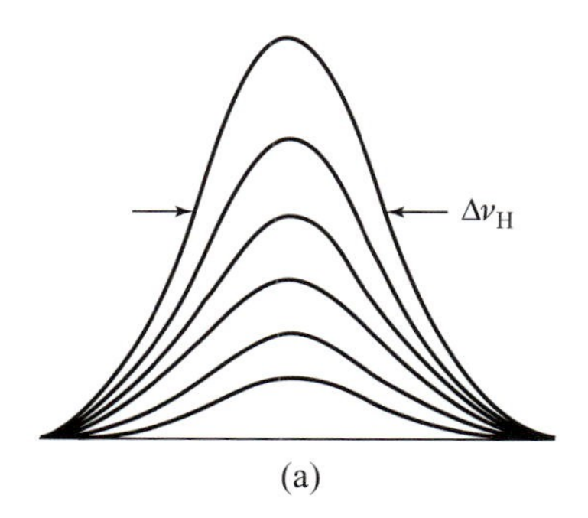

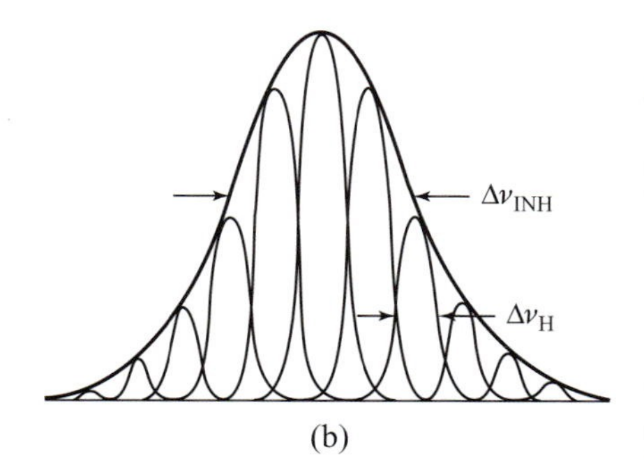

Figure 6.24 Illustration of (a) homogeneous band showing absorption at several concentrations and (b) inhomogeneous band showing absorption at one concentration by numerous sub-populations.

convolving the kind of $I(\omega)$ function that results from Doppler, lifetime, rotational diffusion, and pressure broadening with a Gaussian distribution $P(\Delta E)$ that describes the inhomogeneous distribution of energy level splittings:

$$I(\omega) = \int I^0(\omega; \Delta E) P(\Delta E)\, d\Delta E. \tag{6.161}$$

Here $I^0(\omega; \Delta E)$ is a line shape function such as those described earlier each of which contains a set of frequencies (e.g., $\omega_{\mathrm{fv,iv}} + \Delta E_{\mathrm{i,f}}/\hbar \pm \omega_J + \Delta E/\hbar$) at which absorption or emission occurs and $P(\Delta E)$ is a Gaussian probability function describing the inhomogeneous broadening of the energy splitting ΔE.

A common experimental test to determine whether inhomogeneous broadening is significant involves hole burning. In such experiments, an intense light source (often a laser) is tuned to a frequency ω_{burn} that lies within the spectral line being probed for inhomogeneous broadening. Then, with the intense light source constantly turned on, a second tunable light source is used to scan through the profile of the spectral line, and an absorption spectrum is recorded. Given an absorption profile as shown in Fig. 6.25 in the absence of the intense burning light source, one expects to see a profile such as that shown in Fig. 6.26 if inhomogeneous broadening is operative.

The interpretation of the change in the absorption profile caused by the bright light source proceeds as follows:

(i) In the ensemble of molecules contained in the sample, some molecules will absorb at or near the frequency of the bright light source ω_{burn}; other molecules (those

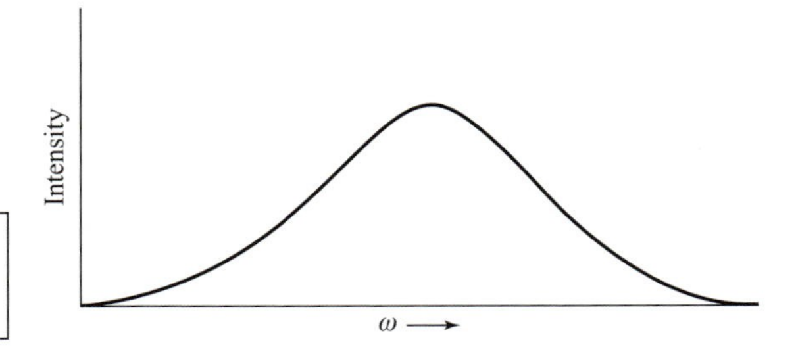

Figure 6.25 Absorption profile in the absence of hole burning.

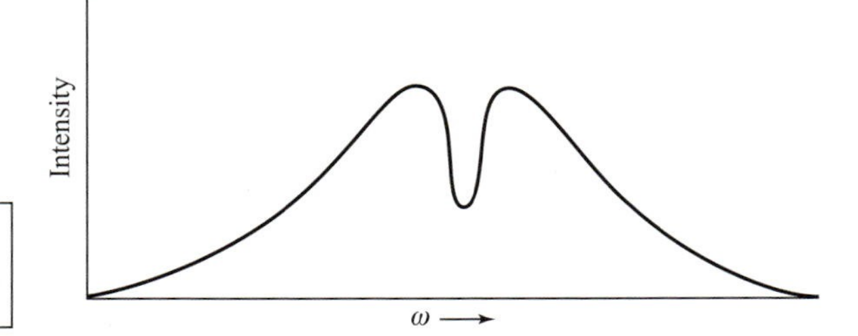

Figure 6.26 Absorption profile with laser turned on to burn a hole.

whose environments do not produce energy level splittings that match ω_{burn}) will not absorb at this frequency.

(ii) Those molecules that do absorb at ω_{burn} will have their transition saturated by the intense light source, thereby rendering this frequency region of the line profile transparent to further absorption.

(iii) When the "probe" light source is scanned over the line profile, it will induce absorptions for those molecules whose local environments did not allow them to be saturated by the ω_{burn} light. The absorption profile recorded by this probe light source's detector thus will match that of the original line profile, until

(iv) the probe light source's frequency matches ω_{burn}, upon which no absorption of the probe source's photons will be recorded because molecules that absorb in this frequency regime have had their transition saturated.

(v) Hence, a "hole" will appear in the absorption spectrum recorded by the probe light source's detector in the region of ω_{burn}.

Unfortunately, the technique of hole burning does not provide a fully reliable method for identifying inhomogeneously broadened lines. If a hole is observed in such a burning experiment, this provides ample evidence, but if one is not seen, the result is not definitive. In the latter case, the transition may not be strong enough (i.e., may not have a large enough "rate of photon absorption") for the intense light source to saturate the transition to the extent needed to form a hole.

6.6 Photoelectron spectroscopy

Photoelectron spectroscopy (PES) is a special kind of electronic spectroscopy. It uses visible or UV light to excite a molecule or ion to a final state in which an electron is ejected. In effect, it induces transitions to final states in which an electron has been promoted to an unbound or so-called continuum orbital. Most PES experiments are carried out using a fixed-frequency light source (usually a laser). This source's photons, when absorbed, eject electrons whose intensity and kinetic energies (*KE*) are then measured. Subtracting the electrons' *KE* from the photon's energy $h\nu$ gives the binding energy (*BE*) of the electron:

$$BE = h\nu - KE. \tag{6.162}$$

If the sample subjected to the PES experiment has molecules in a variety of initial states (e.g., two electronic states or various vibrational–rotational levels of the ground electronic state) having various binding energies BE_k, one will observe a series of "peaks" corresponding to electrons ejected with a variety of kinetic energies KE_k as Fig. 6.27 illustrates and as the energy-balance condition requires:

$$BE_k = h\nu - KE_k. \tag{6.163}$$

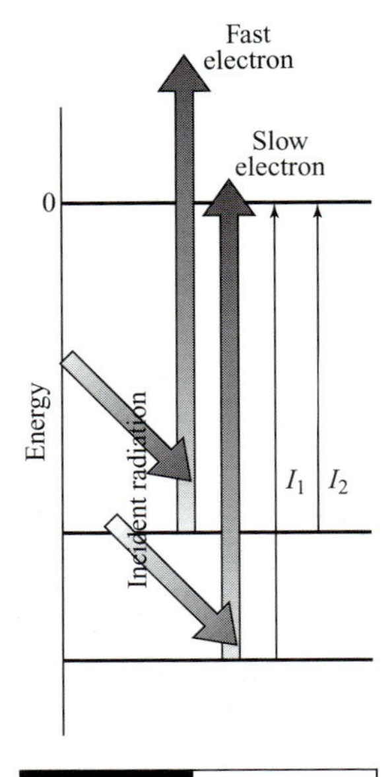

Figure 6.27
Photoelectron spectrum showing absorption from two states of the parent.

The peak of electrons detected with the highest kinetic energy came from the highest-lying state of the parent, while those with low kinetic energy came from the lowest-energy state of the parent. By examining the spacings between these peaks, one learns about the spacings between the energy levels of the parent species that has been subjected to electron loss.

Alternatively, if the parent species exists primarily in its lowest state but the daughter species produced when an electron is removed from the parent has excited (electronic, vibration–rotation) states, one can observe a different progression of peaks. In this case, the electrons with highest kinetic energy arise from transitions leading to the lowest-energy state of the daughter as Fig. 6.28 illustrates. In that figure, the lower-energy surface belongs to the parent and the upper curve to the daughter.

An example of experimental photodetachment data is provided in Fig. 6.29 showing the intensity of electrons detected when Cu_2^- loses an electron vs. the kinetic energy of the ejected electrons. The peak at a kinetic energy of *c.* 1.54 eV, corresponding to a binding energy of 1.0 eV, arises from Cu_2^- in $v = 0$ losing an electron to produce Cu_2 in $v = 0$. The most intense peak corresponds to a $v = 0$ to $v = 4$ transition. As in the visible–UV spectroscopy case, Franck–Condon factors involving the overlap of the Cu_2^- and Cu_2 vibrational wave functions govern the relative intensities of the PES peaks.

Another example is given in Fig. 6.30 where the photodetachment spectrum of $H_2C{=}C^-$ (the anion of the carbene vinylidene) appears. In this spectrum, the peaks having electron binding energies near 0.5 eV correspond to transitions in which ground-state $H_2C{=}C^-$ in $v = 0$ is detached to produce ground-state

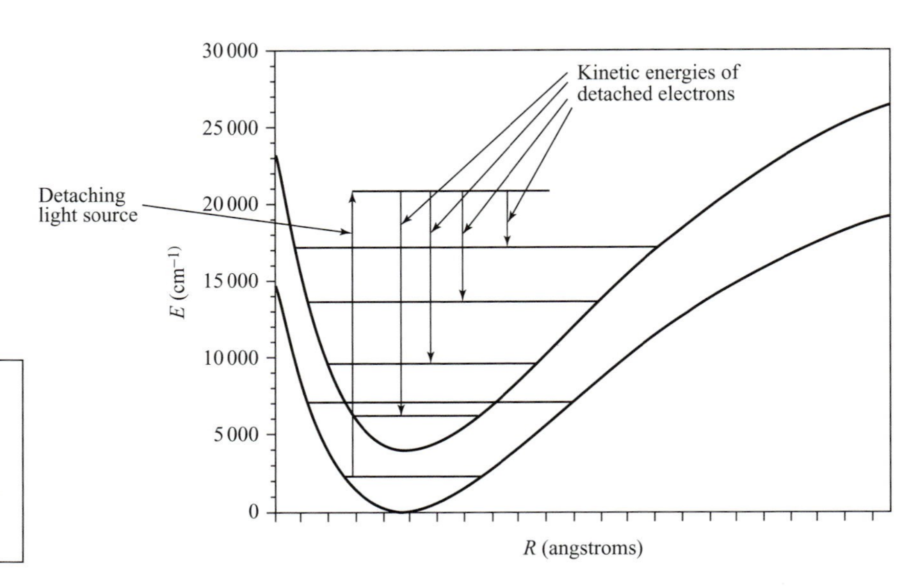

Figure 6.28
Photoelectron events showing detachment from one state of the parent to several states of the daughter. *R* is bond length.

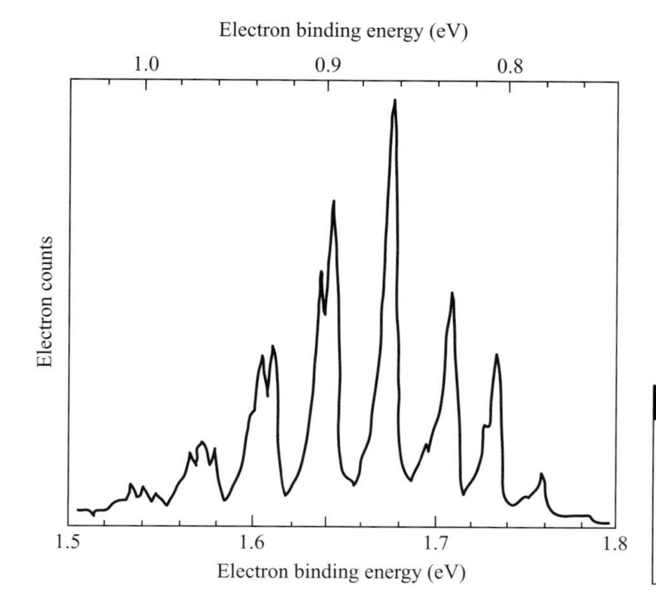

Figure 6.29 Photoelectron spectrum of Cu_2^-. The peaks belong to a Franck–Condon vibrational progression of neutral Cu_2.

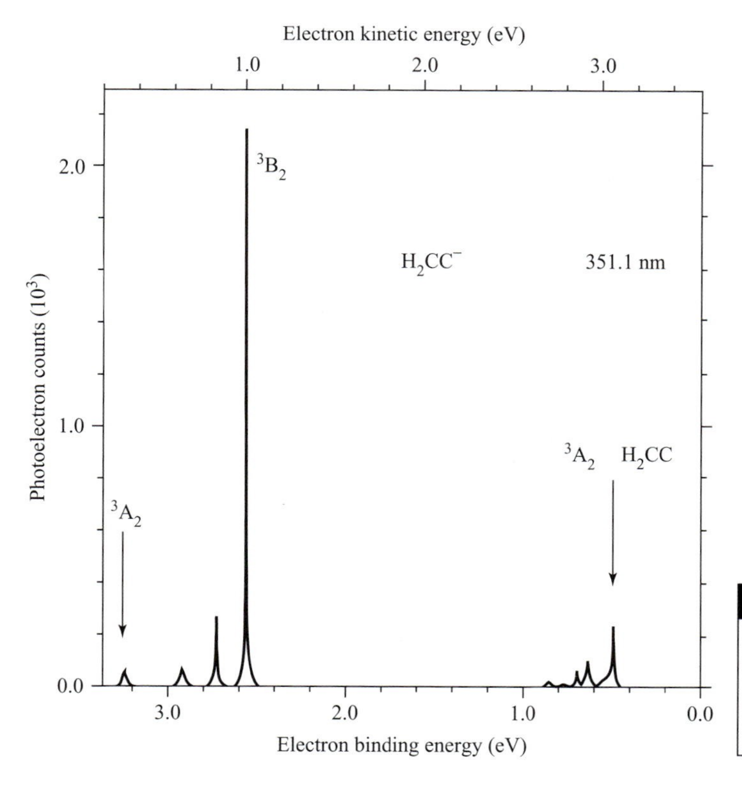

Figure 6.30 Photoelectron spectrum of $H_2C{=}C^-$ showing detachments to two electronic states of the neutral.

($^1A_1$) $H_2C{=}C$ in various v levels. The spacings between this group of peaks relate to the spacings in vibrational states of this $^1A_1$ electronic state. The series of peaks with binding energies near 2.5 eV correspond to transitions in which $H_2C{=}C^-$ is detached to produce $H_2C{=}C$ in its $^3B_2$ excited electronic state. The spacings between peaks in this range relate to spacings in vibrational states of this $^3B_2$ state. The spacing between the peaks near 0.5 eV and those near 2.5 eV relate to the energy difference between the $^3B_2$ and $^1A_1$ electronic states of the neutral $H_2C{=}C$.

Because PES offers a direct way to measure energy differences between anion and neutral or neutral and cation state energies, it is a powerful and widely used means of determining molecular electron affinities (EAs) and ionization potentials (IPs). Because IPs and EAs relate, via Koopmans' theorem, to orbital energies, PES is thus seen to be a way to measure orbital energies. Its vibrational envelopes also offer a good way to probe vibrational energy level spacings, and hence the bonding strengths.

6.7 Probing continuum orbitals

There is another type of spectroscopy that can be used to directly probe the orbitals of a molecule that lie in the continuum (i.e., at energies higher than that of the parent neutral). I ask that you reflect back on our discussion of tunneling and of resonance states that can occur when an electron experiences both attractive and repulsive potentials. In such cases, there exists a special energy at which the electron can be trapped by the attractive potential and have to tunnel through the repulsive barrier to escape. It is these kinds of situations that this spectroscopy probes.

This experiment is called electron-transmission spectroscopy (ETS). In such an experiment a beam of electrons having a known intensity I_0 and narrowly defined range of kinetic energies E is allowed to pass through a sample (usually gaseous) of thickness L. The intensity I of electrons observed to pass through the sample and arrive at a detector lying along the incident beam's direction is monitored, as are the kinetic energies of these electrons E'. Such an experiment is described in qualitative form in Fig. 6.31.

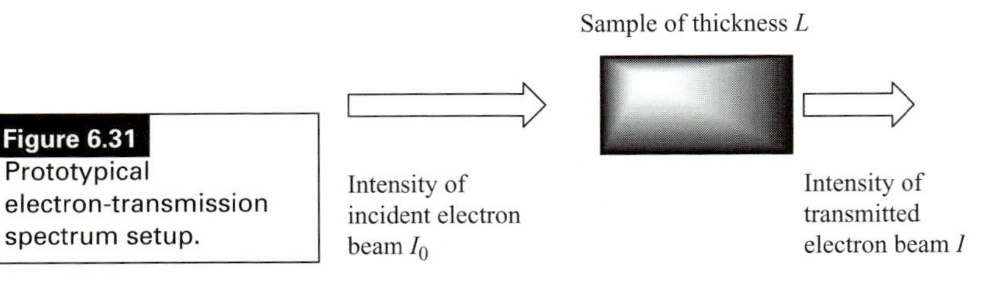

Figure 6.31 Prototypical electron-transmission spectrum setup.

If the molecules in the sample have a resonance orbital whose energy is close to the kinetic energy E of the colliding electrons, it is possible for an electron from the beam to be captured into such an orbital and to exist in this orbital for a considerable time. Of course, in the absence of any collisions or other processes to carry away excess energy, this anion will re-emit an electron at a later time. Hence, such anions are called metastable and their electronic states are called resonance states. If the captured electron remains in this orbital for a length of time comparable to or longer than the time it takes for the nascent molecular anion to undergo vibrational or rotational motion, various events can take place before the electron is re-emitted:

(i) some bond lengths or angles can change (this will happen if the orbital occupied by the beam's electron has bonding or antibonding character) so, when the electron is subsequently emitted, the neutral molecule is left with a change in vibrational energy;
(ii) the molecule may rotate, so when the electron is ejected, it is not emitted in the same direction as the incident beam.

In the former case, one observes electrons emitted with energies E' that differ from that of the incident beam by amounts related to the internal vibrational energy levels of the anion. In the latter, one sees a reduction in the intensity of the beam that is transmitted directly through the sample and electrons that are scattered away from this direction.

Such an ETS spectrum is shown in Fig. 6.32 for a gaseous sample of CO_2 molecules. In this spectrum, the energy of the transmitted beam's electrons is plotted on the horizontal axis and the derivative of the intensity of the transmitted beam is plotted on the vertical axis. It is common to plot such derivatives in ETS-type experiments to allow the variation of the signal with energy to be more clearly identified. In this ETS spectrum of CO_2, the oscillations that appear within the one spectral feature displayed correspond to stretching and bending vibrational levels of the metastable CO_2^- anion. It is the bending vibration that is primarily excited because the beam electron enters the LUMO of CO_2, which is an orbital of the form shown in Fig. 6.33. Occupancy of this antibonding π^* orbital causes both C—O bonds to lengthen and the O—C—O angle to bend away from 180°. The bending allows the antibonding nature of this orbital to be reduced.

Other examples of ETS spectra are shown in Fig. 6.34. Here again a derivative spectrum is shown, and the vertical lines have been added to show where the derivative passes through zero, which is where the ETS signal would have a "peak". These maxima correspond to electrons entering various virtual π^* orbitals of the uracil and DNA base molecules. It is by finding these peaks in the ETS spectrum that one can determine the energies of such continuum orbitals.

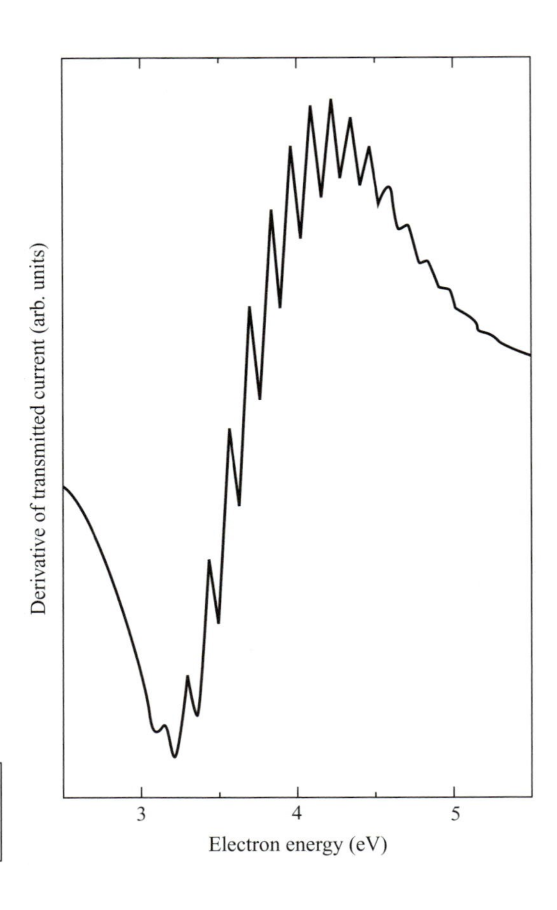

Figure 6.32 ETS spectrum (plotted in derivative form) of CO_2^-.

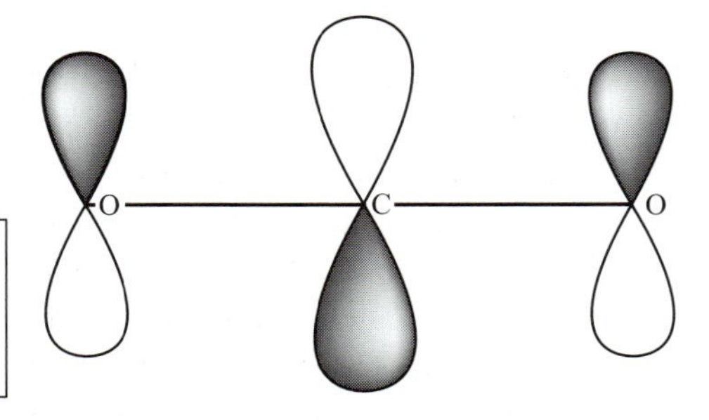

Figure 6.33 Antibonding π^* orbital of CO_2 holding the excess electron in CO_2^-.

Before closing this section, it is important to describe how one uses theory to simulate the metastable states that arise in such ETS experiments. Such calculations are not at all straightforward, and require the introduction of special tools designed to properly model the resonant continuum orbital.

For metastable anions, it is difficult to approximate the potential experienced by the excess electron. For example, singly charged anions in which the excess electron occupies a molecular orbital ϕ that possesses non-zero angular

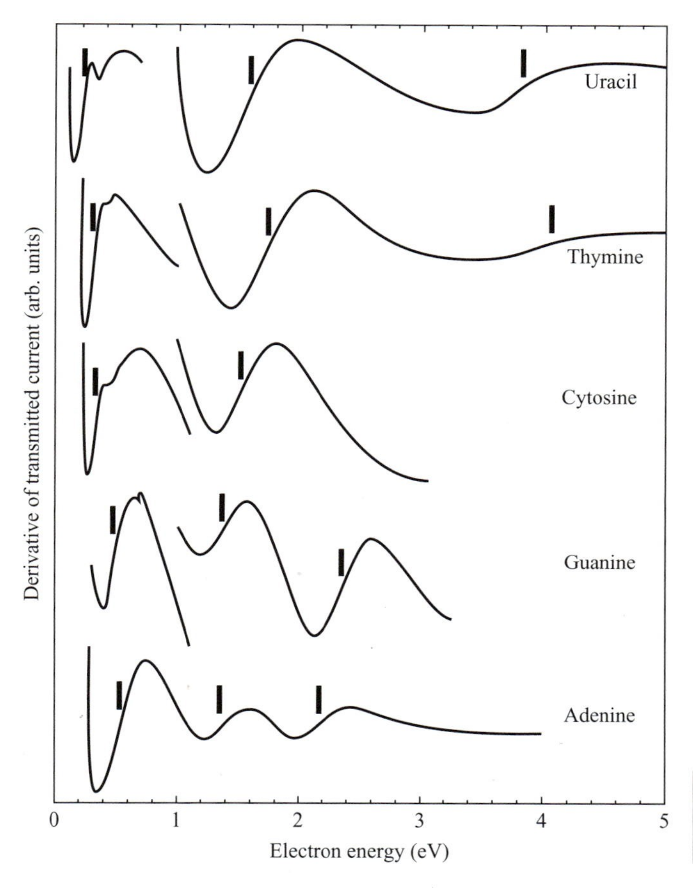

Figure 6.34 ETS spectra of several molecules.

momentum have effective potentials as shown in Fig. 6.35, which depend on the angular momentum L value of the orbital.

For example, the π^* orbital of N_2^- shown in Fig. 6.36 produces two counteracting contributions to the effective radial potential $V_{\text{eff}}(r)$ experienced by an electron occupying it. First, the two nitrogen centers exert attractive potentials on the electron in this orbital. These attractions are strongest when the excess electron is near the nuclei but decay rapidly at larger distances because the other electrons' Coulomb repulsions screen the nuclear attractions. Second, because the π^* molecular orbital is comprised of atomic basis functions of p_π, d_π, etc. symmetry, it possesses non-zero angular momentum. Because the π^* orbital has gerade symmetry, its large-r character is dominated by $L = 2$ angular momentum. As a result, the excess electron has a centrifugal radial potential $L(L+1)/2m_3r^2$ derived largely from its $L = 2$ character.

The attractive short-range valence potentials $V(r)$ and the centrifugal potential combine to produce a net effective potential as illustrated in Fig. 6.35. The energy

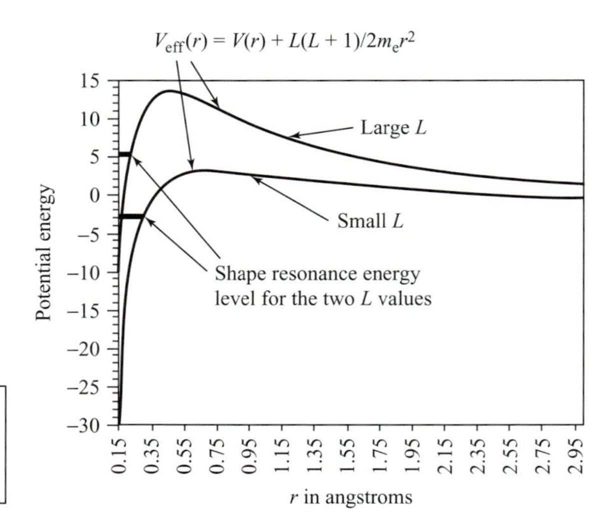

Figure 6.35 Radial potentials and shape resonance energy levels for two L values.

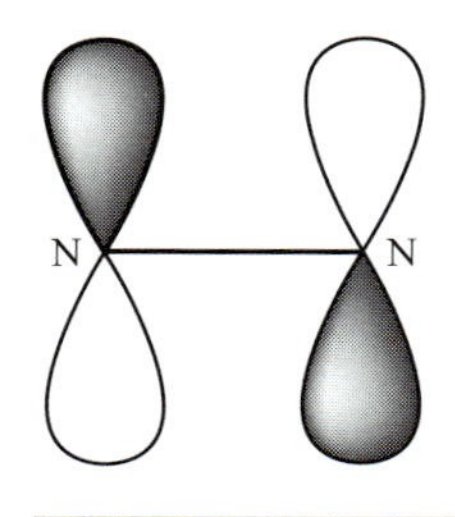

Figure 6.36 Antibonding π^* orbital of N_2^- showing its $L = 2$ character.

of an electron experiencing such a potential may or may not lie below the $r \to \infty$ asymptote. If the attractive potential is sufficiently strong, as it is for O_2^{-1}, the electron in the π^* orbital will be bound and its energy will lie below this asymptote. On the other hand, if the attractive potential is not as strong, as is the case for the less-electronegative nitrogen atoms in N_2^{-1}, the energy of the π^* orbital can lie above the asymptote. In the latter cases, we speak of metastable shape-resonance states. They are metastable because their energies lie above the asymptote so they can decay by tunneling through the centrifugal barrier. They are called shape-resonances because their metastability arises from the shape of their repulsive centrifugal barrier.

If one had in-hand a reasonable approximation to the attractive short-range potential $V(r)$ and if one knew the L-symmetry of the orbital occupied by the excess electron, one could form $V_{eff}(r)$ as above. However, to compute the lifetime of the shape resonance, one has to know the energy E of this state.

The most common and powerful tool for studying such metastable states theoretically is the stabilization method (SM). This method involves embedding the system of interest (e.g., the N_2^{-1} anion) within a finite radial "box" in order to convert the continuum of states corresponding, for example, to $N_2 + e^-$, into discrete states that can be handled using more conventional methods. By then varying the size of the box, one can vary the energies of the discrete states that correspond to $N_2 + e^-$ (i.e., one varies the kinetic energy of the orbital containing the excess electron). As the box size is varied, one eventually notices (e.g., by plotting the orbitals) that one of the $N_2 + e^-$ states possesses a significant amount of valence (i.e., short-range) character. That is, one such state has significant amplitude not only at large-r but also in the region of the two nitrogen centers. It is this state that corresponds to the metastable shape-resonance state, and it is the energy E where significant valence components develop that provides the stabilization estimate of the state energy.

Let us continue using N_2^{-1} as an example for how the SM would be employed, especially how one usually varies the box within which the anion is constrained. One would use a conventional atomic orbital basis set that would likely include s and p functions on each N atom, perhaps some polarization d functions and some conventional diffuse s and p orbitals on each N atom. These basis orbitals serve primarily to describe the motions of the electrons within the usual valence regions of space.

To this basis, one would append an extra set of diffuse π-symmetry orbitals. These orbitals could be p_π (and maybe d_π) functions centered on each nitrogen atom, or they could be d_π orbitals centered at the midpoint of the N—N bond. One usually would not add just one such function; rather several such functions, each with an orbital exponent α_J that characterizes its radial extent, would be used. Let us assume, for example, that K such π functions have been used.

Next, using the conventional atomic orbital basis as well as the K extra π basis functions, one carries out a calculation (most often a variational calculation in which one computes many energy levels) of the N_2^{-1} anion. In this calculation, one tabulates the energies of many (say M) of the electronic states of N_2^{-1}. Of course, because a finite atomic orbital basis set must be used, one finds a discrete "spectrum" of orbital energies and thus of electronic state energies. There are occupied orbitals having negative energy that represent, via Koopmans' theorem, the bound states of the N_2^-. There are also so-called virtual orbitals (i.e., those orbitals that are not occupied) whose energies lie above zero (i.e., do not describe bound states). The latter orbitals offer a discrete approximation to the continuum within which the resonance state of interest lies.

One then scales the orbital exponents $\{\alpha_J\}$ of the K extra π basis orbitals by a factor $\eta : \alpha_J \rightarrow \eta\alpha_J$ and repeats the calculation of the energies of the M lowest energies of N_2^{-1}. This scaling causes the extra π basis orbitals to contract radially (if $\eta > 1$) or to expand radially (if $\eta < 1$). It is this basis orbital expansion and contraction that produces expansion and contraction of the "box" discussed above. That is, one does not employ a box directly; instead, one varies the radial extent of the most diffuse basis orbitals to simulate the box variation.

If the conventional orbital basis is adequate, one finds that the extra π orbitals, whose exponents are being scaled, do not affect appreciably the energy of the neutral N_2 molecule. This can be probed by plotting the N_2 energy as a function of the scaling parameter η; if the energy varies little with η, the conventional basis is adequate.

In contrast to plots of the neutral N_2 energy vs. η, plots of the energies of the M N_2^{-1} states show significant η-dependence as Fig. 6.37 illustrates.

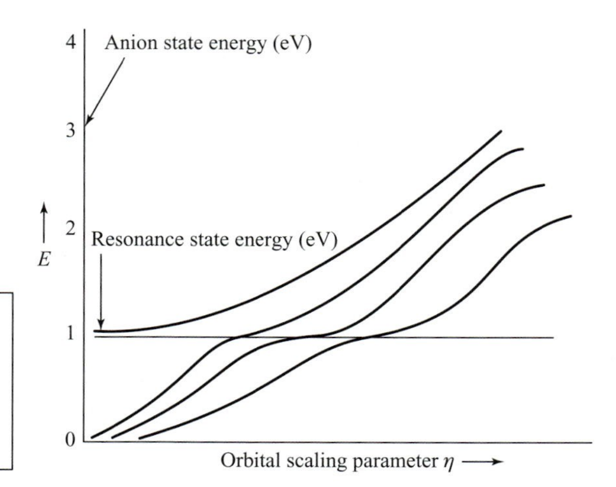

Figure 6.37 Typical stabilization plot showing several levels of the metastable anion and their avoided crossings.

What does such a stabilization plot tell us and what do the various branches of the plot mean? First, one should notice that each of the plots of the energy of an anion state (relative to the neutral molecule's energy, which is independent of η) grows with increasing η. This η-dependence arises from the η-scaling of the extra diffuse π basis orbitals. Because most of the amplitude of such basis orbitals lies outside the valence region, the kinetic energy is the dominant contributor to such orbitals' energy. Because η enters into each orbital as $\exp(-\eta\alpha r^2)$, and because the kinetic energy operator involves the second derivative with respect to r, the kinetic energies of orbitals dominated by the diffuse π basis functions vary as η^2.

For small η, all of the π diffuse basis functions have their amplitudes concentrated at large r and have low kinetic energy. As η grows, these functions become more radially compact and their kinetic energies grow. For example, note the three lowest energies shown above increasing from near zero as η grows.

As η further increases, one reaches a point at which the third and fourth anion-state energies undergo an avoided crossing. At this η value, if one examines the nature of the two wave functions whose energies avoid one another, one finds that one of them contains substantial amounts of both valence and extra diffuse π function character. Just to the left of the avoided crossing, the lower-energy state (the third state for small η) contains predominantly extra diffuse π orbital character, while the higher-energy state (the fourth state) contains largely valence π^* orbital character.

However, at the special value of η where these two states nearly cross, the kinetic energy of the third state (as well as its radial size and de Broglie wavelength) are appropriate to connect properly with the fourth state. By connect properly we mean that the two states have wave function amplitudes, phases, and slopes that match. So, at this special η value, one can achieve a description of the shape-resonance state that correctly describes this state both in the valence region and in the large-r region. Only by tuning the energy of the large-r states using the η-scaling can one obtain this proper boundary condition matching.

In summary, by carrying out a series of anion-state energy calculations for several states and plotting them vs. η, one obtains a stabilization graph. By examining this graph and looking for avoided crossings, one can identify the energies at which metastable resonances occur. It is also possible to use the shapes (i.e., the magnitude of the energy splitting between the two states and the slopes of the two avoiding curves) of the avoided crossings in a stabilization graph to compute the lifetimes of the metastable states. Basically, the larger the avoided crossing energy splitting between the two states, the shorter is the lifetime of the resonance state. So, the ETS and PES experiments offer wonderful probes of the bound and continuum states of molecules and ions that tell us a lot about the electronic nature and chemical bonding of these species. The theoretical study of these phenomena is complicated by the need to properly identify and describe any continuum orbitals and states that are involved. The stabilization technique allows us to achieve a good approximation to resonance states that lie in such continua.

Chapter 7
Statistical mechanics

When one is faced with a condensed-phase system, usually containing many molecules, that is at or near thermal equilibrium, it is not necessary or even wise to try to describe it in terms of quantum wave functions or even classical trajectories of all of the constituent molecules. Instead, the powerful tools of statistical mechanics allow one to focus on quantities that describe the most important features of the many-molecule system. In this chapter, you will learn about these tools and see some important examples of their application.

Collections of molecules at or near equilibrium

As noted in Chapter 5, the approach one takes in studying a system composed of a very large number of molecules at or near thermal equilibrium can be quite different from how one studies systems containing a few isolated molecules. In principle, it is possible to conceive of computing the quantum energy levels and wave functions of a collection of many molecules, but doing so becomes impractical once the number of atoms in the system reaches a few thousand or if the molecules have significant intermolecular interactions. Also, as noted in Chapter 5, following the time evolution of such a large number of molecules can be "confusing" if one focuses on the short-time behavior of any single molecule (e.g., one sees "jerky" changes in its energy, momentum, and angular momentum). By examining, instead, the long-time average behavior of each molecule or, alternatively, the average properties of a significantly large number of molecules, one is often better able to understand, interpret, and simulate such condensed-media systems. This is where the power of statistical mechanics comes into play.

7.1 Distribution of energy among levels

One of the most important concepts of statistical mechanics involves how a specified amount of total energy E can be shared among a collection of molecules and among the internal (translational, rotational, vibrational, electronic) degrees of freedom of these molecules. The primary outcome of asking what is the most probable distribution of energy among a large number N of molecules within a container of volume V that is maintained in equilibrium at a specified temperature T is the most important equation in statistical mechanics, the Boltzmann population formula:

$$P_j = \Omega_j \frac{\exp\left(-\frac{E_j}{kT}\right)}{Q}. \tag{7.1}$$

This equation expresses the probability P_j of finding the system (which, in the case introduced above, is the whole collection of N interacting molecules) in its jth quantum state, where E_j is the energy of this quantum state, k is the Boltzmann constant, T is the temperature in K, Ω_j is the degeneracy of the jth state, and the denominator Q is the so-called partition function:

$$Q = \sum_j \Omega_j \exp\left(-\frac{E_j}{kT}\right). \tag{7.2}$$

The classical mechanical equivalent of the above quantum Boltzmann population formula for a system with M coordinates (collectively denoted q) and M momenta (denoted p) is

$$P(q, p) = h^{-M} \frac{\exp\left(-\frac{H(q,p)}{kT}\right)}{Q}, \tag{7.3}$$

where H is the classical Hamiltonian, h is Planck's constant, and the classical partition function Q is

$$Q = h^{-M} \int \exp\left(-\frac{H(q, p)}{kT}\right) dq dp. \tag{7.4}$$

Notice that the Boltzmann formula does not say that only those states of a given energy can be populated; it gives non-zero probabilities for populating all states from the lowest to the highest. However, it does say that states of higher energy E_j are disfavored by the $\exp(-E_j/kT)$ factor, but if states of higher energy have larger degeneracies Ω_j (which they usually do), the overall population of such states may not be low. That is, there is a competition between state degeneracy Ω_j, which tends to grow as the state's energy grows, and $\exp(-E_j/kT)$ which decreases with increasing energy. If the number of particles N is huge, the degeneracy Ω grows as a high power (let's denote this power as K) of E because the degeneracy is related to the number of ways the energy can be distributed among the N molecules. In fact, K grows at least as fast as N. As a result of Ω growing as E^K, the product function $P(E) = E^K \exp(-E/kT)$ has the form shown in Fig. 7.1 (for $K = 10$). By taking the derivative of this function $P(E)$ with respect to E, and finding the energy at which this derivative vanishes, one can show that this probability function has a peak at $E^* = K\,kT$, and

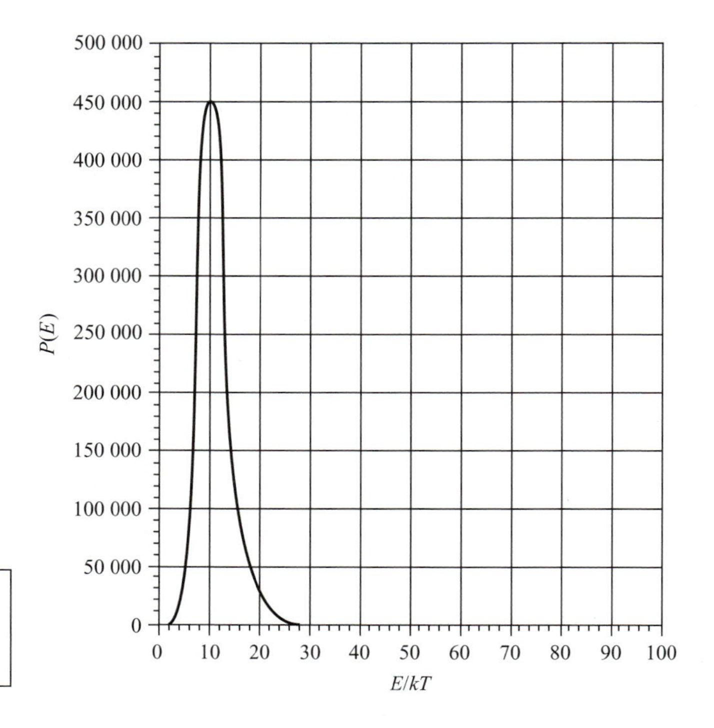

Figure 7.1 Probability weighting factor $P(E)$ as a function of E for $K = 10$.

that at this energy value

$$P(E^*) = (KkT)^K \exp(-K). \tag{7.5}$$

By then asking at what energy E' the function $P(E)$ drops to $\exp(-1)$ of this maximum value $P(E^*)$:

$$P(E') = \exp(-1)P(E^*), \tag{7.6}$$

one finds

$$E' = KkT\left(1 + (2/K)^{1/2}\right). \tag{7.7}$$

So the width of the $P(E)$ graph, measured as the change in energy needed to cause $P(E)$ to drop to $\exp(-1)$ of its maximum value divided by the value of the energy at which $P(E)$ assumes this maximum value, is

$$\frac{(E' - E^*)}{E^*} = \left(\frac{2}{K}\right)^{1/2}. \tag{7.8}$$

This width gets smaller and smaller as K increases. The primary conclusion is that as the number N of molecules in the sample grows, which, as discussed earlier, causes K to grow, the energy probability function becomes more and more sharply peaked about the most probable energy E^*. This, in turn, suggests that we may be able to model, aside from infrequent fluctuations, the behavior of systems with many molecules by focusing on the most probable situation (i.e., having the energy E^*) and ignoring deviations from this case.

It is for the reasons just shown that for so-called macroscopic systems near equilibrium, in which N (and hence K) is extremely large (e.g., $N \sim 10^{10}$ to 10^{24}), only the most probable distribution of the total energy among the N molecules need be considered. This is the situation in which the equations of statistical mechanics are so useful. Certainly, there are fluctuations (as evidenced by the finite width of the above graph) in the energy content of the N-molecule system about its most probable value. However, these fluctuations become less and less important as the system size (i.e., N) becomes larger and larger.

To understand how this narrow Boltzmann distribution of energies arises when the number of molecules N in the sample is large, we consider a system composed of M identical containers, each having volume V, and each made of a material that allows for efficient heat transfer to its surroundings but material that does not allow the N molecules in each container to escape. These containers are arranged into a regular lattice as shown in Fig. 7.2 in a manner that allows their thermally conducting walls to come into contact. Finally, the entire collection of M such containers is surrounded by a perfectly insulating material that assures that the total energy (of all $N \times M$ molecules) can not change. So, this collection of M identical containers each containing N molecules constitutes a closed (i.e., with no molecules coming or going) and isolated (i.e., so total energy is constant) system.

One of the fundamental assumptions of statistical mechanics is that, for a closed isolated system at equilibrium, all quantum states of the system having an energy equal to the energy E with which the system is prepared are equally likely to be occupied. This is called the assumption of equal a priori probability

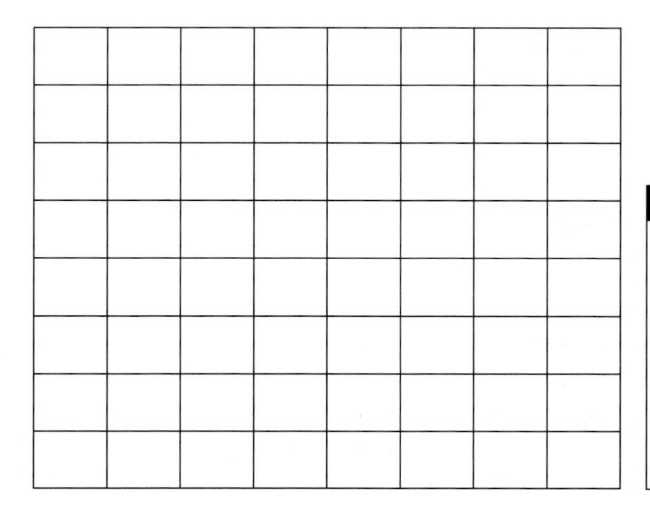

Figure 7.2 Collection of M identical cells having energy conducting walls that do not allow molecules to pass between cells. Each cell contains N molecules in volume V. There are M such cells and the total energy of these M cells is E.

for such energy-allowed quantum states. The quantum states relevant to this case are not the states of individual molecules. Nor are they the states of N of the molecules in one of the containers of volume V. They are the quantum states of the entire system comprised of $N \times M$ molecules. Because our system consists of M identical containers, each with N molecules in it, we can describe the quantum states of the entire system in terms of the quantum states of each such container.

In particular, let's pretend that we know the quantum states that pertain to N molecules in a container of volume V as shown in Fig. 7.2, and let's label these states by an index J. That is $J = 1$ labels the first energy state of N molecules in the container of volume V, $J = 2$ labels the second such state, and so on. I understand that it may seem daunting to think of how one actually finds these N-molecule eigenstates. However, we are just deriving a general framework that gives the probabilities of being in each such state. In so doing, we are allowed to pretend that we know these states. In any actual application, we will, of course, have to use approximate expressions for such energies.

An energy labeling for states of the entire collection of M containers can be realized by giving the number of containers that exist in each single-container J-state. This is possible because the energy of each M-container state is a sum of the energies of the M single-container states that comprise that M-container state. For example, if $M = 9$, the label 1, 1, 2, 2, 1, 3, 4, 1, 2 specifies the energy of this 9-container state in terms of the energies $\{\varepsilon_J\}$ of the states of the 9 containers: $E = 4\varepsilon_1 + 3\varepsilon_2 + \varepsilon_3 + \varepsilon_4$. Notice that this 9-container state has the same energy as several other 9-container states; for example, 1, 2, 1, 2, 1, 3, 4, 1, 2 and 4, 1, 3, 1, 2, 2, 1, 1, 2 have the same energy although they are different individual states. What differs among these distinct states is which box occupies which single-box quantum state.

The above example illustrates that an energy level of the M-container system can have a high degree of degeneracy because its total energy can be achieved by having the various single-container states appear in various orders. That is, which container is in which state can be permuted without altering the total energy E. The formula for how many ways the M container states can be permuted such that (i) there are n_J containers appearing in single-container state J, with (ii) a total of M containers, is

$$\Omega(\mathbf{n}) = \frac{M!}{\{\prod_J n_J!\}}. \tag{7.9}$$

Here $\mathbf{n} = \{n_1, n_2, n_3, \ldots n_J, \ldots\}$ denotes the number of containers existing in single-container states $1, 2, 3, \ldots J, \ldots$. This combinatorial formula reflects the permutational degeneracy arising from placing n_1 containers into state 1, n_2 containers into state 2, etc.

If we imagine an extremely large number of containers and we view M as well as the $\{n_J\}$ as being large numbers (n.b., we will soon see that this is the case), we can ask for what choices of the variables $\{n_1, n_2, n_3, \ldots n_J, \ldots\}$ is this degeneracy function $\Omega(\mathbf{n})$ a maximum. Moreover, we can examine $\Omega(\mathbf{n})$ at its maximum and compare its value at values of the $\{\mathbf{n}\}$ parameters changed only slightly from the values that maximized $\Omega(\mathbf{n})$. As we will see, Ω is very strongly peaked at its maximum and decreases extremely rapidly for values of $\{\mathbf{n}\}$ that differ only slightly from the "optimal" values. It is this property that gives rise to the very narrow energy distribution discussed earlier in this section. So, let's take a closer look at how this energy distribution formula arises.

We want to know what values of the variables $\{n_1, n_2, n_3, \ldots n_J, \ldots\}$ make $\Omega = M!/\{\Pi_J n_J!\}$ a maximum. However, all of the $\{n_1, n_2, n_3, \ldots n_J, \ldots\}$ variables are not independent; they must add up to M, the total number of containers, so we have a constraint

$$\sum_J n_J = M \tag{7.10}$$

that the variables must obey. The $\{n_j\}$ variables are also constrained to give the total energy E of the M-container system when summed as

$$\sum_J n_J \varepsilon_J = E. \tag{7.11}$$

We have two problems: (i) how to maximize Ω and (ii) how to impose these constraints. Because Ω takes on values greater than unity for any choice of the $\{n_j\}$, Ω will experience its maximum where $\ln \Omega$ has its maximum, so we can maximize $\ln \Omega$ if doing so helps. Because the n_J variables are assumed to take on large numbers (when M is large), we can use Stirling's approximation $\ln X! = X \ln X - X$ to approximate $\ln \Omega$ as follows:

$$\ln \Omega = \ln M! - \sum_J \{n_J \ln n_J - n_J). \tag{7.12}$$

This expression will prove useful because we can take its derivative with respect to the n_J variables, which we need to do to search for the maximum of $\ln \Omega$.

To impose the constraints $\sum_J n_J = M$ and $\sum_J n_J \varepsilon_J = E$ we use the technique of Lagrange multipliers. That is, we seek to find values of $\{n_J\}$ that maximize the following function:

$$F = \ln M! - \sum_J \{n_J \ln n_J - n_J) - \alpha \left(\sum_J n_J - M \right) - \beta \left(\sum_J n_J \varepsilon_J - E \right). \tag{7.13}$$

Notice that this function F is exactly equal to the $\ln \Omega$ function we wish to maximize whenever the $\{n_J\}$ variables obey the two constraints. So, the maxima of F and of $\ln \Omega$ are identical if the $\{n_J\}$ have values that obey the constraints. The two Lagrange multipliers α and β are introduced to allow the values of $\{n_J\}$ that maximize F to ultimately obey the two constraints. That is, we will find values of the $\{n_J\}$ variables that make F maximum; these values will depend on α and

β and will not necessarily obey the constraints. However, we will then choose α and β to assure that the two constraints are obeyed. This is how the Lagrange multiplier method works.

Taking the derivative of F with respect to each independent n_K variable and setting this derivative equal to zero gives

$$-\ln n_K - \alpha - \beta\varepsilon_K = 0. \tag{7.14}$$

This equation can be solved to give $n_K = \exp(-\alpha)\exp(-\beta\varepsilon_K)$. Substituting this result into the first constraint equation gives $M = \exp(-\alpha)\sum_J \exp(-\beta\varepsilon_J)$, which allows us to solve for $\exp(-\alpha)$ in terms of M. Doing so, and substituting the result into the expression for n_K gives

$$n_K = M\frac{\exp(-\beta\varepsilon_K)}{Q}, \tag{7.15}$$

where

$$Q = \sum_J \exp(-\beta\varepsilon_J). \tag{7.16}$$

Notice that the n_K are, as we assumed earlier, large numbers if M is large because n_K is proportional to M. Notice also that we now see the appearance of the partition function Q and of exponential dependence on the energy of the state that gives the Boltzmann population of that state.

It is possible to relate the β Lagrange multiplier to the total energy E of the M containers by using

$$\begin{aligned} E &= M\sum_J \varepsilon_J \frac{\exp(-\beta\varepsilon_K)}{Q} \\ &= -M\left(\frac{\partial \ln Q}{\partial \beta}\right)_{N,V}. \end{aligned} \tag{7.17}$$

This shows that the average energy of a container, computed as the total energy E divided by the number M of such containers, can be computed as a derivative of the logarithm of the partition function Q. As we show in the following section, all thermodynamic properties of the N molecules in the container of volume V can be obtained as derivatives of the logarithm of this Q function. This is why the partition function plays such a central role in statistical mechanics.

To examine the range of energies over which each of the M single-container systems ranges with appreciable probability, let us consider not just the degeneracy $\Omega(\mathbf{n}^*)$ of that set of variables $\{\mathbf{n}^*\} = \{n_1^*, n_2^*, \ldots\}$ that makes Ω maximum, but also the degeneracy $\Omega(\mathbf{n})$ for values of $\{n_1, n_2, \ldots\}$ differing by small amounts $\{\delta n_1, \delta n_2, \ldots\}$ from the optimal values $\{\mathbf{n}^*\}$. Expanding $\ln\Omega$ as a Taylor series in the parameters $\{n_1, n_2, \ldots\}$ and evaluating the expansion in the neighborhood of the values $\{\mathbf{n}^*\}$, we find

$$\ln \Omega = \ln \Omega(\{n_1^*, n_2^*, \ldots\}) + \sum_J \left(\frac{\partial \ln \Omega}{\partial n_J}\right) \delta n_J + \frac{1}{2} \sum_{J,K} \left(\frac{\partial^2 \ln \Omega}{\partial n_J\, \partial n_K}\right) \delta n_J \delta n_K + \cdots. \quad (7.18)$$

We know that all of the first derivative terms $(\partial \ln \Omega/\partial n_J)$ vanish because $\ln \Omega$ has been made maximum at $\{\mathbf{n}^*\}$. The first derivative of $\ln \Omega$ as given above is $\partial \ln \Omega/\partial n_J = -\ln n_J$, so the second derivatives needed to complete the Taylor series through second order are

$$\left(\frac{\partial^2 \ln \Omega}{\partial n_J\, \partial n_K}\right) = -\delta_{J,K} n_j^{-1}. \quad (7.19)$$

We can thus express $\Omega(\mathbf{n})$ in the neighborhood of $\{\mathbf{n}^*\}$ as follows:

$$\ln \Omega(\mathbf{n}) = \ln \Omega(\mathbf{n}^*) - 1/2 \sum_J \frac{(\delta n_J)^2}{n_J^*}, \quad (7.20)$$

or, equivalently,

$$\Omega(\mathbf{n}) = \Omega(\mathbf{n}^*) \exp\left[-\frac{1}{2} \sum_J \frac{(\delta n_J)^2}{n_J^*}\right]. \quad (7.21)$$

This result clearly shows that the degeneracy, and hence, by the equal a priori probability hypothesis, the probability of the M-container system occupying a state having $\{n_1, n_2, \ldots\}$ falls off exponentially as the variables n_J move away from their "optimal" values $\{\mathbf{n}^*\}$.

As we noted earlier, the n_J^* are proportional to M (i.e., $n_J^* = M \exp(-\beta \varepsilon_J)/Q = f_J M$), so when considering deviations δn_J away from the optimal n_J^*, we should consider deviations that are also proportional to $M : \delta n_J = M \delta f_J$. In this way, we are treating deviations of specified percentage or fractional amount which we denote f_J. Thus, the ratio $(\delta n_J)^2/n_J^*$ that appears in the above exponential has an M-dependence that allows $\Omega(\mathbf{n})$ to be written as

$$\Omega(\mathbf{n}) = \Omega(\mathbf{n}^*) \exp\left[-\frac{M}{2} \sum_J \frac{(\delta f_J)^2}{f_J^*}\right], \quad (7.22)$$

where f_J^* and δf_J are the fraction and fractional deviation of containers in state $J : f_J^* = n_J^*/M$ and $\delta f_J = \delta n_J/M$. The purpose of writing $\Omega(\mathbf{n})$ in this manner is to explicitly show that, in the so-called thermodynamic limit, when M approaches infinity, only the most probable distribution of energy $\{\mathbf{n}^*\}$ need be considered because only $\{\delta f_J = 0\}$ is important as M approaches infinity.

Let's consider this very narrow distribution issue a bit further by examining fluctuations in the energy of a single container around its average energy $E_{\text{ave}} = E/M$. We already know that the number of containers in a given state K can be written as $n_K = M \exp(-\beta\varepsilon_K)/Q$. Alternatively, we can say that the probability of a container occupying the state J is

$$P_J = \frac{\exp(-\beta\varepsilon_K)}{Q}. \tag{7.23}$$

Using this probability, we can compute the average energy E_{ave} as

$$E_{\text{ave}} = \sum_J P_J\varepsilon_J = \sum_J \varepsilon_J \frac{\exp(-\beta\varepsilon_K)}{Q} = -\left(\frac{\partial \ln Q}{\partial \beta}\right)_{N,V}. \tag{7.24}$$

To compute the fluctuation in energy, we first note that the fluctuation is defined as the average of the square of the deviation in energy from the average:

$$\begin{aligned}(E - E_{\text{ave}})^2_{\text{ave}} &= \sum_J (\varepsilon_J - E_{\text{ave}})^2 P_J = \sum_J P_J\left(\varepsilon_J^2 - 2\varepsilon_J E_{\text{ave}} + E^2_{\text{ave}}\right)\\ &= \sum_J P_J\left(\varepsilon_J^2 - E^2_{\text{ave}}\right).\end{aligned} \tag{7.25}$$

The following identity is now useful for further re-expressing the fluctuations:

$$\begin{aligned}\left(\frac{\partial^2 \ln Q}{\partial \beta^2}\right)_{N,V} &= \frac{\partial\left(-\sum_J \varepsilon_J \frac{\exp(-\beta\varepsilon_J)}{Q}\right)}{\partial \beta}\\ &= \sum_J \varepsilon_J^2 \frac{\exp(-\beta\varepsilon_J)}{Q} - \left\{\sum_J \varepsilon_J \frac{\exp(-\beta\varepsilon_J)}{Q}\right\}\left\{\sum_L \varepsilon_L \frac{\exp(-\beta\varepsilon_L)}{Q}\right\}.\end{aligned} \tag{7.26}$$

Recognizing the first factor immediately above as $\sum_J \varepsilon_J^2 P_J$, and the second factor as E^2_{ave}, and noting that $\sum_J P_J = 1$, allows the fluctuation formula to be rewritten as

$$(E - E_{\text{ave}})^2_{\text{ave}} = \left(\frac{\partial^2 \ln Q}{\partial \beta^2}\right)_{N,V} = -\left(\frac{\partial E_{\text{ave}}}{\partial \beta}\right)_{N,V}. \tag{7.27}$$

Because the parameter β can be shown to be related to the Kelvin temperature T as $\beta = 1/(kT)$, the above expression can be re-written as

$$(E - E_{\text{ave}})^2_{\text{ave}} = -\left(\frac{\partial E_{\text{ave}}}{\partial \beta}\right)_{N,V} = kT^2\left(\frac{\partial E_{\text{ave}}}{\partial T}\right)_{N,V}. \tag{7.28}$$

Recognizing the formula for the constant-volume heat capacity

$$C_V = \left(\frac{\partial E_{\text{ave}}}{\partial T}\right)_{N,V} \tag{7.29}$$

allows the fractional fluctuation in the energy around the mean energy $E_{\text{ave}} = E/M$ to be expressed as

$$\frac{(E - E_{\text{ave}})^2_{\text{ave}}}{E^2_{\text{ave}}} = \frac{kT^2 C_V}{E^2_{\text{ave}}}. \tag{7.30}$$

What does this fractional fluctuation formula tell us? On its left-hand side it gives a measure of the fractional spread of energies over which each of the containers ranges about its mean energy E_{ave}. On the right side, it contains a ratio of two quantities that are extensive properties, the heat capacity and the mean energy. That is, both C_V and E_{ave} will be proportional to the number N of molecules in the container as long as N is reasonably large. However, because the right-hand side involves C_V/E_{ave}^2, it is proportional to N^{-1} and thus will be very small for large N as long as C_V does not become large. As a result, except near so-called critical points where the heat capacity does indeed become extremely large, the fractional fluctuation in the energy of a given container of N molecules will be very small (i.e., proportional to N^{-1}). It is this fact that causes the narrow distribution in energies that we discussed earlier in this section.

7.2 Partition functions and thermodynamic properties

Let us now examine how this idea of the most probable energy distribution being dominant gives rise to equations that offer molecular-level expressions of thermodynamic properties. The first equation is the fundamental Boltzmann population formula that we already examined:

$$P_j = \Omega_j \frac{\exp\left(-\frac{E_j}{kT}\right)}{Q}, \tag{7.31}$$

which expresses the probability for finding the N-molecule system in its jth quantum state having energy E_j and degeneracy Ω_j.

Using this result, it is possible to compute the average energy $\langle E \rangle$ of the system

$$\langle E \rangle = \sum_j P_j E_j, \tag{7.32}$$

and, as we saw earlier in this section, to show that this quantity can be recast as

$$\langle E \rangle = kT^2 \left(\frac{\partial \ln Q}{\partial T}\right)_{N,V}. \tag{7.33}$$

To review how this proof is carried out, we substitute the expressions for P_j and for Q into the expression for $\langle E \rangle$:

$$\langle E \rangle = \left\{\sum_j E_j \Omega_j \exp\left(-\frac{E_j}{kT}\right)\right\} \Big/ \left\{\sum_l \Omega_l \exp\left(-\frac{E_l}{kT}\right)\right\}. \tag{7.34}$$

By noting that $\partial(\exp(-E_j/kT))/\partial T = (1/kT^2)E_j \exp(-E_j/kT)$, we can then rewrite $\langle E \rangle$ as

$$\langle E \rangle = kT^2 \left\{\sum_j \Omega_j \frac{\partial \exp\left(-\frac{E_j}{kT}\right)}{\partial T}\right\} \Big/ \left\{\sum_l \Omega_l \exp\left(\frac{-E_l}{kT}\right)\right\}. \tag{7.35}$$

And then recalling that $\{\partial X/\partial T\}/X = \partial \ln X/\partial T$, we finally obtain

$$\langle E\rangle = kT^2 \left(\frac{\partial \ln Q}{\partial T}\right)_{N,V}. \tag{7.36}$$

All other equilibrium properties can also be expressed in terms of the partition function Q. For example, if the average pressure $\langle p\rangle$ is defined as the pressure of each quantum state

$$p_j = \left(\frac{\partial E_j}{\partial V}\right)_N \tag{7.37}$$

multiplied by the probability P_j for accessing that quantum state, summed over all such states, one can show, realizing that only E_j (not T or Ω) depend on the volume V, that

$$\begin{aligned}\langle p\rangle &= \sum_j \left(\frac{\partial E_j}{\partial V}\right)_N \Omega_j \frac{\exp\left(-\frac{E_j}{kT}\right)}{Q} \\ &= kT\left(\frac{\partial \ln Q}{\partial V}\right)_{N,T}.\end{aligned} \tag{7.38}$$

Without belaboring the point further, it is possible to express all of the usual thermodynamic quantities in terms of the partition function Q. The average energy and average pressure are given above; the average entropy is given as

$$\langle S\rangle = k \ln Q + kT\left(\frac{\partial \ln Q}{\partial N}\right)_{V,T}, \tag{7.39}$$

the Helmholtz free energy A is

$$A = -kT \ln Q \tag{7.40}$$

and the chemical potential μ is expressed as

$$\mu = -kT\left(\frac{\partial \ln Q}{\partial N}\right)_{T,V}. \tag{7.41}$$

As we saw earlier, it is also possible to express fluctuations in thermodynamic properties in terms of derivatives of partition functions and, thus, as derivatives of other properties. For example, the fluctuation in the energy $\langle(E - \langle E\rangle)^2\rangle$ was shown above to be given by

$$\langle(E - \langle E\rangle)^2\rangle = kT^2 C_V. \tag{7.42}$$

The *Statistical Mechanics* text by McQuarrie has an excellent treatment of these topics and shows how all of these expressions are derived.

So, if one were able to evaluate the partition function Q for N molecules in a volume V at a temperature T, either by summing the quantum-state degeneracy and $\exp(-E_j/kT)$ factors

$$Q = \sum_j \Omega_j \exp\left(-\frac{E_j}{kT}\right), \tag{7.43}$$

or by carrying out the phase-space integral over all M of the coordinates and momenta of the system

$$Q = h^{-M} \int \exp\left(\frac{-H(q, p)}{kT}\right) dq dp, \tag{7.44}$$

one could then use the above formulas to evaluate any thermodynamic properties as derivatives of $\ln Q$.

What do these partition functions mean? They represent the thermal-average number of quantum states that are accessible to the system. This can be seen best by again noting that, in the quantum expression,

$$Q = \sum_j \Omega_j \exp\left(\frac{-E_j}{kT}\right), \tag{7.45}$$

the partition function is equal to a sum of (i) the number of quantum states in the jth energy level multiplied by (ii) the Boltzmann population factor $\exp(-E_j/kT)$ of that level. So, Q is dimensionless and is a measure of how many states the system can access at temperature T. Another way to think of Q is suggested by rewriting the Helmholtz free energy definition given above as $Q = \exp(-A/kT)$. This identity shows that Q can be viewed as the Boltzmann population, not of a given energy E, but of a specified amount of free energy A.

Keep in mind that the energy levels E_j and degeneracies Ω_j are those of the full N-molecule system. In the special case for which the interactions among the molecules can be neglected (i.e., in the dilute ideal-gas limit), each of the energies E_j can be expressed as a sum of the energies of each individual molecule: $E_j = \sum_{k=1,N} \varepsilon_j(k)$. In such a case, the above partition function Q reduces to a product of individual-molecule partition functions:

$$Q = (N!)^{-1} q^N, \tag{7.46}$$

where the $N!$ factor arises as a degeneracy factor having to do with the permutational indistinguishability of the N molecules, and q is the partition function of an individual molecule,

$$q = \sum_l \omega_l \exp\left(\frac{-\varepsilon_l}{kT}\right). \tag{7.47}$$

Here, ε_l is the energy of the lth level of the molecule and ω_l is its degeneracy.

The molecular partition functions q, in turn, can be written as products of translational, rotational, vibrational, and electronic partition functions if the molecular energies ε_l can be approximated as sums of such energies. The following equations give explicit expressions for these individual contributions to q in the most usual case of a non-linear polyatomic molecule.

Translational

$$q_t = (2\pi mkT/h^2)^{3/2}V, \tag{7.48}$$

where m is the mass of the molecule and V is the volume to which its motion is constrained. For molecules constrained to a surface of area A, the corresponding result is $q_t = (2\pi mkT/h^2)^{2/2}A$, and for molecules constrained to move along a single axis over a length L, the result is $q_t = (2\pi mkT/h^2)^{1/2}L$. The magnitudes of these partition functions can be computed, using m in amu, T in kelvin, and L, A, or V in cm, cm^2 or cm^3, as

$$q = (3.28 \times 10^{13} mT)^{1/2,2/2,3/2} L,\ A,\ V. \tag{7.49}$$

Rotational

$$q_{\text{rot}} = (\pi^{1/2}/\sigma)(8\pi^2 I_A kT/h^2)^{1/2}(8\pi^2 I_B kT/h^2)^{1/2}(8\pi^2 I_C kT/h^2)^{1/2}, \tag{7.50}$$

where I_A, I_B, and I_C are the three principal moments of inertia of the molecule (i.e., eigenvalues of the moment of inertia tensor). σ is the symmetry number of the molecule defined as the number of ways the molecule can be rotated into a configuration that is indistinguishable from its original configuration. For example, σ is 2 for H_2 or D_2, 1 for HD, 3 for NH_3, and 12 for CH_4. The magnitudes of these partition functions can be computed using bond lengths in Å and masses in amu and T in K, using

$$\left(\frac{8\pi^2 I_A kT}{h^2}\right)^{1/2} = 9.75 \times 10^6 (IT)^{1/2}. \tag{7.51}$$

Vibrational

$$q_{\text{vib}} = \prod_{k=1,3N-6} \left\{\exp\left(-\frac{h\nu_j}{2kT}\right)\Big/\left[1 - \exp\left(-\frac{h\nu_j}{kT}\right)\right]\right\}, \tag{7.52}$$

where ν_j is the frequency of the jth harmonic vibration of the molecule, of which there are $3N - 6$.

Electronic

$$q_{\text{e}} = \sum_J \omega_J \exp\left(\frac{-\varepsilon_J}{kT}\right), \tag{7.53}$$

where ε_J and ω_J are the energies and degeneracies of the Jth electronic state; the sum is carried out for those states for which the product $\omega_J \exp(-\varepsilon_J/kT)$ is numerically significant. It is conventional to define the energy of a molecule or ion with respect to that of its atoms. So, the first term above is usually written as $\omega_e \exp(D_e/kT)$, where ω_e is the degeneracy of the ground electronic state and D_e is the energy required to dissociate the molecule into its constituent atoms, all in their ground electronic states. Notice that the magnitude of the translational partition function is much larger than that of the rotational partition function, which, in turn, is larger than that of the vibrational function. Moreover, note that the three-dimensional translational partition function is larger than the two-dimensional, which is larger than the one-dimensional. These orderings are simply reflections of the average number of quantum states that are accessible to the respective degrees of freedom at the temperature T.

The above partition function and thermodynamic equations form the essence of how statistical mechanics provides the tools for connecting molecule-level properties such as energy levels and degeneracies, which ultimately determine the E_j and the Ω_j, to the macroscopic properties such as $\langle E\rangle$, $\langle S\rangle$, $\langle p\rangle$, μ, etc.

If one has a system for which the quantum energy levels are not known, it is possible to express all of the thermodynamic properties in terms of the classical partition function. This partition function is computed by evaluating the following classical phase-space integral (phase space is the collection of coordinates q and conjugate momenta p)

$$Q = h^{-NM}(N!)^{-1} \int \exp\left(-\frac{H(q,p)}{kT}\right) dq dp. \tag{7.54}$$

In this integral, one integrates over the internal (e.g., bond lengths and angles), orientational, and translational coordinates and momenta of the N molecules. If each molecule has K internal coordinates, 3 translational coordinates, and 3 orientational coordinates, the total number of such coordinates per molecule is $M = K + 6$. One can then compute all thermodynamic properties of the system using this Q in place of the quantum Q in the equations given above for $\langle E\rangle$, $\langle p\rangle$, etc.

The classical partition functions discussed above are especially useful when substantial intermolecular interactions are present (and, thus, where knowing the quantum energy levels of the N-molecule system is highly unlikely). In such cases, the classical Hamiltonian is usually written in terms of H^0 which contains all of the kinetic energy factors as well as all of the potential energies other than the intermolecular potentials, and the intermolecular potential U, which depends only on a subset of the coordinates: $H = H^0 + U$. For example, let us assume that U depends only on the relative distances between molecules (i.e., on the $3N$ translational degrees of freedom which we denote r). Denoting all of the remaining coordinates as y, the classical partition function integral can be re-expressed as follows:

$$Q = h^{-NM}(N!)^{-1} \int \exp\left(\frac{-H^0(y,p)}{kT}\right) dy dp \left\{\int \exp\left(\frac{-U(r)}{kT}\right) dr\right\}. \tag{7.55}$$

The factor

$$Q_{\text{ideal}} = h^{-NM}(N!)^{-1} \int \exp\left(\frac{-H^0(y,p)}{kT}\right) dy\, dp V^N \tag{7.56}$$

would be the partition function if the Hamiltonian H contained no intermolecular interactions U. The V^N factor would arise from the integration over all of the translational coordinates if $U(r)$ were absent (i.e., if $U = 0$). The other factor

$$Q_{\text{inter}} = \frac{1}{V^N}\left\{\int \exp\left(\frac{-U(r)}{kT}\right) dr\right\} \tag{7.57}$$

contains all of the effects of intermolecular interactions and reduces to unity if the potential U vanishes. If, as the example considered here assumes, U only depends on the positions of the centers of mass of the molecules (i.e., not on molecular orientations or internal geometries), the Q_{ideal} partition function can be written in terms of the molecular translational, rotational, and vibrational partition functions shown earlier:

$$\begin{aligned} Q_{\text{ideal}} = \frac{1}{N!}\Bigg\{&\left(\frac{2\pi mkT}{h^2}\right)^{3/2} V \frac{\pi^{1/2}}{\sigma}\left(\frac{8\pi^2 I_A kT}{h^2}\right)^{1/2}\left(\frac{8\pi^2 I_B kT}{h^2}\right)^{1/2}\left(\frac{8\pi^2 I_C kT}{h^2}\right)^{1/2} \\ &\times \prod_{k=1,3N-6}\left\{\exp\left(\frac{-h\nu_j}{2kT}\right)\Big/\left[1-\exp\left(\frac{-h\nu_j}{kT}\right)\right]\right\}\sum_J \omega_J \exp\left(\frac{-\varepsilon_J}{kT}\right)\Bigg\}^N. \end{aligned} \tag{7.58}$$

Because all of the equations that relate thermodynamic properties to partition functions contain ln Q, all such properties will decompose into a sum of two parts, one coming from ln Q_{ideal} and one coming from ln Q_{inter}. The latter contains all of the effects of the intermolecular interactions. This means that all of the thermodynamic equations can, in this case, be written as an "ideal" component plus a part that arises from the intermolecular forces. Again, the *Statistical Mechanics* text by McQuarrie is a good source for reading more details on these topics.

7.3 Equilibrium constants in terms of partition functions

One of the most important and useful applications of statistical thermodynamics arises in the relation giving the equilibrium constant of a chemical reaction or for a physical transformation in terms of molecular partition functions. Specifically, for any chemical or physical equilibrium (e.g., the former could be the HF $\leftrightarrow$ $H^+ + F^-$ equilibrium; the latter could be $H_2O(l) \leftrightarrow H_2O(g)$), one can relate the equilibrium constant (expressed in terms of numbers of molecules per unit volume) in terms of the partition functions of these molecules. For example, in the hypothetical chemical equilibrium A + B $\leftrightarrow$ C, the equilibrium constant K

can be written, neglecting the effects of intermolecular potentials, as

$$K = (N_C/V)/[(N_A/V)(N_B/V)] = (q_C/V)/[(q_A/V)(q_B/V)]. \quad (7.59)$$

Here, q_J is the partition function for molecules of type J confined to volume V at temperature T. Alternatively, for an isomerization reaction involving the normal (N) and zwitterionic (Z) forms of arginine that were discussed in Chapter 5, the pertinent equilibrium constant would be

$$K = (N_Z/V)/[(N_N/V)] = (q_Z/V)/[(q_N/V)]. \quad (7.60)$$

So, if one can evaluate the partition functions q for reactant and product molecules in terms of the translational, electronic, vibrational, and rotational energy levels of these species, one can express the equilibrium constant in terms of these molecule-level properties.

Notice that the above equilibrium constant expressions equate ratios of species concentrations (in numbers of molecules per unit volume) to ratios of corresponding partition functions per unit volume. Because partition functions are a count of the thermal-average number of quantum states available to the system at temperature T (i.e., the average density of quantum states), this means that we equate species number densities to quantum state densities when we use the above expressions for the equilibrium constant.

7.4 Monte-Carlo evaluation of properties

A tool that has proven extremely powerful in statistical mechanics since computers became fast enough to permit simulations of complex systems is the Monte-Carlo (MC) method. This method allows one to evaluate the classical partition function described above by generating a sequence of configurations (i.e., locations of all of the molecules in the system as well as of all the internal coordinates of these molecules) and assigning a weighting factor to these configurations. By introducing an especially efficient way to generate configurations that have high weighting, the MC method allows us to simulate extremely complex systems that may contain millions of molecules.

To illustrate how this process works, let us consider carrying out a MC simulation representative of liquid water at some density ρ and temperature T. One begins by placing N water molecules in a "box" of volume V with V chosen such that N/V reproduces the specified density. To effect the MC process, we must assume that the total (intramolecular and intermolecular) potential energy E of these N water molecules can be computed for any arrangement of the N molecules within the box and for any values of the internal bond lengths and angles of the water molecules. Notice that E does not include the kinetic energy of the molecules; it is only the potential energy. Usually, this energy E is expressed as a sum of intramolecular bond stretching and bending contributions, one for each molecule, plus a pairwise additive intermolecular potential:

$$E = \sum_J E(\text{internal})_J + \sum_{J,K} E(\text{intermolecular})_{J,K}. \tag{7.61}$$

However, the energy E could be computed in other ways, if appropriate. For example, E might be evaluated as the Born–Oppenheimer energy if an *ab initio* electronic structure calculation on the full N-molecule system were feasible. The MC process does not depend on how E is computed, but, most commonly, it is evaluated as above.

In each "step" of the MC process, this potential energy E is evaluated for the current positions of the N water molecules. In its most common and straightforward implementation, a single water molecule is then chosen at random and one of its internal (bond lengths or angle) or external (position or orientation) coordinates is selected at random. This one coordinate (q) is then altered by a small amount ($q \rightarrow q + \delta q$) and the potential energy E is evaluated at the "new" configuration ($q + \delta q$). The amount δq by which coordinates are varied is usually chosen to make the fraction of MC steps that are accepted (see below) approximately 50%. This has been shown to optimize the performance of the MC algorithm.

Note that, when the intermolecular energy is pairwise additive as suggested above, evaluation of the energy change $E(q + \delta q) - E(q) = \delta E$ accompanying the change in q requires computational effort that is proportional to the number N of molecules in the system because only those factors $E(\text{intermolecular})_{J,K}$, with J or K equal to the single molecule that "moved" need be computed. This is why pairwise additive forms for E are often employed.

If the energy change δE is negative (i.e., if the potential energy is lowered by the "move"), the change in coordinate δq is allowed to occur and the resulting "new" configuration is counted among the MC "accepted" configurations. On the other hand, if δE is positive, the candidate move from q to $q + \delta q$ is not simply rejected (to do so would produce an algorithm directed toward finding a minimum on the energy landscape, which is not the goal). Instead, the quantity $P = \exp(-\delta E/kT)$ is used to compute the probability for accepting this energy-increasing move. In particular, a random number between, for example, 0.000 and 1.000 is selected. If the number is greater than P (expressed in the same decimal format), then the move is accepted and included in the list of accepted MC configurations. If the random number is less than P, the move is not accepted. Instead, a new water molecule and its internal or external coordinate are chosen at random and the entire process is restarted.

In this manner, one generates a sequence of "accepted moves" that generate a series of configurations for the system of N water molecules. This set of configurations has been shown to be properly representative of the geometries that the system will experience as it moves around at equilibrium at the specified

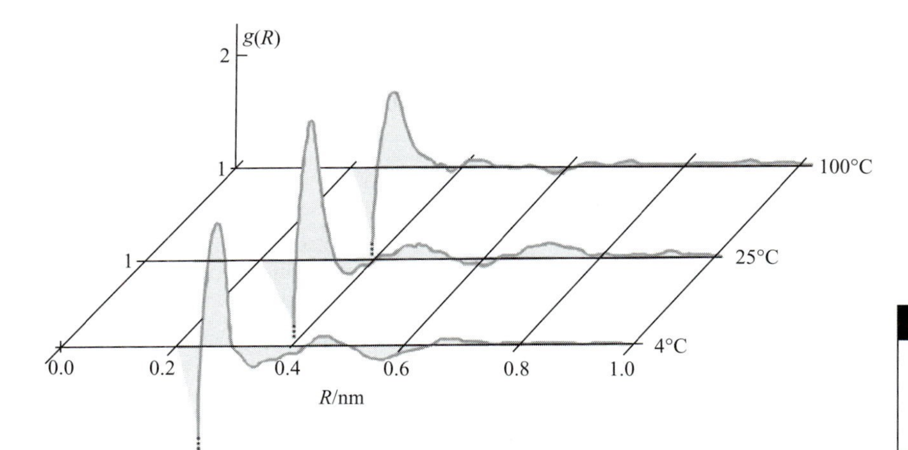

Figure 7.3 Radial distribution functions between pairs of oxygen atoms in H_2O at three different temperatures.

temperature T (n.b., T is the only way that the molecules' kinetic energy enters the MC process). As the series of accepted steps is generated, one can keep track of various geometrical and energetic data for each accepted configuration. For example, one can monitor the distances R among all pairs of oxygen atoms and then average this data over all of the accepted steps to generate an oxygen–oxygen radial distribution function $g(R)$ as shown in Fig. 7.3. Alternatively, one might accumulate the intermolecular interaction energies between pairs of water molecules and average this over all accepted configurations to extract the cohesive energy of the liquid water.

The MC procedure thus allows us to compute the equilibrium average of any property $A(q)$ that depends on the coordinates of the N molecules. Such an average would be written in terms of the normalized coordinate probability distribution function $P(q)$ as

$$\langle A \rangle = \int P(q) A(q)\, dq = \frac{\int \exp[-\beta E(q)] A(q)\, dq}{\int \exp[-\beta E(q)]\, dq}. \tag{7.62}$$

The denominator in the definition of $P(q)$ is, of course, proportional to the coordinate contribution to the partition function Q.

In the MC process, this same average is computed by forming the following sum over the M accepted MC configurations q_J:

$$\langle A \rangle = (1/M) \sum_J A(q_J). \tag{7.63}$$

In most MC simulations, millions of accepted steps contribute to the above averages. At first glance, it may seem that such a large number of steps represent an extreme computational burden. However, consider what might be viewed as an alternative procedure. Namely, suppose that the N molecules' three translational coordinates are the only variables to be treated (this certainly is a lower limit) and suppose one divides the range of each of these $3N$ coordinates into only 10 values. To compute an integral such as

$$\int \exp[-\beta E(q)] A(q)\, dq \tag{7.64}$$

in terms of such a 10-site discretization of the $3N$ coordinates would require the evaluation of the following $3N$-fold sum:

$$\sum_{q_1, q_2, \ldots q_{3N}} A(q_1, q_2, \ldots q_{3N}) \exp[-\beta E(q_1, \ldots q_{3N})]. \tag{7.65}$$

This sum contains 10^{3N} terms! Clearly, even for $N = 6$ (i.e., six molecules), the sum would require as much computer effort as the one million MC steps mentioned above, and MC simulations are often performed on thousands and even millions of molecules.

So, how do MC simulations work? That is, how can one handle thousands or millions of coordinates when the above analysis would suggest that performing an integral over so many coordinates would require 10^{1000} or $10^{1\,000\,000}$ computations? The main thing to understand is that the 10-site discretization of the $3N$ coordinates is a "stupid" way to perform the above integral because there are many (in fact, most) coordinate values where $A \exp(-\beta E)$ is negligible. On the other hand, the MC algorithm is designed to select (as accepted steps) those coordinates for which $\exp(-\beta E)$ is non-negligible. So, it avoids configurations that are "stupid" and focuses on those for which the probability factor is largest. This is why the MC method works!

It turns out that the MC procedure as outlined above is a highly efficient method for computing multidimensional integrals of the form

$$\int P(q) A(q)\, dq, \tag{7.66}$$

where $P(q)$ is a normalized (positive) probability distribution and $A(q)$ is any property that depends on the multidimensional variable q.

There are, however, cases where this conventional MC approach needs to be modified by using so-called umbrella sampling. To illustrate how this is done, suppose that one wanted to use the MC process to compute an average, with the above $\exp[-\beta E(q)]$ as the weighting factor, of a function $A(q)$ that is large whenever two or more molecules have high (i.e., repulsive) intermolecular potentials. For example, one could have

$$A(q) = \sum_{I<J} \frac{a}{|\mathbf{R}_I - \mathbf{R}_J|^n}. \tag{7.67}$$

Such a function could, for example, be used to monitor when pairs of molecules, with center-of-mass coordinates $\mathbf{R}_J$ and $\mathbf{R}_I$, approach closely enough to undergo reaction.

The problem with using conventional MC methods to compute

$$\langle A \rangle = \int A(q)P(q)\,dq \tag{7.68}$$

in such cases is that:

(i) $P(q) = \exp[-\beta E(q)]/\int \exp(-\beta E)\,dq$ favors those coordinates for which the total potential energy E is low. So, coordinates with high $E(q)$ are very infrequently accepted.
(ii) $A(q)$ is designed to identify events in which pairs of molecules approach closely and thus have high $E(q)$ values.

So, there is a competition between $P(q)$ and $A(q)$ that renders the MC procedure ineffective in such cases.

What is done to overcome this competition is to introduce a so-called umbrella weighting function $U(q)$ that:

(i) attains its largest values where $A(q)$ is large, and
(ii) is positive and takes on values between 0 and 1.

One then replaces $P(q)$ in the MC algorithm by the product $P(q)U(q)$. To see how this replacement works, we re-write the average that needs to be computed as follows:

$$\begin{aligned}\langle A \rangle &= \int P(q)A(q)\,dq = \frac{\int A(q)\exp(-\beta E(q))\,dq}{\int \exp(-\beta E(q))\,dq} \\ &= \frac{\int \frac{A(q)}{U(q)}U(q)\exp(-\beta E(q))\,dq}{\int U(q)\exp(-\beta E(q))\,dq} \Bigg/ \frac{\int \frac{1}{U(q)}U(q)\exp(-\beta E(q))\,dq}{\int U(q)\exp(-\beta E(q))\,dq}.\end{aligned} \tag{7.69}$$

The interpretation of the last identity is that $\langle A \rangle$ can be computed by:

(i) using the MC process to evaluate the average of $[A(q)/U(q)]$ but with a probability weighting factor of $U(q)\exp[-\beta E(q)]$ to accept or reject coordinate changes, and
(ii) also using the MC process to evaluate the average of $[1/U(q)]$ again with $U(q)\exp[-\beta E(q)]$ as the weighting factor, and finally
(iii) taking the average of (A/U) divided by the average of $(1/U)$ to obtain the final result.

The secret to the success of umbrella sampling is that the product $U\exp(-\beta E)$ causes the MC process to focus on those coordinates for which both $\exp(-\beta E)$ and U (and hence A) are significant.

7.5 Molecular dynamics simulations of properties

One thing that the MC process does not address directly is information about the time evolution of the system. That is, the "steps" one examines in the MC algorithm are not straightforward to associate with a time-duration, so it is not designed to compute the rates at which events take place. If one is interested in simulating such dynamical processes, even when the N-molecule system is at or near equilibrium, it is more appropriate to carry out a classical molecular dynamics (MD) simulation. In such a MD calculation, one usually assigns to each of the internal and external coordinates of each of the N molecules an initial amount of kinetic energy (proportional to T). However, whether one assigns this initial kinetic energy equally to each coordinate or not does not matter much because, as time evolves and the molecules interact, the energy becomes more or less randomly shared in any event and eventually properly simulates the dynamics of the equilibrium system. Moreover, one usually waits until such energy randomization has occurred before beginning to use data extracted from the simulation to compute properties. Hence, any effects caused by improper specifications of the initial conditions can be removed.

With each coordinate having its initial velocity $(dq/dt)_0$ and its initial value q_0 specified as above, one then uses Newton's equations written for a time step of duration δt to propagate q and dq/dt forward in time according, for example, to the following first order propagation formula:

$$q(t+\delta t) = q_0 + \left(\frac{dq}{dt}\right)_0 \delta t \tag{7.70}$$

$$\frac{dq}{dt}(t+\delta t) = \left(\frac{dq}{dt}\right)_0 - \delta t \left[\frac{\left(\frac{\partial E}{\partial q}\right)_0}{m_q}\right]. \tag{7.71}$$

Here m_q is the mass factor connecting the velocity dq/dt and the momentum p_q conjugate to the coordinate q:

$$p_q = m_q \frac{dq}{dt}, \tag{7.72}$$

and $-(\partial E/\partial q)_0$ is the force along the coordinate q at the "initial" geometry q_0. In most modern MD simulations, more sophisticated numerical methods can be used to propagate the coordinates and momenta. However, what I am outlining here provides you with the basic idea of how MD simulations are performed. The forces can be obtained from gradients of a Born–Oppenheimer electronic energy surface if this is computationally feasible. Alternatively, they can be computed from derivatives of an empirical force field. In the latter case, the system's potential energy E is expressed in terms of analytical functions of:

(i) intramolecular bond lengths, angles, and torsional angles, as well as
(ii) intermolecular distances and orientations.

The parameters appearing in such force fields have usually been determined from electronic structure calculations on molecular fragments, spectroscopic determination of vibrational force constants, and experimental measurements of intermolecular forces.

By applying this time-propagation to all of the coordinates and momenta of the N molecules, one generates a set of "new" coordinates $q(t + \delta t)$ and new velocities $dq/dt(t + \delta t)$ appropriate to the system at time $t + \delta t$. Using these new coordinates and momenta as q_0 and $(dq/dt)_0$ and evaluating the forces $-(\partial E/\partial q)_0$ at these new coordinates, one can again use the Newton equations to generate another finite-time-step set of new coordinates and velocities. Through the sequential application of this process, one generates a sequence of coordinates and velocities that simulate the system's behavior. By following these coordinates and momenta, one can interrogate any dynamical properties that one is interested in.

In Chapter 8, I again discuss using Newtonian dynamics to follow the time evolution of a chemical system. There is a fundamental difference between the situation just described and the case treated in Chapter 8. In the former, one allows the N-molecule system to reach equilibrium (i.e., by waiting until the dynamics has randomized the energy) before monitoring the subsequent time evolution. In the problem of Chapter 8, we use MD to follow the time progress of a system representing a single bimolecular collision in two crossed beams of molecules. Each such beam contains molecules whose initial velocities are narrowly defined rather than Maxwell–Boltzmann distributed. In this case, we do not allow the system to equilibrate because we are not trying to model an equilibrium system. Instead, we select initial conditions that represent the two beams and we then follow the Newtonian dynamics to monitor the outcome (e.g., reaction or non-reactive collision).

Unlike the MC method, which is very amenable to parallel computation, MD simulations are more difficult to carry out in a parallel manner. One can certainly execute many different classical trajectories on many different computer nodes; however, to distribute one trajectory over many nodes is difficult. The primary difficulty is that, for each time step, all N of the molecules undergo moves to new coordinates and momenta. To compute the forces on all N molecules requires of the order of N^2 calculations (e.g., when pairwise additive potentials are used). In contrast, each MC step requires that one evaluate the potential energy change accompanying the displacement of only one molecule. This uses only of the order of N computational steps (again, for pairwise additive potentials).

Another factor that complicates MD simulations has to do with the wide range of times scales that may be involved. For example, for one to use a time step δt short enough to follow high-frequency motions (e.g., O—H stretching) in a simulation of an ion or polymer in water solvent, δt must be of the order of 10^{-15} s. To then simulate the diffusion of an ion or the folding of a polymer in the liquid state, which might require 10^{-4} s, one would have to carry out 10^{11} MD steps. This likely would render the simulation not feasible. For such reasons, when carrying out long-time MD simulations, it is common to ignore the high-frequency intramolecular motions by, for example, simply not including these coordinates and momenta in the Netwonian dynamics. Of course, this is an approximation whose consequences must be tested and justified.

In summary, MD simulations are not difficult to implement if one has available a proper representation of the intramolecular and intermolecular potential energy E. Such calculations are routinely carried out on large biomolecules or condensed-media systems containing thousands to millions of atomic centers. There are, however, difficulties primarily connected to the time scales over which molecular motions change and over which the process being simulated changes that limit the success of this method.

7.6 Time correlation functions

One of the most active research areas in statistical mechanics involves the evaluation of so-called equilibrium time correlation functions such as we encountered in Chapter 6. The correlation function $C(t)$ is defined in terms of two physical operators A and B, a time dependence that is carried by a Hamiltonian H via $\exp(-iHt/\hbar)$, and an equilibrium average over a Boltzmann population $\exp(-\beta H)/Q$.

The quantum mechanical expression for $C(t)$ is

$$C(t) = \sum_j \langle \Phi_j | A \exp\left(\frac{iHt}{\hbar}\right) B \exp\left(\frac{-iHt}{\hbar}\right) |\Phi_j\rangle \frac{\exp(-\beta E_j)}{Q}, \tag{7.73}$$

while the classical mechanical expression is

$$C(t) = \int dq \int dp A(q(0), p(0)) B(q(t), p(t)) \exp[-\beta H(q(0), p(0))]/Q, \tag{7.74}$$

where $q(0)$ and $p(0)$ are the values of all the coordinates and momenta of the system at $t = 0$ and $q(t)$ and $p(t)$ are their values, according to Newtonian mechanics, at time t. As shown above, an example of a time correlation function that relates to molecular spectroscopy is the dipole–dipole correlation function that we discussed in Chapter 6:

$$C(t) = \sum_j \langle \Phi_j | e \cdot \mu \exp\left(\frac{iHt}{\hbar}\right) e \cdot \mu \exp\left(\frac{-iHt}{\hbar}\right) |\Phi_j\rangle \frac{\exp(-\beta E_j)}{Q}, \tag{7.75}$$

for which A and B are both the electric dipole interaction $e \cdot \mu$ between the photon's electric field and the molecule's dipole operator. The Fourier transform of this particular $C(t)$ relates to the absorption intensity for light of frequency ω:

$$I(\omega) = \int dt C(t) \exp(i\omega t). \tag{7.76}$$

It turns out that many physical properties (e.g., absorption line shapes, Raman scattering intensities) and transport coefficients (e.g., diffusion coefficients, viscosity) can be expressed in terms of time correlation functions. It is beyond the scope of this text to go much further in this direction, so I will limit my discussion to the optical spectroscopy case at hand which now requires that we discuss how the time-evolution aspect of this problem is dealt with. The *Statistical Mechanics* text by McQuarrie has a nice treatment of other such correlation functions, so the reader is directed to that text for further details.

The computation of correlation functions involves propagating either wave functions or classical trajectories which produce the $q(t)$, $p(t)$ values entering into the expression for $C(t)$. In the classical case, one carries out a large number of Newtonian trajectories with initial coordinates $q(0)$ and momenta $p(0)$ chosen to represent the equilibrium condition of the N-molecule system. For example, one could use the MC method to select these variables employing $\exp(-\beta H(p, q))$ as the probability function for accepting or rejecting initial q and p values. In this case, the weighting function contains not just the potential energy but also the kinetic energy (and thus the total Hamiltonian H) because now we need to also select proper initial values for the momenta. So, with many (e.g., M) selections of the initial q and p variables of the N-molecules being made, one would allow the Newtonian dynamics of each set of initial conditions to proceed. During each such trajectory, one would monitor the initial value of the $A(q(0),p(0))$ property and the time progress of the $B(q(t),p(t))$ property. One would then compute the MC average to obtain the correlation function:

$$C(t) = (1/M) \sum_{J=1,M} A(q_J(0),p_J(0))B(q_J(t),p_J(t)) \exp[-\beta H(q_J(0),p_J(0))]. \tag{7.77}$$

In the quantum case, the time propagation is especially challenging and is somewhat beyond the scope of this text. However, I want to give you some idea of the steps that are involved, realizing that this remains an area of very active research development. As noted in the Background Material, it is possible to time-propagate a wave function Φ that is known at $t = 0$ if one is able to expand Φ in terms of the eigenfunctions of the Hamiltonian H. However, for systems comprised of many molecules, which are most common in statistical mechanics studies, it is impossible to compute (or realistically approximate) these eigenfunctions. Thus, it is not productive to try to express $C(t)$ in terms of these eigenfunctions. Therefore, an entirely new set of tools has been introduced to handle time propagation in the quantum case, and it is these new devices that I now attempt to describe in a manner much like we saw in the Background Material's discussion of time propagation of wave functions.

To illustrate, consider the time propagation issue contained in the quantum definition of $C(t)$ shown above. One is faced with:

(i) propagating $|\Phi_j\rangle$ from $t = 0$ up to time t, using $\exp(-iHt/\hbar)\,|\Phi_j\rangle$ and then acting with the operator B;
(ii) acting with the operator A^+ on $|\Phi_j\rangle$ and then propagating $A^+|\Phi_j\rangle$ from $t = 0$ up to time t, using $\exp(-iHt/\hbar)A^+|\Phi_j\rangle$;
(iii) $C(t)$ then requires that these two time-propagated functions be multiplied together and integrated over the coordinates that Φ depends on.

The $\exp(-\beta H)$ operator that also appears in the definition of $C(t)$ can be combined, for example, with the first time propagation step and actually handled as part of the time propagation as follows:

$$\exp\left(\frac{-iHt}{\hbar}\right)|\Phi_j\rangle\exp(-\beta E_j) = \exp\left(\frac{-iHt}{\hbar}\right)\exp(-\beta H)|\Phi_j\rangle = \exp\left[\frac{-i\left(t+\frac{\beta\hbar}{i}\right)_H}{\hbar}\right]|\Phi_j\rangle. \tag{7.78}$$

The latter expression can be viewed as involving a propagation in complex time from $t = 0$ to $t = t + \beta\hbar/i$. Although having a complex time may seem unusual, as I will soon point out, it turns out that it can have a stabilizing influence on the success of these tools for computing quantum correlation functions.

Much like we saw earlier in the Background Material, so-called Feynman path integral techniques can be used to carry out the above time propagations. One begins by dividing the time interval into P discrete steps (this can be the real time interval or the complex interval)

$$\exp\left(\frac{-iHt}{\hbar}\right) = \left\{\exp\left(\frac{-iH\delta t}{\hbar}\right)\right\}^P. \tag{7.79}$$

The number P will eventually be taken to be very large, so each time step $\delta t = t/P$ has a very small magnitude. This fact allows us to use approximations to the exponential operator appearing in the propagator that are valid only for short time steps. For each of these short time steps one then approximates the propagator in the most commonly used so-called split symmetric form:

$$\exp\left(\frac{-iH\delta t}{\hbar}\right) = \exp\left(\frac{-iV\delta t}{2\hbar}\right)\exp\left(\frac{-iT\delta t}{\hbar}\right)\exp\left(\frac{-iV\delta t}{2\hbar}\right). \tag{7.80}$$

Here, V and T are the potential and kinetic energy operators that appear in $H = T + V$. It is possible to show that the above approximation is valid up to terms of order $(\delta t)^4$, whereas the form used in the Background Material is valid only to order δt^2. So, for short times (i.e., small δt), these symmetric split operator approximations to the propagator should be accurate.

The time evolved wave function $\Phi(t)$ can then be expressed as

$$\Phi(t) = \left\{\exp\left(\frac{-iV\delta t}{2\hbar}\right)\exp\left(\frac{-iT\delta t}{\hbar}\right)\exp\left(\frac{-iV\delta t}{2\hbar}\right)\right\}^P \Phi(t=0). \tag{7.81}$$

The potential V is (except when external magnetic fields are present) a function only of the coordinates $\{q_j\}$ of the system, while the kinetic term T is a function of the momenta $\{p_j\}$ (assuming Cartesian coordinates are used). By making use of the completeness relations for eigenstates of the coordinate operator

$$1 = \int dq|q_j\rangle\langle q_j| \tag{7.82}$$

and inserting this identity P times (once between each combination of $\exp(-iV\delta t/2\hbar)\exp(-iT\delta t/\hbar)\exp(-iV\delta t/2\hbar)$ factors), the expression given above for $\Phi(t)$ can be rewritten as follows:

$$\Phi(q_P, t) = \int dq_{P-1}\, dq_{P-2}\ldots dq_1 dq_0 \prod_{j=1,P} \exp\left\{\left(\frac{-i\delta t}{2\hbar}\right)[V(q_j)+V(q_{j-1})]\right\}$$
$$\times\langle q_j|\exp\left(\frac{-i\delta t T}{\hbar}\right)|q_{j-1}\rangle\Phi(q_0, 0). \tag{7.83}$$

Then, by using the analogous completeness identity for the momentum operator

$$1 = (1/\hbar)\int dp_j|p_j\rangle\langle p_j| \tag{7.84}$$

one can write

$$\langle q_j|\exp\left(\frac{-i\delta t T}{\hbar}\right)|q_{j-1}\rangle = \left(\frac{1}{\hbar}\right)\int dp\langle q_j \mid p\rangle\exp\left(\frac{-ip^2\delta t}{2m\hbar}\right)\langle p \mid q_{j-1}\rangle. \tag{7.85}$$

Finally, by using the fact (recall this from the Background Material) that the momentum eigenfunctions $|p\rangle$, when expressed as functions of coordinates q, are given by

$$\langle q_j \mid p\rangle = \left(\frac{1}{2\pi}\right)^{1/2}\exp\left(\frac{ipq}{\hbar}\right), \tag{7.86}$$

the above integral becomes

$$\langle q_j|\exp\left(\frac{-i\delta t T}{\hbar}\right)|q_{j-1}\rangle = \left(\frac{1}{2\pi\hbar}\right)\int dp\exp\left(\frac{-ip^2\delta t}{2m\hbar}\right)\exp\left[\frac{ip(q_j-q_{j-1})}{\hbar}\right]. \tag{7.87}$$

This integral over p can be carried out analytically to give

$$\langle q_j|\exp\left(\frac{-i\delta t T}{\hbar}\right)|q_{j-1}\rangle = \left(\frac{m}{2\pi i\hbar\delta t}\right)^{1/2}\exp\left[\frac{im(q_j-q_{j-1})^2}{2\hbar\delta t}\right]. \tag{7.88}$$

When substituted back into the multidimensional integral for $\Phi(q_P, t)$, we obtain

$$\Phi(q_P, t) = \left(\frac{m}{2\pi i\hbar\delta t}\right)^{P/2} \int dq_{P-1}\, dq_{P-2} \ldots dq_1\, dq_0 \times \prod_{j=1,P} \exp\left\{\left(\frac{-i\delta t}{2\hbar}\right)[V(q_j) + V(q_{j-1})]\right\} \exp\left[\frac{im(q_j - q_{j-1})^2}{2\hbar\delta t}\right] \Phi(q_0, 0) \tag{7.89}$$

or

$$\Phi(q_P, t) = \left(\frac{m}{2\pi i\hbar\delta t}\right)^{P/2} \int dq_{P-1}\, dq_{P-2} \ldots dq_1\, dq_0 \times \exp\left\{\sum_{j=1,P}\left[\left(\frac{-i\delta t}{2\hbar}\right)[V(q_j) + V(q_{j-1})] + \left(\frac{im(q_j - q_{j-1})^2}{2\hbar\delta t}\right)\right]\right\} \Phi(q_0, 0). \tag{7.90}$$

Why are such multidimensional integrals called path integrals? Because the sequence of positions $q_1, \ldots q_{P-1}$ describes a "path" connecting q_0 to q_P. By integrating over all of the intermediate positions $q_1, q_2, \ldots q_{P-1}$ for any given q_0 and q_P one is integrating over all paths that connect q_0 to q_P. Further insight into the meaning of the above is gained by first realizing that

$$\left(\frac{m}{2\delta t}\right)(q_j - q_{j-1})^2 = \left(\frac{m}{2(\delta t)^2}\right)(q_j - q_{j-1})^2 \delta t = \int T dt \tag{7.91}$$

is the representation, within the P discrete time steps of length δt, of the integral of $T dt$ over the jth time step, and that

$$\left(\frac{\delta t}{2}\right)[V(q_j) + V(q_{j-1})] = \int V(q)\, dt \tag{7.92}$$

is the representation of the integral of $V dt$ over the jth time step. So, for any particular path (i.e., any specific set of $q_0, q_1, \ldots q_{P-1}, q_P$ values), the sum over all P such terms $\sum_{j=1,P}[m(q_j - q_{j-1})^2/2\delta t - \delta t(V(q_j) + V(q_{j-1}))/2]$ represents the integral over all time from $t = 0$ until $t = t$ of the so-called Lagrangian $L = T - V$:

$$\sum_{j=1,P}\left[\frac{m(q_j - q_{j-1})^2}{2\delta t} - \delta t\frac{(V(q_j) + V(q_{j-1}))}{2}\right] = \int L\, dt. \tag{7.93}$$

This time integral of the Lagrangian is called the "action" S in classical mechanics (recall that in the Background Material we used quantization of the action in the particle-in-a-box problem). Hence, the N-dimensional integral in terms of which $\Phi(q_P, t)$ is expressed can be written as

$$\Phi(q_P, t) = \left(\frac{m}{2\pi i\hbar\delta t}\right)^{P/2} \sum_{\text{all paths}} \exp\left\{\frac{i}{\hbar}\int dt L\right\} \Phi(q_0, t = 0). \tag{7.94}$$

Here, the notation "all paths" is realized in the earlier version of this equation by dividing the time axis from $t = 0$ to $t = t$ into P equal divisions, and denoting the coordinates of the system at the jth time step by q_j. By then allowing each q_j to assume all possible values (i.e., integrating over all possible values of q_j using, for example, the Monte-Carlo method discussed earlier), one visits all possible paths that begin at q_0 at $t = 0$ and end at q_P at $t = t$. By forming the classical action S

$$S = \int dt L \tag{7.95}$$

for each path and then summing $\exp(iS/\hbar)\Phi(q_0, t = 0)$ over all paths and multiplying by $(m/2\pi\hbar\delta t)^{P/2}$, one is able to form $\Phi(q_P, t)$.

The difficult step in implementing this Feynman path integral method in practice involves how one identifies all paths connecting $q_0, t = 0$ to q_P, t. Each path contributes an additive term involving the complex exponential of the quantity

$$\sum_{j=1,P} \left[\frac{m(q_j - q_{j-1})^2}{2\delta t} - \delta t \frac{(V(q_j) + V(q_{j-1}))}{2} \right]. \tag{7.96}$$

Because the time variable $\delta t = t/P$ appearing in each action component can be complex (recall that, in one of the time evolutions, t is really $t + \beta\hbar/i$), the exponentials of these action components can have both real and imaginary parts. The real parts, which arise from the $\exp(-\beta H)$, cause the exponential terms to be damped (i.e., to undergo exponential decay), but the real parts give rise (in $\exp(iS/\hbar)$) to oscillations. The sum of many, many (actually, an infinite number of) oscillatory $\exp(iS/\hbar) = \cos(S/\hbar) + i \sin(S/\hbar)$ terms is extremely difficult to evaluate because of the tendency of contributions from one path to cancel those of another path. The practical evaluation of such sums remains a very active research subject.

The most commonly employed approximation to this sum involves finding the path(s) for which the action

$$S = \sum_{j=1,P} \left[\frac{m(q_j - q_{j-1})^2}{2\delta t} - \delta t \frac{(V(q_j) + V(q_{j-1}))}{2} \right] \tag{7.97}$$

is smallest because such paths produce the lowest frequency oscillations in $\exp(iS/\hbar)$, and thus may be less subject to cancellation by contributions from other paths. The path(s) that minimize the action S are, in fact, the classical paths. That is, they are the paths that the system whose quantum wave function is being propagated would follow if the system were undergoing classical Newtonian mechanics subject to the conditions that the system be at q_0 at $t = 0$ and at q_P at $t = t$. In this so-called semi-classical approximation to the propagation of the initial wave function using Feynman path integrals, one finds all classical paths that connect q_0 at $t = 0$ and q_P at $t = t$, and one evaluates the action S for each such path. One then applies the formula

$$\Phi(q_P, t) = \left(\frac{m}{2\pi i\hbar\delta t}\right)^{P/2} \sum_{\text{all paths}} \exp\left\{\frac{i}{\hbar}\int dt L\right\} \Phi(q_0, t = 0) \tag{7.98}$$

but includes in the sum only the contribution from the classical path(s). In this way, one obtains an approximate quantum propagated wave function via a procedure that requires knowledge of only classical propagation paths.

Clearly, the quantum propagation of wave functions, even within the semi-classical approximation discussed above, is a rather complicated affair. However, keep in mind the alternative that one would face in evaluating, for example, spectroscopic line shapes if one adopted a time-independent approach. One would have to know the energies and wave functions of a system comprised of many interacting molecules. This knowledge is simply not accessible for any but the simplest molecules. For this reason, the time-dependent framework in which one propagates classical trajectories or uses path-integral techniques to propagate initial wave functions offers the most feasible way to evaluate the correlation functions that ultimately produce spectral line shapes and other time correlation functions for complex molecules in condensed media.

Some important chemical applications of statistical mechanics

7.7 Gas-molecule thermodynamics

The equations relating the thermodynamic variables to the molecular partition functions can be employed to obtain the following expressions for the energy E, heat capacity C_V, Helmholz free energy A, entropy S, and chemical potential μ in the case of a gas (i.e., in the absence of intermolecular interactions) of polyatomic molecules:

$$\frac{E}{NkT} = \frac{3}{2} + \frac{3}{2} + \sum_{J=1,3N-6}\left[\frac{h\nu_J}{2kT} + \frac{h\nu_J}{kT}\left[\exp\left(\frac{h\nu_J}{kT}\right) - 1\right]^{-1}\right] - \frac{D_e}{kT}, \tag{7.99}$$

$$\frac{C_V}{Nk} = \frac{3}{2} + \frac{3}{2} + \sum_{J=1,3N-6}\left(\frac{h\nu_J}{kT}\right)^2 \exp\left(\frac{h\nu_J}{kT}\right)\left[\exp\left(\frac{h\nu_J}{kT}\right) - 1\right]^{-2}, \tag{7.100}$$

$$\begin{aligned}-A/NkT = {} & \ln\left\{\left[\frac{2\pi mkT}{h^2}\right]^{3/2}\left(\frac{Ve}{N}\right)\right\} + \ln\left[\left(\frac{\pi^{1/2}}{\sigma}\right)\left(\frac{8\pi^2 I_A kT}{h^2}\right)^{1/2}\right. \\ & \left.\times\left(\frac{8\pi^2 I_B kT}{h^2}\right)^{1/2}\left(\frac{8\pi^2 I_C kT}{h^2}\right)^{1/2}\right] \\ & - \sum_{J=1,3N-6}\left[\frac{h\nu_J}{2kT} + \ln\left(1 - \exp\left(\frac{-h\nu_J}{kT}\right)\right)\right] + \frac{D_e}{kT} + \ln\omega_e,\end{aligned} \tag{7.101}$$

$$\frac{S}{Nk} = \ln\left[\left(\frac{2\pi mkT}{h^2}\right)^{3/2}\left(\frac{Ve^{5/2}}{N}\right)\right] + \ln\left[\left(\frac{\pi^{1/2}}{\sigma}\right)\left(\frac{8\pi^2 I_A kT}{h^2}\right)^{1/2}\right.$$
$$\left.\times\left(\frac{8\pi^2 I_B kT}{h^2}\right)^{1/2}\left(\frac{8\pi^2 I_C kT}{h^2}\right)^{1/2}\right] + \sum_{J=1,3N-6}\left\{\frac{h\nu_J}{kT}\left[\exp\left(\frac{h\nu_J}{kT}\right) - 1\right]^{-1}\right.$$
$$\left. - \ln\left[1 - \exp\left(\frac{-h\nu_\vartheta}{kT}\right)\right]\right\} + \ln\omega_e, \tag{7.102}$$

$$\frac{\mu}{kT} = -\ln\left\{\left[\frac{2\pi mkT}{h^2}\right]^{3/2}\left(\frac{kT}{p}\right)\right\} - \ln\left[\left(\frac{\pi^{1/2}}{\sigma}\right)\left(\frac{8\pi^2 I_A kT}{h^2}\right)^{1/2}\right.$$
$$\left.\times\left(\frac{8\pi^2 I_B kT}{h^2}\right)^{1/2}\left(\frac{8\pi^2 I_C kT}{h^2}\right)^{1/2}\right]$$
$$+ \sum_{J=1,3N-6}\left\{\frac{h\nu_J}{2kT} + \ln\left[1 - \exp\left(\frac{-h\nu_J}{kT}\right)\right]\right\} - \frac{D_e}{kT} - \ln\omega_e. \tag{7.103}$$

Notice that, except for μ, all of these quantities are extensive properties that depend linearly on the number of molecules in the system N. Except for the chemical potential μ and the pressure p, all of the variables appearing in these expressions have been defined earlier when we showed the explicit expressions for the translational, vibrational, rotational, and electronic partition functions. These are the working equations that allow one to compute thermodynamic properties of stable molecules, ions, and even reactive species such as radicals in terms of molecular properties such as geometries, vibrational frequencies, electronic state energies and degeneracies, and the temperature, pressure, and volume.

7.8 Einstein and Debye models of solids

These two models deal with the vibrations of crystals that involve motions among the neighboring atoms, ions, or molecules that comprise the crystal. These inter-fragment vibrations are called phonons. In the Einstein model of a crystal, one assumes that:

(i) Each atom, ion, or molecule from which the crystal is constituted is trapped in a potential well formed by its interactions with neighboring species. This potential is denoted $\phi(V/N)$ with the V/N ratio written to keep in mind that it likely depends on the packing density (i.e., the distances among neighbors) within the crystal, and that ϕ represents the interaction of any specific atom, ion, or molecule with the $N - 1$ other such species. So, $N\phi/2$, not $N\phi$ is the total of interaction energies among all of the species; the factor of 1/2 is necessary to avoid double counting.

(ii) Each such species is assumed to undergo local harmonic motions about its equilibrium position (q_J^0) within the local well that traps it. If the crystal is isotropic, the force constants k_J that characterize the harmonic potential $1/2\ k_J(q_J - q_J^0)^2$ along the x, y, and z directions are equal; if not, these k_J parameters may be unequal. It is these force constants, along with the masses m of the atoms, ions, or molecules, that determine the harmonic frequencies $\nu_J = 1/2\pi(k_J/m)^{1/2}$ of the crystal.

(iii) The inter-species phonon vibrational partition function of the crystal is then assumed to be a product of N partition functions, one for each species in the crystal, with each partition function taken to be of the harmonic vibrational form:

$$Q = \exp\left(\frac{-N\phi}{2kT}\right)\left\{\prod_{J=1,3}\exp\left(\frac{-h\nu_J}{2kT}\right)\left[1-\exp\left(\frac{-h\nu_J}{kT}\right)\right]^{-1}\right\}^N. \quad (7.104)$$

There is no factor of $N!$ in the denominator because, unlike a gas of N species, each of these N species (atoms, ions, or molecules) are constrained to stay put (i.e., not free to roam independently) in the trap induced by their neighbors. In this sense, the N species are distinguishable rather than indistinguishable. The $N\phi/2kT$ factor arises when one asks what the total energy of the crystal is, aside from its vibrational energy, relative to N separated species; in other words, what is the total cohesive energy of the crystal. This energy is N times the energy of any single species ϕ, but, as noted above, divided by 2 to avoid double counting the inter-species interaction energies.

This partition function can be subjected to the thermodynamic equations discussed earlier to compute various thermodynamic properties. One of the most useful to discuss for crystals is the heat capacity C_V, which is given by

$$C_V = Nk\sum_{J=1,3}\left(\frac{h\nu_J}{kT}\right)^2\exp\left(\frac{h\nu_J}{kT}\right)\left[\exp\left(\frac{h\nu_J}{kT}\right)-1\right]^{-2}. \quad (7.105)$$

At very high temperatures, this function can be shown to approach $3Nk$, which agrees with the experimental observation know as the law of Dulong and Petit. However, at very low temperatures, this expression approaches

$$C_V \to \sum_{J=1,3} Nk\left(\frac{h\nu_J}{kT}\right)^2\exp\left(\frac{-h\nu_J}{kT}\right), \quad (7.106)$$

which goes to zero as T approaches zero, but not in a way that is consistent with experimental observation. That is, careful experimental data shows that all crystal heat capacities approach zero proportional to T^3 at low temperature; the Einstein model's C_V does not.

So, although the Einstein model offers a very useful model of how a crystal's stability relates to $N\phi$ and how its C_V depends on vibrational frequencies of the phonon modes, it does not work well at low temperatures. Nevertheless, it remains a widely used model in which to understand the phonons' contributions to thermodynamic properties as long as one does not attempt to extrapolate its predictions to low T.

In the Debye model of phonons in crystals, one abandons the view in which each atom, ion, or molecule vibrates independently about its own equilibrium position and replaces this with a view in which the constituent species vibrate collectively in wave-like motions. Each such wave has a wavelength λ and a frequency ν that are related to the speed c of propagation of such waves in the crystal by

$$c = \lambda\nu. \tag{7.107}$$

The speed c is a characteristic of the crystal's inter-species forces; it is large for "stiff" crystals and small for "soft" crystals.

In a manner much like we used to determine the density of quantum states $\Omega(E)$ within a three-dimensional box, one can determine how many waves can fit within a cubic crystalline "box" having frequencies between ν and $\nu + d\nu$. The approach to this problem is to express the allowed wavelengths and frequencies as

$$\lambda_n = 2L/n, \tag{7.108}$$

$$\nu_n = nc/2L, \tag{7.109}$$

where L is the length of the box on each of its sides and n is an integer $1, 2, 3, \ldots$. This prescription forces all wavelengths to match the boundary condition for vanishing at the box boundaries.

Then, carrying out a count of how many $(\Omega(\nu))$ waves have frequencies between ν and $\nu + d\nu$ for a box whose sides are all equal gives the following expression:

$$\Omega(\nu) = 12\pi V\nu^2/c^3. \tag{7.110}$$

The primary observation to be made is that the density of waves is proportional to ν^2:

$$\Omega(\nu) = a\nu^2. \tag{7.111}$$

It is conventional to define the parameter a in terms of the maximum frequency ν_m that one obtains by requiring that the integral of $\Omega(\nu)$ over all allowed ν add up to $3N$, the total number of inter-species vibrations that can occur:

$$3N = \int \Omega(\nu)d\nu = a\nu_m^3/3. \tag{7.112}$$

This then gives the constant a in terms of ν_m and N and allows $\Omega(\nu)$ to be written as

$$\Omega(\nu) = 9N\nu^2/\nu_m^3. \tag{7.113}$$

The Debye model uses this wave picture and computes the total energy E of the crystal much as done in the Einstein model, but with the sum over $3N$ vibrational modes replaced by a continuous integral over the frequencies ν weighted by the density of such states $\Omega(\nu)$ (see Eq. (7.99)):

$$E = \frac{N\phi}{2} + \left(\frac{9NkT}{\nu_m^3}\right)\int\left[\frac{h\nu}{2kT} + \left(\frac{h\nu}{kT}\right)\left(\exp\left(\frac{h\nu}{kT}\right) - 1\right)^{-1}\right]\nu^2 d\nu, \quad (7.114)$$

where the integral over ν ranges from 0 to ν_m. It turns out that the C_V heat capacity obtained by taking the temperature derivative of this expression for E can be written as follows:

$$C_V = 3Nk\left[4D\left(\frac{h\nu_m}{kT}\right) - 3\left(\frac{h\nu_m}{kT}\right)\left(\exp\left(\frac{h\nu_m}{kT}\right) - 1\right)^{-1}\right], \quad (7.115)$$

where the so-called Debye function $D(u)$ is defined by

$$D(u) = 3u^{-3}\int x^3(\exp(x) - 1)^{-1}dx, \quad (7.116)$$

and the integral is taken from $x = 0$ to $x = u$.

The important thing to be noted about the Debye model is that the heat capacity, as defined above, extrapolates to $3Nk$ at high temperatures, thus agreeing with the law of Dulong and Petit, and varies at low temperature as

$$C_V \to \frac{12}{5}Nk\pi^4\left(\frac{kT}{h\nu_\mathrm{m}}\right)^3. \quad (7.117)$$

So, the Debye heat capacity does indeed vary as T^3 at low T as careful experiments indicate. For this reason, it is appropriate to use the Debye model whenever one is interested in properly treating the energy, heat capacity, and other thermodynamic properties of crystals at temperatures for which $kT/h\nu_\mathrm{m}$ is small. At higher temperatures, it is appropriate to use either the Debye or Einstein models. The major difference between the two lies in how they treat the spectrum of vibrational frequencies that occur in a crystal. The Einstein model says that only one (or at most three, if three different k_J values are used) frequency occurs $\nu_J = 1/2\pi(k_J/\mu)^{1/2}$; each species in the crystal is assumed to vibrate at this frequency. In contrast, the Debye model says that the species vibrate collectively and with frequencies ranging from $\nu = 0$ up to $\nu = \nu_\mathrm{m}$, the so-called Debye frequency, which is proportional to the speed c at which phonons propagate in the crystal. In turn, this speed depends on the stiffness (i.e., the inter-species potentials) within the crystal.

7.9 Lattice theories of surfaces and liquids

This kind of theory can be applied to a wide variety of chemical and physical problems, so it is a very useful model to be aware of. The starting point of the model is to consider a lattice containing M sites, each of which has c nearest neighbor sites (n.b., clearly, c will depend on the structure of the lattice) and to imagine that each of these sites can exist in either of two “states” that we label A and B. Before deriving the basic equations of this model, let me explain how the concepts of sites and A and B states are used to apply the model to various problems. For example:

(i) The sites can represent binding sites on the surface of a solid and the two states A and B can represent situations in which the site is either occupied (A) or unoccupied (B) by a molecule that is chemi-sorbed or physi-sorbed to the site. This point of view is taken when one applies lattice models to adsorption of gases or liquids to solid surfaces.

(ii) The sites can represent individual spin $= 1/2$ molecules or ions within a lattice, and the states can denote the α and β spin states of these species. This point of view allows the lattice models to be applied to magnetic materials.

(iii) The sites can represent positions that either of two kinds of molecules A and B might occupy in a liquid or solid in which case A and B are used to label whether each site contains an A or a B molecule. This is how we apply the lattice theories to liquid mixtures.

(iv) The sites can represent cis- and trans-conformations in linkages within a polymer, and A and B can be used to label each such linkage as being either cis- or trans-. This is how we use these models to study polymer conformations.

In Fig. 7.4 I show a two-dimensional lattice having 25 sites of which 16 are occupied by dark (A) species and 9 are occupied by lighter (B) species. The partition function for such a lattice is written in terms of a degeneracy Ω and an energy E, as usual. The degeneracy is computed by considering the number of

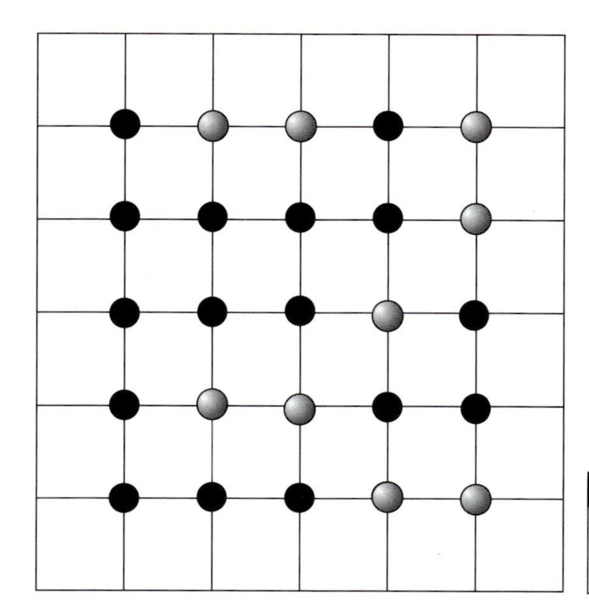

Figure 7.4 Two-dimensional lattice having 25 sites with 16 A and 9 B species.

ways a total of $N_A + N_B$ species can be arranged on the lattice:

$$\Omega = (N_A + N_B)!/[N_A!N_B!]. \tag{7.118}$$

The interaction energy among the A and B species for any arrangement of the A and B on the lattice is assumed to be expressed in terms of pairwise interaction energies. In particular, if only nearest neighbor interaction energies are considered, one can write the total interaction energy E_{int} of any arrangement as

$$E_{int} = N_{AA}E_{AA} + N_{BB}E_{BB} + N_{AB}E_{AB}, \tag{7.119}$$

where N_{IJ} is the number of nearest neighbor pairs of type I–J and E_{IJ} is the interaction energy of an I–J pair. The example shown in Fig. 7.4 has $N_{AA} = 16$, $N_{BB} = 4$ and $N_{AB} = 19$.

The three parameters N_{IJ} that characterize any such arrangement can be re-expressed in terms of the numbers N_A and N_B of A and B species and the number of nearest neighbors per site c as follows:

$$2N_{AA} + N_{AB} = cN_A, \tag{7.120}$$

$$2N_{BB} + N_{AB} = cN_B. \tag{7.121}$$

Note that the sum of these two equations states the obvious fact that twice the sum of AA, BB, and AB pairs must equal the number of A and B species multiplied by the number of neighbors per species, c.

Using the above relationships among N_{AA}, N_{BB}, and N_{AB}, we can rewrite the interaction energy as

$$\begin{aligned} E_{int} &= E_{AA}(cN_A - N_{AB})/2 + E_{BB}(cN_B - N_{AB})/2 + E_{AB}N_{AB} \\ &= (N_A E_{AA} + N_B E_{BB})c/2 + (2E_{AB} - E_{AA} - E_{BB})N_{AB}/2. \end{aligned} \tag{7.122}$$

The reason it is helpful to write E_{int} in this manner is that it allows us to express things in terms of two variables over which one has direct experimental control, N_A and N_B, and one variable N_{AB} that characterizes the degree of disorder among the A and B species. That is, if N_{AB} is small, the A and B species are arranged on the lattice in a phase-separated manner; whereas, if N_{AB} is large, the A and B are well mixed.

The total partition function of the A and B species arranged on the lattice is written as follows:

$$Q = q_A^{NA} q_B^{NB} \sum_{N_{AB}} \Omega(N_A, N_B, N_{AB}) \exp(-E_{int}/kT). \tag{7.123}$$

Here, q_A and q_B are the partition functions (electronic, vibrational, etc.) of the A and B species as they sit bound to a lattice site and $\Omega(N_A, N_B, N_{AB})$ is the number of ways that N_A species of type A and N_B of type B can be arranged on the lattice such that there are N_{AB} A–B type nearest neighbors. Of course, E_{int} is the interaction energy discussed earlier. The sum occurs because a partition function is a sum over all possible states of the system. There are no $(1/N_J!)$ factors because, as in the Einstein and Debye crystal models, the A and B species are not free to roam but are tied to lattice sites and thus are distinguishable.

This expression for Q can be rewritten in a manner that is more useful by employing the earlier relationships for N_{AA} and N_{BB}:

$$Q = \left(q_{\mathrm{A}} \exp\left(-\frac{cE_{\mathrm{AA}}}{2kT}\right)\right)^{N_{\mathrm{A}}} \left(q_{\mathrm{B}} \exp\left(-\frac{cE_{\mathrm{BB}}}{2kT}\right)\right)^{N_{\mathrm{B}}} \times \sum_{N_{\mathrm{AB}}} \Omega(N_{\mathrm{A}}, N_{\mathrm{B}}, N_{\mathrm{AB}}) \exp\left(\frac{N_{\mathrm{AB}}X}{2kT}\right), \tag{7.124}$$

where

$$X = (-2E_{\mathrm{AB}} + E_{\mathrm{AA}} + E_{\mathrm{BB}}). \tag{7.125}$$

The quantity X plays a central role in all lattice theories because it provides a measure of how different the A–B interaction energy is from the average of the A–A and B–B interaction energies. As we will soon see, if X is large and negative (i.e, if the A–A and B–B interactions are highly attractive), phase separation can occur; if X is positive, phase separation will not occur.

The problem with the above expression for the partition function is that no one has yet determined an analytical expression for the degeneracy $\Omega(N_{\mathrm{A}}, N_{\mathrm{B}}, N_{\mathrm{AB}})$ factor. Therefore, in the most elementary lattice theory, known as the Bragg–Williams approximation, one approximates the sum over N_{AB} by taking the following average value of N_{AB}:

$$N^{*}_{\mathrm{AB}} = \frac{N_{\mathrm{A}}(cN_{\mathrm{B}})}{(N_{\mathrm{A}} + N_{\mathrm{B}})} \tag{7.126}$$

in the expression for Ω. This then produces

$$Q = \left[q_{\mathrm{A}} \exp\left(-\frac{cE_{\mathrm{AA}}}{2kT}\right)\right]^{N_{\mathrm{A}}} \left[q_{\mathrm{B}} \exp\left(-\frac{cE_{\mathrm{BB}}}{2kT}\right)\right]^{N_{\mathrm{B}}} \times \exp\left(\frac{N^{*}_{\mathrm{AB}}X}{2kT}\right) \sum_{N_{\mathrm{AB}}} \Omega(N_{\mathrm{A}}, N_{\mathrm{B}}, N_{\mathrm{AB}}). \tag{7.127}$$

Finally, we realize that the sum $\sum_{N_{\mathrm{AB}}} \Omega(N_{\mathrm{A}}, N_{\mathrm{B}}, N_{\mathrm{AB}})$ is equal to the number of ways of arranging N_{A} A species and N_{B} B species on the lattice regardless of how many A–B neighbor pairs there are. This number is, of course, $(N_{\mathrm{A}} + N_{\mathrm{B}})!/[(N_{\mathrm{A}}!)(N_{\mathrm{B}}!)]$.

So, the Bragg–Williams lattice model partition function reduces to

$$Q = \left[q_{\mathrm{A}} \exp\left(-\frac{cE_{\mathrm{AA}}}{2kT}\right)\right]^{N_{\mathrm{A}}} \left[q_{\mathrm{B}} \exp\left(-\frac{cE_{\mathrm{BB}}}{2kT}\right)\right]^{N_{\mathrm{B}}} \frac{(N_{\mathrm{A}} + N_{\mathrm{B}})!}{(N_{\mathrm{A}}!)(N_{\mathrm{B}}!)} \exp\left(\frac{N^{*}_{\mathrm{AB}}X}{2kT}\right). \tag{7.128}$$

The most common connection one makes to experimental measurements using this partition function arises by computing the chemical potentials of the A and B species on the lattice and equating these to the chemical potentials of the A and B as they exist in the gas phase. In this way, one uses the equilibrium conditions (equal chemical potentials in two phases) to relate the vapor pressures of A and B, which arise through the gas-phase chemical potentials, to the interaction energy X.

Let me now show you how this is done. First, we use

$$\mu_J = -kT\left(\frac{\partial \ln Q}{\partial N_J}\right)_{T,V} \tag{7.129}$$

to compute the A and B chemical potentials on the lattice. This gives

$$\begin{aligned}\mu_{\mathrm{A}} = -kT\Bigg\{&\ln\left[q_{\mathrm{A}}\exp\left(-\frac{cE_{\mathrm{AA}}}{2kT}\right)\right] - \ln\left[\frac{N_{\mathrm{A}}}{(N_{\mathrm{A}}+N_{\mathrm{B}})}\right]\\ &+\left(1-\left[\frac{N_{\mathrm{A}}}{(N_{\mathrm{A}}+N_{\mathrm{B}})}\right]\right)^2\frac{cX}{2kT}\Bigg\}\end{aligned} \tag{7.130}$$

and an analogous expression for μ_{B} with N_{B} replacing N_{A}. The expression for the gas-phase chemical potentials μ_{A}^g and μ_{B}^g given earlier has the form

$$\begin{aligned}\mu = &-kT\ln\left[\left(\frac{2\pi mkT}{h^2}\right)^{3/2}\left(\frac{kT}{p}\right)\right] - kT\ln\left[\left(\frac{\pi^{1/2}}{\sigma}\right)\left(\frac{8\pi^2 I_{\mathrm{A}}kT}{h^2}\right)^{1/2}\right.\\ &\times\left.\left(\frac{8\pi^2 I_{\mathrm{B}}kT}{h^2}\right)^{1/2}\left(\frac{8\pi^2 I_{\mathrm{C}}kT}{h^2}\right)^{1/2}\right] + kT\sum_{J=1,3N-6}\\ &\times\left[\frac{h\nu_J}{2kT} + \ln\left(1-\exp\left(-\frac{h\nu_J}{kT}\right)\right)\right] - D_{\mathrm{e}} - kT\ln\omega_{\mathrm{e}},\end{aligned} \tag{7.131}$$

within which the vapor pressure appears. The pressure dependence of this gas-phase expression can be factored out to write each μ as

$$\mu_{\mathrm{A}}^g = \mu_{\mathrm{A}}^0 + kT\ln p_{\mathrm{A}}, \tag{7.132}$$

where p_{A} is the vapor pressure of A (in atmosphere units) and μ_{A}^0 denotes all of the other factors in μ_{A}^g. Likewise, the lattice-phase chemical potentials can be written as a term that contains the N_{A} and N_{B} dependence and a term that does not:

$$\mu_{\mathrm{A}} = -kT\left\{\ln\left[q_{\mathrm{A}}\exp\left(-\frac{cE_{\mathrm{AA}}}{2kT}\right)\right] - \ln X_{\mathrm{A}} + (1-X_{\mathrm{A}})^2\frac{cX}{2kT}\right\}, \tag{7.133}$$

where X_{A} is the mole fraction of A ($N_{\mathrm{A}}/(N_{\mathrm{A}}+N_{\mathrm{B}})$). Of course, an analogous expression holds for μ_{B}.

We now perform two steps:

(i) We equate the gas-phase and lattice-phase chemical potentials of species A in a case where the mole fraction of A is unity. This gives

$$\mu_A^0 + kT \ln\left(p_A^0\right) = -kT\left\{\ln\left[q_A \exp\left(-\frac{cE_{AA}}{2kT}\right)\right]\right\}, \tag{7.134}$$

where p_A^0 is the vapor pressure of A that exists over the lattice in which only A species are present.

(ii) We equate the gas- and lattice-phase chemical potentials of A for an arbitrary chemical potential X_A and obtain

$$\mu_A^0 + kT \ln p_A = -kT\left\{\ln\left[q_A \exp\left(-\frac{cE_{AA}}{2kT}\right)\right] - \ln X_A + (1 - X_A)^2 \frac{cX}{2kT}\right\}, \tag{7.135}$$

which contains the vapor pressure p_A of A over the lattice covered by A and B with X_A being the mole fraction of A.

Subtracting these two equations and rearranging, we obtain an expression for how the vapor pressure of A depends on X_A:

$$p_A = p_A^0 X_A \exp\left(-\frac{cX(1 - X_A)^2}{2kT}\right). \tag{7.136}$$

Recall that the quantity X is related to the interaction energies among various species as

$$X = -2E_{AB} + E_{AA} + E_{BB}. \tag{7.137}$$

Let us examine that physical meaning of the above result for the vapor pressure. First, if one were to totally ignore the interaction energies (i.e., by taking $X = 0$), one would obtain the well known Raoult's law expression for the vapor pressure of a mixture:

$$p_A = p_A^0 X_A, \tag{7.138}$$

$$p_B = p_B^0 X_B. \tag{7.139}$$

In Fig. 7.5, I plot the A and B vapor pressures vs. X_A. The two straight lines are, of course, just the Raoult's law findings. I also plot the p_A vapor pressure for three values of the X interaction energy parameter. When X is positive, meaning that the A–B interactions are more energetically favorable than the average of the A–A and B–B interactions, the vapor pressure of A is found to deviate negatively from the Raoult's law prediction. This means that the observed vapor pressure is lower than that expected based solely on Raoult's law. On the other hand, when X is negative, the vapor pressure deviates positively from Raoult's law.

An especially important and interesting case arises when the X parameter is negative and has a value that makes $cX/2kT$ be more negative than -4. It

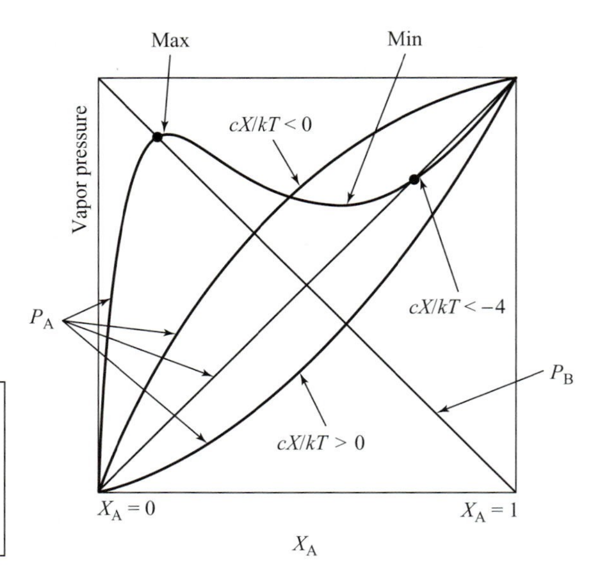

Figure 7.5 Plots of vapor pressures in an A, B mixture as predicted in the lattice model with the Bragg–Williams approximation.

turns out that in such cases, the function p_A suggested in this Bragg–Williams model displays a behavior that suggests a phase transition may occur. Hints of this behavior are clear in Fig. 7.5 where one of the plots displays both a maximum and a minimum, but the plots for $X > 0$ and for $cX/2kT > -4$ do not. Let me now explain this further by examining the derivative of p_A with respect to X_A:

$$\frac{dp_A}{dX_A} = p_A^0 \left\{1 + X_A(1 - X_A)\frac{2cX}{2kT}\right\} \exp\left(-\frac{cX(1 - X_A)^2}{2kT}\right). \tag{7.140}$$

Setting this derivative to zero (in search of a maximum or minimum), and solving for the values of X_A that make this possible, one obtains:

$$X_A = \frac{1}{2}\left\{1 \pm \left(1 + \frac{4kT}{cX}\right)^{12}\right\}. \tag{7.141}$$

Because X_A is a mole fraction, it must be less than unity and greater than zero. The above result giving the mole fraction at which $dp_A/dX_A = 0$ will not produce a realistic value of X_A unless

$$\frac{cX}{kT} < -4. \tag{7.142}$$

If $cX/kT = -4$, there is only one value of X_A (i.e., $X_A = 1/2$) that produces a zero slope; for $cX/kT < -4$, there will be two such values given by $X_A = 1/2\{1 \pm (1 + 4kT/cX)^{12}\}$, which is what we see in Fig. 7.5 where the plot displays both a maximum and a minimum.

What does it mean for cX/kT to be less than -4 and why is this important? For X to be negative, it means that the average of the A–A and B–B interactions is more energetically favorable than of the A–B interactions. It is for this reason that a phase separation may be favored in such cases (i.e., the A species "prefer" to be near other A species more than to be near B species, and similarly for the B species). However, thermal motion can overcome a slight preference for such separation. That is, if X is not large enough, kT can overcome this slight preference. This is why cX must be less than $-4kT$, not just less than zero.

So, the bottom line is that if the A–A and B–B interactions are more attractive, on average, than are the A–B interactions, one can experience a phase separation in which the A and B species do not remain mixed on the lattice but instead gather into two distinct kinds of domains. One domain will be rich in the A species, having an X_A value equal to that shown in the right dot in Fig. 7.5. The other domains will be rich in B and have an X_A value of that shown by the left dot.

As I noted in the introduction to this section, lattice models can be applied to a variety of problems. We just analyzed how it is applied, within the Bragg–Williams approximation, to mixtures of two species. In this way, we obtain expressions for how the vapor pressures of the two species in the liquid or solid mixture display behavior that reflects their interaction energies. Let me now briefly show you how the lattice model is applied in some other areas.

In studying adsorption of gases to sites on a solid surface, one imagines a surface containing M sites per unit area A with N_{ad} molecules (that have been adsorbed from the gas phase) bound to these sites. In this case, the interaction energy E_{int} introduced earlier involves only interactions among neighboring adsorbed molecules; there are no interactions among empty surface sites or between empty surface sites and adsorbed molecules. So, we can make the following replacements in our earlier equations:

$$N_A \rightarrow N_{ad}, \tag{7.143}$$

$$N_B \rightarrow M - N_{ad}, \tag{7.144}$$

$$E_{int} = E_{ad,ad} N_{ad,ad}, \tag{7.145}$$

where $N_{ad,ad}$ is the number of nearest neighbor pairs of adsorbed species and $E_{ad,ad}$ is the pairwise interaction energy between such a pair. The primary result obtained by equating the chemical potentials of the gas-phase and adsorbed molecules is

$$p = kT \left(\frac{q_{gas}}{V}\right)\left(\frac{1}{q_{ad}}\right)\left[\frac{\theta}{(1-\theta)}\right]\exp\left(\frac{E_{ad}c\theta}{kT}\right). \tag{7.146}$$

Here q_{gas}/V is the partition function of the gas-phase molecules per unit volume, q_{ad} is the partition function of the adsorbed molecules (which contains the adsorption energy as $\exp(-\phi/kT)$) and θ is called the coverage (i.e., the fraction of surface sites to which molecules have adsorbed). Clearly, θ plays the role that the mole fraction X_A played earlier. This so-called adsorption isotherm equation allows one to connect the pressure of the gas above the solid surface to the coverage.

As in our earlier example, something unusual occurs when the quantity $E_{ad}c\theta/kT$ is negative and beyond a critical value. In particular, differentiating the expression for p with respect to θ and finding for what θ value(s) $dp/d\theta$ vanishes, one finds

$$\theta = \frac{1}{2}\left[1 \pm \left(1 + \frac{4kT}{cE_{\text{ad}}}\right)^{1/2}\right]. \tag{7.147}$$

Since θ is a positive fraction, this equation can only produce useful values if

$$\frac{cE_{\text{ad}}}{kT} < -4. \tag{7.148}$$

In this case, this means that if the attraction between neighboring adsorbed molecules is strong enough, it can overcome thermal factors to cause phase separation to occur. The kind of phase separation one observes is the formation of islands of adsorbed molecules separated by regions where the surface has few or no adsorbed molecules.

There is another area where this kind of lattice model is widely used. When studying magnetic materials one often uses the lattice model to describe the interactions among pairs of neighboring spins (e.g., unpaired electrons on neighboring molecules or nuclear spins on neighboring molecules). In this application, one assumes that "up" or "down" spin states are distributed among the lattice sites, which represent where the molecules are located. N_α and N_β are the total numbers of such spins, so $(N_\alpha - N_\beta)$ is a measure of what is called the net magnetization of the sample. The result of applying the Bragg–Williams approximation in this case is that one again observes a critical condition under which strong spin pairings occur. In particular, because the interactions between α and α spins, denoted $-J$, and between α and β spins, denoted $+J$, are equal and opposite, the X variable characteristic of all lattice models reduces to

$$X = -2E_{\alpha,\beta} + E_{\alpha,\alpha} + E_{\beta,\beta} = -4J. \tag{7.149}$$

The critical condition under which one expects like spins to pair up and thus to form islands of α-rich centers and other islands of β-rich centers is

$$\frac{-4cJ}{kT} < -4 \tag{7.150}$$

or

$$\frac{cJ}{kT} > 1. \tag{7.151}$$

7.10 Virial corrections to ideal-gas behavior

Recall from our earlier treatment of the classical partition function that one can decompose the total partition function into a product of two factors:

$$Q = h^{-NM}(N!)^{-1}\int \exp\left(\frac{-H^0(y,p)}{kT}\right) dy\, dp \int \exp\left(-\frac{U(r)}{kT}\right) dr, \tag{7.152}$$

one of which,

$$Q_{\text{ideal}} = h^{-NM}(N!)^{-1} \int \exp\left(-\frac{H^0(y,p)}{kT}\right) dy\, dp V^N, \tag{7.153}$$

is the result if no intermolecular potentials are operative. The second factor,

$$Q_{\text{inter}} = \left(\frac{1}{V^N}\right)\left\{\int \exp\left(-\frac{U(r)}{kT}\right) dr\right\}, \tag{7.154}$$

thus contains all of the effects of intermolecular interactions. Recall also that all of the equations relating partition functions to thermodynamic properties involve taking ln Q and derivatives of ln Q. So, all such equations can be cast into sums of two parts; that arising from ln Q_{ideal} and that arising from ln Q_{inter}. In this section, we will be discussing the contributions of Q_{inter} to such equations.

The first thing that is done to develop the so-called cluster expansion of Q_{inter} is to assume that the total intermolecular potential energy can be expressed as a sum of pairwise additive terms:

$$U = \sum_{I<J} U(r_{IJ}), \tag{7.155}$$

where r_{IJ} labels the distance between molecule I and molecule J. This allows the exponential appearing in Q_{inter} to be written as a product of terms, one for each pair of molecules:

$$\exp\left(-\frac{U}{kT}\right) = \exp\left(-\sum_{I<J}\frac{U(r_{IJ})}{kT}\right) = \prod_{I<J}\exp\left(-\frac{U(r_{IJ})}{kT}\right). \tag{7.156}$$

Each of the exponentials $\exp(-U(r_{IJ})/kT)$ is then expressed as follows:

$$\exp\left(-\frac{U(r_{IJ})}{kT}\right) = 1 + \left(\exp\left(-\frac{U(r_{IJ})}{kT}\right) - 1\right) = 1 + f_{IJ}, \tag{7.157}$$

the last equality being what defines f_{IJ}. These f_{IJ} functions are introduced because, whenever the molecules I and J are distant from one another and thus not interacting, $U(r_{IJ})$ vanishes, so $\exp(-U(r_{IJ})/kT)$ approaches unity, and thus f_{IJ} vanishes. In contrast, whenever molecules I and J are close enough to experience strong repulsive interactions, $U(r_{IJ})$ is large and positive, so f_{IJ} approaches -1. These properties make f_{IJ} a useful measure of how molecules are interacting; if they are not, $f = 0$, if they are repelling strongly, $f = -1$, and if they are strongly attracting, f is large and positive.

Inserting the f_{IJ} functions into the product expansion of the exponential, one obtains

$$\exp(-U/kT) = \prod_{I<J}(1 + f_{IJ}) = 1 + \sum_{I<J} f_{IJ} + \sum_{I<J}\sum_{K<L} f_{IJ} f_{KL} + \cdots, \tag{7.158}$$

which is called the cluster expansion in terms of the f_{IJ} pair functions. When this expansion is substituted into the expression for Q_{inter}, we find

$$Q_{\text{inter}} = V^{-N} \int \left(1 + \sum_{I<J} f_{IJ} + \sum_{I<J} \sum_{K<L} f_{IJ} f_{KL} + \cdots \right) dr, \tag{7.159}$$

where the integral is over all $3N$ of the N molecule's center of mass coordinates.

The integrals involving only one f_{IJ} function are all equal (i.e., for any pair I, J, the molecules are identical in their interaction potentials) and reduce to

$$[N(N-1)/2]V^{-2} \int f(r_{1,2}) dr_1\, dr_2. \tag{7.160}$$

The integrals over $dr_3 \ldots dr_N$ produce V^{N-2}, which combines with V^{-N} to produce the V^{-2} seen. Finally, because $f(r_{1,2})$ depends only on the relative positions of molecules 1 and 2, the six-dimensional integral over $dr_1 dr_2$ can be replaced by integrals over the relative location of the two molecules r, and the position of their center of mass R. The integral over R gives one more factor of V, and the above cluster integral reduces to

$$4\pi[N(N-1)/2]V^{-1} \int f(r) r^2 dr. \tag{7.161}$$

with the 4π coming from the angular integral over the relative coordinate r. Because the total number of molecules N is very large, it is common to write the $N(N-1)/2$ factor as $N^2/2$.

The cluster integrals containing two $f_{IJ} f_{KL}$ factors can also be reduced. However, it is important to keep track of different kinds of such factors (depending on whether the indices I, J, K, L are all different or not). For example, terms of the form

$$V^{-N} \int f_{IJ} f_{KL} dr_1 dr_2 \ldots dr_N \qquad \text{with } I, J, K, \text{and } L \text{ all unique} \tag{7.162}$$

reduce (again using the equivalence of the molecules and the fact that f_{IJ} depends only on the relative positions of I and J) to

$$1/4 N^4 (4\pi)^2 V^{-2} \int f_{12} r_{12}^2 dr_{12} \int f_{34} r_{34}^2 dr_{34}, \tag{7.163}$$

where again I used the fact that N is very large to replace $N(N-1)/2(N-2)(N-3)/2$ by $N^4/4$.

On the other hand, cluster integrals with, for example, $I = K$ but J and L different reduce as follows:

$$V^{-N} \int f_{12} f_{13} dr_1 dr_2 \ldots dr_N = 1/2 V^{-3} N^3 \int f_{12} f_{13} dr_1 dr_2 dr_3. \tag{7.164}$$

Because f_{12} depends only on the relative positions of molecules 1 and 2 and f_{13} depends on the relative positions of 1 and 3, the nine-dimensional integral over $dr_1 dr_2 dr_3$ can be changed to a six-dimensional integral over $dr_{12} dr_{13}$ and an integral over the location of molecule 1; the latter integral produces a factor of V when carried out. Thus, the above cluster integral reduces to

$$(4\pi)^2 1/2V^{-2}N^3 \int f_{12}f_{13}r_{12}^2 r_{13}^2 dr_{12}dr_{13}. \tag{7.165}$$

There is a fundamental difference between cluster integrals of the type $f_{12}f_{34}$ and those involving $f_{12}f_{13}$. The former are called unlinked clusters because they involve the interaction of molecules 1 and 2 and a separate interaction of molecules 3 and 4. The latter are called linked because they involve molecule 1 interacting simultaneously with molecules 2 and 3 (although 2 and 3 need not be close enough to cause f_{23} to be non-zero). The primary differences between unlinked and linked cluster contributions are:

(i) The total number of unlinked terms is proportional to N^4, while the number of linked terms is proportional to N^3. This causes the former to be more important than the latter.
(ii) The linked terms only become important at densities where there is a significant probability that three molecules occupy nearby regions of space. The linked terms, on the other hand, do not require that molecules 1 and 2 be anywhere near molecules 3 and 4. This also causes the unlinked terms to dominate especially at low and moderate densities.

I should note that a similar observation was made in Chapter 6 when we discussed the configuration interaction and coupled-cluster expansion of electronic wave functions. That is, we noted that doubly excited configurations (analogous to f_{IJ}) are the most important contributions beyond the single determinant, and that quadruple excitations in the form of unlinked products of double excitations were next most important, not triple excitations. The unlinked nature in this case was related to the amplitudes of the quadruple excitations being products of the amplitudes of two double excitations. So, both in electronic structures and in liquid structure, one finds that pair correlations followed by unlinked pair correlations are the most important to consider.

Clearly, the cluster expansion approach to Q_{inter} can be carried to higher and higher level clusters (e.g., involving $f_{12}f_{34}f_{56}$ or $f_{12}f_{13}f_{34}$, etc.). Generally, one finds that the unlinked terms (e.g., $f_{12}f_{34}f_{56}$ in this example) are most important (because they are proportional to higher powers of N and because they do not require more than binary collisions). It is most common, however, to employ a severely truncated expansion and to retain only the linked terms. Doing so for Q_{inter} produces at the lower levels:

$$\begin{aligned} Q_{\text{inter}} &= 1 + \frac{1}{2}\left(\frac{N}{V}\right)^2 4\pi V \int fr^2 dr + \frac{1}{4}\left(\frac{N}{V}\right)^4 \left[4\pi V \int fr^2 dr\right]^2 \\ &+ \frac{1}{2}\left(\frac{N}{V}\right)^3 V(4\pi)^2 \int f_{12}f_{13}r_{12}^2 r_{13}^2 dr_{12}dr_{13}. \end{aligned} \tag{7.166}$$

One of the most common properties to compute using a partition function that includes molecular interactions in the cluster manner is the pressure, which is calculated as

$$p = kT \left(\frac{\partial \ln Q}{\partial V} \right)_{N,T}. \qquad (7.167)$$

Using $Q = Q_{\text{ideal}} Q_{\text{inter}}$ and inserting the above expression for Q_{inter} produces the following result for the pressure:

$$\frac{pV}{NkT} = 1 + B_2 \left(\frac{N}{V} \right) + B_3 \left(\frac{N}{V} \right)^2 + \cdots, \qquad (7.168)$$

where the so-called virial coefficients B_2 and B_3 are defined as the factors proportional to (N/V) and $(N/V)^2$, respectively. The second virial coefficient's expression in terms of the cluster integrals is

$$B_2 = -2\pi \int f r^2 dr = -2\pi \int [\exp(-U(r)/kT) - 1] r^2 dr. \qquad (7.169)$$

The third virial coefficient involves higher order cluster integrals.

The importance of such cluster analyses is that they allow various thermodynamic properties (e.g., the pressure above) to be expressed as one contribution that would occur if the system consisted of non-interacting molecules and a second contribution that arises from the intermolecular forces. It thus allows experimental measurements of the deviation from ideal (i.e., non-interacting) behavior to provide a direct way to determine intermolecular potentials. For example, by measuring pressures at various N/V values and various temperatures, one can determine B_2 and thus gain valuable information about the intermolecular potential U.

Chapter 8
Chemical dynamics

Chemical dynamics is a field in which scientists study the rates and mechanisms of chemical reactions. It also involves the study of how energy is transferred among molecules as they undergo collisions in gas-phase or condensed-phase environments. Therefore, the experimental and theoretical tools used to probe chemical dynamics must be capable of monitoring the chemical identity and energy content (i.e., electronic, vibrational, and rotational state populations) of the reacting species. Moreover, because the rates of chemical reactions and energy transfer are of utmost importance, these tools must be capable of doing so on time scales over which these processes, which are often very fast, take place. Let us begin by examining many of the most commonly employed theoretical models for simulating and understanding the processes of chemical dynamics.

Theoretical tools for studying chemical change and dynamics

8.1 Transition state theory

The most successful and widely employed theoretical approach for studying reaction rates involving species that are undergoing reaction at or near thermal-equilibrium conditions is the transition state theory (TST) of Eyring. This would not be a good way to model, for example, photochemical reactions in which the reactants do not reach thermal equilibrium before undergoing significant reaction progress. However, for most thermal reactions, it is remarkably successful.

In this theory, one views the reactants as undergoing collisions that act to keep all of their degrees of freedom (translational, rotational, vibrational, electronic) in thermal equilibrium. Among the collection of such reactant molecules, at any instant of time, some will have enough internal energy to access a transition state (TS) on the Born–Oppenheimer ground-state potential energy surface. Within TST, the rate of progress from reactants to products is then expressed in terms of the concentration of species that exist near the TS multiplied by the rate at which these species move through the TS region of the energy surface.

The concentration of species at the TS is, in turn, written in terms of the equilibrium constant expression of statistical mechanics discussed in Chapter 7. For example, for a bimolecular reaction A + B → C passing through a TS denoted AB*, one writes the concentration (in molecules per unit volume) of AB* species in terms of the concentrations of A and of B and the respective partition functions as

$$[\mathrm{AB}^*] = \left(\frac{q_{\mathrm{AB}^*}}{V}\right) \Big/ \left\{\left(\frac{q_{\mathrm{A}}}{V}\right)\left(\frac{q_{\mathrm{B}}}{V}\right)\right\} [\mathrm{A}]\,[\mathrm{B}]\,. \tag{8.1}$$

There is, however, one aspect of the partition function of the TS species that is specific to this theory. The q_{AB^*} contains all of the usual translational, rotational, vibrational, and electronic partition functions that one would write down, as we did in Chapter 7, for a conventional AB molecule except for one modification. It does not contain an $\{\exp(-h\nu_j/2kT)/(1-\exp(-h\nu_j/kT))\}$ vibrational contribution for motion along the one internal coordinate corresponding to the reaction path. In the vicinity of the TS, the reaction path can be identified as that direction along which the PES has negative curvature; along all other directions, the energy surface is positively curved. For example, in Fig. 8.1, a reaction path begins at transition structure B and is directed "downhill". More specifically, if one knows the gradients $\{(\partial E/\partial s_k)\}$ and Hessian matrix elements $\{H_{j,k} = \partial^2 E/\partial s_j\,\partial s_k\}$ of the energy surface at the TS, one can express the variation of the potential energy along the $3N$ Cartesian coordinates $\{s_k\}$ of the molecule as follows:

$$E(s_k) = E(0) + \sum_k \left(\frac{\partial E}{\partial s_k}\right) s_k + 1/2 \sum_{j,k} s_j H_{j,k} s_k + \cdots, \tag{8.2}$$

where $E(0)$ is the energy at the TS, and the $\{s_k\}$ denote displacements away

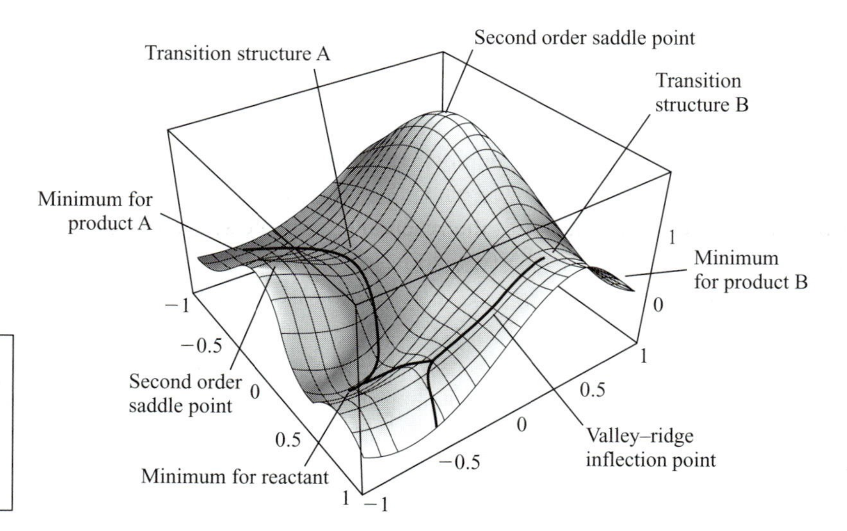

Figure 8.1 Typical potential energy surface in two dimensions showing local minima, transition states and paths connecting them.

from the TS geometry. Of course, at the TS, the gradients all vanish because this geometry corresponds to a stationary point. As we discussed in the Background Material, the Hessian matrix $H_{j,k}$ has 6 zero eigenvalues whose eigenvectors correspond to overall translation and rotation of the molecule. This matrix has $3N - 7$ positive eigenvalues whose eigenvectors correspond to the vibrations of the TS species, as well as one negative eigenvalue. The latter has an eigenvector whose components $\{s_k\}$ along the $3N$ Cartesian coordinates describe the direction of the reaction path as it begins its journey from the TS backward to reactants (when followed in one direction) and onward to products (when followed in the opposite direction). Once one moves a small amount along the direction of negative curvature, the reaction path is subsequently followed by taking infinitesimal "steps" downhill along the gradient vector **g** whose $3N$ components are $(\partial E/\partial s_k)$. Note that once one has moved downhill away from the TS by taking the initial step along the negatively curved direction, the gradient no longer vanishes because one is no longer at the stationary point.

Returning to the TST rate calculation, one therefore is able to express the concentration [AB*] of species at the TS in terms of the reactant concentrations and a ratio of partition functions. The denominator of this ratio contains the conventional partition functions of the reactant molecules and can be evaluated as discussed in Chapter 7. However, the numerator contains the partition function of the TS species but with one vibrational component missing (i.e., $q_{\text{vib}} = \Pi_{k=1,3N-7}\{\exp(-h\nu_j/2kT)/(1 - \exp[-h\nu_j/kT])\}$. Other than the one missing q_{vib}, the TS's partition function is also evaluated as in Chapter 7. The motion along the reaction path coordinate contributes to the rate expression in terms of the frequency (i.e., how often) with which reacting flux crosses the TS region given that the system is in near-thermal equilibrium at temperature T.

To compute the frequency with which trajectories cross the TS and proceed onward to form products, one imagines the TS as consisting of a narrow region along the reaction coordinate s; the width of this region we denote δs. We next ask what the classical weighting factor is for a collision to have momentum p_s along the reaction coordinate. Remembering our discussion of such matters in Chapter 7, we know that the momentum factor entering into the classical partition function for translation along the reaction coordinate is $(1/h)\exp(-p_s^2/2\mu kT)dp_s$. Here, μ is the mass factor associated with the reaction coordinate s. We can express the rate or frequency at which such trajectories pass through the narrow region of width δs as $(p_s/\mu\delta s)$, with p_s/μ being the speed of passage (cm s^{-1}) and $1/\delta s$ being the inverse of the distance that defines the TS region. So, $(p_s/\mu\delta s)$ has units of s^{-1}. In summary, we expect the rate of trajectories moving through the TS region to be

$$\left(\frac{1}{h}\right)\exp\left(\frac{-p_s^2}{2\mu kT}\right)dp_s\left(\frac{p_s}{\mu\delta s}\right). \tag{8.3}$$

However, we still need to integrate this over all values of p_s that correspond to enough energy $p_s^2/2\mu$ to access the TS's energy, which we denote E^*. Moreover, we have to account for the fact that it may be that not all trajectories with kinetic energy equal to E^* or greater pass on to form product molecules; some trajectories may pass through the TS but later recross the TS and return to produce reactants. Moreover, it may be that some trajectories with kinetic energy along the reaction coordinate less than E^* can react by tunneling through the barrier.

The way we account for the facts that a reactive trajectory must have at least E^* in energy along s and that not all trajectories with this energy will react is to integrate over only values of p_s greater than $(2\mu E^*)^{1/2}$ and to include in the integral a so-called transmission coefficient κ that specifies the fraction of trajectories crossing the TS that eventually proceed onward to products. Putting all of these pieces together, we carry out the integration over p_s just described to obtain

$$\int\int\left(\frac{1}{h}\right)\kappa\exp\left(\frac{-p_s^2}{2\mu kT}\right)\left(\frac{p_s}{\mu\delta s}\right)ds\,dp_s, \tag{8.4}$$

where the momentum is integrated from $p_s = (2\mu E^*)^{1/2}$ to ∞ and the s-coordinate is integrated only over the small region δs. If the transmission coefficient is factored out of the integral (treating it as a multiplicative factor), the integral over p_s can be done and yields the following:

$$\kappa\left(\frac{kT}{h}\right)\exp\left(\frac{-E^*}{kT}\right). \tag{8.5}$$

The exponential energy dependence is usually then combined with the partition function of the TS species that reflect this species' other $3N-7$ vibrational coordinates and momenta and the reaction rate is then expressed as

$$\text{Rate} = \kappa\left(\frac{kT}{h}\right)[\text{AB}^*] = \kappa\left(\frac{kT}{h}\right)\left(\frac{q_{\text{AB}^*}}{V}\right)\Big/\left\{\left(\frac{q_A}{V}\right)\left(\frac{q_\text{B}}{V}\right)\right\}[\text{A}][\text{B}]. \tag{8.6}$$

This implies that the rate coefficient k_{rate} for this bimolecular reaction is given in terms of molecular partition functions by

$$k_{\text{rate}} = \kappa\frac{kT}{h}\left(\frac{q_{\text{AB}^*}}{V}\right)\Big/\left\{\left(\frac{q_\text{A}}{V}\right)\left(\frac{q_\text{B}}{V}\right)\right\} \tag{8.7}$$

which is the fundamental result of TST. Once again we notice that ratios of partition functions per unit volume can be used to express ratios of species concentrations (in number of molecules per unit volume), just as appeared in earlier expressions for equilibrium constants as in Chapter 7.

The above rate expression undergoes only minor modifications when unimolecular reactions are considered. For example, in the hypothetical reaction A $\rightarrow$ B via the TS (A^*), one obtains

$$k_{\text{rate}} = \kappa\frac{kT}{h}\left\{\left(\frac{q_{\text{A}^*}}{V}\right)\Big/\left(\frac{q_\text{A}}{V}\right)\right\}, \tag{8.8}$$

where again q_{A^*} is a partition function of A* with one missing vibrational component.

Before bringing this discussion of TST to a close, I need to stress that this theory is not exact. It assumes that the reacting molecules are nearly in thermal equilibrium, so it is less likely to work for reactions in which the reactant species are prepared in highly non-equilibrium conditions. Moreover, it ignores tunneling by requiring all reactions to proceed through the TS geometry. For reactions in which a light atom (i.e., an H or D atom) is transferred, tunneling can be significant, so this conventional form of TST can provide substantial errors in such cases. Nevertheless, TST remains the most widely used and successful theory of chemical reaction rates and can be extended to include tunneling and other corrections as we now illustrate.

8.2 Variational transition state theory

Within the TST expression for the rate constant of a bimolecular reaction, $k_{rate} = \kappa kT/h(q_{AB^*}/V)/\{(q_A/V)(q_B/V)\}$ or of a unimolecular reaction, $k_{rate} = \kappa kT/h\{(q_{A^*}/V)/(q_A/V)\}$, the height ($E^*$) of the barrier on the potential energy surface appears in the TS species' partition function q_{AB^*} or q_{A^*}, respectively. In particular, the TS partition function contains a factor of the form $\exp(-E^*/kT)$ in which the Born–Oppenheimer electronic energy of the TS relative to that of the reactant species appears. This energy E^* is the value of the potential energy $E(S)$ at the TS geometry, which we denote S_0.

It turns out that the conventional TS approximation to k_{rate} over-estimates reaction rates because it assumes all trajectories that cross the TS proceed onward to products unless the transmission coefficient is included to correct for this. In the variational transition state theory (VTST), one does not evaluate the ratio of partition functions appearing in k_{rate} at the TS S_0, but one first determines at what geometry (S^*) the TS partition function (i.e., q_{AB^*} or q_{A^*}) is smallest. Because this partition function is a product of (i) the $\exp(-E(S)/kT)$ factor as well as (ii) 3 translational, 3 rotational, and $3N - 7$ vibrational partition functions (which depend on S), the value of S for which this product is smallest need not be the conventional TS value S_0. What this means is that the location (S^*) along the reaction path at which the free energy reaches a saddle point is not the same as the location S_0 where the Born–Oppenheimer electronic energy $E(S)$ has its saddle. This interpretation of how S^* and S_0 differ can be appreciated by recalling that partition functions are related to the Helmholtz free energy A by $q = \exp(-A/kT)$; so determining the value of S where q reaches a minimum is equivalent to finding that S where A is at a maximum.

So, in VTST, one adjusts the "dividing surface" (through the location on the reaction coordinate S) to first find that value S^* where k_{rate} has a minimum. One then evaluates both $E(S^*)$ and the other components of the TS species partition functions at this value S^*. Finally, one then uses the k_{rate} expressions given above, but with S taken at S^*. This is how VTST computes reaction rates in a somewhat different manner than does the conventional TST. As with TST, the VTST, in the form outlined above, does not treat tunneling and the fact that not all trajectories crossing S^* proceed to products. These corrections still must be incorporated as an "add-on" to this theory (i.e., in the κ factor) to achieve high accuracy for reactions involving light species (recall from the Background Material that tunneling probabilities depend exponentially on the mass of the tunneling particle).

8.3 Reaction path Hamiltonian theory

Let us review what the reaction path is as defined above. It is a path that:

(i) begins at a transition state (TS) and evolves along the direction of negative curvature on the potential energy surface (as found by identifying the eigenvector of the Hessian matrix $H_{j,k} = \partial^2 E/\partial s_k \partial s_j$ that belongs to the negative eigenvalue);
(ii) moves further downhill along the gradient vector $\mathbf{g}$ whose components are $g_k = \partial E/\partial s_{k'}$;
(iii) terminates at the geometry of either the reactants or products (depending on whether one began moving away from the TS forward or backward along the direction of negative curvature).

The individual "steps" along the reaction coordinate can be labeled $S_0, S_1, S_2, \ldots S_{\text{P}}$ as they evolve from the TS to the products (labeled S_{P}) and $S_{-\text{R}}, S_{-\text{R}+1}, \ldots S_0$ as they evolve from reactants ($S_{-\text{R}}$) to the TS. If these steps are taken in very small (infinitesimal) lengths, they form a continuous path and a continuous coordinate that we label S.

At any point S along a reaction path, the Born–Oppenheimer potential energy surface $E(S)$, its gradient components $g_k(S) = (\partial E(S)/\partial s_k)$ and its Hessian components $H_{k,j}(S) = (\partial^2 E(S)/\partial s_k \partial s_j)$ can be evaluated in terms of derivatives of E with respect to the $3N$ Cartesian coordinates of the molecule. However, when one carries out reaction path dynamics, one uses a different set of coordinates for reasons that are similar to those that arise in the treatment of normal modes of vibration as given in the Background Material. In particular, one introduces $3N$ mass-weighted coordinates $x_j = s_j(m_j)^{1/2}$ that are related to the $3N$ Cartesian coordinates s_j in the same way as we saw in the Background Material.

The gradient and Hessian matrices along these new coordinates $\{x_j\}$ can be evaluated in terms of the original Cartesian counterparts:

$$g'_k(S) = g_k(S)(m_k)^{-1/2}, \tag{8.9}$$
$$H'_{j,k} = H_{j,k}(m_j m_k)^{-1/2}. \tag{8.10}$$

The eigenvalues $\{\omega_k^2\}$ and eigenvectors $\{\mathbf{v}_k\}$ of the mass-weighted Hessian H' can then be determined. Upon doing so, one finds:

(i) 6 zero eigenvalues whose eigenvectors describe overall rotation and translation of the molecule;
(ii) $3N - 7$ positive eigenvalues $\{\omega_K^2\}$ and eigenvectors $\mathbf{v}_K$ along which the gradient $\mathbf{g}$ has zero (or nearly so) components;
(iii) and one eigenvalue ω_S^2 (that may be positive, zero, or negative) along whose eigenvector $\mathbf{v}_S$ the gradient $\mathbf{g}$ has its largest component.

The one unique direction along $\mathbf{v}_S$ gives the direction of evolution of the reaction path (in these mass-weighted coordinates). All other directions (i.e., within the space spanned by the $3N - 7$ other vectors $\{\mathbf{v}_K\}$) possess zero gradient component and positive curvature. This means that at any point S on the reaction path being discussed:

(i) one is at a local minimum along all $3N - 7$ directions $\{\mathbf{v}_K\}$ that are transverse to the reaction path direction (i.e., the gradient direction);
(ii) one can move to a neighboring point on the reaction path by moving a small (infinitesimal) amount along the gradient.

In terms of the $3N - 6$ mass-weighted Hessian's eigenmode directions ($\{\mathbf{v}_K\}$ and $\mathbf{v}_S$), the potential energy surface can be approximated, in the neighborhood of each such point on the reaction path S, by expanding it in powers of displacements away from this point. If these displacements are expressed as components:

(i) δX_k along the $3N - 7$ eigenvectors $\mathbf{v}_K$ and
(ii) δS along the gradient direction $\mathbf{v}_S$,

one can write the Born–Oppenheimer potential energy surface locally as

$$E = E(S) + \mathbf{v}_S \delta S + 1/2\omega_S^2\, \delta S^2 + \sum_{K=1,3N-7} 1/2\omega_K^2\, \delta X_K^2. \tag{8.11}$$

Within this local quadratic approximation, E describes a sum of harmonic potentials along each of the $3N - 7$ modes transverse to the reaction path direction. Along the reaction path, E appears with a non-zero gradient and a curvature that may be positive, negative, or zero.

The eigenmodes of the local (i.e., in the neighborhood of any point S along the reaction path) mass-weighted Hessian decompose the $3N - 6$ internal coordinates into $3N - 7$ along which E is harmonic and one (S) along which the reaction evolves. In terms of these same coordinates, the kinetic energy T can also be written and thus the classical Hamiltonian $H = T + V$ can be constructed. Because the coordinates we use are mass-weighted, in Cartesian form the kinetic energy T contains no explicit mass factors:

$$T = \frac{1}{2}\sum_j m_j \left(\frac{ds_j}{dt}\right)^2 = \frac{1}{2}\sum_j \left(\frac{dx_j}{dt}\right)^2. \tag{8.12}$$

This means that the momenta conjugate to each (mass-weighted) coordinate x_j, obtained in the usual way as $p_j = \partial[T - V]/\partial(dx_j/dt) = dx_j/dt$, all have identical (unit) mass factors associated with them.

To obtain the working expression for the reaction path Hamiltonian (RPH), one must transform the above equation for the kinetic energy T by replacing the $3N$ Cartesian mass-weighted coordinates $\{x_j\}$ by:

(i) the $3N - 7$ eigenmode displacement coordinates δX_j,
(ii) the reaction path displacement coordinate δS, and
(iii) 3 translation and 3 rotational coordinates.

The 3 translational coordinates can be separated and ignored (because center-of-mass energy is conserved) in further consideration. The 3 rotational coordinates do not enter into the potential E, but they do appear in T. However, it is most common to ignore their effects on the dynamics that occurs in the internal coordinates; this amounts to ignoring the effects of overall centrifugal forces on the reaction dynamics. We will proceed with this approximation in mind.

Although it is tedious to perform the coordinate transformation of T outlined above, it has been done and results in the following form for the RPH:

$$H = \sum_{K=1,3N-7} 1/2\left[p_K^2 + \omega_K^2(S)\right] + E(S) + 1/2\left[p_S - \sum_{K,K'=1,3N-7} p_K \delta X_{K'} B_{K,K'}\right]^2 \Big/ 2(1+F), \quad (8.13)$$

where

$$(1+F) = \left[1 + \sum_{K=1,3N-7} \delta X_K B_{K,S}\right]^2. \quad (8.14)$$

In the absence of the so-called dynamical coupling factors $B_{K,K'}$ and $B_{K,S}$, this expression for H describes:

(i) $3N - 7$ harmonic oscillator Hamiltonians $1/2[p_K^2 + \omega_K^2(S)]$ each of which has a locally defined frequency $\omega_K(S)$ that varies along the reaction path (i.e., is S-dependent);
(ii) a Hamiltonian $1/2p_S^2 + E(S)$ for motion along the reaction coordinate S with $E(S)$ serving as the potential.

In this limit (i.e., with the B factors "turned off"), the reaction dynamics can be simulated in what is termed a "vibrationally adiabatic" manner by:

(i) placing each transverse oscillator into a quantum level v_K that characterizes the reactant's population of this mode;
(ii) assigning an initial momentum $p_S(0)$ to the reaction coordinate that is characteristic of the collision to be simulated (e.g., $p_S(0)$ could be sampled from a

Maxwell–Boltzmann distribution if a thermal reaction is of interest, or $p_S(0)$ could be chosen equal to the mean collision energy of a beam-collision experiment);

(iii) time evolving the S and p_S, coordinate and momentum using the above Hamiltonian, assuming that each transverse mode remains in the quantum state v_K that it had when the reaction began.

The assumption that v_K remains fixed, which is why this model is called vibrationally adiabatic, does not mean that the energy content of the Kth mode remains fixed because the frequencies $\omega_K(S)$ vary as one moves along the reaction path. As a result, the kinetic energy along the reaction coordinate $1/2p_S^2$ will change both because $E(S)$ varies along S and because $\sum_{K=1,3N-7} \hbar\omega_K^2(S)[v_K + 1/2]$ varies along S.

Let's return now to the RPH theory in which the dynamical couplings among motion along the reaction path and the modes transverse to it are included. In the full RPH, the terms $B_{K,K'}(S)$ couple modes K and K', while $B_{K,S}(S)$ couples the reaction path to mode K. These couplings are how energy can flow among these various degrees of freedom. Explicit forms for the $B_{K,K'}$ and $B_{K,S}$ factors are given in terms of the eigenvectors $\{\mathbf{v}_K, \mathbf{v}_S\}$ of the mass-weighted Hessian matrix as follows:

$$B_{K,K'} = \left\langle \frac{d\mathbf{v}_K}{dS} \mid \mathbf{v}_{K'} \right\rangle; \qquad B_{K,S} = \left\langle \frac{d\mathbf{v}_K}{dS} \mid \mathbf{v}_S \right\rangle, \tag{8.15}$$

where the derivatives of the eigenvectors $\{d\mathbf{v}_K/dS\}$ are usually computed by taking the eigenvectors at two neighboring points S and S' along the reaction path:

$$\frac{d\mathbf{v}_K}{dS} = \frac{\{\mathbf{v}_K(S') - \mathbf{v}_K(S)\}}{(S' - S)}. \tag{8.16}$$

In summary, once a reaction path has been mapped out, one can compute, along this path, the mass-weighted Hessian matrix and the potential $E(S)$. Given these quantities, all terms in the RPH

$$H = \sum_{K=1,3N-7} \frac{1}{2}\left[p_K^2 + \omega_K^2(S)\right] + E(S) + \frac{1}{2}\frac{\left[p_S - \sum_{K,K'=1,3N-7} p_K \delta X_{K'} B_{K,K'}\right]^2}{(1+F)} \tag{8.17}$$

are in hand.

This knowledge can, subsequently, be used to perform the propagation of a set of classical coordinates and momenta forward in time. For any initial (i.e., $t = 0$) momenta p_S and p_K, one can use the above form for H to propagate the coordinates $\{\delta X_K, \delta S\}$ and momenta $\{p_K, p_S\}$ forward in time. In this manner, one can use the RPH theory to follow the time evolution of a chemical reaction that begins ($t = 0$) with coordinates and moment characteristic of reactants under specified laboratory conditions and moves through a TS and onward to products. Once time has evolved long enough for product geometries to be realized, one can interrogate the values of $1/2[p_K^2 + \omega_K^2(S)]$ to determine how much energy has been deposited into various product-molecule vibrations and of $1/2p_S^2$ to see what the final kinetic energy of the product fragments is. Of course, one also monitors what fraction of the trajectories, whose initial conditions are chosen to represent some experimental situation, progress to product geometries vs. returning to reactant geometries. In this way, one can determine the overall reaction probability.

8.4 Classical dynamics simulation of rates

One can perform classical dynamics simulations of reactive events without using the reaction path Hamiltonian. Following a procedure like that outlined in Chapter 7 where condensed-media MD simulations were discussed, one can time-evolve the Newton equations of motion of the molecular reaction species using, for example, the Cartesian coordinates of each atom in the system and with either a Born–Oppenheimer surface or a parameterized functional form. Of course, it is essential for whatever function one uses to accurately describe the reactive surface, especially near the transition state.

With each such coordinate having an initial velocity $(dq/dt)_0$ and an initial value q_0, one then uses the Newton equations written for a time step of duration δt to propagate q and dq/dt forward in time according, for example, to the following first order propagation formula:

$$q(t+\delta t) = q_0 + \left(\frac{dq}{dt}\right)_0 \delta t, \tag{8.18}$$

$$\frac{dq}{dt}(t+\delta t) = \left(\frac{dq}{dt}\right)_0 - \delta t \left[\frac{(\partial E/\partial q)_0}{m_q}\right]. \tag{8.19}$$

Here m_q is the mass factor connecting the velocity dq/dt and the momentum p_q conjugate to the coordinate q:

$$p_q = m_q \frac{dq}{dt}, \tag{8.20}$$

and $-(\partial E/\partial q)_0$ is the force along the coordinate q at the "initial" geometry q_0. Again, as in Chapter 7, I should note that the above formulas for propagating q and p forward in time represent only the most elementary approach to this problem. There are other, more sophisticated, numerical methods for effecting more accurate and longer-time propagations, but I will not go into them here. Rather, I wanted to focus on the basics of how these simulations are carried out.

By applying the time-propagation process, one generates a set of "new" coordinates $q(t+\delta t)$ and new velocities $dq/dt(t+\delta t)$ appropriate to the system at time $t+\delta t$. Using these new coordinates and momenta as q_0 and $(dq/dt)_0$ and evaluating the forces $-(\partial E/\partial q)_0$ at these new coordinates, one can again use the Newton equations to generate another finite-time-step set of new coordinates and velocities. Through the sequential application of this process, one generates a sequence of coordinates and velocities that simulate the system's dynamical behavior.

In using this kind of classical trajectory approach to study chemical reactions, it is important to choose the initial coordinates and momenta in a way that is representative of the experimental conditions that one is attempting to simulate. The tools of statistical mechanics discussed in Chapter 7 guide us in making these choices. When one attempts, for example, to simulate the reactive collisions of an A atom with a BC molecule to produce AB + C, it is not appropriate to consider a single classical (or quantal) collision between A and BC. Why? Because in any laboratory setting:

(i) The A atoms are probably moving toward the BC molecules with a distribution of relative speeds. That is, within the sample of molecules (which likely contains 10^{10} or more molecules), some A + BC pairs have low relative kinetic energies when they collide, and others have higher relative kinetic energies. There is a probability distribution $P(E_{KE})$ for this relative kinetic energy that must be properly sampled in choosing the initial conditions.
(ii) The BC molecules may not all be in the same rotational (J) or vibrational (v) state. There is a probability distribution function $P(J, v)$ describing the fraction of BC molecules that are in a particular J state and a particular v state. Initial values of the BC molecule's internal vibrational coordinate and momentum as well as its orientation and rotational angular momentum must be selected to represent this $P(J, v)$.
(iii) When the A and BC molecules collide with a relative motion velocity vector $\boldsymbol{v}$, they do not all hit "head on". Some collisions have small impact parameter b (the closest distance from A to the center of mass of BC if the collision were to occur with no attractive or repulsive forces), and some have large b-values (see Fig. 8.2).

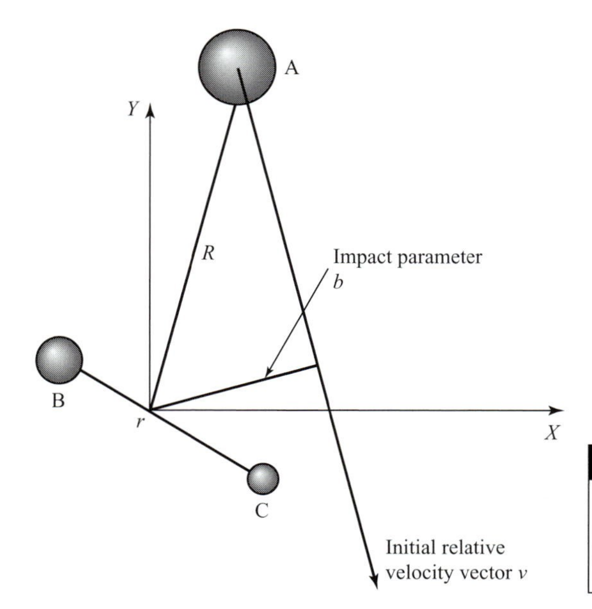

Figure 8.2 Coordinates needed to characterize an atom–diatom collision showing the impact parameter *b*.

The probability function for these impact parameters is $P(b) = 2\pi b\, db$, which is simply a statement of the geometrical fact that larger b-values have more geometrical volume element than smaller b-values.

So, to simulate the entire ensemble of collisions that occur between A atoms and BC molecules in various J, v states and having various relative kinetic energies E_{KE} and impact parameters b, one must:

(i) run classical trajectories (or quantum propagations) for a large number of J, v, E_{KE}, and b values,
(ii) with each such trajectory assigned an overall weighting (or importance) of

$$P_{total} = P(E_{KE})P(J, v)2\pi b\, db. \tag{8.21}$$

After such an ensemble of trajectories representative of an experimental condition has been carried out, one has available a great deal of data. This data includes knowledge of what fraction of the trajectories produced final geometries characteristic of products, so the net reaction probability can be calculated. In addition, the kinetic and potential energy content of the internal (vibrational and rotational) modes of the product molecules can be interrogated and used to compute probabilities for observing products in these states. This is how classical dynamics simulations allow us to study chemical reactions and/or energy transfer.

8.5 RRKM theory

Another theory that is particularly suited for studying unimolecular decomposition reactions is named after the four scientists who developed it – Rice, Ramsperger, Kassel, and Marcus. To use this theory, one imagines an ensemble of molecules that have been "activated" to a state in which they possess a specified total amount of internal energy E of which an amount E^*_{rot} exists as rotational energy and the remainder as internal vibrational energy.

The mechanism by which the molecules become activated could involve collisions or photochemistry. It does not matter as long as enough time has passed to permit one to reasonably assume that these molecules have the energy $E - E^*_{rot}$ distributed randomly among all their internal vibrational degrees of freedom. When considering thermally activated unimolecular decomposition of a molecule, the implications of such assumptions are reasonably clear. For photochemically activated unimolecular decomposition processes, one usually also assumes that the molecule has undergone radiationless relaxation and returned to its ground electronic state but in a quite vibrationally "hot" situation. That is, in this case, the molecule contains excess vibrational energy equal to the energy of the optical photon used to excite it. Finally, when applied to bimolecular reactions, one assumes that collision between the two fragments results in a long-lived complex. The lifetime of this intermediate must be long enough to allow the energy $E - E^*_{rot}$, which is related to the fragments' collision energy, to be randomly distributed among all vibrational modes of the collision complex.

For bimolecular reactions that proceed "directly" (i.e., without forming a long-lived intermediate), one does not employ RRKM-type theories because their primary assumption of energy randomization almost certainly would not be valid in such cases.

The RRKM expression of the unimolecular rate constant for activated molecules A* (i.e., either a long-lived complex formed in a bimolecular collision or a "hot" molecule) dissociating to products through a transition state, $A^* \to TS \to P$, is

$$k_{rate} = \frac{G(E - E_0 - E'_{tot})}{N(E - E^*_{rot})h}. \tag{8.22}$$

Here, the total energy E is related to the energies of the activated molecules by

$$E = E^*_{rot} + E^*_{vib} \tag{8.23}$$

where E^*_{rot} is the rotational energy of the activated molecule and E^*_{vib} is the vibrational energy of this molecule. This same energy E must, of course, appear in the transition state where it is decomposed as an amount E_0 needed to move from A* to the TS (i.e, the energy needed to reach the barrier) and vibrational (E'_{vib}), translational (E'_{trans} along the reaction coordinate), and rotational (E'_{rot}) energies:

$$E = E_0 + E'_{vib} + E'_{trans} + E'_{rot}. \tag{8.24}$$

In the rate coefficient expression, $G(E - E_0 - E'_{rot})$ is the total sum of internal vibrational quantum states that the transition state possesses having energies up to and including $E - E_0 - E'_{rot}$. This energy is the total energy E but with the activation energy E_0 removed and the overall rotational energy E'_{rot} of the TS removed. The quantity $N(E - E^*_{rot})$ is the density of internal vibrational quantum states (excluding the mode describing the reaction coordinate) that the activated molecule possesses having an energy between $E - E^*_{rot}$ and $E - E^*_{rot} + dE$. In this expression, the energy $E - E^*_{rot}$ is the total energy E with the rotational energy E^*_{rot} of the activated species removed.

In the most commonly employed version of RRKM theory, the rotational energies of the activated molecules E^*_{rot} and of the TS E'_{rot} are assumed to be related by

$$E^*_{rot} - E'_{rot} = J(J+1)h^2/8\pi^2\{1/I^* - 1/I'\} = E^*_{rot}\{1 - I^*/I'\}. \tag{8.25}$$

Here I^* and I' are the average (taken over the three eigenvalues of the moment inertia tensors) moments of inertia of the activated molecules and TS species, respectively. The primary assumption embodied in the above relationship is that the rotational angular momenta of the activated and TS species are the same, so their rotational energies can be related, as expressed in the equation, to changes in geometries as reflected in their moments of inertia. Because RRKM theory assumes that the vibrational energy is randomly distributed, its fundamental rate coefficient equation $k_{\rm rate} = G(E - E_0 - E'_{\rm tot})/(N(E - E^*_{\rm rot})h)$ depends on the total energy E, the energy E_0 required to access the TS, and the amount of energy contained in the rotational degrees of freedom that is thus not available to the vibrations.

To implement a RRKM rate coefficient calculation, one must know:

(i) the total energy E available,
(ii) the barrier energy E_0,
(iii) the geometries (and hence the moments of inertia I^* and I') of the activated molecules and of the TS, respectively,
(iv) the rotational energy $E^*_{\rm rot}$ of the activated molecules, as well as
(v) all $3N - 6$ vibrational energies of the activated molecules and all $3N - 7$ vibrational energies of the TS (i.e., excluding the reaction coordinate).

The rotational energy of the TS species can then be related to that of the activated molecules through $E^*_{\rm rot} - E'_{\rm rot} = E^*_{\rm rot}\{1 - I^*/I'\}$.

To simulate an experiment in which the activated molecules have a thermal distribution of rotational energies, the RRKM rate constant is computed for a range of $E^*_{\rm rot}$ values and then averaged over $E^*_{\rm rot}$ using the thermal Boltzmann population $(2J + 1)\exp[-J(J + 1)h^2/(8\pi^2 I^* kT)]$ as a weighting factor. This can be carried out, for example, using the MC process for selecting rotational J values. This then produces a rate constant for any specified total energy E. Alternatively, to simulate experiments in which the activated species are formed in bimolecular collisions at a specified energy E, the RRKM rate coefficient is computed for a range of $E^*_{\rm rot}$ values with each $E^*_{\rm rot}$ related to the collisional impact parameter b that we discussed earlier. In that case, the collisional angular momentum J is given as $J = \mu v b$, where v is the collision speed (related to the collision energy) and μ is the reduced mass of the two colliding fragments. Again using $E^*_{\rm rot} - E'_{\rm rot} = E^*_{\rm rot}\{1 - I^*/I'\}$ the TS rotational energy can be related to that of the activated species. Finally, the RRKM rate coefficient is evaluated by averaging the result over a series of impact parameters b (each of which implies a J value and thus an $E^*_{\rm rot}$) with $2\pi b\, db$ as the weighting factor.

The evaluation of the sum of states $G(E - E_0 - E'_{\rm tot})$ and the density of states $N(E - E^*_{\rm rot})$ that appear in the RRKM expression is usually carried out using a state-counting algorithm such as that implemented by Beyer and Swinehart (*Commun. Assoc. Comput. Machin.* **16**, 372 (1973)). This algorithm uses knowledge of the $3N - 6$ harmonic vibrational frequencies of the activated molecules and the $3N - 7$ frequencies of the TS and determines how many ways a given amount of energy can be distributed among these modes. By summing over all such distributions for energy varying from zero to E, the algorithm determines $G(E)$. By taking the difference $G(E + \delta E) - G(E)$, it determines $N(E)\,\delta E$.

8.6 Correlation function expressions for rates

Recall from Chapter 6 that rates of photon absorption can, in certain circumstances, be expressed either (a) in terms of squares of transition dipole matrix elements connecting each initial state Φ_i to each final state Φ_f:

$$|\mathbf{E}_0 \cdot \langle \Phi_\mathrm{f} | \mu | \Phi_\mathrm{i} \rangle|^2 \tag{8.26}$$

or (b) in terms of the equilibrium average of the product of a transition dipole vector at time $t = 0$ dotted into this same vector at another time t:

$$\sum_\mathrm{i} \rho_\mathrm{i} \langle \Phi_\mathrm{i} | \mathbf{E}_0 \cdot \mu \mathbf{E}_0 \cdot \mu(t) | \Phi_\mathrm{i} \rangle. \tag{8.27}$$

That is, these rates can be expressed either in a state-to-state manner or in a time-dependent correlation function framework. In Chapter 7, this same correlation function approach was examined further.

In an analogous fashion, it is possible to express chemical reaction rate constants in a time-domain language again using time correlation functions. The TST (or VTST) and RRKM expressions for the rate constant k_rate all involve, through the partition functions or state densities, the reactant and transition-state energy levels and degeneracies. These theories are therefore analogs of the state-to-state photon-absorption rate equations.

To make the connection between the state-to-state and time-correlation function expressions, one can begin with a classical expression for the rate constant given below:

$$k(T) = Q_\mathrm{r}^{-1}(2\pi\hbar)^{-L} \int dp\, dq\, e^{-\beta H(q,p)} F(p,q)\, \chi(p,q). \tag{8.28}$$

Here Q_r is the partition function of the reactant species, L is the number of coordinates and momenta upon which the Hamiltonian $H(\mathbf{p}, \mathbf{q})$ depends, and β is $1/kT$. The flux factor F and the reaction probability χ are defined in terms of a dividing surface which could, for example, be a plane perpendicular to the reaction coordinate S and located along the reaction path that was discussed earlier in this chapter in Section 8.3. Points on such a surface can be defined by specifying one condition that the L coordinates $\{q_j\}$ must obey, and we write this condition as

$$f(\mathbf{q}) = 0. \tag{8.29}$$

Points lying where $f(\mathbf{q}) < 0$ are classified as lying in the reactant region of coordinate space, while those lying where $f > 0$ are in the product region. For example, if the dividing surface is defined as mentioned above as being a plane perpendicular to the reaction path, the function f can be written as:

$$f(\mathbf{q}) = (S - S_0). \tag{8.30}$$

Here, S is the reaction coordinate (which, of course, depends on all of the $\mathbf{q}$ variables) and S_0 is the value of S at the dividing surface. If the dividing surface is placed at the transition state on the energy surface, S_0 vanishes because the transition state is then, by convention, the origin of the reaction coordinate.

So, now we see how the dividing surface can be defined, but how are the flux F and probability χ constructed? The flux factor F is defined in terms of the dividing surface function $f(\mathbf{q})$ as follows:

$$\begin{aligned} F(\mathbf{p}, \mathbf{q}) &= \frac{dh(f(\mathbf{q}))}{dt} \\ &= \left(\frac{dh}{df}\right)\left(\frac{df}{dt}\right) \\ &= \left(\frac{dh}{df}\right)\sum_j \frac{\partial f}{\partial q_j}\left(\frac{dq_j}{dt}\right) \\ &= \delta(f(\mathbf{q}))\sum_j \frac{\partial f}{\partial q_j}\left(\frac{dq_j}{dt}\right). \end{aligned} \tag{8.31}$$

Here, $h(f(\mathbf{q}))$ is the Heaviside step function ($h(x) = 1$ if $x > 0; h(x) = 0$ if $x < 0$), whose derivative $dh(x)/dx$ is the Dirac delta function $\delta(x)$, and the other identities follow by using the chain rule. When the dividing surface is defined in terms of the reaction path coordinate S as introduced earlier (i.e., $f(\mathbf{q}) = (S - S_0)$), the factor $\sum_j \partial f/\partial q_j\,(dq_j/dt)$ contains only one term when the L coordinates $\{q_j\}$ are chosen, as in the reaction path theory, to be the reaction coordinate S and $L - 1$ coordinates $\{q'_j\} = \mathbf{q}'$ perpendicular to the reaction path. For such a choice, one obtains

$$\sum_j \frac{\partial f}{\partial q_j}\left(\frac{dq_j}{dt}\right) = \frac{dS}{dt} = \frac{P_S}{m_S}, \tag{8.32}$$

where P_S is the momentum along S and m_S is the mass factor associated with S in the reaction path Hamiltonian. So, in this case, the total flux factor F reduces to

$$F(\mathbf{p}, \mathbf{q}) = \delta(S - S_0)P_S/m_S. \tag{8.33}$$

We have seen exactly this construct before in Section 8.1 where the TST expression for the rate coefficient was developed.

The reaction probability factor $\chi(\mathbf{p}, \mathbf{q})$ is defined in terms of those trajectories that evolve, at long time $t \to \infty$, onto the product side of the dividing surface; such trajectories obey

$$\chi(\mathbf{p}, \mathbf{q}) = \lim_{t\to\infty} h(f(\mathbf{q}(t))) = 1. \tag{8.34}$$

This long-time limit can, in turn, be expressed in a form where the flux factor again occurs:

$$\lim_{t\to\infty} h(f(\mathbf{q}(t))) = \int_0^\infty \frac{dh(f(\mathbf{q}(t)))}{dt} dt = \int_0^\infty F dt. \tag{8.35}$$

In this expression, the flux $F(t)$ pertains to coordinates $\mathbf{q}(t)$ and momenta $\mathbf{p}(t)$ at $t > 0$. Because of time reversibility, the integral can be extended to range from $t = -\infty$ to $t = \infty$.

Using the expressions for χ and for F as developed above in the equation for the rate coefficient given at the beginning of this section allows the rate coefficient $k(T)$ to be rewritten as follows:

$$\begin{aligned} k(T) &= Q_{\mathrm{r}}^{-1}(2\pi\hbar)^{-L} \int d\mathbf{p}\, d\mathbf{q}\, e^{-\beta H(\mathbf{q},\mathbf{p})} F(\mathbf{p},\mathbf{q})\chi(\mathbf{p},\mathbf{q}) \\ &= Q_{\mathrm{r}}^{-1}(2\pi\hbar)^{-L} \int_{-\infty}^{\infty} dt \int d\mathbf{p}\, d\mathbf{q} e^{-\beta H(\mathbf{q},\mathbf{p})}\, F(\mathbf{p},\mathbf{q}) F(\mathbf{p}(t),\mathbf{q}(t)). \end{aligned} \tag{8.36}$$

In this form, the rate constant $k(T)$ appears as an equilibrium average (represented by the integral over the initial values of the variables $\mathbf{p}$ and $\mathbf{q}$ with the $Q_{\mathrm{r}}^{-1}(2\pi\hbar)^{-L} \exp(-\beta H)$ weighting factor) of the time correlation function of the flux F:

$$\int_{-\infty}^{\infty} dt\ F(\mathbf{p},\mathbf{q}) F(\mathbf{p}(t),\mathbf{q}(t)). \tag{8.37}$$

To evaluate the rate constant in this time-domain framework for a specific chemical reaction, one would proceed as follows.

(i) Run an ensemble of trajectories whose initial coordinates and momenta $\{\mathbf{q},\mathbf{p}\}$ are selected (e.g., using Monte-Carlo methods discussed in Chapter 7) from a distribution with $\exp(-\beta H)$ as its weighting factor.
(ii) Make sure that the initial coordinates $\{\mathbf{q}\}$ lie on the dividing surface because the flux (xpression contains the $\delta(f(\mathbf{q}))$ factor.
(iii) Monitor each trajectory to observe when it again crosses the dividing surface (i.e., when $\{\mathbf{q}(t)\}$ again obeys $f(\mathbf{q}(t)) = 0$ at which time the quantity
(iv) $F(\mathbf{p}(t),\mathbf{q}(t))$ can be evaluated as $F(\mathbf{p},\mathbf{q}) = \delta(f(\mathbf{q}))\sum_j \partial f/\partial q_j (dq_j/dt)$, using the coordinates and momenta at time t to compute these quantities.

Using a planar dividing surface attached to the reaction path at $S = S_0$ as noted earlier allows $F(\mathbf{q},\mathbf{p})$ to be calculated in terms of the initial ($t = 0$) momentum lying along the reaction path direction as $F(\mathbf{p},\mathbf{q}) = \delta(S - S_0)P_S/m_S$ and permits $F(\mathbf{p}(t),\mathbf{q}(t))$ to be computed when the trajectory again crosses this surface at time t as $F(\mathbf{p}(t),\mathbf{q}(t)) = \delta(S - S_0)P_S(t)/m_S$. So, all that is really needed if the dividing surface is defined in this manner is to start trajectories with $S = S_0$; to keep track of the initial momentum along S; to determine at what times t the trajectory returns to $S = S_0$; and to form the product $(P_S/m_S)(P_S(t)/m_S)$ for each such time. It is in this manner that one can compute flux–flux correlation functions and, thus, the rate coefficient.

Notice that trajectories that undergo surface re-crossings contribute negative terms to the flux–flux correlation function computed as discussed above. That is, a trajectory with a positive initial value of (P_S/m_S) can, at some later time t, cross the dividing surface with a negative value of $(P_S(t)/m_S)$ (i.e, be directed back toward reactants). This re-crossing will contribute a negative value, via the product $(P_S/m_S)(P_S(t)/m_S)$, to the total correlation function, which integrates over all times. Of course, if this same trajectory later undergoes yet another crossing of the dividing surface at t' with positive $P_S(t')$, it will contribute a positive term to the correlation function via $(P_S/m_S)(P_S(t')/m_S)$. Thus, the correlation function approach to computing the rate coefficient can properly account for surface re-crossings, unlike the TST which requires one to account for such effects (and tunneling) in the transmission coefficient κ.

8.7 Wave packet propagation

The preceding discussion should have made it clear that it is very difficult to propagate wave functions rigorously using quantum mechanics. On the other hand, to propagate a classical trajectory is relatively straightforward. There exists a powerful tool that allows one to retain much of the computational ease and convenient interpretation of the classical trajectory approach while also incorporating quantum effects that are appropriate under certain circumstances. In this wave packet propagation approach, one begins with a quantum mechanical wave function that is characterized by two parameters that give the average value of the position and momentum along each coordinate. One then propagates not the quantum wave function but the values of these two parameters, which one assumes will evolve according to Newtonian dynamics. Let's see how these steps are taken in more detail and try to understand when such an approach is expected to work or to fail.

First, the form of the so-called wave packet quantum function is written as follows:

$$\Psi(\mathbf{q}, \mathbf{Q}, \mathbf{P}) = \prod_{J=1,N} \left(2\pi \langle \delta q_J^2 \rangle\right)^{-1/2} \exp\left[(i P_J q_J/\hbar) - (q_J - Q_J)^2/4\langle \delta q_J^2 \rangle\right]. \quad (8.38)$$

Here, we have a total of N coordinates that we denote $\{q_J : J = 1, N\}$. It is these coordinates that the quantum wave function depends upon. The total wave function is a product of terms, one for each coordinate. Notice that this wave function has two distinct ways in which the coordinates q_J appear. First, it has a Gaussian ($\exp[-(q_J - Q_J)^2/4\langle\delta q_J^2\rangle]$) dependence centered at the values Q_J and having Gaussian width factors related to $\langle\delta q_J^2\rangle$. This dependence tends to make the wave function's amplitude largest when q_J is close to Q_J. Second, it has a form $\exp[(i P_J q_J/\hbar)]$ that looks like the traveling wave that we encountered in the Background Material in which the coordinate q_J moves with momentum P_J. So, these wave packet functions have built into them characteristics that allow them to describe motion (via the P_J) of an amplitude that is centered at Q_J with a width given by the parameter $\langle \delta q_J^2 \rangle$.

The parameters P_J and Q_J we assume in this approach to chemical dynamics will undergo classical time evolution according to the Newton equations:

$$\frac{dQ_J}{dt} = \frac{P_J}{m_J}, \tag{8.39}$$

$$\frac{dP_J}{dt} = -\frac{\partial E}{\partial Q_J}, \tag{8.40}$$

where E is the potential energy surface (Born–Oppenheimer or force field) upon which we wish to propagate the wave packet, and m_J is the mass associated with coordinate q_J. The Q_J and P_J parameters can be shown to be the expectation values of the coordinates q_J and momenta $-i\hbar\partial/\partial q_J$ for the above function:

$$Q_J = \int \Psi^* q_J \Psi d\mathbf{q}, \tag{8.41}$$

$$P_J = \int \Psi^*(-i\hbar\partial/\partial q_J)\Psi d\mathbf{q}. \tag{8.42}$$

Moreover, the $\langle \delta q_J^2 \rangle$ parameter appearing in the Gaussian part of the function can be shown to equal the dispersion or "spread" of this wave function along the coordinate q_J:

$$\langle \delta q_J^2 \rangle = \int \Psi^*(q_J - Q_J)^2 \Psi d\mathbf{q}. \tag{8.43}$$

There is an important characteristic of the above Gaussian wave packet functions that we need to point out. It turns out that functions of this form:

$$\Psi(\mathbf{q}, \mathbf{Q(t)}, \mathbf{P(t)}) = \prod_{J=1,N} \left(2\pi \langle \delta q_J^2 \rangle\right)^{-1/2} \exp\left[(iP_J(t)q_J/\hbar) - (q_J - Q_J(t))^2/4 \langle \delta q_J^2 \rangle\right] \tag{8.44}$$

can be shown to have uncertainties in q_J and in $-i\hbar\partial/\partial q_J$ that are as small as possible:

$$\langle (q_J - Q_J)^2 \rangle \langle (-i\hbar\partial/\partial q_J - P_J)^2 \rangle = \hbar^2/4. \tag{8.45}$$

The Heisenberg uncertainty relation, which is discussed in many texts dealing with quantum mechanics, says that this product of coordinate and momentum dispersions must be greater than or equal to $\hbar^2/4$. In a sense, the Gaussian wave packet function is the most classical function that one can have because its uncertainty product is as small as possible (i.e, equals $\hbar^2/4$). We say this is the most classical possible quantum function because in classical mechanics, both the coordinate and the momentum can be known precisely. So, whatever quantum wave function allows these two variables to be least uncertain is the most classical.

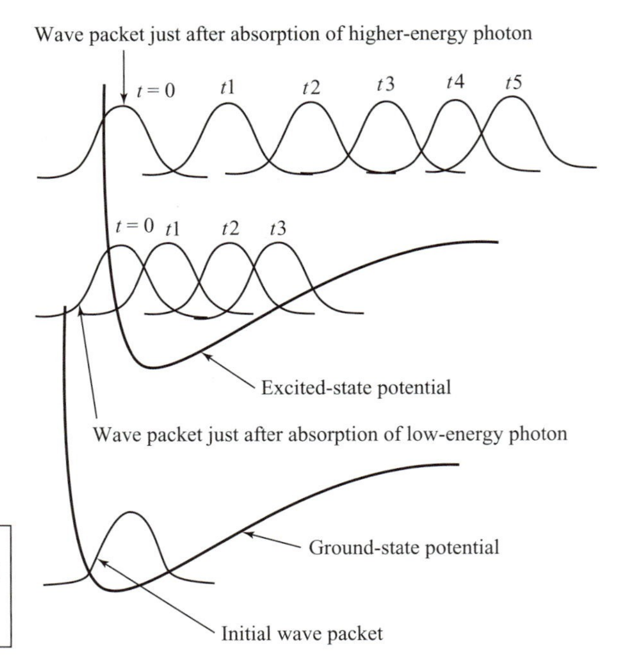

Figure 8.3 Propagation of wave packet prepared by absorption of two different photons.

To use wave packet propagation to simulate a chemical dynamics event, one begins with a set of initial classical coordinates and momenta $\{Q_J(0), P_J(0)\}$ as well as a width $\langle \delta q_J^2 \rangle$ or uncertainty for each coordinate. Each width must be chosen to represent the range of that coordinate in the experiment that is to be simulated. For example, assume one were to represent the dynamics of a wave function that is prepared by photon absorption of a $v = 0$ vibrational state of the HCl molecule from the ground $^1\sum$ state to an excited-state energy surface ($E(R)$). Such a situation is described qualitatively in Fig. 8.3. In this case, one could choose $\langle \delta R^2 \rangle$ to be the half-width of the $v = 0$ harmonic (or Morse) oscillator wave function $\chi_0(R)$ of HCl, and take $P(0) = 0$ (because this is the average value of the momentum for χ_0) and $R(0) = R_{\text{eq}}$, the equilibrium bond length.

For such initial conditions, classical Newtonian dynamics would then be used to propagate the Q_J and P_J. In the HCl example, introduced above, this propagation would be performed using the excited-state energy surface for E since, for $t > 0$, the molecule is assumed to be on this surface. The total energy at which the initial wave packet it delivered to the upper surface would be dictated by the energy of the photon used to perform the excitation. In Fig. 8.3, two such examples are shown.

Once the packet is on the upper surface, its position Q and momentum P begin to change according to the Newton equations. This, in turn, causes the packet to move as shown for several equally spaced time "steps" in Fig. 8.3 for the two different photons' cases. At such subsequent times, the quantum wave function is then assumed, within this model, to be given by

$$\Psi(\mathbf{q}, \mathbf{Q(t)}, \mathbf{P(t)}) = \prod_{J=1,N} \left(2\pi \langle \delta q_J^2 \rangle\right)^{-1/2} \exp\left[(i P_J(t) q_J/\hbar) - (q_J - Q_J(t))^2/4\langle \delta q_J^2 \rangle\right]. \tag{8.46}$$

That is, it is taken to be of the same form as the initial wave function but to have simply moved its center from $Q(0)$ to $Q(t)$ with a momentum that has changed from $P(0)$ to $P(t)$. It should be noticed that the time evolution of the wave packet shown in Fig. 8.3 displays clear classical behavior. For example, as time evolves, it moves to large R-values and its speed (as evidenced by the spacings between neighboring packets for equal time steps) is large when the potential is low and small when the potential is higher. As we learned in the Background Material and in Chapter 6, the time correlation function

$$C(t) = \langle \Psi(q, Q(0), P(0)) \mid \Psi(q, Q(t), P(t)) \rangle \tag{8.47}$$

can be used to extract spectral information by Fourier transformation. For the HCl example considered here, this correlation function will be large at $t = 0$ but will decay in magnitude as the wave packet $\Psi(q, Q(t), P(t))$ moves to the right (at $t1$, $t2$, etc.) because its overlap with $\Psi(q, Q(0), P(0))$ becomes smaller and smaller as time evolves. This decay in $C(t)$ will occur more rapidly for the high-energy photon case because $\Psi(q, Q(t), P(t))$ moves to the right more quickly because the classical momentum $P(t)$ grows more rapidly. These dynamics will induce exponential decays in $C(t)$ (i.e., $C(t)$ will vary as $\exp(-t/\tau_1)$) at short times.

In fact, the decay of $C(t)$ discussed above produces, when $C(t)$ is Fourier transformed, the primary characteristic of the correlation function for the higher-energy photon case where dissociation ultimately occurs. In such photodissociation spectra, one observes a Lorentzian line shape whose width is characterized by the decay rate $(1/\tau_1)$, which, in turn, relates to the total energy of the packet and the steepness of the excited-state surface. This steepness determines how fast $P(t)$ grows, which then determines how fast the H—Cl bond fragments.

In the lower-energy photon case shown in Fig. 8.3, a qualitatively different behavior occurs in $C(t)$ and thus in the spectrum. The packet's movement to larger R causes $C(t)$ to initially undergo $\exp(-t/\tau_1)$ decay. However, as the packet moves to its large-R turning point (shortly after time $t3$), it strikes the outer wall of the surface where it is reflected. Subsequently, it undergoes motion to smaller R, eventually returning to its initial location. Such recurrences, which occur on time scales that we denote τ_2, are characteristic of bound motion in contrast to the directly dissociative motion discussed earlier. This recurrence will cause $C(t)$ to again achieve a large amplitude, but $C(t)$ will subsequently again undergo $\exp(-t/\tau_1)$ decay as the packet once again departs. Clearly, the correlation function will display a series of recurrences followed by exponential decays. The frequency of the recurrences is determined by the frequency with which the packet traverses from its inner to outer turning points and back again, which is proportional to $1/\tau_2$. This, of course, is the vibrational period of the H—Cl bond. So, in such bound-motion cases, the spectrum (i.e., the Fourier transform of $C(t)$) will display a series of peaks spaced by $(1/\tau_2)$ with the envelope of such peaks having a width determined by $1/\tau_1$.

In more complicated multi-mode cases (e.g., in molecules containing several coordinates), the periodic motion of the wave packet usually shows another feature that we have not yet discussed. Let us, for simplicity, consider a case in which only two coordinates are involved. For the wave packet to return to (or near) its initial location enough time must pass for both coordinates to have undergone an excursion to their turning points and back. For example, consider the situation in which one coordinate's vibrational frequency is *ca.* 1000 cm^{-1} and the other's is 300 cm^{-1}; these two modes then require *ca.* 1/30 ps and 1/9 ps, respectively, to undergo one complete oscillation. At $t = 0$, the wave packet, which is a product of two packets, $\prod_{J=1,2}(2\pi\langle\delta q_J^2\rangle)^{-1/2}\exp[(iP_J(t)q_J/\hbar) - (q_J - Q_J(t))^2/4\langle\delta q_J^2\rangle]$, one for each mode, produces a large $C(t)$. After 1/30 ps, the first mode's coordinate has returned to its initial location, but the second mode is only 9/30 of the way along in its periodic motion. Moreover, after 1/9 ps, the second mode's coordinate has returned to near where it began, but now the first mode has moved away. So, at both 1/30 ps and 1/9 ps, the correlation function will not be large because one of the mode contributions to $C(t) = \langle\Psi(q, Q(0), P(0)) \mid \Psi(q, Q(t), P(t))\rangle$ will be small. However, after 1/3 ps, both modes' coordinates will be in positions to produce a large value of $C(t)$; the high-frequency mode will have undergone 10 oscillations, and the lower-frequency mode will have undergone 3 oscillations. My point in discussing this example has been to illustrate that molecules having many coordinates can produce spectra that display rather complicated patterns but which, in principle, can be related to the time evolution of these coordinates using the correlation function's connection to the spectrum.

Of course, there are problems that arise in using the wave packet function to describe the time evolution of a molecule (or any system that should be treated using quantum mechanics). One of the most important limitations of the wave packet approach to be aware of relates to its inability to properly treat wave reflections. It is well known that when a wave strikes a hard wall it is reflected by the wall. However, when, for example, a water wave moves suddenly from a region of deep water to a much more shallow region, one observes both a reflected and a transmitted wave. In the discussion of tunneling resonances given in the Background Material, we also encountered reflected and transmitted waves. Furthermore, when a wave strikes a barrier that has two or more holes or openings in it, one observes wave fronts coming out of these openings. The problem with the most elementary form of wave packets presented above is that each packet contains only one "piece". It therefore can not break into two or more "pieces" as it, for example, reflects from turning points or passes through barriers with holes. Because such wave packets can not fragment into two or more packets that subsequently undergo independent dynamical evolution, they are not able to describe dynamical processes that require multiple-fragmentation events. It is primarily for this reason that wave packet approaches to simulating dynamics are usually restricted to treating short-time dynamics where such fragmentation of the wave packet is less likely to occur. Prompt molecular photodissociation processes such as we discussed above are a good example of such a short-time phenomenon.

8.8 Surface hopping dynamics

There are, of course, chemical reactions and energy transfer collisions in which two or more Born–Oppenheimer energy surfaces are involved. Under such circumstances, it is essential to have available the tools needed to describe the coupled electronic and nuclear-motion dynamics appropriate to this situation.

The way this problem is addressed is by returning to the Schrödinger equation before the single-surface Born–Oppenheimer approximation was made and expressing the electronic wave function $\Psi(\mathbf{r} \mid \mathbf{R})$, which depends on the electronic coordinates $\{\mathbf{r}\}$ and the nuclear coordinates $\{\mathbf{R}\}$, as

$$\Psi(\mathbf{r} \mid \mathbf{R}) = \sum_J a_J(t)\psi_J(\mathbf{r} \mid \mathbf{R}) \exp\left[-i/\hbar \int H_{J,J}(\mathbf{R})dt\right]. \tag{8.48}$$

Here, $\psi_J(\mathbf{r} \mid \mathbf{R})$ is the Born–Oppenheimer electronic wave function belonging to the Jth electronic state, and $H_{J,J}(\mathbf{R})$ is the expectation value of H for this state. The $a_J(t)$ are amplitudes that will eventually relate to the probability that the system is "on" the Jth surface. Substituting this expansion into the time-dependent Schrödinger equation

$$i\hbar\frac{\partial\Psi}{\partial t} = H\Psi \tag{8.49}$$

followed by multiplying on the left by $\psi_K^*(\mathbf{r} \mid \mathbf{R})$ and integrating over the electronic coordinates $\{\mathbf{r}\}$, gives an equation for the $a_K(t)$ amplitudes:

$$i\hbar\frac{da_K}{dt} = \sum_{J\neq K}\left\{H_{K,J} - i\hbar\left\langle\psi_K \mid \frac{d\psi_J}{dt}\right\rangle\right\}\exp\left[\frac{-i}{\hbar}\int[H_{J,J}(\mathbf{R}) - H_{K,K}(\mathbf{R})]\,dt\right]a_J. \tag{8.50}$$

Here, $H_{K,J}$ is the Hamiltonian matrix that couples ψ_K to ψ_J. The integral appearing in the exponential is taken from an initial time t_i when one is assumed to know that the system resides on a particular Born–Oppenheimer surface (K) up to the time t, at which the amplitude for remaining on this surface a_K as well as the amplitudes for hopping to other surfaces $\{a_J\}$ are needed. This differential equation for the amplitudes is solved numerically by starting at t_i with $a_K = 1$ and $a_{J\neq K} = 0$ and propagating the amplitudes' values forward in time.

The next step is to express $\langle\psi_K \mid d\psi_J/dt\rangle$, using the chain rule, in terms of derivatives with respect to the nuclear coordinates $\{\mathbf{R}\}$ and the time rate of change of these coordinates:

$$\left\langle \psi_K \mid \frac{d\psi_J}{dt} \right\rangle = \sum_a \left\langle \psi_K \mid \frac{d\psi_J}{dR_a} \right\rangle \frac{dR_a}{dt}. \tag{8.51}$$

This is how a classical dynamical treatment of the nuclear motions will be introduced (i.e., by assuming that the nuclear coordinates R_a and dR_a/dt obey the Newton equations). So, now the equations for the $a_K(t)$ read as follows:

$$i\hbar \frac{da_K}{dt} = \sum_{J \neq K} \left\{ H_{K,J} - i\hbar \sum_a \frac{dR_a}{dt} \left\langle \psi_K \mid \frac{d\psi_J}{dR_a} \right\rangle \right\} \times \exp\left[\frac{-i}{\hbar} \int [H_{J,J}(\mathbf{R}) - H_{K,K}(\mathbf{R})]\, dt \right] a_J. \tag{8.52}$$

The $\langle \psi_K | d\psi_J / dR_a \rangle$ are called non-adiabatic coupling matrix elements, and it is their magnitudes that play a central role in determining how efficient surface hoppings are. These matrix elements are becoming more commonly available in widely utilized quantum chemistry and dynamics computer packages (although their efficient evaluation remains a challenge that is undergoing significant study).

In addition to the above prescription for calculating amplitudes (the probabilities then being computed as $|a_J|^2$), one also needs to identify (using, perhaps the kind of strategy discussed in Chapter 3) the seam at which the surfaces of interest intersect. Let us, for example, assume that there are two surfaces that undergo such an intersection in a sub-space of the $3N - 6$ dimensional energy surfaces $H_{1,1}$ and $H_{2,2}$, and let us denote nuclear coordinates $\{R_a\}$ that lie on this seam as obeying

$$F(R_a) = 0, \tag{8.53}$$

following the discussion of Chapter 3.

To utilize the most basic form of surface hopping theory, one proceeds as follows:

(i) One begins with initial values of the nuclear coordinates $\{R_a\}$ and their velocities $\{dR_a/dt\}$ that properly characterize the kind of collision or reaction one wishes to simulate. Of course, one may have to perform many such surface hopping trajectories with an ensemble of initial conditions chosen to properly describe such an experimental situation. In addition, one specifies which electronic surface (say the Kth surface) the system is initially on.

(ii) For each such set of initial conditions, one propagates a classical trajectory describing the time evolution of the $\{R_a\}$ and $\{dR_a/dt\}$ on this initial surface, until such a trajectory approaches a geometry lying on the intersection seam.

(iii) As one is propagating the classical trajectory from t_i up to the time t_1 when it approaches the seam, one also propagates the (in this example, two) differential equations $i\hbar\, da_K/dt = \sum_{J \neq K} \{H_{K,J} - i\hbar \Sigma_a\, dR_a/dt \langle \psi_K \mid d\psi_J/dR_a \rangle\}$ $\exp[-i/\hbar \int [H_{J,J}(\mathbf{R}) - H_{K,K}(\mathbf{R})] dt] a_J$ that produce the time evolution of the $\{a_J\}$ amplitudes.

(iv) Once the classical trajectory reaches the seam, the initial trajectory is halted and two (in the case of M coupled surfaces, M) new trajectories are initiated, one beginning (at t_1) on each of the (two in our case) coupled surfaces. Each of these new trajectories (two) is assigned a probability weighting factor computed as the square of the amplitude belonging to the surface upon which the trajectory now resides: $|a_J|^2$.

(v) Each of the new trajectories is subsequently allowed to propagate from t_1 (where the nuclear coordinates are what they were just prior to dividing the one initial trajectory into two (or M) trajectories) but now on different Born–Oppenheimer surfaces. This propagation is continued until one of these trajectories again approaches a geometry on the intersection seam, at which time the trajectory hopping scheme is repeated.

Clearly, it is possible that the one initial trajectory will generate a large number of subsequent trajectories each of which is assigned an overall weighting factor taken to be the product of all of the $|a_J|^2$ probabilities associated with trajectories from which this particular trajectory was descended. It is these overall weightings that one uses to predict, once this process has been carried out long enough to allow the reaction or collision to proceed to completion, the final yield of products in each Born–Oppenheimer state. This surface hopping algorithm remains one of the most widely used approaches to treating such coupled-state dynamics.

Experimental probes of reaction dynamics

8.9 Spectroscopic methods

To follow the rate of any chemical reaction, one must have a means of monitoring the concentrations of reactant or product molecules as time evolves. In the majority of current experiments that relate to reaction dynamics, one uses some form of spectroscopic or alternative physical probe (e.g., an electrochemical signature) to monitor these concentrations as functions of time. Of course, in all such measurements, one must know how the intensity of the signal detected relates to the concentration of the molecules that cause the signal. For example, in most absorption experiments, as illustrated in Fig. 8.4, light is passed through a sample of "thickness" L and the intensity of the light beam in the absence of the

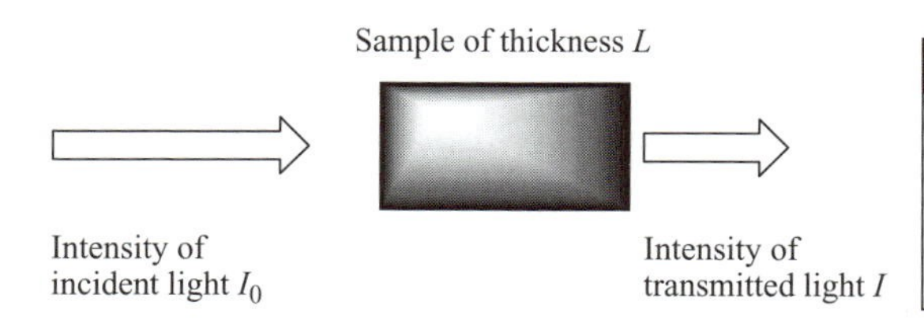

Figure 8.4 Typical Beer's law experiment in which a light beam of intensity I_0 is passed through a sample of thickness L.

sample I_0 and with the sample present I are measured. The Beer–Lambert law:

$$\log(I_0/I) = \varepsilon[A]L \tag{8.54}$$

then allows the concentration $[A]$ of the absorbing molecules to be determined, given the path length L over which absorption occurs and given the extinction coefficient ε of the absorbing molecules.

These extinction coefficients, which relate to the electric dipole matrix elements that are discussed in Chapter 6, are usually determined empirically by preparing a known concentration of the absorbing molecules and measuring the I_0/I ratio that this concentration produces in a cell of length L. For molecules and ions that are extremely reactive, this "calibration" approach to determining ε is often not feasible because one can not prepare a sample with a known and stable concentration. In such cases, one often must resort to using the theoretical expressions given in Chapter 6 (and discussed in most textbooks on molecular spectroscopy) to compute ε in terms of the wave functions of the absorbing species. In any event, one must know how the strength of the signal relates to the concentrations of the species if one wishes to monitor chemical reaction or energy transfer rates.

Because modern experimental techniques are capable of detecting molecules in particular electronic and vibration–rotation states, it has become common to use such tools to examine chemical reaction dynamics on a state-to-state level and to follow energy transfer processes, which clearly require such state-specific data. In such experiments, one seeks to learn the rate at which reactants in a specific state Φ_i react to produce products in some specific state Φ_f. One of the most common ways to monitor such state-specific rates is through a so-called "pump–probe" experiment in which:

(i) A short-duration light source is used to excite reactant molecules to some specified initial state Φ_i. Usually a tunable laser is used because its narrow frequency spread allows specific states to be pumped. The time at which this "pump" laser thus prepares the excited reactant molecules in state Φ_ι defines $t = 0$.
(ii) After a "delay time" of duration τ, a second light source is used to "probe" the product molecules that have been formed in various final states, Φ_f. Usually, the frequency of this probe source is scanned so that one can examine populations of many such final states.

The concentrations of reactant and product molecules in the initial and final states, Φ_i and Φ_f, are determined by the Beer–Lambert relation assuming that the extinction coefficients ε_i and ε_f for these species and states are known. In the former case, the extinction coefficient ε_ι relates to absorption of the pump photons to prepare reactant molecules in the specified initial state. In the latter, ε_f refers to absorption of the product molecules that are created in the state Φ_f.

Carrying out a series of such final-state absorption measurements at various delay times τ allows one to determine the concentration of these states as a function of time.

This kind of laser pump–probe experiment is used not only to probe specific electronic or vibration–rotation states of the reactants and products but also when the reaction is fast (i.e., complete in 10^{-4} s or less). In these cases, one is not using the high frequency resolution of the laser but its fast time response. Because laser pulses of quite short duration can be generated, these tools are well suited in such fast chemical reaction studies. The reactions can be in the gas phase (e.g., fast radical reactions in the atmosphere or in explosions) or in solution (e.g., photo-induced electron transfer reactions in biological systems).

8.10 Beam methods

Another approach to probing chemical reaction dynamics is to use a beam of reactant molecules A that collides with other reactants B that may also be in a beam or in a "bulb" in equilibrium at some temperature T. Such crossed-beam and beam–bulb experiments are illustrated in Fig. 8.5. Almost always, these beam and bulb samples contain molecules, radicals, or ions in the gas phase, so these techniques are most prevalent in gas-phase dynamics studies.

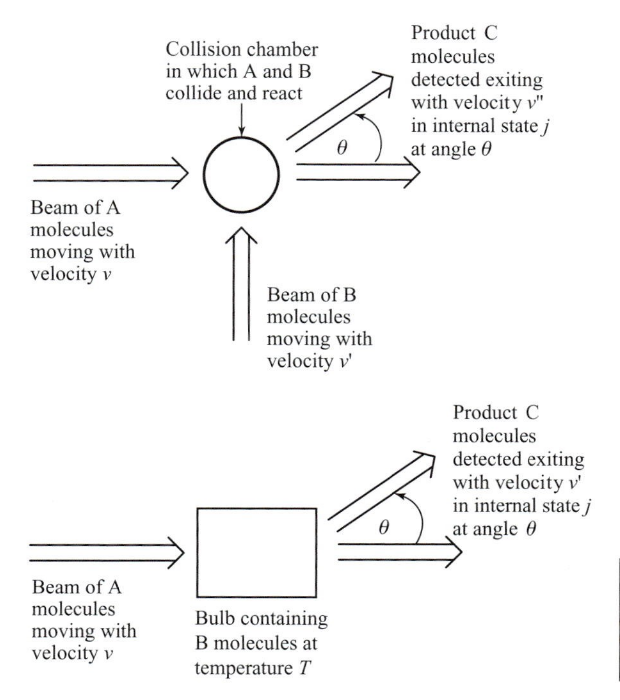

Figure 8.5 Typical crossed-beam and beam–bulb experimental setups.

The advantages of the crossed-beam type experiments are that:

(i) one can control the velocities, and hence the collision energies, of both reagents,
(ii) one can examine the product yield as a function of the angle θ through which the products are scattered,
(iii) one can probe the velocity of the products and,
(iv) by using spectroscopic methods, one can determine the fraction of products generated in various internal (electronic/vibrational/rotational) states.

Such measurements allow one to gain very detailed information about how the reaction rate coefficient depends on collisional (kinetic) energy and where the total energy available to the products is deposited (i.e., into product translational energy or product internal energy). The angular distribution of product molecules can also give information about the nature of the reaction process. For example, if the A + B collision forms a long-lived (i.e., on rotational time scales) collision complex, the product C molecules display a very isotropic angular distribution.

In beam–bulb experiments, one is not able to gain as much detailed information because one of the reactant molecules B is not constrained to be moving with a known fixed velocity in a specified direction when the A + B → C collisions occur. Instead, the B molecules collide with A molecules in a variety of orientations and with a distribution of collision energies whose range depends on the Maxwell–Boltzmann distribution of kinetic energies of the B molecules in the bulb. The advantage of beam–bulb experiments is that one can achieve much higher collision densities than in crossed-beam experiments because the density of B molecules inside the bulb is usually much higher than are the densities achievable in a beam of B molecules.

There are cases in which the beam–bulb experiments can be used to determine how the reaction rate depends on collision energy even though the molecules in the bulb have a distribution of kinetic energies. That is, if the species in the beam have much higher kinetic energies than most of the B molecules, then the A + B collision energy is primarily determined by the beam energy. An example of this situation is provided by so-called guided-ion beam experiments in which a beam of ions having well-specified kinetic energy E impinges on molecules in a bulb having a temperature T for which $kT \ll E$. Figure 8.6 illustrates data that can be extracted from such an experiment. In Fig. 8.6, we illustrate the cross-section σ (related to the bimolecular rate constant k by $\sigma v = k$, where v is the relative collision speed) for production of Na^+ ions when a beam of Na^+ (uracil) complexes having energy E (the horizontal axis) collides with a bulb containing Xe atoms at room temperature. In this case, the reaction is simply the collision-induced dissociation (CID) process in which the complex undergoes unimolecular decomposition after gaining internal energy in its collsions with Xe atoms:

$$Na^+(\text{uracil}) \rightarrow Na^+ + \text{uracil}. \tag{8.55}$$

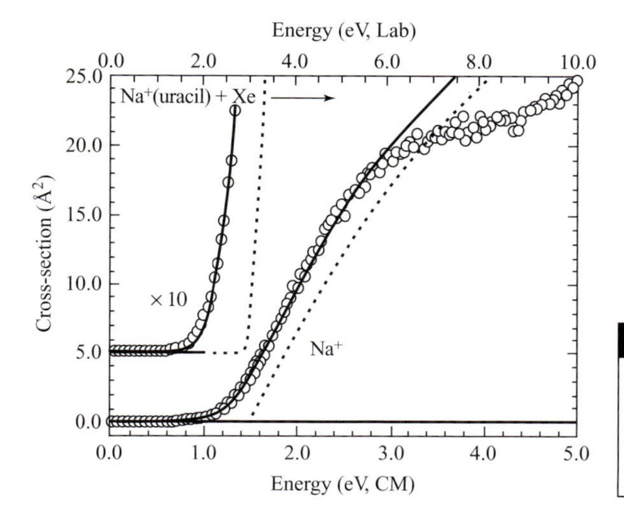

Figure 8.6 Collision-induced dissociation data showing cross-section as a function of collision energy.

The primary knowledge gained in this CID experiment is the threshold energy E^*; that is, the minimum collision energy needed to effect dissociation of the Na^+(uracil) complex. This kind of data has proven to offer some of the most useful information about bond dissociation energies of a wide variety of species. In addition, the magnitude of the reaction cross-section σ as a function of collision energy is a valuable product of such experiments. These CID beam–bulb experiments offer one of the most powerful and widely used means of determining such bond-rupture energies and reaction rate constants.

8.11 Other methods

Of course, not all chemical reactions occur so quickly that they require the use of fast lasers to follow concentrations of reacting species or pump–probe techniques to generate and probe these molecules. For slower chemical reactions, one can use other methods for monitoring the relevant concentrations. These methods include electrochemistry (where the redox potential is the species' signature) and NMR spectroscopy (where the chemical shifts of substituents are the signatures), both of whose instrumental response times would be too slow for probing fast reactions.

In addition, when the reactions under study do not proceed to completion but exist in equilibrium with a back reaction, alternative approaches can be used. The example discussed in Chapter 5 is one such case. Let us briefly review it here and again consider the reaction of an enzyme E and a substrate S to form the enzyme–substrate complex ES:

$$E + S \Leftrightarrow ES. \tag{8.56}$$

In the perturbation-type experiments, the equilibrium concentrations of the species are "shifted" by a small amount δ by application of the perturbation, so that

$$[\mathrm{ES}] = [\mathrm{ES}]_{\mathrm{eq}} - \delta, \tag{8.57}$$

$$[\mathrm{E}] = [\mathrm{E}]_{\mathrm{eq}} + \delta, \tag{8.58}$$

$$[\mathrm{S}] = [\mathrm{S}]_{\mathrm{eq}} + \delta. \tag{8.59}$$

Subsequently, the following rate law will govern the time evolution of the concentration change δ:

$$-d\delta/dt = -k_r([\mathrm{ES}]_{\mathrm{eq}} - \delta) + k_f([\mathrm{E}]_{\mathrm{eq}} + \delta)([\mathrm{S}]_{\mathrm{eq}} + \delta). \tag{8.60}$$

Assuming that δ is very small (so that the term involving δ^2 can be neglected) and using the fact that the forward and reverse rates balance at equilibrium, this equation for the time evolution of δ can be reduced to

$$-d\delta/dt = (k_r + k_f[\mathrm{S}]_{\mathrm{eq}} + k_f[\mathrm{E}_{\mathrm{eq}}])\delta. \tag{8.61}$$

So, the concentration deviations from equilibrium will return to equilibrium exponentially with an effective rate coefficient that is equal to a sum of terms:

$$k_{\mathrm{eff}} = k_r + k_f[\mathrm{S}]_{\mathrm{eq}} + k_f[\mathrm{E}_{\mathrm{eq}}]. \tag{8.62}$$

So, by following the concentrations of the reactants or products as they return to their equilibrium values, one can extract the effective rate coefficient k_{eff}. Doing this at a variety of different initial equilibrium concentrations (e.g., $[\mathrm{S}]_{\mathrm{eq}}$ and $[\mathrm{E}]_{\mathrm{eq}}$), and seeing how k_{eff} changes, one can then determine both the forward and reverse rate constants.

Problems

The following are some problems that will help you refresh your memory about material you should have learned in undergraduate chemistry classes and that allow you to exercise the material taught in this text.

Suggestions about what you should be able to do relative to the background material in the chapters of Part I

1 You should be able to set up and solve the one- and two-dimensional particle-in-a-box Schrödinger equations. I suggest you now try this and make sure you see:

a. How the second order differential equations have two independent solutions, so the most general solution is a sum of these two.
b. How the two boundary conditions reduce the number of acceptable solutions from two to one and limit the values of E that can be "allowed".
c. How the wave function is continuous even at the box boundaries, but $d\Psi/dx$ is not. In general $d\Psi/dx$, which relates to the momentum because $-i\hbar d/dx$ is the momentum operator, is continuous except at points where the potential $V(x)$ undergoes an infinite jump as it does at the box boundaries. The infinite jump in V, when viewed classically, means that the particle would undergo an instantaneous reversal in momentum at this point, so its momentum would not be continuous. Of course, in any realistic system, V does not have infinite jumps, so momentum will vary smoothly and thus $d\Psi/dx$ will be continuous.
d. How the energy levels grow with quantum number n as n^2.
e. What the wave functions look like when plotted.

2 You should go through the various wave functions treated in Part I (e.g., particles in boxes, rigid rotor, harmonic oscillator) and make sure you see how the $|\Psi|^2$ probability plots of such functions are not at all like the classical probability distributions except when the quantum number is very large.

3 You should make sure you understand how the time evolution of an eigenstate Ψ produces a simple $\exp(-itE/\hbar)$ multiple of Ψ so that $|\Psi|^2$ does not depend on time. However, when Ψ is not an eigenstate (e.g., when it is a combination of such states), its time propagation produces a Ψ whose $|\Psi|^2$ probability distribution changes with time.

4 You should notice that the densities of states appropriate to the one-, two-, and three-dimensional particle-in-a-box problems (which relate to translations in these dimensions) depend of different powers of E for the different dimensions.

5 You should be able to solve 2×2 and 3×3 Hückel matrix eigenvalue problems to obtain both the orbital energies and the normalized eigenvectors. For practice, try to do so for

a. the allyl radical's three π orbitals

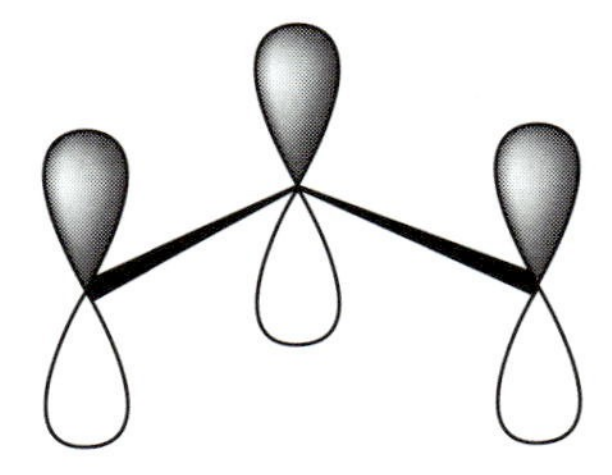

b. the cyclopropenyl radical's three π orbitals.

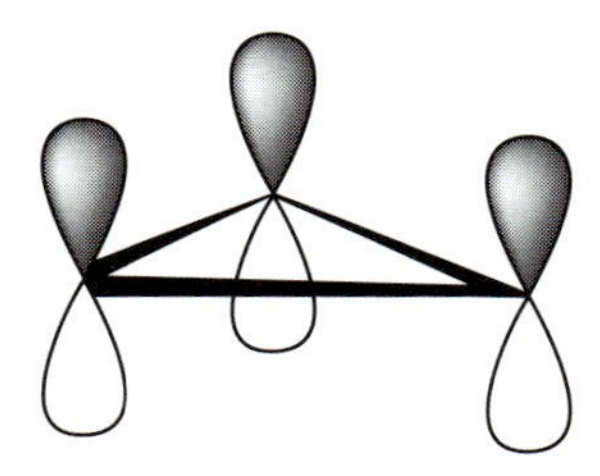

Do you see that the algebra needed to find the above sets of orbitals is exactly the same as is needed when we treat the linear and triangular sodium trimer?

6 You should be able to follow the derivation of the tunneling probability. Doing this offers a good test of your ability to apply the boundary conditions properly, so I suggest you do this task. You should appreciate how the tunneling probability decays exponentially with the "thickness" of the tunneling barrier and with the "height" of this barrier and that tunneling for heavier particles is less likely than for light particles. This is why tunneling usually is considered only for electrons, protons, and neutrons.

7 I do not expect that you could carry off a full solution to the Schrödinger equation for the hydrogenic atom. However, I think you need to pay attention to:

a. How separation of variables leads to a radial and two angular second order differential equations.

b. How the boundary condition that ϕ and $\phi + 2\pi$ are equivalent points in space produces the m quantum number.
c. How the l quantum number arises from the θ equation.
d. How the condition that the radial wave function not "explode" (i.e., go to infinity) as the coordinate r becomes large gives rise to the equation for the energy E.
e. The fact that the angular parts of the wave functions are spherical harmonics, and that these are exactly the same wave functions for the rotational motion of a linear molecule.
f. How the energy E depends on the n quantum number as n^{-2} and on the nuclear charge Z as Z^2, and that the bound state energies are negative. (Do you understand what this means? That is, what is the zero or reference point of energy?)

8 You should make sure that you are familiar with how the rigid-rotor and harmonic oscillator energies vary with quantum numbers (J, M in the former case, v in the latter). You should also know how these energies depend on the molecular geometry (in the former) and on the force constant and reduced mass (in the latter). You should note that E depends quadratically on J but linearly on v.

9 You should know what the Morse potential is and what its parameters mean. You should understand that the Morse potential displays anharmonicity but the harmonic potential does not.

10 You should be able to follow how the mass-weighted Hessian matrix can be used to approximate the vibrational motions of a polyatomic molecule. You should understand how the eigenvalues of this matrix produce the harmonic vibrational frequencies and the corresponding eigenvectors describe the motions of the molecule associated with these frequencies.

Practice with matrices and operators

1 Find the eigenvalues and corresponding normalized eigenvectors of the following matrices:

a.

$$\begin{bmatrix} -1 & 2 \\ 2 & 2 \end{bmatrix}$$

b.

$$\begin{bmatrix} -2 & 0 & 0 \\ 0 & -1 & 2 \\ 0 & 2 & 2 \end{bmatrix}$$

2 Replace the following classical mechanical expressions with their corresponding quantum mechanical operators:

a. $\text{KE} = \frac{mv^2}{2}$ in three-dimensional space.
b. $\mathbf{p} = m\mathbf{v}$, a three-dimensional Cartesian vector.
c. y-component of angular momentum: $L_y = zp_x - xp_z$.

3 Transform the following operators into the specified coordinates:

a. $\mathbf{L}_x = \frac{\hbar}{\mathrm{i}}\{y\frac{\partial}{\partial z} - z\frac{\partial}{\partial y}\}$ from Cartesian to spherical polar coordinates.
b. $\boldsymbol{L}_z = \frac{\hbar}{i}\frac{\partial}{\partial \phi}$ from spherical polar to Cartesian coordinates.

4 Match the eigenfunctions in column B to their operators in column A. What is the eigenvalue for each eigenfunction?

	Column A	Column B
(i)	$(1 - x^2)\frac{d^2}{dx^2} - x\frac{d}{dx}$	$4x^4 - 12x^2 + 3$
(ii)	$\frac{d^2}{dx^2}$	$5x^4$
(iii)	$x\frac{d}{dx}$	$e^{3x} + e^{-3x}$
(iv)	$\frac{d^2}{dx^2} - 2x\frac{d}{dx}$	$x^2 - 4x + 2$
(v)	$x\frac{d^2}{dx^2} + (1 - x)\frac{d}{dx}$	$4x^3 - 3x$

Review of shapes of orbitals

5 Draw qualitative shapes of the (1) s, (3) p and (5) d atomic orbitals (note that these orbitals represent only the angular portion and do not contain the radial portion of the hydrogen-like atomic wave functions). Indicate with light and dark shading the relative signs of the wave functions and the position(s) (if any) of any nodes.

6 Plot the radial portions of the 4s, 4p, 4d, and 4f hydrogen-like atomic wave functions.

7 Plot the radial portions of the 1s, 2s, 2p, 3s, and 3p hydrogen-like atomic wave functions for the Si atom using screening concepts for any inner electrons.

Labeling orbitals using point group symmetry

8 Define the symmetry adapted "core" and "valence" atomic orbitals of the following systems:

a. NH_3 in the C_{3v} point group,
b. H_2O in the C_{2v} point group,
c. H_2O_2 (cis) in the C_2 point group,
d. N in $D_{\infty h}$, D_{2h}, C_{2v}, and C_s point groups,
e. N_2 in $D_{\infty h}$, D_{2h}, C_{2v}, and C_s point groups.

A problem to practice the basic tools of the Schrödinger equation

9 A particle of mass m moves in a one-dimensional box of length L, with boundaries at $x = 0$ and $x = L$. Thus, $V(x) = 0$ for $0 \le x \le L$, and $V(x) = \infty$

elsewhere. The normalized eigenfunctions of the Hamiltonian for this system are given by $\Psi_n(x) = (\frac{2}{L})^{1/2} \sin \frac{n\pi x}{L}$, with $E_n = \frac{n^2\pi^2\hbar^2}{2mL^2}$, where the quantum number n can take on the values $n = 1, 2, 3, \ldots$.

a. Assuming that the particle is in an eigenstate, $\Psi_n(x)$, calculate the probability that the particle is found somewhere in the region $0 \le x \le \frac{L}{4}$. Show how this probability depends on n.
b. For what value of n is there the largest probability of finding the particle in $0 \le x \le \frac{L}{4}$?
c. Now assume that Ψ is a superposition of two eigenstates,

$$\Psi = a\Psi_n + b\Psi_m, \text{ at time } t = 0.$$

What is Ψ at time t? What energy expectation value does Ψ have at time t and how does this relate to its value at $t = 0$?
d. For an experimental measurement which is capable of distinguishing systems in state Ψ_n from those in Ψ_m, what fraction of a large number of systems each described by Ψ will be observed to be in Ψ_n? What energies will these experimental measurements find and with what probabilities?
e. For those systems originally in $\Psi = a\Psi_n + b\Psi_m$ which were observed to be in Ψ_n at time t, what state (Ψ_n, Ψ_m, or whatever) will they be found in if a second experimental measurement is made at a time t' later than t?
f. Suppose by some method (which need not concern us at this time) the system has been prepared in a non-stationary state (that is, it is not an eigenfunction of **H**). At the time of a measurement of the particle's energy, this state is specified by the normalized wave function $\Psi = (\frac{30}{L^5})^{1/2} x(L - x)$ for $0 \le x \le L$, and $\Psi = 0$ elsewhere. What is the probability that a measurement of the energy of the particle will give the value $E_n = \frac{n^2\pi^2\hbar^2}{2mL^2}$ for any given value of n?
g. What is the expectation value of **H**, i.e. the average energy of the system, for the wave function Ψ given in part (f)?

A problem on the properties of non-stationary states

10 Show that for a system in a non-stationary state, $\Psi = \sum_j C_j \Psi_j e^{-iE_j t/\hbar}$, the average value of the energy does not vary with time but the expectation values of other properties do vary with time.

A problem about Jahn–Teller distortion

11 The energy states and wave functions for a particle in a three-dimensional box whose lengths are L_1, L_2, and L_3 are given by

$$E(n_1, n_2, n_3) = \frac{h^2}{8m}\left[\left(\frac{n_1}{L_1}\right)^2 + \left(\frac{n_2}{L_2}\right)^2 + \left(\frac{n_3}{L_3}\right)^2\right]$$

and $$\Psi(n_1, n_2, n_3) = \left(\frac{2}{L1}\right)^{\frac{1}{2}} \left(\frac{2}{L_2}\right)^{\frac{1}{2}} \left(\frac{2}{L_3}\right)^{\frac{1}{2}} \sin\left(\frac{n_1\pi x}{L_1}\right) \sin\left(\frac{n_2\pi y}{L_2}\right) \sin\left(\frac{n_3\pi z}{L_3}\right).$$

These wave functions and energy levels are sometimes used to model the motion of electrons in a central metal atom (or ion) which is surrounded by six ligands in an octahedral manner.

a. Show that the lowest energy level is non-degenerate and the second energy level is triply degenerate if $L_1 = L_2 = L_3$. What values of $n_{1,} n_2$, and n_3 characterize the states belonging to the triply degenerate level?
b. For a box of volume $V = L_1 L_2 L_3$, show that for three electrons in the box (two in the non-degenerate lowest "orbital", and one in the next), a lower total energy will result if the box undergoes a rectangular distortion ($L_1 = L_2 \neq L_3$) which preserves the total volume than if the box remains undistorted. (Hint: if V is fixed and $L_1 = L_2$, then $L_3 = \frac{V}{L_1^2}$ and L_1 is the only "variable".)
c. Show that the degree of distortion (ratio of L_3 to L_1) which will minimize the total energy is $L_3 = \sqrt{2}\, L_1$. How does this problem relate to Jahn–Teller distortions? Why (in terms of the property of the central atom or ion) do we do the calculation with fixed volume?
d. By how much (in eV) will distortion lower the energy (from its value for a cube, $L_1 = L_2 = L_3$) if $V = 8$ Å^3 and $\frac{h^2}{8m} = 6.01 \times 10^{-27}$ erg cm^2. 1 eV $= 1.6 \times 10^{-12}$ erg.

A particle on a ring model for electrons moving in cyclic compounds

12 The π-orbitals of benzene, C_6H_6, may be modeled very crudely using the wave functions and energies of a particle on a ring. Let's first treat the particle on a ring problem and then extend it to the benzene system.

a. Suppose that a particle of mass m is constrained to move on a circle (of radius r) in the xy plane. Further assume that the particle's potential energy is constant (choose zero as this value). Write down the Schrödinger equation in the normal Cartesian coordinate representation. Transform this Schrödinger equation to cylindrical coordinates where $x = r\cos\phi$, $y = r\sin\phi$, and $z = z$ ($z = 0$ in this case). Taking r to be held constant, write down the general solution, $\Phi(\phi)$, to this Schrödinger equation. The "boundary" conditions for this problem require that $\Phi(\phi) = \Phi(\phi + 2\pi)$. Apply this boundary condition to the general solution. This results in the quantization of the energy levels of this system. Write down the final expression for the normalized wave functions and quantized energies. What is the physical significance of these quantum numbers that can have both positive and negative values? Draw an energy diagram representing the first five energy levels.
b. Treat the six π-electrons of benzene as particles free to move on a ring of radius 1.40 Å, and calculate the energy of the lowest electronic transition. Make sure the

Pauli principle is satisfied! What wavelength does this transition correspond to? Suggest some reasons why this differs from the wavelength of the lowest observed transition in benzene, which is 2600 Å.

A non-stationary state wave function

13 A diatomic molecule constrained to rotate on a flat surface can be modeled as a planar rigid rotor (with eigenfunctions, $\Phi(\phi)$, analogous to those of the particle on a ring of Problem 12) with fixed bond length r. At $t = 0$, the rotational (orientational) probability distribution is observed to be described by a wave function $\Psi(\phi, 0) = \sqrt{\frac{4}{3\pi}} \cos^2 \phi$. What values, and with what probabilities, of the rotational angular momentum, $(-i\hbar\frac{\partial}{\partial\phi})$, could be observed in this system? Explain whether these probabilities would be time dependent as $\Psi(\phi, 0)$ evolves into $\Psi(\phi, t)$.

A problem about Franck–Condon factors

14 Consider an N_2 molecule, in the ground vibrational level of the ground electronic state, which is bombarded by 100 eV electrons. This leads to ionization of the N_2 molecule to form N_2^+. In this problem we will attempt to calculate the vibrational distribution of the newly formed N_2^+ ions, using a somewhat simplified approach.

a. Calculate (according to classical mechanics) the velocity (in cm s^{-1}) of a 100 eV electron, ignoring any relativistic effects. Also calculate the amount of time required for a 100 eV electron to pass an N_2 molecule, which you may estimate as having a length of 2 Å.
b. The radial Schrödinger equation for a diatomic molecule treating vibration as a harmonic oscillator can be written as

$$-\frac{\hbar^2}{2\mu r^2}\left(\frac{\partial}{\partial r}\left(r^2\frac{\partial\Psi}{\partial r}\right)\right) + \frac{k}{2}(r - r_e)^2\Psi = E\Psi.$$

Substituting $\Psi(r) = \frac{F(r)}{r}$, this equation can be rewritten as

$$-\frac{\hbar^2}{2\mu}\frac{\partial^2}{\partial r^2}F(r) + \frac{k}{2}(r - r_e)^2 F(r) = EF(r).$$

The vibrational Hamiltonian for the ground electronic state of the N_2 molecule within this approximation is given by

$$H(N_2) = -\frac{\hbar^2}{2\mu}\frac{d^2}{dr^2} + \frac{k_{N_2}}{2}\left(r - r_{N_2}\right)^2,$$

where r_{N_2} and k_{N_2} have been measured experimentally to be 1.097 69 Å and 2.294×10^6 g s^{-2}, respectively. The vibrational Hamiltonian for the N_2^+ ion, however, is given by

$$H(N_2) = -\frac{\hbar^2}{2\mu}\frac{d^2}{dr^2} + \frac{k_{N_2^+}}{2}\left(r - r_{N_2^+}\right)^2,$$

where $r_{N_2^+}$ and $k_{N_2^+}$ have been measured experimentally to be 1.116 42 Å and 2.009×10^6 g s^{-2}, respectively. In both systems the reduced mass is $\mu = 1.1624 \times 10^{-23}$ g. Use the above information to write out the ground-state vibrational wave functions of the N_2 and N_2^+ molecules, giving explicit values for any constants which appear in them. The $v = 0$ harmonic oscillator function is $\Psi_0 = (\alpha/\pi)^{1/4}\exp(-\alpha x^2/2)$.

c. During the time scale of the ionization event (which you calculated in part (a)), the vibrational wave function of the N_2 molecule has effectively no time to change. As a result, the newly formed N_2^+ ion finds itself in a vibrational state which is not an eigenfunction of the new vibrational Hamiltonian, $\mathbf{H}(N_2^+)$. Assuming that the N_2 molecule was originally in its $v = 0$ vibrational state, calculate the probability that the N_2^+ ion will be produced in its $v = 0$ vibrational state.

Vibration of a diatomic molecule

15 The force constant, k, of the C—O bond in carbon monoxide is 1.87×10^6 g s^{-2}. Assume that the vibrational motion of CO is purely harmonic and use the reduced mass $\mu = 6.857$ amu.

a. Calculate the spacing between vibrational energy levels in this molecule, in units of ergs and cm^{-1}.
b. Calculate the uncertainty in the internuclear distance in this molecule, assuming it is in its ground vibrational level. Use the ground-state vibrational wave function ($\Psi_{v=0}$; recall that I gave you this function in Problem 14), and calculate $\langle x \rangle$, $\langle x^2 \rangle$, and $\Delta x = (\langle x^2 \rangle - \langle x \rangle^2)^{1/2}$.
c. Under what circumstances (i.e., large or small values of k; large or small values of μ) is the uncertainty in internuclear distance large? Can you think of any relationship between this observation and the fact that helium remains a liquid down to absolute zero?

A variational method problem

16 A particle of mass m moves in a one-dimensional potential whose Hamiltonian is given by

$$H = -\frac{\hbar^2}{2m}\frac{d^2}{dx^2} + a|x|,$$

where the absolute value function is defined by $|x| = x$ if $x \geq 0$ and $|x| = -x$ if $x \leq 0$.

a. Use the normalized trial wave function $\phi = (\frac{2b}{\pi})^{\frac{1}{4}} e^{-bx^2}$ to estimate the energy of the ground state of this system, using the variational principle to evaluate $W(b)$, the variational expectation value of H.
b. Optimize b to obtain the best approximation to the ground-state energy of this system, using a trial function of the form of ϕ, as given above. The numerically calculated exact ground-state energy is $0.808\,616\,\hbar^{\frac{2}{3}} m^{-\frac{1}{3}} a^{-\frac{2}{3}}$. What is the percentage error in your value?

Another variational method problem

17 The harmonic oscillator is specified by the Hamiltonian

$$H = -\frac{\hbar^2}{2m}\frac{d^2}{dx^2} + \frac{1}{2}kx^2.$$

Suppose the ground-state solution to this problem were unknown, and that you wish to approximate it using the variational theorem. Choose as your trial wave function

$$\phi = \sqrt{\frac{15}{16}} a^{-\frac{5}{2}}(a^2 - x^2) \quad \text{for} \quad -a < x < a,$$

$$\phi = 0 \quad \text{for } |x| \geq a,$$

where a is an arbitrary parameter which specifies the range of the wave function. Note that ϕ is properly normalized as given.

a. Calculate $\int_{-\infty}^{+\infty} \phi^* H\phi\, dx$ and show that it is given by

$$\int_{-\infty}^{+\infty} \phi^* H\phi\, dx = \frac{5}{4}\frac{\hbar^2}{ma^2} + \frac{ka^2}{14}.$$

b. Calculate $\int_{-\infty}^{+\infty} \phi^* H\phi\, dx$ for $a = b(\frac{\hbar^2}{km})^{\frac{1}{4}}$.
c. To find the best approximation to the true wave function and its energy, find the minimum of $\int_{-\infty}^{+\infty} \phi^* H\phi\, dx$ by setting $\frac{d}{da}\int_{-\infty}^{+\infty} \phi^* H\phi\, dx = 0$ and solving for a. Substitute this value into the expression for $\int_{-\infty}^{+\infty} \phi^* H\phi\, dx$ given in part (a) to obtain the best approximation for the energy of the ground state of the harmonic oscillator.
d. What is the percentage error in your calculated energy of part (c)?

A perturbation theory problem

18 Consider an electron constrained to move on the surface of a sphere of radius r_0. The Hamiltonian for such motion consists of a kinetic energy term only, $H_0 = \frac{\mathbf{L}^2}{2m_e r_0^2}$, where $\mathbf{L}$ is the orbital angular momentum operator involving derivatives with respect to the spherical polar coordinates (θ, ϕ). H_0 has the complete set of eigenfunctions $\psi_{lm}^{(0)} = Y_{l,m}(\theta, \phi)$.

a. Compute the zeroth order energy levels of this system.
b. A uniform electric field is applied along the z-axis, introducing a perturbation $V = -e\varepsilon z = -e\varepsilon r_0 \cos\theta$, where ε is the strength of the field. Evaluate the correction to the energy of the lowest level through second order in perturbation theory, using the identity

$$\cos\theta\, Y_{l,m}(\theta, \phi) = \sqrt{\frac{(l+m+1)(l-m+1)}{(2l+1)(2l+3)}}\, Y_{l+1,m}(\theta, \phi) + \sqrt{\frac{(l+m)(l-m)}{(2l+1)(2l-1)}}\, Y_{l-1,m}(\theta, \phi).$$

Note that this identity enables you to utilize the orthonormality of the spherical harmonics.
c. The electric polarizability α gives the response of a molecule to an externally applied electric field, and is defined by $\alpha = -\frac{\partial^2 E}{\partial^2 \varepsilon}|_{\varepsilon=0}$ where E is the energy in the presence of the field and ε is the strength of the field. Calculate α for this system.
d. Use this problem as a model to estimate the polarizability of a hydrogen atom, where $r_0 = a_0 = 0.529$ Å, and a cesium atom, which has a single 6s electron with $r_0 \approx 2.60$ Å. The corresponding experimental values are $\alpha_{\mathrm{H}} = 0.6668$ Å^3 and $\alpha_{\mathrm{Cs}} = 59.6$ Å^3.

A Hartree–Fock problem you can do by hand

19 Given the following orbital energies (in hartrees) for the N atom and the coupling elements between two like atoms (these coupling elements are the Fock matrix elements from standard *ab initio* minimum-basis SCF calculations), calculate the molecular orbital energy levels and orbital expansion coefficients. Draw the orbital correlation diagram for formation of the N_2 molecule. Indicate the symmetry of each atomic and each molecular orbital. Designate each of the molecular orbitals as bonding, non-bonding, or antibonding.

$$N_{1s} = -15.31^*$$

$$N_{2s} = -0.86^*$$

$$N_{2p} = -0.48^*$$

N_2 σ_g Fock matrix*

$$\begin{bmatrix} -6.52 & & \\ -6.22 & -7.06 & \\ 3.61 & 4.00 & -3.92 \end{bmatrix}$$

N_2 π_g Fock matrix*

$$[0.28]$$

N_2 σ_u Fock matrix*

$$\begin{bmatrix} 1.02 & & \\ -0.60 & -7.59 & \\ 0.02 & 7.42 & -8.53 \end{bmatrix}$$

N_2 π_u Fock matrix*

$$[-0.58]$$

*The Fock matrices (and orbital energies) were generated using standard minimum basis set SCF calculations. The Fock matrices are in the orthogonal basis formed from these orbitals.

An orbital correlation diagram problem

20 Given the following valence orbital energies for the C atom and H_2 molecule, draw the orbital correlation diagram for formation of the CH_2 molecule (via a C_{2v} insertion of C into H_2 resulting in bent CH_2). Designate the symmetry of each atomic and molecular orbital both in their highest point group symmetry and in that of the reaction path (C_{2v}).

$$C_{1s} = -10.91^* \quad H_2\,\sigma_g = -0.58^*$$
$$C_{2s} = -0.60^* \quad H_2\,\sigma_u = 0.67^*$$
$$C_{2p} = -0.33^*$$

*The orbital energies were generated using standard STO3G minimum basis set SCF calculations.

Practice using point group symmetry

21 Qualitatively analyze the electronic structure (orbital energies and orbitals) of PF_5. Analyze only the 3s and 3p electrons of P and the one 2p bonding electron of each F. Proceed with a D_{3h} analysis in the following manner:

a. Symmetry adapt the top and bottom F atomic orbitals.
b. Symmetry adapt the three (trigonal) F atomic orbitals.

c. Symmetry adapt the P 3s and 3p atomic orbitals.
d. Allow these three sets of D_{3h} orbitals to interact and draw the resultant orbital energy diagram.
e. Symmetry label each of these molecular energy levels. Fill this energy diagram with 10 "valence" electrons.

Practice with term symbols and determinental wave functions for atoms and molecules

22 For the given orbital occupations (configurations) of the following systems, determine all possible states (all possible allowed combinations of spin and space states). There is no need to form the determinental wave functions, simply label each state with its proper term symbol.

a. CH_2 $1a_1^2 2a_1^2 1b_2^2 3a_1^1 1b_1^1$
b. B_2 $1\sigma_g^2 1\sigma_u^2 2\sigma_g^2 2\sigma_u^2 1\pi_u^1 2\pi_u^1$
c. O_2 $1\sigma_g^2 1\sigma_u^2 2\sigma_g^2 2\sigma_u^2 1\pi_u^4 3\sigma_g^2 1\pi_g^2$
d. Ti $1s^2 2s^2 2p^6 3s^2 3p^6 4s^2 3d^1 4d^1$
e. Ti $1s^2 2s^2 2p^6 3s^2 3p^6 4s^2 3d^2$

23 Construct Slater determinant wave functions for each of the following states of CH_2:

a. ${}^1B_1(1a_1^2 2a_1^2 1b_2^2 3a_1^1 1b_1^1)$
b. ${}^3B_1(1a_1^2 2a_1^2 1b_2^2 3a_1^1 1b_1^1)$
c. ${}^1A_1(1a_1^2 2a_1^2 1b_2^2 3a_1^2)$

A Woodward–Hoffmann rules problem

24 Let us investigate the reactions:

(i) $CH_2({}^1A_1) \rightarrow H_2 + C$, and
(ii) $CH_2({}^3B_1) \rightarrow H_2 + C$,

under an assumed C_{2v} reaction pathway utilizing the following information:

$$\text{C atom:} \quad {}^3P \xrightarrow{29.2 \text{ kcal mol}^{-1}} {}^1D \xrightarrow{32.7 \text{ kcal mol}^{-1}} {}^1S$$

$$C({}^3P) + H_2 \rightarrow CH_2({}^3B_1) \quad \Delta E = -78.8 \text{ kcal mol}^{-1}$$

$$C({}^1D) + H_2 \rightarrow CH_2({}^1A_1) \quad \Delta E = -97.0 \text{ kcal mol}^{-1}$$

$$IP(H_2) > IP \text{ (2s carbon)}.$$

a. Write down (first in terms of $2p_{1,0,-1}$ orbitals and then in terms of $2p_{x,y,z}$ orbitals) the:
   (i) three Slater determinant (SD) wave functions belonging to the ${}^3P$ states all of which have $M_S = 1$,
   (ii) five ${}^1D$ SD wave functions, and
   (iii) one ${}^1S$ SD wave function.

b. Using the coordinate system shown below, label the hydrogen orbitals σ_g, σ_u and the carbon 2s, $2p_x$, $2p_y$, $2p_z$ orbitals as a_1, $b_1(x)$, $b_2(y)$, or a_2. Do the same for the $\sigma, \sigma, \sigma^*, \sigma^*, n$, and p_π orbitals of CH_2.

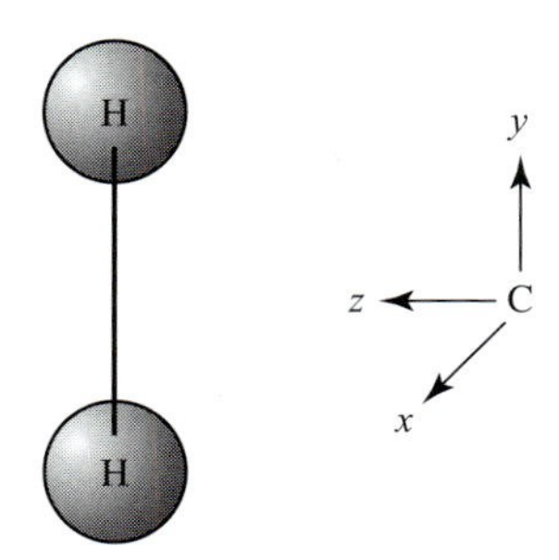

c. Draw an orbital correlation diagram for the $CH_2 \rightarrow H_2 + C$ reactions. Try to represent the relative energy orderings of the orbitals correctly.

d. Draw a configuration correlation diagram for $CH_2(^3B_1) \rightarrow H_2 + C$ showing *all* configurations which arise from the $C(^3P) + H_2$ products. You can assume that doubly excited configurations lie much ($\sim$100 kcal mol^{-1}) above their parent configurations.

e. Repeat step (d) for $CH_2(^1A_1) \rightarrow H_2 + C$ again showing all configurations which arise from the $C(^1D) + H_2$ products.

f. Do you expect the reaction $C(^3P) + H_2 \rightarrow CH_2$ to have a large activation barrier? About how large? What state of CH_2 is produced in this reaction? Would distortions away from C_{2v} symmetry be expected to raise or lower the activation barrier? Show how one could estimate where along the reaction path the barrier top occurs.

g. Would $C(^1D) + H_2 \rightarrow CH_2$ be expected to have a larger or smaller barrier than you found for the ^3P C reaction?

Another Woodward–Hoffmann rule problem

25 The decomposition of the ground-state singlet carbene,

to produce acetylene and ^1D carbon is known to occur with an activation energy equal to the reaction endothermicity. However, when the corresponding triplet carbene decomposes to acetylene and ground-state (triplet) carbon, the activation energy exceeds this reaction's endothermicity. Construct orbital, configuration, and state correlation diagrams that permit you to explain the above observations. Indicate whether single configuration or configuration interaction wave functions would be required to describe the above singlet and triplet decomposition processes.

Practice with rotational spectroscopy and its relation to molecular structure

26 Consider the molecules CCl_4, $CHCl_3$, and CH_2Cl_2.

a. What kind of rotor are they (symmetric top, etc.; do not bother with oblate, or near-prolate, etc.)
b. Will they show pure rotational (i.e., microwave) spectra?

27 Assume that ammonia shows a pure rotational spectrum. If the rotational constants are 9.44 and 6.20 cm^{-1}, use the energy expression

$$E = (A - B)K^2 + BJ(J+1)$$

to calculate the energies (in cm^{-1}) of the first three lines (i.e., those with lowest K, J quantum number for the absorbing level) in the absorption spectrum (ignoring higher order terms in the energy expression).

A problem on vibration–rotation spectroscopy

28 The molecule $^{11}B^{16}O$ has a vibrational frequency $\omega_e = 1885$ cm^{-1}, a rotational constant $B_e = 1.78$ cm^{-1}, and a bond energy from the bottom of the potential well of $D_e^0 = 8.28$ eV. Use integral atomic masses in the following:

a. In the approximation that the molecule can be represented as a Morse oscillator, calculate the bond length, R_e in angstroms, the anharmonicity constant, $\omega_e x_e$ in cm^{-1}, the zero-point corrected bond energy, D_0^0 in eV, the vibration–rotation interaction constant, α_e in cm^{-1}, and the vibrational state specific rotation constants, B_0 and B_1 in cm^{-1}. Use the vibration–rotation energy expression for a Morse oscillator:

$$E = \hbar\omega_e(v + 1/2) - \hbar\omega_e x_e(v + 1/2)^2 + B_v J(J+1) - D_e J^2(J+1)^2,$$

where $B_v = B_e - \alpha_e(v + 1/2)$, $\alpha_e = \dfrac{-6B_e^2}{\hbar\omega_e} + \dfrac{6\sqrt{B_e^3\hbar\omega_e x_e}}{\hbar\omega_e}$, and $D_e = \dfrac{4B_e^3}{\hbar\omega_e^2}$.

b. Will this molecule show a pure rotation spectrum? A vibration–rotation spectrum? Assuming that it does, what are the energies (in cm^{-1}) of the first three lines in the P branch ($\Delta v = +1$, $\Delta J = -1$) of the fundamental absorption?

A problem labeling vibrational modes by symmetry

29 Consider trans-$C_2H_2Cl_2$. The vibrational normal modes of this molecule are shown below. What is the symmetry of the molecule? Label each of the modes with the appropriate irreducible representation.

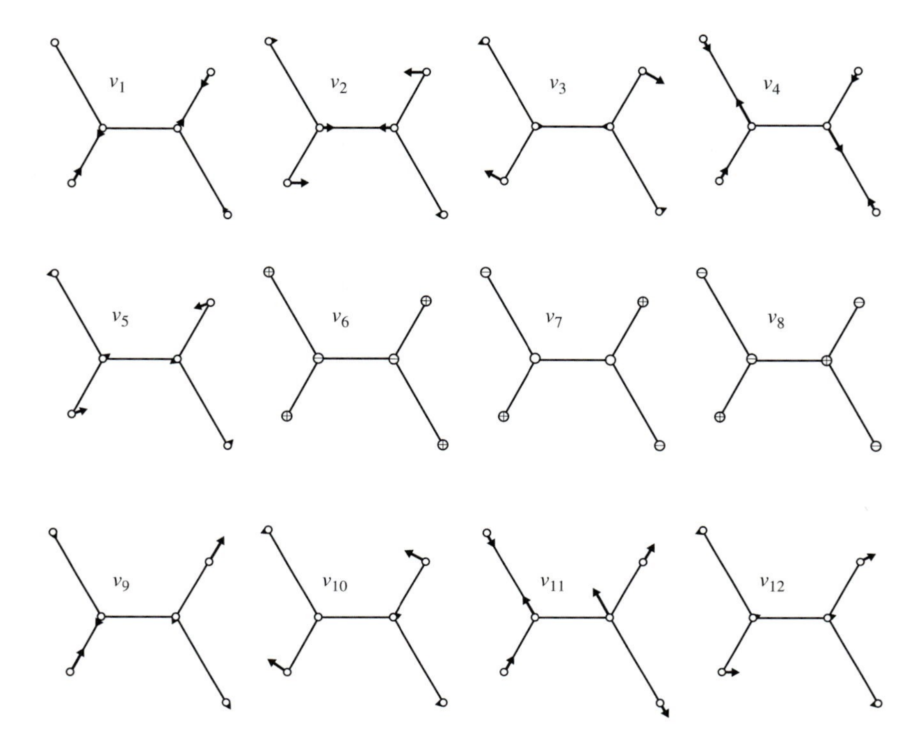

A problem in rotational spectroscopy

30 Suppose you are given two molecules (one is CH_2 and the other is CH_2^- but you don't know which is which). Both molecules have C_{2v} symmetry. The CH bond length of molecule I is 1.121 Å and for molecule II it is 1.076 Å. The bond angle of molecule I is 104° and for molecule II it is 136°.

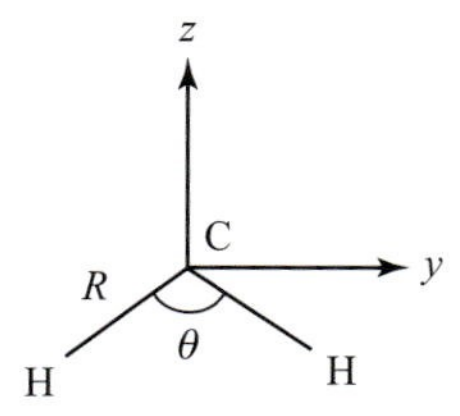

a. Using a coordinate system centered on the C nucleus as shown above (the molecule is in the yz plane), compute the moment of inertia tensors of both species (I and II). The definitions of the components of the tensor are, for example:

$$I_{xx} = \sum_j m_j \left(y_j^2 + z_j^2\right) - M(Y^2 + Z^2),$$

$$I_{xy} = -\sum_j m_j x_j y_j - MXY.$$

Here, m_j is the mass of the nucleus j, M is the mass of the entire molecule, and X, Y, Z are the coordinates of the center of mass of the molecule. Use Å for distances and amu for masses.

b. Find the principal moments of inertia $I_a < I_b < I_c$ for both compounds (in amu Å^2 units) and convert these values into rotational constants A, B, and C in cm^{-1} using, for example,

$$A = h(8\pi^2 c I_a)^{-1}.$$

c. Both compounds are "nearly prolate tops" whose energy levels can be well approximated using the prolate top formula:

$$E = (A - B)K^2 + BJ(J + 1),$$

if one uses for the B constant the average of the B and C values determined earlier. Thus, take the B and C values (for each compound) and average them to produce an effective B constant to use in the above energy formula. Write down (in cm^{-1} units) the energy formula for both species. What values are J and K allowed to assume? What is the degeneracy of the level labeled by a given J and K?

d. Draw pictures of both compounds and show the directions of the three principal axes (a, b, c). On these pictures, show the kind of rotational motion associated with the quantum number K.

e. Suppose you are given the photoelectron spectrum of CH_2^-. In this spectrum $J_j = J_i + 1$ transitions are called R-branch absorptions and those obeying $J_j = J_i - 1$ are called P-branch transitions. The spacing between lines can increase or decrease as functions of J_i depending on the changes in the moment of inertia for the transition. If spacings grow closer and closer, we say that the spectrum exhibits a so-called band head formation. In the photoelectron spectrum that you are given, a rotational analysis of the vibrational lines in this spectrum is carried out and it is found that the R-branches show band head formation but the P-branches do not. Based on this information, determine which compound, I or II, is the CH_2^- anion. Explain you reasoning.

f. At what J value (of the absorbing species) does the band head occur and at what rotational energy difference?

Using point group symmetry in vibrational spectroscopy

31 Let us consider the vibrational motions of benzene. To consider all of the vibrational modes of benzene we should attach a set of displacement vectors in the x, y, and z directions to each atom in the molecule (giving 36 vectors in all), and evaluate how these transform under the symmetry operations of D_{6h}. For this problem, however, let's only inquire about the C—H stretching vibrations.

a. Represent the C—H stretching motion on each C—H bond by an outward-directed vector on each H atom, designated r_i:

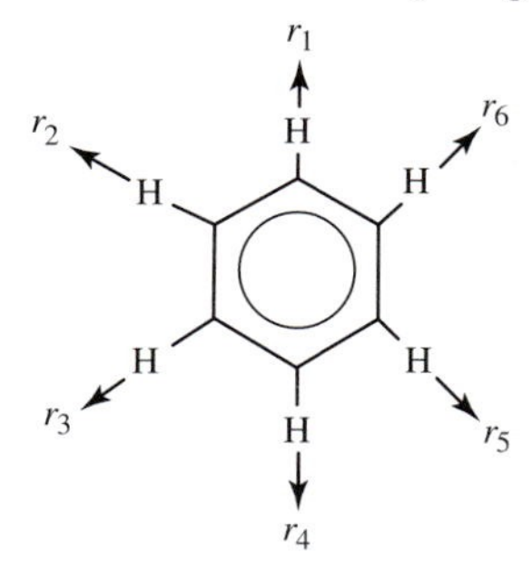

b. These vectors form the basis for a reducible representation. Evaluate the characters for this reducible representation under the symmetry operations of the D_{6h} group.

c. Decompose the reducible representation you obtained in part (b) into its irreducible components. These are the symmetries of the various C—H stretching vibrational modes in benzene.

d. The vibrational state with zero quanta in each of the vibrational modes (the ground vibrational state) of any molecule always belongs to the totally symmetric representation. For benzene, the ground vibrational state is therefore of A_{1g} symmetry. An excited state which has one quantum of vibrational excitation in a mode which is of a given symmetry species has the same symmetry species as the mode which is excited (because the vibrational wave functions are given as Hermite polynomials in the stretching coordinate). Thus, for example, excitation (by one quantum) of a vibrational mode of A_{2u} symmetry gives a wave function of A_{2u} symmetry. To resolve the question of what vibrational modes may be excited by the absorption of infrared radiation we must examine the x, y, and z components of the transition dipole integral for initial and final state wave functions ψ_i and ψ_f, respectively:

$$|\langle\psi_f|x|\psi_i\rangle|, \quad |\langle\psi_f|y|\psi_i\rangle|, \quad \text{and} \quad |\langle\psi_f|z|\psi_i\rangle|.$$

Using the information provided above, which of the C—H vibrational modes of benzene will be infrared-active, and how will the transitions be polarized? How many C—H vibrations will you observe in the infrared spectrum of benzene?

e. A vibrational mode will be active in Raman spectroscopy only if one or more of the following integrals is non-zero:

$$|\langle\psi_f|xy|\psi_i\rangle|, \quad |\langle\psi_f|xz|\psi_i\rangle|, \quad |\langle\psi_f|yz|\psi_i\rangle|,$$
$$|\langle\psi_f|x^2|\psi_i\rangle|, \quad |\langle\psi_f|y^2|\psi_i\rangle|, \quad \text{and} \quad |\langle\psi_f|z^2|\psi_i\rangle|.$$

Using the fact that the quadratic operators transform according to the irreducible representations:

$$(x^2 + y^2, z^2) \Rightarrow A_{1g}$$
$$(xz, yz) \Rightarrow E_{1g}$$
$$(x^2 - y^2, xy) \Rightarrow E_{2g}$$

determine which of the C—H vibrational modes will be Raman active.

A problem on electronic spectra and lifetimes

32 Time-dependent perturbation theory provides an expression for the radiative lifetime of an excited electronic state, given by τ_R:

$$\tau_R = \frac{3\hbar^4 c^3}{4(E_i - E_f)^3 |\mu_{fi}|^2},$$

where i refers to the excited state, f refers to the lower state, and μ_{fi} is the transition dipole.

a. Evaluate the z-component of the transition dipole for the $2p_z \rightarrow 1s$ transition in a hydrogenic atom of nuclear charge Z, given

$$\psi_{1s} = \frac{1}{\sqrt{\pi}}\left(\frac{Z}{a_0}\right)^{\frac{3}{2}} e^{\frac{-Zr}{a_0}}, \qquad \text{and} \quad \psi_{2p_z} = \frac{1}{4\sqrt{2\pi}}\left(\frac{Z}{a_0}\right)^{\frac{5}{2}} r\cos\theta\, e^{\frac{-Zr}{2a_0}}.$$

   Express your answer in units of ea_0.

b. Use symmetry to demonstrate that the x- and y-components of μ_{fi} are zero, i.e.,

$$\langle 2p_z|ex|1s\rangle = \langle 2p_z|ey|1s\rangle = 0.$$

c. Calculate the radiative lifetime τ_R of a hydrogen-like atom in its $2p_z$ state. Use the relation $e^2 = \frac{\hbar^2}{m_e a_0}$ to simplify your results.

The difference between slowly and quickly turning on a perturbation

33 Consider a case in which the complete set of states $\{\phi_k\}$ for a Hamiltonian is known.

a. If the system is initially in the state m at time $t = 0$ when a constant perturbation V is suddenly turned on, find the probability amplitudes $C_k^{(2)}(t)$ and $C_m^{(2)}(t)$, to second order in V, that describe the system being in a different state k or the same state m at time t.
b. If the perturbation is turned on adiabatically (i.e., very slowly), what are $C_k^{(2)}(t)$ and $C_m^{(2)}(t)$? Here, consider that the initial time is $t_0 \rightarrow -\infty$, and the potential is $Ve^{\eta}t$, where the positive parameter η is allowed to approach zero, $\eta \rightarrow 0$, in order to describe the adiabatically turned on perturbation.
c. Compare the results of parts (a) and (b) and explain any differences.
d. Ignore first order contributions (assume they vanish) and evaluate the transition rates $\frac{d}{dt}|C_k^{(2)}(t)|^2$ for the results of part (b) by taking the limit $\eta \rightarrow 0^+$, to obtain the adiabatic results.

An example of quickly turning on a perturbation – the sudden approximation

34 Consider an interaction or perturbation which is carried out suddenly (instantaneously, e.g., within an interval of time Δt which is small compared to the natural period ω_{nm}^{-1} corresponding to the transition from state m to state n), and after that is turned off adiabatically (i.e., extremely slowly as $V e^{\eta t}$). The transition probability in this case is given as

$$T_{nm} \approx \frac{|\langle n|V|m\rangle|^2}{\hbar^2 \omega_{nm}^2},$$

where V corresponds to the maximum value of the interaction when it is turned on. This formula allows one to calculate the transition probabilities under the action of sudden perturbations which are small in absolute value whenever perturbation theory is applicable.

Let's use this "sudden approximation" to calculate the probability of excitation of an electron under a sudden change of the charge of the nucleus. Consider the reaction

$$^3_1\text{H} \to {}^3_2\text{He}^+ + \text{e}^-,$$

and assume the tritium atom has its electron initially in a 1s orbital.

a. Calculate the transition probability for the transition 1s $\to$ 2s for this reaction using the above formula for the transition probability.
b. Suppose that at time $t = 0$ the system is in a state which corresponds to the wave function φ_m, which is an eigenfunction of the operator H_0. At $t = 0$, the sudden change of the Hamiltonian occurs (now denoted as H and remains unchanged). Calculate the same 1s $\to$ 2s transition probability as in part (a), only this time as the square of the magnitude of the coefficient $A_{1\text{s},2\text{s}}$ using the expansion

$$\Psi(r,0) = \varphi_m(r) = \sum_n A_{nm}\psi_n(r), \qquad \text{where} \quad A_{nm} = \int \varphi_m(r)\psi_n(r)d^3r.$$

Note that the eigenfunctions of H are ψ_n with eigenvalues E_n. Compare this value with that obtained by perturbation theory in part (a).

A symmetric top rotational spectrum problem

35 The methyl iodide molecule is studied using microwave (pure rotational) spectroscopy. The following integral governs the rotational selection rules for transitions labeled $J, M, K \to J', M', K'$:

$$I = \langle D^{J'}_{M'K'} \,|\, \boldsymbol{\varepsilon} \cdot \boldsymbol{\mu} \,|\, D^{J}_{MK} \rangle.$$

The dipole moment $\boldsymbol{\mu}$ lies along the molecule's C_3 symmetry axis. Let the electric field of the light $\boldsymbol{\varepsilon}$ define the lab-fixed z-direction.

a. Using the fact that $\cos\beta = D^{1*}_{00}$, show that

$$I = 8\pi^2 \mu\varepsilon(-1)^{(M+K)} \begin{pmatrix} J' & 1 & J \\ M & 0 & M \end{pmatrix} \begin{pmatrix} J' & 1 & J \\ K & 0 & K \end{pmatrix} \delta_{M'M}\delta_{K'K}.$$

b. What restrictions does this result place on $\Delta J = J' - J$? Explain physically why the K quantum number can not change.

A problem in electronic and photoelectron spectroscopy

36 Consider the molecule BO.

a. What is the total number of possible electronic states that can be formed by combination of ground-state B and O atoms?
b. What electron configurations of the molecule are likely to be low in energy? Consider all reasonable orderings of the molecular orbitals. What are the states corresponding to these configurations?
c. What are the bond orders in each of these states?
d. The true ground state of BO is $^2\Sigma$. Specify the $+/-$ and u/g symmetries for this state.
e. Which of the excited states you derived above will radiate to the $^2\Sigma$ ground state? Consider electric dipole radiation only.
f. Does ionization of the molecule to form a cation lead to a stronger, weaker, or equivalent bond strength?
g. Assuming that the energies of the molecular orbitals do not change upon ionization, what are the ground state, the first excited state, and the second excited state of the positive ion?
h. Considering only these states, predict the structure of the photoelectron spectrum you would obtain for ionization of BO.

A problem on vibration–rotation spectroscopy

37

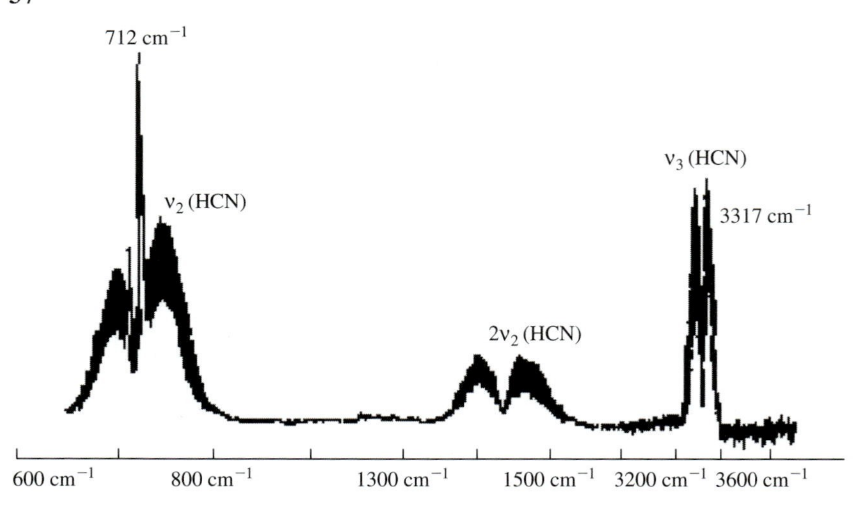

The above figure shows part of the infrared absorption spectrum of HCN gas. The molecule has a CH stretching vibration, a bending vibration, and a CN stretching vibration.

a. Are any of the vibrations of linear HCN degenerate?
b. To which vibration does the group of peaks between 600 cm^{-1} and 800 cm^{-1} belong?
c. To which vibration does the group of peaks between 3200 cm^{-1} and 3400 cm^{-1} belong?
d. What are the symmetries (σ, π, δ) of the CH stretch, CN stretch, and bending vibrational motions?
e. Starting with HCN in its 0, 0, 0 vibrational level, which fundamental transitions would be infrared active under parallel polarized light (i.e., z-axis polarization):

$000 \rightarrow 001$?

$000 \rightarrow 100$?

$000 \rightarrow 010$?

f. Why does the 712 cm^{-1} transition have a Q-branch, whereas that near 3317 cm^{-1} has only P- and R-branches?

A problem in which you can practice deriving equations This is important because a theory scientist does derivations as part of her/his job

38 By expanding the molecular orbitals $\{\phi_k\}$ as linear combinations of atomic orbitals $\{\chi_\mu\}$,

$$\phi_k = \sum_\mu c_{\mu k} \chi_\mu,$$

show how the canonical Hartree–Fock (HF) equations,

$$F\phi_i = \varepsilon_i \phi_i,$$

reduce to the matrix eigenvalue-type equation of the form

$$\sum_\nu F_{\mu\nu} C_{\nu i} = \varepsilon_i \sum_\nu S_{\mu\nu} C_{\nu i},$$

where

$$F_{\mu\nu} = \langle \chi_\mu | h | \chi_\nu \rangle + \sum_{\delta\kappa} \left[\gamma_{\delta\kappa} \langle \chi_\mu \chi_\delta | g | \chi_\nu \chi_\kappa \rangle - \gamma_{\delta\kappa}^{\text{ex}} \langle \chi_\mu \chi_\delta | g | \chi_\kappa \chi_\nu \rangle \right],$$

$$S_{\mu\nu} = \langle \chi_\mu \mid \chi_\nu \rangle, \qquad \gamma_{\delta\kappa} = \sum_{i=\text{occ}} C_{\delta i} C_{\kappa i},$$

and

$$\gamma_{\delta\kappa}^{\text{ex}} = \sum_{i=\substack{\text{occ and}\\\text{same spin}}} C_{\delta i} C_{\kappa i}.$$

Note that the sum over i in $\gamma_{\delta\kappa}$ and $\gamma_{\delta\kappa}^{\text{ex}}$ is a sum over spin orbitals. In addition, show that this Fock matrix can be further reduced for the closed-shell case to

$$F_{\mu\nu} = \langle \chi_\mu | h | \chi_\nu \rangle + \sum_{\delta\kappa} P_{\delta\kappa} \left[\langle \chi_\mu \chi_\delta | g | \chi_\nu \chi_\kappa \rangle - \frac{1}{2} \langle \chi_\mu \chi_\delta | g | \chi_\kappa \chi_\nu \rangle \right],$$

where the charge bond order matrix, P, is defined to be

$$P_{\delta\kappa} = \sum_{i=\text{occ}} 2 C_{\delta i} C_{\kappa i},$$

where the sum over i here is a sum over orbitals not spin orbitals.

Another derivation practice problem

39 Show that the HF total energy for a closed-shell system may be written in terms of integrals over the orthonormal HF orbitals as

$$E(\text{SCF}) = 2 \sum_k^{\text{occ}} \langle \phi_k | h | \phi_k \rangle + \sum_{kl}^{\text{occ}} [2 \langle kl | g | kl \rangle - \langle kl | g | lk \rangle] + \sum_{\mu > \nu} \frac{Z_\mu Z_\nu}{R_{\mu\nu}}.$$

One more derivation problem

40 Show that the HF total energy may alternatively be expressed as

$$E(\text{SCF}) = \sum_k^{\text{occ}} (\varepsilon_k + \langle \phi_k | h | \phi_k \rangle) + \sum_{\mu > \nu} \frac{Z_\mu Z_\nu}{R_{\mu\nu}},$$

where the ε_k refer to the HF orbital energies.

A molecular Hartree–Fock SCF problem

41 This problem will be concerned with carrying out an SCF calculation for the HeH^+ molecule in the ${}^1\sum_g^+(1\sigma^2)$ ground state. The one- and two-electron integrals (in atomic units) needed to carry out this SCF calculation at $R = 1.4$ a.u. using Slater type orbitals with orbital exponents of 1.6875 and 1.0 for the He and H, respectively, are:

$$S_{11} = 1.0, \quad S_{22} = 1.0, \quad S_{12} = 0.5784,$$
$$h_{11} = -2.6442, \quad h_{22} = -1.7201, \quad h_{12} = -1.5113,$$
$$g_{1111} = 1.0547, \quad g_{1121} = 0.4744, \quad g_{1212} = 0.5664,$$
$$g_{2211} = 0.2469, \quad g_{2221} = 0.3504, \quad g_{2222} = 0.6250,$$

where 1 refers to $1s_{\text{He}}$ and 2 to $1s_{\text{H}}$. The two-electron integrals are given in Dirac notation. Parts (a)–(d) should be done by hand. Any subsequent parts can make use of the *QMIC* software that can be found at http://www.emsl.pnl.gov:2080/docs/tms/quantummechanics.

a. Using $\phi_1 \approx 1s_{\text{He}}$ for the initial guess of the occupied molecular orbital, form a 2×2 Fock matrix. Use the equation derived above in Problem 38 for $F_{\mu\nu}$.
b. Solve the Fock matrix eigenvalue equations given above to obtain the orbital energies and an improved occupied molecular orbital. In so doing, note that $\langle \phi_1 \mid \phi_1 \rangle = 1 = C_1^{\text{T}} S C_1$ gives the needed normalization condition for the expansion coefficients of the ϕ_1 in the atomic orbital basis.
c. Determine the total SCF energy using the expression of Problem 39 at this step of the iterative procedure. When will this energy agree with that obtained by using the alternative expression for $E(\text{SCF})$ given in Problem 40?
d. Obtain the new molecular orbital, ϕ_1, from the solution of the matrix eigenvalue problem (part (b)).
e. A new Fock matrix and related total energy can be obtained with this improved choice of molecular orbital, ϕ_1. This process can be continued until a convergence criterion has been satisfied. Typical convergence criteria include: no significant change in the molecular orbitals or the total energy (or both) from one iteration to the next. Perform this iterative procedure for the HeH^+ system until the difference in total energy between two successive iterations is less than 10^{-5} a.u.
f. Show, by comparing the difference between the SCF total energy at one iteration and the converged SCF total energy, that the convergence of the above SCF approach is primarily linear (or first order).
g. Is the SCF total energy calculated at each iteration of the above SCF procedure, as in Problem 39, an upper bound to the exact ground-state total energy?
h. Does this SCF wave function give rise (at $R \to \infty$) to proper dissociation products?

A configuration interaction problem

42 This problem will continue to address the same HeH^+ molecular system as above, extending the analysis to include correlation effects. We will use the one- and two-electron integrals (same geometry) in the converged (to 10^{-5} au) SCF molecular orbital basis which we would have obtained after seven iterations above. The converged MOs you should have obtained in Problem 41 are

$$\phi_1 = \begin{bmatrix} -0.899\,977\,92 \\ -0.158\,430\,12 \end{bmatrix} \qquad \phi_2 = \begin{bmatrix} -0.832\,331\,80 \\ 1.215\,580\,30 \end{bmatrix}.$$

a. Carry out a two-configuration CI calculation using the $1\sigma^2$ and $2\sigma^2$ configurations first by obtaining an expression for the CI matrix elements $H_{I,J}(I, J = 1\sigma^2, 2\sigma^2)$ in terms of one- and two-electron integrals, and secondly by showing that the resultant CI matrix is (ignoring the nuclear repulsion energy)

$$\begin{bmatrix} -4.2720 & 0.1261 \\ 0.1261 & -2.0149 \end{bmatrix}.$$

b. Obtain the two CI energies and eigenvectors for the matrix found in part (a).
c. Show that the lowest energy CI wave function is equivalent to the following two-determinant (single configuration) wave function:

$$\frac{1}{2}\left[\left|\left(a^{\frac{1}{2}}\phi_1 + b^{\frac{1}{2}}\phi_2\right)\alpha\left(a^{\frac{1}{2}}\phi_1 - b^{\frac{1}{2}}\phi_2\right)\beta\right| + \left|\left(a^{\frac{1}{2}}\phi_1 - b^{\frac{1}{2}}\phi_2\right)\alpha\left(a^{\frac{1}{2}}\phi_1 + b^{\frac{1}{2}}\phi_2\right)\beta\right|\right]$$

involving the polarized orbitals: $a^{\frac{1}{2}}\phi_1 \pm b^{\frac{1}{2}}\phi_2$, where $a = 0.9984$ and $b = 0.0556$.

d. Expand the CI list to three configurations by adding $1\sigma 2\sigma$ to the original $1\sigma^2$ and $2\sigma^2$ configurations of part (a) above. First, express the proper singlet spin-coupled $1\sigma 2\sigma$ configuration as a combination of Slater determinants and then compute all elements of this 3×3 CI matrix.
e. Obtain all eigenenergies and corresponding normalized eigenvectors for this CI problem.
f. Determine the excitation energies and transition moments for HeH^+ using the full CI result of part (e) above. The non-vanishing matrix elements of the dipole operator $\mathbf{r}(x, y, z)$ in the atomic basis are

$$\langle 1s_H|z|1s_{He}\rangle = 0.2854 \qquad \text{and} \qquad \langle 1s_H|z|1s_H\rangle = 1.4.$$

First determine the matrix elements of $\mathbf{r}$ in the SCF orbital basis then determine the excitation energies and transition moments from the ground state to the two excited singlet states of HeH^+.

g. Now turning to perturbation theory, carry out a perturbation theory calculation of the first order wave function $|1\sigma^2 >^{(1)}$ for the case in which the zeroth order wave function is taken to be the $1\sigma^2$ Slater determinant. Show that the first order wave function is given by

$$|1\sigma^2\rangle^{(1)} = -0.0442|2\sigma^2\rangle.$$

h. Why does the $|1\sigma 2\sigma\rangle$ configuration not enter into the first order wave function?

i. Normalize the resultant wave function that contains zeroth plus first order parts and compare it to the wave function obtained in the two-configuration CI study of part (b).
j. Show that the second order RSPT correlation energy, $E^{(2)}$, of HeH^+ is -0.0056 a.u. How does this compare with the correlation energy obtained from the two-configuration CI study of part (b)?

Repeating the SCF problem but with a computer program

43 Using either programs that are available to you or the *QMIC* programs that you can find at the web site http://www.emsl.pnl.gov:2080/docs/tms/quantummechanics calculate the SCF energy of HeH^+ using the same geometry as in Problem 42 and the STO3G basis set provided in the *QMIC* basis set library. How does this energy compare to that found in Problem 42? Run the calculation again with the 3-21G basis set provided. How does this energy compare to the STO3G and the energy found using STOs in Problem 42?

A series of SCF calculations to produce a potential energy curve

44 Generate SCF potential energy surfaces for HeH^+ and H_2 using the *QMIC* software or your own programs. Use the 3-21G basis set and generate points for geometries of $R = 1.0, 1.2, 1.4, 1.6, 1.8, 2.0, 2.5,$ and $10.0\ a_0$. Plot the energies vs. geometry for each system. Which system dissociates properly?

Configuration interaction potential curves for several states

45 Generate CI potential energy surfaces for the four states of H_2 resulting from a calculation with two electrons occupying the lowest two SCF orbitals ($1\sigma_g$ and $1\sigma_u$) in all possible ways. Use the same geometries and basis set as in Problem 44. Plot the energies vs. geometry for each system. Properly label and characterize each of the states (e.g., repulsive, dissociate properly, etc.).

A problem on partition functions and thermodynamic properties

46 F atoms have $1s^2 2s^2 2p^5\ {}^2P$ ground electronic states that are split by spin–orbit coupling into ${}^2P_{3/2}$ and ${}^2P_{1/2}$ states that differ by only 0.05 eV in energy.

a. Write the electronic partition function (take the energy of the ${}^2P_{3/2}$ state to be zero and that of the ${}^2P_{1/2}$ state to be 0.05 eV and ignore all other states) for each F atom.
b. Using $\bar{E} = kT^2(\frac{\partial \ln Q}{\partial T})_{N,V}$, derive an expression for the average electronic energy $\bar{E}$ of N gaseous F atoms.

c. Using the fact that $kT = 0.03$ eV at $T = 300$ K, make a (qualitative) graph of $\bar{E}/N$ vs. T for T ranging from 100 K to 3000 K.

A problem using transition state theory

47 Suppose that we used transition state theory to study the reaction $NO(g) + Cl_2(g) \rightarrow NOCl(g) + Cl(g)$ assuming it to proceed through a bent transition state, and we obtained an expression for the rate coefficient:

$$k_{\text{bent}} = \frac{kT}{h} e^{-E^{\neq}/kT} \frac{(q^{\neq}/V)}{\frac{(q_{\text{NO}})}{V} \frac{(q_{\text{Cl}_2})}{V}}.$$

a. Now, let us consider what differences would occur if the transition state structure were linear rather than bent. Assuming that the activation energy $E^{\neq}$ and electronic state degeneracies are not altered, derive an expression for the ratio of the rate coefficients for the linear and bent transition state cases:

$$\frac{k_{\text{linear}}}{k_{\text{bent}}} =$$

b. Using the following order of magnitude estimates of translational, rotational, and vibrational partition functions per degree of freedom at 300 K:

$$q_{\text{t}} \sim 10^8, \qquad q_{\text{r}} \sim 10^2, \qquad q_{\text{v}} \sim 1,$$

what ratio would you expect for $k_{\text{linear}}/k_{\text{bent}}$?

A problem with Slater determinants

48 Show that the configuration (determinant) corresponding to the Li^+ $1s(\alpha)1s(\alpha)$ state vanishes.

Another problem with Slater determinants and angular momenta

49 Construct the three triplet and one singlet wave functions for the Li^+ $1s^12s^1$ configuration. Show that each state is a proper eigenfunction of S^2 and S_z (use raising and lowering operators for S^2).

A problem with Slater determinants for a linear molecule

50 Construct determinant wave functions for each state of the $1\sigma^2 2\sigma^2 3\sigma^2 1\pi^2$ configuration of NH.

A problem with Slater determinants for an atom

51 Construct determinant wave functions for each state of the $1s^1 2s^1 3s^1$ configurations of Li.

A problem on angular momentum of an atom

52 Determine all term symbols that arise from the $1s^2 2s^2 2p^2 3d^1$ configuration of the excited N atom.

Practice with the Slater–Condon rules

53 Calculate the energy (using Slater–Condon rules) associated with the 2p valence electrons for the following states of the C atom.

a. $^3P(M_L = 1, M_S = 1)$,
b. $^3P(M_L = 0, M_S = 0)$,
c. $^1S(M_L = 0, M_S = 0)$, and
d. $^1D(M_L = 0, M_S = 0)$.

More practice with the Slater–Condon rules

54 Calculate the energy (using Slater–Condon rules) associated with the π valence electrons for the following states of the NH molecule.

a. $^1\Delta(M_L = 2, M_S = 0)$,
b. $^1\sum(M_L = 0, M_S = 0)$, and
c. $^3\sum(M_L = 0, M_S = 0)$.

Practice with the equations of statistical mechanics

55 Match each of the equations below with the proper phrase A–K.

$$B_2 = -2\pi \int_0^\infty r^2\left(e^{-u(r)/kT} - 1\right)dr$$ ☐

$$\bar{E}^2 - (\bar{E})^2 = kT^2 \left(\frac{\partial E}{\partial T}\right)_{N,V}$$ ☐

$$\left(\frac{2\pi mkT}{h^2}\right)$$ ☐

$$Q = e^{-\frac{N\phi}{2kT}} \left(\frac{e^{-\theta/2T}}{1-e^{-\theta/T}}\right)^{3N}$$ ☐

$$g(\nu) = \alpha\nu^2$$ ☐

$Q = \frac{M!}{N!(M-N)!} q^N$ ☐

$\Theta = \frac{q e^{\mu 0/kT} p}{1 + q e^{\mu 0/kT} p}$ ☐

$p_{\mathrm{A}} = p_{\mathrm{A}}^0 X_{\mathrm{A}}$ ☐

$\frac{c\omega}{kT} = -4$ ☐

$W = W_{\mathrm{AA}} N_{\mathrm{AA}} + W_{\mathrm{BB}} N_{\mathrm{BB}} + W_{\mathrm{AB}} N_{\mathrm{AB}}$ ☐

$N_{\mathrm{AB}} \cong \frac{N_{\mathrm{A}}\, \mathrm{c}\, N_{\mathrm{B}}}{N_{\mathrm{A}} + N_{\mathrm{B}}}$ ☐

A. Raoult's law

B. Debye solid

C. Critical point

D. Ideal adsorption

E. Langmuir isotherm

F. Bragg–Williams

G. Partition function for surface translation

H. Concentrated solution

I. Fluctuation

J. Virial coefficient

K. Einstein solid

A problem dealing with the second virial coefficient

56 The van der Waals equation of state is

$$\left(p + \left(\frac{N}{V}\right)^2 a\right)(V - Nb) = NkT.$$

Solve this equation for p, and then obtain an expression for $\frac{pV}{NkT}$. Finally, expand $\frac{pV}{NkT}$ in powers of $(\frac{N}{V})$ and obtain an expression for the second virial coefficient of this van der Waals gas in terms of b, a, and T.

A problem to make you think about carrying out Monte-Carlo and molecular dynamics simulations

57 Briefly answer each of the following:
For which of the following would you be wisest to use Monte-Carlo (MC) simulation and for which should you use molecular dynamics (MD)?

a. Determining the rate of diffusion of CH_4 in liquid Kr.
b. Determining the equilibrium radial distribution of Kr atoms relative to the CH_4 in the above example.
c. Determining the mean square end-to-end distance for a floppy hydrocarbon chain in the liquid state.

Suppose you are carrying out a Monte-Carlo simulation involving 1000 Ar atoms. Further suppose that the potentials are pairwise additive and that your computer requires approximately 50 floating point operations (FPOs) (e.g., multiply, add, divide, etc.) to compute the interaction potential between *any pair* of atoms.

d. For each MC trial move, how many FPOs are required? Assuming your computer has a speed of 100 MFlops (i.e., 100 million FPOs per second), how long will it take you to carry out 1 000 000 MC moves?
e. If the fluctuations observed in the calculation of question (d) are too large, and you wish to make a longer MC calculation to reduce the statistical "noise", how long will your new calculation require if you wish to cut the noise in half?
f. How long would the calculation of question (d) require if you were to use 1 000 000 Ar atoms (with the same potential and the same computer)?
g. Assuming that the evaluation of the forces between pairs of Ar atoms ($\partial V/\partial r$) requires approximately the same number of FPOs (50) as for computing the pair potential, how long (in seconds) would it take to carry out a molecular dynamics simulation involving 1000 Ar atoms using a time step (δt) of 10^{-15} s and persisting for a total time duration of one nanosecond (10^{-9} s) using the 100 MFlop computer?
h. How long would a 10^{-6} MFlop (i.e., 1 FPO per sec) Ph.D. student take to do the calculation in part (d)?

A problem to practice using partition functions

58 In this problem, you will compute the pressure-unit equilibrium constant K_p for the equilibrium

$$2\ \mathrm{Na} \leftrightarrow \mathrm{Na}_2$$

in the gas phase at a temperature of 1000 K. Your final answer should be expressed in units of atm^{-1}. In doing so, you need to consider the electronic term symbols of Na and of Na_2, and you will need to use the following data:

(i) Na has no excited electronic states that you need to consider.
(ii) $(h^2/8\pi^2 Ik) = 0.221$ K for Na_2.

(iii) $(h\nu/k) = 229$ K for Na_2.
(iv) 1 atm $= 1.01 \times 10^6$ dynes cm^{-2}.
(v) The dissociation energy of Na_2 from the $v = 0$ to dissociation is $D_0 = 17.3$ kcal mol^{-1}.

a. First, write the expressions for the Na and Na_2 partition functions showing their translational, rotational, vibrational and electronic contributions.
b. Next, substitute the data and compute K_p, and change units to atm^{-1}.

A problem using transition state theory

59 Looking back at the NO + Cl_2 reaction treated using transition state theory in Problem 47, let us assume that this same reaction (via the bent transition state) were to occur while two reagents NO and Cl_2 were adsorbed to a surface in the following manner:

(i) Both NO and Cl_2 lie flat against the surface with both of their atoms touching the surface.
(ii) Both NO and Cl_2 move freely along the surface (i.e., they can translate parallel to the surface).
(iii) Both NO and Cl_2 are tightly bound to the surface in a manner that causes their movements perpendicular to the surface to become high-frequency vibrations.

Given this information, and again assuming the following order of magnitude estimates of partition functions,

$$q_t \sim 10^8, \qquad q_r \sim 10^2, \qquad q_v \sim 1,$$

calculate the ratio of the TS rate constants for this reaction occurring in the surface adsorbed state and in the gas phase. In doing so, you may assume that the activation energy and all properties of the transition state are identical in the gas and adsorbed state, except that the TS species is constrained to lie flat on the surface just as are NO and Cl_2.

Solutions

1 a. First determine the eigenvalues:

$$\det\begin{bmatrix} -1-\lambda & 2 \\ 2 & 2-\lambda \end{bmatrix} = (-1-\lambda)(2-\lambda) - 2^2$$
$$= -2 + \lambda - 2\lambda + \lambda^2 - 4$$
$$= \lambda^2 - \lambda - 6$$
$$= (\lambda - 3)(\lambda + 2)$$
$$= 0$$
$$\lambda = 3 \quad \text{or} \quad \lambda = -2.$$

Next, determine the eigenvectors. First, the eigenvector associated with eigenvalue −2:

$$\begin{bmatrix} -1 & 2 \\ 2 & 2 \end{bmatrix}\begin{bmatrix} C_{11} \\ C_{21} \end{bmatrix} = -2\begin{bmatrix} C_{11} \\ C_{21} \end{bmatrix}$$
$$-C_{11} + 2C_{21} = -2C_{11}$$
$$C_{11} = -2C_{21} \quad \text{(Note: the second row offers no new information, e.g. } 2C_{11} + 2C_{21} = -2C_{21}\text{)}$$
$$C_{11}^2 + C_{21}^2 = 1 \quad \text{(from normalization)}$$
$$(-2C_{21})^2 + C_{21}^2 = 1$$
$$4C_{21}^2 + C_{21}^2 = 1$$
$$5C_{21}^2 = 1$$
$$C_{21}^2 = 0.2$$
$$C_{21} = \sqrt{0.2} \quad \text{and therefore} \quad C_{11} = -2\sqrt{0.2}.$$

For the eigenvector associated with eigenvalue 3:

$$\begin{bmatrix} -1 & 2 \\ 2 & 2 \end{bmatrix}\begin{bmatrix} C_{12} \\ C_{22} \end{bmatrix} = 3\begin{bmatrix} C_{12} \\ C_{22} \end{bmatrix}$$

$$-C_{12} + 2C_{22} = 3C_{12}$$
$$-4C_{12} = -2C_{22}$$
$$C_{12} = 0.5C_{22} \quad \text{(again the second row offers no new information)}$$
$$C_{12}^2 + C_{22}^2 = 1 \quad \text{(from normalization)}$$
$$(0.5C_{22})^2 + C_{22}^2 = 1$$
$$0.25C_{22}^2 + C_{22}^2 = 1$$
$$1.25C_{22}^2 = 1$$
$$C_{22}^2 = 0.8$$
$$C_{22} = \sqrt{0.8} = 2\sqrt{0.2}, \quad \text{and therefore} \quad C_{12} = \sqrt{0.2}.$$

Therefore the eigenvector matrix becomes:

$$\begin{bmatrix} -2\sqrt{0.2} & \sqrt{0.2} \\ \sqrt{0.2} & 2\sqrt{0.2} \end{bmatrix}$$

b. First determine the eigenvalues:

$$\det \begin{bmatrix} -2-\lambda & 0 & 0 \\ 0 & -1-\lambda & 2 \\ 0 & 2 & 2-\lambda \end{bmatrix} = 0$$

$$\det\,[-2-\lambda]\,\det \begin{bmatrix} -1-\lambda & 2 \\ 2 & 2-\lambda \end{bmatrix} = 0$$

From 1a, the solutions then become −2, −2, and 3. Next, determine the eigenvectors. First the eigenvector associated with eigenvalue 3 (the third root):

$$\begin{bmatrix} -2 & 0 & 0 \\ 0 & -1 & 2 \\ 0 & 2 & 2 \end{bmatrix} \begin{bmatrix} C_{11} \\ C_{21} \\ C_{31} \end{bmatrix} = 3 \begin{bmatrix} C_{11} \\ C_{21} \\ C_{31} \end{bmatrix}$$
$$-2C_{13} = 3C_{13} \quad \text{(row one)}$$
$$C_{13} = 0$$
$$-C_{23} + 2C_{33} = 3C_{23} \quad \text{(row two)}$$
$$2C_{33} = 4C_{23}$$
$$C_{33} = 2C_{23} \quad \text{(again the third row offers no new information)}$$
$$C_{13}^2 + C_{23}^2 + C_{33}^2 = 1 \quad \text{(from normalization)}$$
$$0 + C_{23}^2 + (2C_{23})^2 = 1$$
$$5C_{23}^2 = 1$$
$$C_{23} = \sqrt{0.2}, \quad \text{and therefore } C_{33} = 2\sqrt{0.2}.$$

Next, find the pair of eigenvectors associated with the degenerate eigenvalue of −2. First, root one eigenvector one:

$-2C_{11} = -2C_{11}$ (no new information from row one)

$-C_{21} + 2C_{31} = -2C_{21}$ (row two)

$C_{21} = -2C_{31}$ (again the third row offers no new information)

$C_{11}^2 + C_{21}^2 + C_{31}^2 = 1$ (from normalization)

$C_{11}^2 + (-2C_{31})^2 + C_{31}^2 = 1$

$C_{11}^2 + 5C_{31}^2 = 1$

$C_{11} = \sqrt{1 - 5C_{31}^2}$ (Note: there are now two equations with three unknowns.)

Second, root two eigenvector two:

$-2C_{12} = -2C_{12}$ (no new information from row one)

$-C_{22} + 2C_{32} = -2C_{22}$ (row two)

$C_{22} = -2C_{32}$ (again the third row offers no new information)

$C_{12}^2 + C_{22}^2 + C_{32}^2 = 1$ (from normalization)

$C_{12}^2 + (-2C_{32})^2 + C_{32}^2 = 1$

$C_{12}^2 + 5C_{32}^2 = 1$

$C_{12} = \left(1 - 5C_{32}^2\right)^{1/2}$
(Note: again, two equations in three unknowns)

$C_{11}C_{12} + C_{21}C_{22} + C_{31}C_{32} = 0$ (from orthogonalization).

Now there are five equations with six unknowns.

Arbitrarily choose $C_{11} = 0$.

(Whenever there are degenerate eigenvalues, there are not unique eigenvectors because the degenerate eigenvectors span a two- or more-dimensional space, not two unique directions. One always is then forced to choose one of the coefficients and then determine all the rest; different choices lead to different final eigenvectors but to identical spaces spanned by these eigenvectors.)

$C_{11} = 0 = \sqrt{1 - 5C_{31}^2}$

$5C_{31}^2 = 1$

$C_{31} = \sqrt{0.2}$

$C_{21} = -2\sqrt{0.2}$

$$C_{11}C_{12} + C_{21}C_{22} + C_{31}C_{32} = 0 \quad \text{(from orthogonalization)}$$
$$0 + (-2\sqrt{0.2}(-2C_{32})) + \sqrt{0.2}\, C_{32} = 0$$
$$5C_{32} = 0$$
$$C_{32} = 0, \qquad C_{22} = 0, \qquad \text{and} \qquad C_{12} = 1.$$

Therefore the eigenvector matrix becomes:

$$\begin{bmatrix} 0 & 1 & 0 \\ -2\sqrt{0.2} & 0 & \sqrt{0.2} \\ \sqrt{0.2} & 0 & 2\sqrt{0.2} \end{bmatrix}$$

2 a.

$$\begin{aligned} \text{KE} &= \frac{mv^2}{2} = \left(\frac{m}{m}\right)\frac{mv^2}{2} = \frac{(mv)^2}{2m} = \frac{p^2}{2m} \\ &= \frac{1}{2m}\left(p_x^2 + p_y^2 + p_z^2\right) \\ &= \frac{1}{2m}\left\{\left(\frac{\hbar\partial}{i\partial x}\right)^2 + \left(\frac{\hbar\partial}{i\partial y}\right)^2 + \left(\frac{\hbar\partial}{i\partial z}\right)^2\right\} \\ &= \frac{-\hbar^2}{2m}\left\{\frac{\partial^2}{\partial x^2} + \frac{\partial^2}{\partial y^2} + \frac{\partial^2}{\partial z^2}\right\}. \end{aligned}$$

b.

$$\mathbf{p} = m\mathbf{v} = \mathbf{i}p_x + \mathbf{j}p_y + \mathbf{k}p_z,$$
$$p = \left\{\mathbf{i}\left(\frac{\hbar\partial}{i\partial x}\right) + \mathbf{j}\left(\frac{\hbar\partial}{i\partial y}\right) + \mathbf{k}\left(\frac{\hbar\partial}{i\partial z}\right)\right\},$$

where **i**, **j**, and **k** are unit vectors along the x, y, and z axes.

c.

$$\begin{aligned} L_y &= zp_x - xp_z \\ &= z\left(\frac{\hbar\partial}{i\partial x}\right) - x\left(\frac{\hbar\partial}{i\partial z}\right). \end{aligned}$$

3 First derive the general formulas for $\frac{\partial}{\partial x}$, $\frac{\partial}{\partial y}$, $\frac{\partial}{\partial z}$ in terms of r, θ, and ϕ, and $\frac{\partial}{\partial r}$, $\frac{\partial}{\partial \theta}$, and $\frac{\partial}{\partial \phi}$ in terms of x, y, and z. The general relationships are as follows:

$$x = r\sin\theta\cos\phi \qquad r^2 = x^2 + y^2 + z^2$$
$$y = r\sin\theta\sin\phi \qquad \sin\theta = \frac{\sqrt{x^2+y^2}}{\sqrt{x^2+y^2+z^2}}$$
$$z = r\cos\theta \qquad \cos\theta = \frac{z}{\sqrt{x^2+y^2+z^2}}$$
$$\tan\phi = \frac{y}{x}$$

Next, $\frac{\partial}{\partial x}$, $\frac{\partial}{\partial y}$, and $\frac{\partial}{\partial z}$ from the chain rule:

$$\frac{\partial}{\partial x} = \left(\frac{\partial r}{\partial x}\right)_{y,z}\frac{\partial}{\partial r} + \left(\frac{\partial \theta}{\partial x}\right)_{y,z}\frac{\partial}{\partial \theta} + \left(\frac{\partial \phi}{\partial x}\right)_{y,z}\frac{\partial}{\partial \phi},$$

$$\frac{\partial}{\partial y} = \left(\frac{\partial r}{\partial y}\right)_{x,z}\frac{\partial}{\partial r} + \left(\frac{\partial \theta}{\partial y}\right)_{x,z}\frac{\partial}{\partial \theta} + \left(\frac{\partial \phi}{\partial y}\right)_{x,z}\frac{\partial}{\partial \phi},$$

$$\frac{\partial}{\partial z} = \left(\frac{\partial r}{\partial z}\right)_{x,y}\frac{\partial}{\partial r} + \left(\frac{\partial \theta}{\partial z}\right)_{x,y}\frac{\partial}{\partial \theta} + \left(\frac{\partial \phi}{\partial z}\right)_{x,y}\frac{\partial}{\partial \phi}.$$

Evaluation of the many "coefficients" gives the following:

$$\left(\frac{\partial r}{\partial x}\right)_{y,z} = \sin\theta\cos\phi, \quad \left(\frac{\partial \theta}{\partial x}\right)_{y,z} = \frac{\cos\theta\cos\phi}{r}, \quad \left(\frac{\partial \phi}{\partial x}\right)_{y,z} = -\frac{\sin\phi}{r\sin\theta},$$

$$\left(\frac{\partial r}{\partial y}\right)_{x,z} = \sin\theta\sin\phi, \quad \left(\frac{\partial \theta}{\partial y}\right)_{x,z} = \frac{\cos\theta\sin\phi}{r}, \quad \left(\frac{\partial \phi}{\partial y}\right)_{x,z} = \frac{\cos\phi}{r\sin\theta},$$

$$\left(\frac{\partial r}{\partial z}\right)_{x,y} = \cos\theta, \quad \left(\frac{\partial \theta}{\partial z}\right)_{x,y} = -\frac{\sin\theta}{r}, \quad \text{and} \quad \left(\frac{\partial \phi}{\partial z}\right)_{x,y} = 0.$$

Upon substitution of these "coefficients":

$$\frac{\partial}{\partial x} = \sin\theta\cos\phi\frac{\partial}{\partial r} + \frac{\cos\theta\cos\phi}{r}\frac{\partial}{\partial \theta} - \frac{\sin\phi}{r\sin\theta}\frac{\partial}{\partial \phi},$$

$$\frac{\partial}{\partial y} = \sin\theta\sin\phi\frac{\partial}{\partial r} + \frac{\cos\theta\sin\phi}{r}\frac{\partial}{\partial \theta} + \frac{\cos\phi}{r\sin\theta}\frac{\partial}{\partial \phi},$$

$$\text{and} \quad \frac{\partial}{\partial z} = \cos\theta\frac{\partial}{\partial r} - \frac{\sin\theta}{r}\frac{\partial}{\partial \theta} + 0\frac{\partial}{\partial \phi}.$$

Next $\frac{\partial}{\partial r}$, $\frac{\partial}{\partial \theta}$, and $\frac{\partial}{\partial \phi}$ from the chain rule:

$$\frac{\partial}{\partial r} = \left(\frac{\partial x}{\partial r}\right)_{\theta,\phi}\frac{\partial}{\partial x} + \left(\frac{\partial y}{\partial r}\right)_{\theta,\phi}\frac{\partial}{\partial y} + \left(\frac{\partial z}{\partial r}\right)_{\theta,\phi}\frac{\partial}{\partial z},$$

$$\frac{\partial}{\partial \theta} = \left(\frac{\partial x}{\partial \theta}\right)_{r,\phi}\frac{\partial}{\partial x} + \left(\frac{\partial y}{\partial \theta}\right)_{r,\phi}\frac{\partial}{\partial y} + \left(\frac{\partial z}{\partial \theta}\right)_{r,\phi}\frac{\partial}{\partial z},$$

$$\text{and} \quad \frac{\partial}{\partial \phi} = \left(\frac{\partial x}{\partial \phi}\right)_{r,\theta}\frac{\partial}{\partial x} + \left(\frac{\partial y}{\partial \phi}\right)_{r,\theta}\frac{\partial}{\partial y} + \left(\frac{\partial z}{\partial \phi}\right)_{r,\theta}\frac{\partial}{\partial z}.$$

Again evaluation of the the many "coefficients" results in:

$$\left(\frac{\partial x}{\partial r}\right)_{\theta,\phi} = \frac{x}{\sqrt{x^2+y^2+z^2}}, \quad \left(\frac{\partial y}{\partial r}\right)_{\theta,\phi} = \frac{y}{\sqrt{x^2+y^2+z^2}},$$

$$\left(\frac{\partial z}{\partial r}\right)_{\theta,\phi} = \frac{z}{\sqrt{x^2+y^2+z^2}}, \quad \left(\frac{\partial x}{\partial \theta}\right)_{r,\phi} = \frac{xz}{\sqrt{x^2+y^2}}, \quad \left(\frac{\partial y}{\partial \theta}\right)_{r,\phi} = \frac{yz}{\sqrt{x^2+y^2}},$$

$$\left(\frac{\partial z}{\partial \theta}\right)_{r,\phi} = -\sqrt{x^2+y^2}, \quad \left(\frac{\partial x}{\partial \phi}\right)_{r,\theta} = -y, \quad \left(\frac{\partial y}{\partial \phi}\right)_{r,\theta} = x,$$

$$\text{and} \quad \left(\frac{\partial z}{\partial \phi}\right)_{r,\theta} = 0.$$

Upon substitution of these "coefficients":

$$\frac{\partial}{\partial r} = \frac{x}{\sqrt{x^2+y^2+z^2}}\frac{\partial}{\partial x} + \frac{y}{\sqrt{x^2+y^2+z^2}}\frac{\partial}{\partial y} + \frac{z}{\sqrt{x^2+y^2+z^2}}\frac{\partial}{\partial z},$$

$$\frac{\partial}{\partial \theta} = \frac{xz}{\sqrt{x^2+y^2}}\frac{\partial}{\partial x} + \frac{yz}{\sqrt{x^2+y^2}}\frac{\partial}{\partial y} - \sqrt{x^2+y^2}\frac{\partial}{\partial z},$$

$$\frac{\partial}{\partial \phi} = -y\frac{\partial}{\partial x} + x\frac{\partial}{\partial y} + 0\frac{\partial}{\partial z}.$$

Note, these many "coefficients" are the elements which make up the Jacobian matrix used whenever one wishes to transform a function from one coordinate representation to another. One very familiar result should be in transforming the volume element $dx\ dy\ dz$ to $r^2 \sin\theta\ dr\ d\theta\ d\phi$. For example:

$$\int f(x,y,z)dx\ dy\ dz$$

$$= \int f(x(r,\theta,\phi),\ y(r,\theta,\phi), z(r,\theta,\phi)) \begin{vmatrix} \left(\frac{\partial x}{\partial r}\right)_{\theta,\phi} & \left(\frac{\partial x}{\partial \theta}\right)_{r,\phi} & \left(\frac{\partial x}{\partial \phi}\right)_{r,\theta} \\ \left(\frac{\partial y}{\partial r}\right)_{\theta,\phi} & \left(\frac{\partial y}{\partial \theta}\right)_{r,\phi} & \left(\frac{\partial y}{\partial \phi}\right)_{r,\theta} \\ \left(\frac{\partial z}{\partial r}\right)_{\theta,\phi} & \left(\frac{\partial z}{\partial \theta}\right)_{r,\phi} & \left(\frac{\partial z}{\partial \phi}\right)_{r,\theta} \end{vmatrix} dr\ d\theta\ d\phi.$$

a.

$$\begin{aligned} \mathbf{L}_x &= \frac{\hbar}{i}\left\{y\frac{\partial}{\partial z} - z\frac{\partial}{\partial y}\right\} \\ &= \frac{\hbar}{i}\left(r\sin\theta\sin\phi\left(\cos\theta\frac{\partial}{\partial r} - \frac{\sin\theta}{r}\frac{\partial}{\partial \theta}\right)\right) \\ &\quad -\frac{\hbar}{i}\left(r\cos\theta\left(\sin\theta\sin\phi\frac{\partial}{\partial r} + \frac{\cos\theta\sin\phi}{r}\frac{\partial}{\partial \theta} + \frac{\cos\phi}{r\sin\theta}\frac{\partial}{\partial \phi}\right)\right) \\ &= -\frac{\hbar}{i}\left(\sin\phi\frac{\partial}{\partial \theta} + \cot\theta\cos\phi\frac{\partial}{\partial \phi}\right). \end{aligned}$$

b.

$$\begin{aligned} \mathbf{L}_z &= \frac{\hbar}{i}\frac{\partial}{\partial \phi} = -i\hbar\frac{\partial}{\partial \phi} \\ &= \frac{\hbar}{i}\left(-y\frac{\partial}{\partial x} + x\frac{\partial}{\partial y}\right). \end{aligned}$$

4

	B	dB/dx	d^2B/dx^2
(i)	$4x^4 - 12x^2 + 3$	$16x^3 - 24x$	$48x^2 - 24$
(ii)	$5x^4$	$20x^3$	$60x^2$
(iii)	$e^{3x} + e^{-3x}$	$3(e^{3x} - e^{-3x})$	$9(e^{3x} + e^{-3x})$
(iv)	$x^2 - 4x + 2$	$2x - 4$	2
(v)	$4x^3 - 3x$	$12x^2 - 3$	$24x$

B(v) is an eigenfunction of A(i):

$$\begin{aligned}(1-x^2)\frac{d^2}{dx^2} - x\frac{d}{dx}\,\mathrm{B(v)} &= (1-x^2)(24x) - x(12x^2 - 3)\\ &= 24x - 24x^3 - 12x^3 + 3x\\ &= -36x^3 + 27x\\ &= -9(4x^3 - 3x) \quad \text{(eigenvalue is } -9).\end{aligned}$$

B(iii) is an eigenfunction of A(ii):

$$\frac{d^2}{dx^2}\,\mathrm{B(iii)} = 9(e^{3x} + e^{-3x}) \quad \text{(eigenvalue is 9).}$$

B(ii) is an eigenfunction of A(iii):

$$\begin{aligned}x\frac{d}{dx}\,\mathrm{B(ii)} &= x(20x^3)\\ &= 20x^4\\ &= 4(5x^4) \quad \text{(eigenvalue is 4).}\end{aligned}$$

B(i) is an eigenfunction of A(vi):

$$\begin{aligned}\frac{d^2}{dx^2} - 2x\frac{d}{dx}\,\mathrm{B(i)} &= (48x^2 - 24) - 2x(16x^3 - 24x)\\ &= 48x^2 - 24 - 32x^4 + 48x^2\\ &= -32x^4 + 96x^2 - 24\\ &= -8(4x^4 - 12x^2 + 3) \quad \text{(eigenvalue is } -8).\end{aligned}$$

B(iv) is an eigenfunction of A(v):

$$\begin{aligned}x\frac{d^2}{dx^2} + (1-x)\frac{d}{dx}\,\mathrm{B(iv)} &= x(2) + (1-x)(2x - 4)\\ &= 2x + 2x - 4 - 2x^2 + 4x\\ &= -2x^2 + 8x - 4\\ &= -2(x^2 - 4x + 2) \quad \text{(eigenvalue is } -2).\end{aligned}$$

5

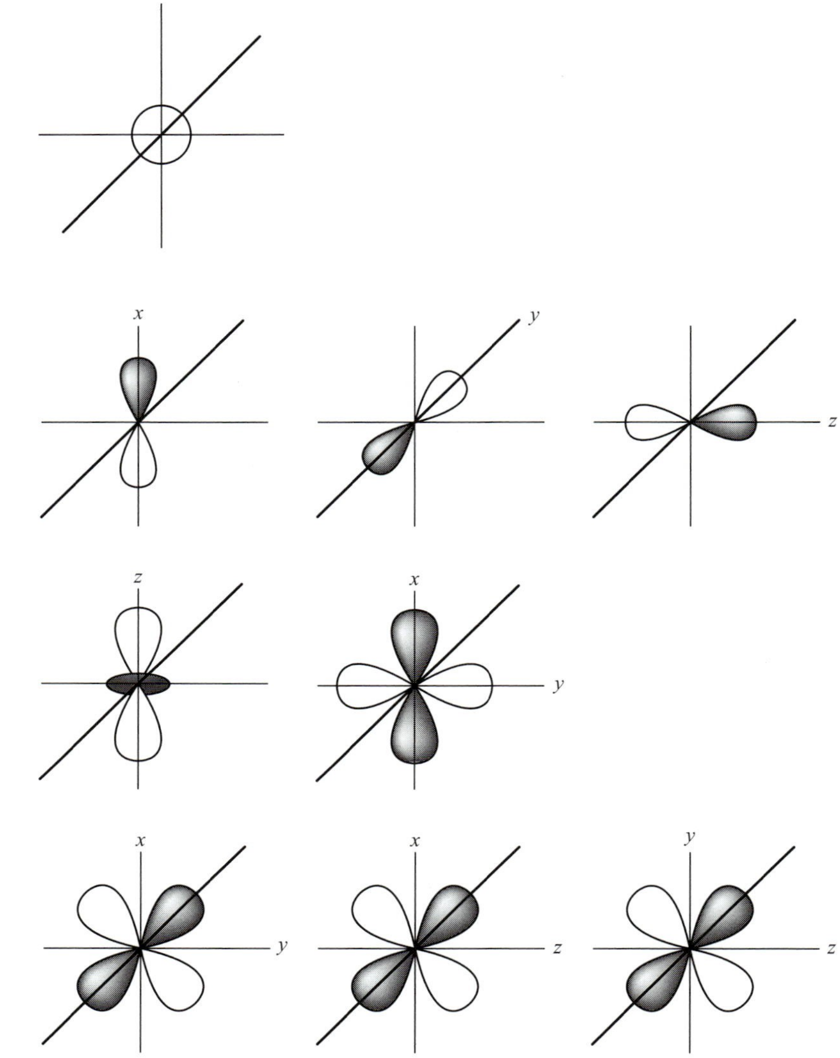

6

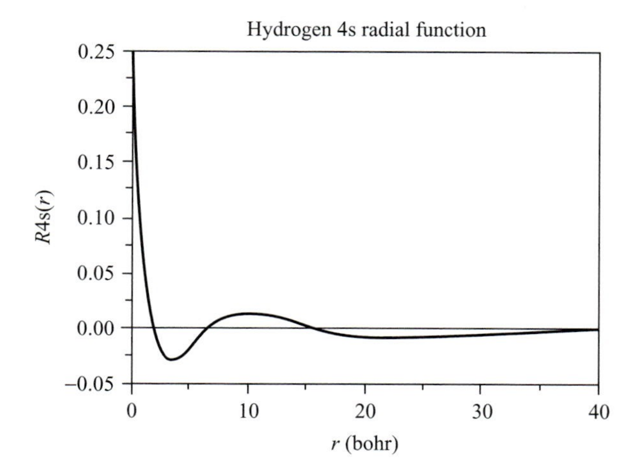
Hydrogen 4s radial function
0.25
0.20
0.15
0.10
0.05
0.00
−0.05
R4s(r)
0
10
20
30
40
r (bohr)

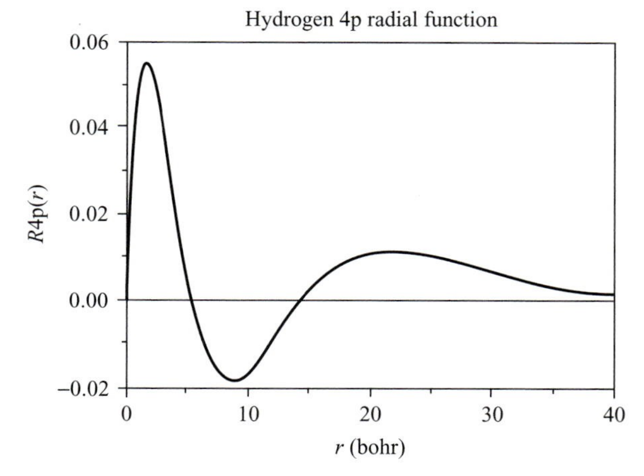
Hydrogen 4p radial function
0.06
0.04
0.02
0.00
−0.02
R4p(r)
0
10
20
30
40
r (bohr)

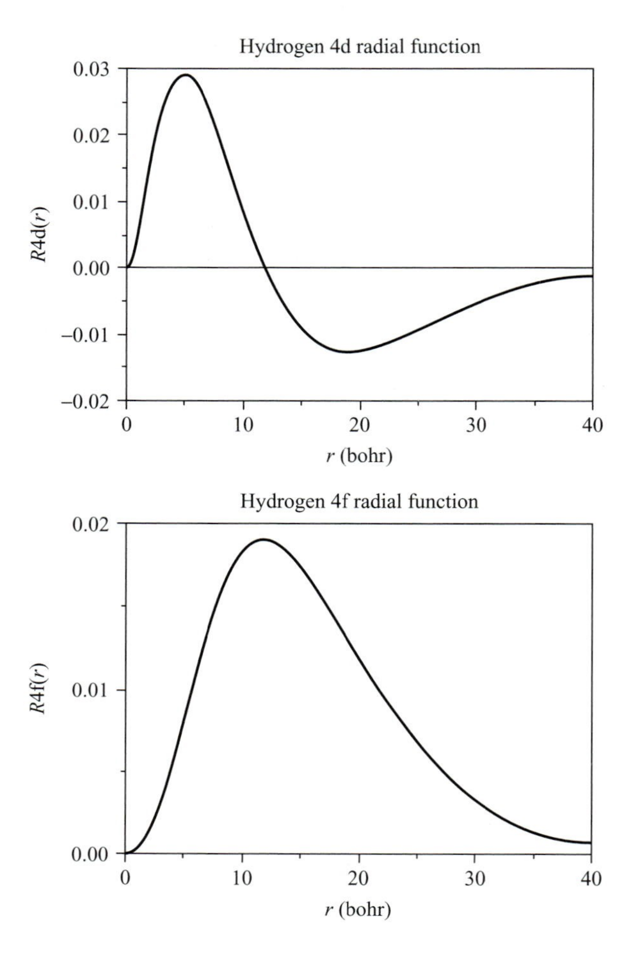
Hydrogen 4d radial function
R4d(r)
0.03
0.02
0.01
0.00
−0.01
−0.02
0
10
20
30
40
r (bohr)
Hydrogen 4f radial function
R4f(r)
0.02
0.01
0.00
0
10
20
30
40
r (bohr)

7

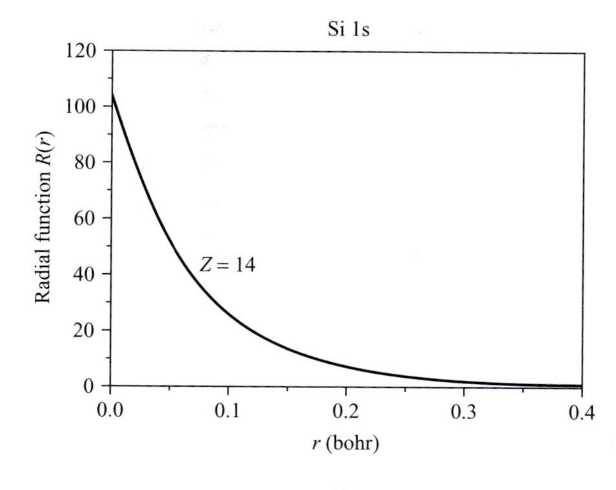
Si 1s
Radial function R(r)
Z = 14
r (bohr)
120
100
80
60
40
20
0
0.0
0.1
0.2
0.3
0.4

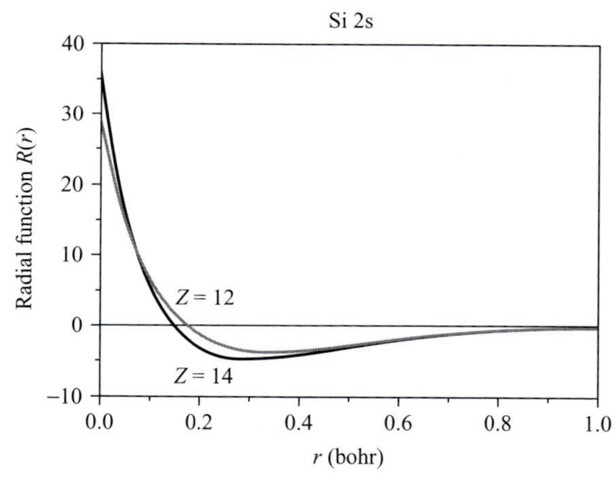
Si 2s
Radial function R(r)
Z = 12
Z = 14
r (bohr)
40
30
20
10
0
−10
0.0
0.2
0.4
0.6
0.8
1.0

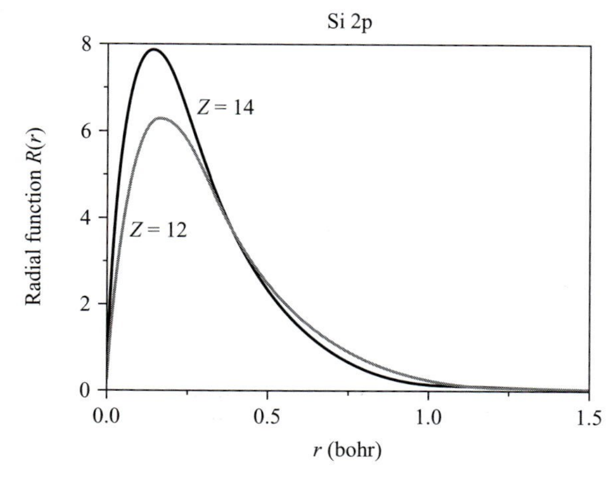
Si 2p
Radial function R(r)
Z = 14
Z = 12
r (bohr)
8
6
4
2
0
0.0
0.5
1.0
1.5

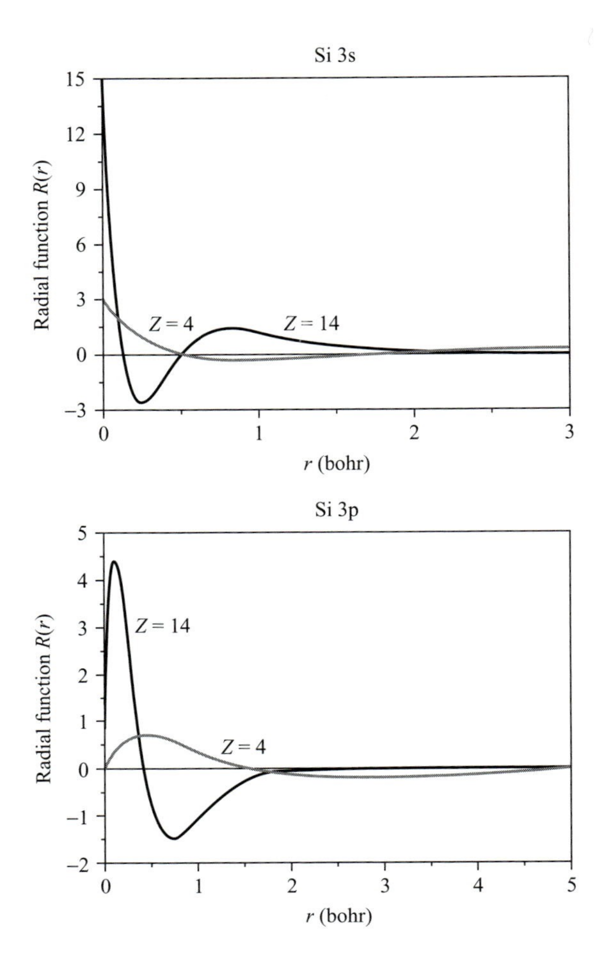

8 a. In ammonia, the only "core" orbital is the N 1s and this becomes an a_1 orbital in C_{3v} symmetry. The N 2s orbitals and 3 H 1s orbitals become $2a_1$ and an e set of orbitals. The remaining N 2p orbitals also become $1a_1$ and a set of e orbitals. The total valence orbitals in C_{3v} symmetry are $3a_1$ and 2e orbitals.

b. In water, the only core orbital is the O 1s and this becomes an a_1 orbital in C_{2v} symmetry. Placing the molecule in the yz plane allows us to further analyze the remaining valence orbitals as: O $2p_z = a_1$, O $2p_y = b_2$, and O $2p_x = b_1$. The (H 1s + H 1s) combination is an a_1 whereas the (H 1s − H 1s) combination is a b_2.

c. Placing the oxygens of H_2O_2 in the yz plane (z bisecting the oxygens) and the (cis) hydrogens distorted slightly in $+x$ and $-x$ directions allows us to analyze the orbitals as follows. The core O 1s + O 1s combination is an a orbital whereas the O 1s − O 1s combination is a b orbital. The valence orbitals are: O 2s + O 2s = a, O 2s − O 2s = b, O $2p_x$ + O $2p_x$ = b, O $2p_x$ − O $2p_x$ = a, O $2p_y$ + O $2p_y$ = a, O $2p_y$ − O $2p_y$ = b, O $2p_z$ + O $2p_z$ = b, O $2p_z$ − O $2p_z$ = a, H 1s + H 1s = a, and finally the H 1s − H 1s = b.

d. For the next two problems we will use the convention of choosing the z axis as principal axis for the $D_{\infty h}$, D_{2h}, and C_{2v} point groups and the xy plane as the horizontal reflection plane in C_s symmetry.

	$D_{\infty h}$	D_{2h}	C_{2v}	C_s
N 1s	σ_g	a_g	a_1	a′
N 2s	σ_g	a_g	a_1	a′
N $2p_x$	π_{xu}	b_{3u}	b_1	a′
N $2p_y$	π_{yu}	b_{2u}	b_2	a′
N $2p_z$	σ_u	b_{1u}	a_1	a''

9 a.

$$\Psi_n(x) = \left(\frac{2}{L}\right)^{\frac{1}{2}} \sin\frac{n\pi x}{L},$$
$$P_n(x)dx = |\Psi_n|^2(x)dx.$$

The probability that the particle lies in the interval $0 \leq x \leq \frac{L}{4}$ is given by

$$P_n = \int_0^{\frac{L}{4}} P_n(x)dx = \left(\frac{2}{L}\right)\int_0^{\frac{L}{4}} \sin^2\left(\frac{n\pi x}{L}\right)dx.$$

This integral can be integrated to give

$$\begin{aligned}
P_n &= \left(\frac{L}{n\pi}\right)\left(\frac{2}{L}\right)\int_0^{\frac{n\pi}{4}} \sin^2\left(\frac{n\pi x}{L}\right)d\left(\frac{n\pi x}{L}\right)\\
&= \left(\frac{L}{n\pi}\right)\left(\frac{2}{L}\right)\int_0^{\frac{n\pi}{4}} \sin^2\theta d\theta\\
&= \frac{2}{n\pi}\left(-\frac{1}{4}\sin 2\theta + \frac{\theta}{2}\Big|_0^{\frac{n\pi}{4}}\right)\\
&= \frac{2}{n\pi}\left(-\frac{1}{4}\sin\frac{2n\pi}{4} + \frac{n\pi}{(2)(4)}\right)\\
&= \frac{1}{4} - \frac{1}{2\pi n}\sin\left(\frac{n\pi}{2}\right).
\end{aligned}$$

b. If n is even, $\sin(\frac{n\pi}{2}) = 0$ and $P_n = \frac{1}{4}$.
If n is odd and $n = 1, 5, 9, 13, \ldots$, $\sin(\frac{n\pi}{2}) = 1$ and $P_n = \frac{1}{4} - \frac{1}{2\pi n}$.
If n is odd and $n = 3, 7, 11, 15, \ldots$, $\sin(\frac{n\pi}{2}) = -1$ and $P_n = \frac{1}{4} + \frac{1}{2\pi n}$.
The highest P_n is when $n = 3$. Then $P_n = \frac{1}{4} + \frac{1}{2\pi 3} = \frac{1}{4} + \frac{1}{6\pi} = 0.303$.

c.
$$\Psi(t) = e^{\frac{-iHt}{\hbar}}[a\Psi_n + b\Psi_m] = a\Psi_n e^{\frac{-iE_nt}{\hbar}} + b\Psi_m e^{\frac{-iE_mt}{\hbar}},$$
$$H\Psi = a\Psi_n E_n e^{\frac{-iE_nt}{\hbar}} + b\Psi_m E_m e^{\frac{iE_mt}{\hbar}},$$
$$\langle\Psi|H|\Psi\rangle = |a|^2E_n + |b|^2E_m + a^*be^{\frac{i(E_n-E_m)t}{\hbar}}\langle\Psi_n|H|\Psi_m\rangle$$
$$+b^*ae^{\frac{-i(E_m-E_n)t}{\hbar}}\langle\Psi_m|H|\Psi_n\rangle.$$

Since $\langle\Psi_n|H|\Psi_m\rangle$ and $\langle\Psi_m|H|\Psi_n\rangle$ are zero, $\langle\Psi|H|\Psi\rangle = |a|^2E_n + |b|^2E_m$ (note the time independence).

d. The fraction of systems observed in Ψ_n is $|a|^2$. The possible energies measured are E_n and E_m. The probabilities of measuring each of these energies are $|a|^2$ and $|b|^2$.

e. Once the system is observed in Ψ_n, it stays in Ψ_n.

f. $P(E_n) = |\langle\Psi_n|\Psi\rangle|^2 = |c_n|^2$

$$c_n = \int_0^L \sqrt{\frac{2}{L}}\sin\left(\frac{n\pi x}{L}\right)\sqrt{\frac{30}{L^5}}x(L-x)dx$$
$$= \sqrt{\frac{60}{L^6}}\int_0^L x(L-x)\sin\left(\frac{n\pi x}{L}\right)dx$$
$$= \sqrt{\frac{60}{L^6}}\left[L\int_0^L x\sin\left(\frac{n\pi x}{L}\right)dx - \int_0^L x^2\sin\left(\frac{n\pi x}{L}\right)dx\right].$$

These integrals can be evaluated to give

$$c_n = \sqrt{\frac{60}{L^6}}\left[L\left(\frac{L^2}{n^2\pi^2}\sin\left(\frac{n\pi x}{L}\right) - \frac{Lx}{n\pi}\cos\left(\frac{n\pi x}{L}\right)\right)\Big|_0^L\right]$$
$$-\sqrt{\frac{60}{L^6}}\left[\left(\frac{2xL^2}{n^2\pi^2}\sin\left(\frac{n\pi x}{L}\right) - \left(\frac{n^2\pi^2x^2}{L^2} - 2\right)\frac{L^3}{n^3\pi^3}\cos\left(\frac{n\pi x}{L}\right)\right)\Big|_0^L\right]$$
$$= \sqrt{\frac{60}{L^6}}\left\{\frac{L^3}{n^2\pi^2}[\sin(n\pi) - \sin(0)] - \frac{L^2}{n\pi}[L\cos(n\pi) - 0\cos(0)]\right.$$
$$-\left(\frac{2L^2}{n^2\pi^2}[L\sin(n\pi) - 0\sin(0)] - (n^2\pi^2 - 2)\frac{L^3}{n^3\pi^3}\cos(n\pi)\right.$$
$$\left.\left.+\left[\frac{n^2\pi^2(0}{L^2} - 2\right]\frac{L^3}{n^3\pi^3}\cos(0)\right)\right\}$$
$$= L^{-3}\sqrt{60}\left\{-\frac{L^3}{n\pi}\cos(n\pi) + (n^2\pi^2 - 2)\frac{L^3}{n^3\pi^3}\cos(n\pi) + \frac{2L^3}{n^3\pi^3}\right\}$$
$$= \sqrt{60}\left(-\frac{1}{n\pi}(-1)^n + (n^2\pi^2 - 2)\frac{1}{n^3\pi^3}(-1)^n + \frac{2}{n^3\pi^3}\right)$$
$$= \sqrt{60}\left(\left(\frac{-1}{n\pi} + \frac{1}{n\pi} - \frac{2}{n^3\pi^3}\right)(-1)^n + \frac{2}{n^3\pi^3}\right)$$

$$= \frac{2\sqrt{60}}{n^3\pi^3}(-(-1)^n + 1).$$

$$|c_n|^2 = \frac{4(60)}{n^6\pi^6}(-(-1)^n + 1)^2.$$

If n is even then $c_n = 0$.
If n is odd then $c_n = \frac{(4)(60)(4)}{n^6\pi^6} = \frac{960}{n^6\pi^6}$.
The probability of making a measurement of the energy and obtaining one of the eigenvalues, given by

$$E_n = \frac{n^2\pi^2\hbar^2}{2mL^2}, \quad \text{is}$$

$$P(E_n) = 0 \quad \text{if } n \text{ is even},$$

$$P(E_n) = \frac{960}{n^6\pi^6} \quad \text{if } n \text{ is odd}.$$

g.

$$\begin{aligned}
\langle\Psi|H|\Psi\rangle &= \int_0^L \left(\frac{30}{L^5}\right)^{\frac{1}{2}} x(L-x)\left(\frac{-\hbar^2}{2m}\frac{d^2}{dx^2}\right)\left(\frac{30}{L^5}\right)^{\frac{1}{2}} x(L-x)dx \\
&= \left(\frac{30}{L^5}\right)\left(\frac{-\hbar^2}{2m}\right)\int_0^L x(L-x)\left(\frac{d^2}{dx^2}\right)(xL-x^2)dx \\
&= \left(\frac{-15\hbar^2}{mL^5}\right)\int_0^L x(L-x)(-2)dx \\
&= \left(\frac{30\hbar^2}{mL^5}\right)\int_0^L (xL-x^2)dx \\
&= \left(\frac{30\hbar^2}{mL^5}\right)\left(L\frac{x^2}{2}-\frac{x^3}{3}\right)\Big|_0^L \\
&= \left(\frac{30\hbar^2}{mL^5}\right)\left(\frac{L^3}{2}-\frac{L^3}{3}\right) \\
&= \left(\frac{30\hbar^2}{mL^2}\right)\left(\frac{1}{2}-\frac{1}{3}\right) \\
&= \frac{30\hbar^2}{6mL^2} = \frac{5\hbar^2}{mL^2}.
\end{aligned}$$

10

$$\langle\Psi|H|\Psi\rangle = \sum_{ij} C_i^* e^{\frac{iE_it}{\hbar}}\langle\Psi_i|H|\Psi_j\rangle e^{\frac{-iE_jt}{\hbar}} C_j.$$

Since $\langle\Psi_i|H|\Psi_j\rangle = E_j\delta_{ij}$,

$$\langle\Psi|H|\Psi\rangle = \sum_j C_j^* C_j E_j e^{\frac{i(E_j-E_j)t}{\hbar}}$$

$$= \sum_j C_j^* C_j E_j \quad \text{(not time dependent)}.$$

For other properties:

$$\langle\Psi|A|\Psi\rangle = \sum_{ij} C_i^* e^{\frac{iE_it}{\hbar}}\langle\Psi_i|A|\Psi_j\rangle e^{\frac{-iE_jt}{\hbar}} C_j$$

but $\langle\Psi_i|A|\Psi_j\rangle$ does not necessarily $= a_j\delta_{ij}$ because the Ψ_j are not eigenfunctions of A unless $[A, H] = 0$.

$$\langle\Psi|A|\Psi\rangle = \sum_{ij} C_i^* C_j e^{\frac{i(E_i - E_j)t}{\hbar}} \langle\Psi_i|A|\Psi_j\rangle.$$

Therefore, in general, other properties are time dependent.

11 a. The lowest energy level for a particle in a three-dimensional box is when $n_1 = 1$, $n_2 = 1$, and $n_3 = 1$. The total energy (with $L_1 = L_2 = L_3$) will be

$$E_{\text{total}} = \frac{h^2}{8mL^2}\left(n_1^2 + n_2^2 + n_3^2\right) = \frac{3h^2}{8mL^2}.$$

Note that $n = 0$ is not possible. The next lowest energy level is when one of the three quantum numbers equals 2 and the other two equal 1:

$$n_1 = 1, \quad n_2 = 1, \quad n_3 = 2$$
$$n_1 = 1, \quad n_2 = 2, \quad n_3 = 1$$
$$n_1 = 2, \quad n_2 = 1, \quad n_3 = 1.$$

All of these three states have the same energy:

$$E_{\text{total}} = \frac{h^2}{8mL^2}\left(n_1^2 + n_2^2 + n_3^2\right) = \frac{6h^2}{8mL^2}.$$

Note that these three states are only degenerate if $L_1 = L_2 = L_3$.

b.

distortion →

$L_1 = L_2 = L_3$ $L_3 \neq L_1 = L_2$

For $L_1 = L_2 = L_3$, $V = L_1L_2L_3 = L_1^3$,

$$\begin{aligned} E_{\text{total}}(L_1) &= 2\varepsilon_1 + \varepsilon_2 \\ &= \frac{2h^2}{8m}\left(\frac{1^2}{L_1^2} + \frac{1^2}{L_2^2} + \frac{1^2}{L_3^2}\right) + \frac{1h^2}{8m}\left(\frac{1^2}{L_1^2} + \frac{1^2}{L_2^2} + \frac{2^2}{L_3^2}\right) \\ &= \frac{2h^2}{8m}\left(\frac{3}{L_1^2}\right) + \frac{1h^2}{8m}\left(\frac{6}{L_1^2}\right) = \frac{h^2}{8m}\left(\frac{12}{L_1^2}\right). \end{aligned}$$

For $L_3 \neq L_1 = L_2$, $V = L_1L_2L_3 = L_1^2L_3$, $L_3 = V/L_1^2$,

$$\begin{aligned} E_{\text{total}}(L_1) &= 2\varepsilon_1 + \varepsilon_2 \\ &= \frac{2h^2}{8m}\left(\frac{1^2}{L_1^2} + \frac{1^2}{L_2^2} + \frac{1^2}{L_3^2}\right) + \frac{1h^2}{8m}\left(\frac{1^2}{L_1^2} + \frac{1^2}{L_2^2} + \frac{2^2}{L_3^2}\right) \\ &= \frac{2h^2}{8m}\left(\frac{2}{L_1^2} + \frac{1}{L_3^2}\right) + \frac{1h^2}{8m}\left(\frac{2}{L_1^2} + \frac{4}{L_3^2}\right) \end{aligned}$$

$$= \frac{2h^2}{8m}\left(\frac{2}{L_1^2} + \frac{1}{L_3^2} + \frac{1}{L_1^2} + \frac{2}{L_3^2}\right)$$
$$= \frac{2h^2}{8m}\left(\frac{3}{L_1^2} + \frac{3}{L_3^2}\right) = \frac{h^2}{8m}\left(\frac{6}{L_1^2} + \frac{6}{L_3^2}\right).$$

In comparing the total energy at constant volume of the undistorted box ($L_1 = L_2 = L_3$) versus the distorted box ($L_3 \neq L_1 = L_2$) it can be seen that

$$\frac{h^2}{8m}\left(\frac{6}{L_1^2} + \frac{6}{L_3^2}\right) \leq \frac{h^2}{8m}\left(\frac{12}{L_1^2}\right) \quad \text{as long as} \quad L_3 \geq L_1.$$

c. In order to minimize the total energy expression, take the derivative of the energy with respect to L_1 and set it equal to zero:

$$\frac{\partial E_{\text{total}}}{\partial L_1} = 0 \qquad \frac{\partial}{\partial L_1}\left[\frac{h^2}{8m}\left(\frac{6}{L_1^2} + \frac{6}{L_3^2}\right)\right] = 0.$$

But since $V = L_1 L_2 L_3 = L_1^2 L_3$, then $L_3 = V/L_1^2$. This substitution gives

$$\frac{\partial}{\partial L_1}\left(\frac{h^2}{8m}\left(\frac{6}{L_1^2} + \frac{6L_1^4}{V^2}\right)\right) = 0$$
$$\left(\frac{h^2}{8m}\left(\frac{(-2)6}{L_1^3} + \frac{(4)6L_1^3}{V^2}\right)\right) = 0$$
$$\left(-\frac{12}{L_1^3} + \frac{24L_1^3}{V^2}\right) = 0$$
$$\left(\frac{24L_1^3}{V^2}\right) = \left(\frac{12}{L_1^3}\right)$$
$$24L_1^6 = 12V^2$$
$$L_1^6 = \frac{1}{2}V^2 = \frac{1}{2}\left(L_1^2 L_3\right)^2 = \frac{1}{2}L_1^4 L_3^2$$
$$L_1^2 = \frac{1}{2}L_3^2$$
$$L_3 = \sqrt{2}L_1.$$

d. Calculate energy upon distortion:

cube: $V = L_1^3$, $L_1 = L_2 = L_3 = (V)^{\frac{1}{3}}$.

distorted: $V = L_1^2 L_3 = L_1^2\sqrt{2}L_1 = \sqrt{2}L_1^3$,

$$L_3 = \sqrt{2}\left(\frac{V}{\sqrt{2}}\right)^{\frac{1}{3}} \neq L_1 = L_2 = \left(\frac{V}{\sqrt{2}}\right)^{\frac{1}{3}}.$$

$$\Delta E = E_{\text{total}}(L_1 = L_2 = L_3) - E_{\text{total}}(L_3 \neq L_1 = L_2)$$
$$= \frac{h^2}{8m}\left(\frac{12}{L_1^2}\right) - \frac{h^2}{8m}\left(\frac{6}{L_1^2} + \frac{6}{L_3^2}\right)$$

$$= \frac{h^2}{8m}\left(\frac{12}{V^{2/3}} - \frac{6(2)^{1/3}}{V^{2/3}} + \frac{6(2)^{1/3}}{2V^{2/3}}\right)$$
$$= \frac{h^2}{8m}\left(\frac{12 - 9(2)^{1/3}}{V^{2/3}}\right).$$

Since $V = 8\,\mathring{A}^3$, $V^{2/3} = 4\,\mathring{A}^2 = 4 \times 10^{-16}\,\text{cm}^2$, and $\frac{h^2}{8m} = 6.01 \times 10^{-27}$ erg cm^2:

$$\Delta E = 6.01 \times 10^{-27}\,\text{erg cm}^2\left(\frac{12 - 9(2)^{1/3}}{4 \times 10^{-16}\,\text{cm}^2}\right)$$
$$= 6.01 \times 10^{-27}\,\text{erg cm}^2\left(\frac{0.66}{4 \times 10^{-16}\,\text{cm}^2}\right)$$
$$= 0.99 \times 10^{-11}\,\text{erg}$$
$$= 0.99 \times 10^{-11}\,\text{erg}\left(\frac{1\,\text{eV}}{1.6 \times 10^{-12}\,\text{erg}}\right)$$
$$= 6.19\,\text{eV}.$$

12 a.
$$\mathbf{H} = \frac{-\hbar^2}{2m}\left\{\frac{\partial^2}{\partial x^2} + \frac{\partial^2}{\partial y^2}\right\} \quad \text{(Cartesian coordinates).}$$

Finding $\frac{\partial}{\partial x}$ and $\frac{\partial}{\partial y}$ from the chain rule gives:

$$\frac{\partial}{\partial x} = \left(\frac{\partial r}{\partial x}\right)_y \frac{\partial}{\partial r} + \left(\frac{\partial \phi}{\partial x}\right)_y \frac{\partial}{\partial \phi}, \qquad \frac{\partial}{\partial y} = \left(\frac{\partial r}{\partial y}\right)_x \frac{\partial}{\partial r} + \left(\frac{\partial \phi}{\partial y}\right)_x \frac{\partial}{\partial \phi}.$$

Evaluation of the "coefficients" gives the following:

$$\left(\frac{\partial r}{\partial x}\right)_y = \cos\phi, \qquad \left(\frac{\partial \phi}{\partial x}\right)_y = -\frac{\sin\phi}{r},$$
$$\left(\frac{\partial r}{\partial y}\right)_x = \sin\phi, \qquad \text{and} \qquad \left(\frac{\partial \phi}{\partial y}\right)_x = \frac{\cos\phi}{r}.$$

Upon substitution of these "coefficients":

$$\frac{\partial}{\partial x} = \cos\phi\frac{\partial}{\partial r} - \frac{\sin\phi}{r}\frac{\partial}{\partial \phi} = -\frac{\sin\phi}{r}\frac{\partial}{\partial \phi} \quad \text{at fixed } r.$$
$$\frac{\partial}{\partial y} = \sin\phi\frac{\partial}{\partial r} + \frac{\cos\phi}{r}\frac{\partial}{\partial \phi} = \frac{\cos\phi}{r}\frac{\partial}{\partial \phi} \quad \text{at fixed } r.$$
$$\frac{\partial^2}{\partial x^2} = \left(-\frac{\sin\phi}{r}\frac{\partial}{\partial \phi}\right)\left(-\frac{\sin\phi}{r}\frac{\partial}{\partial \phi}\right)$$
$$= \frac{\sin^2\phi}{r^2}\frac{\partial^2}{\partial \phi^2} + \frac{\sin\phi\cos\phi}{r^2}\frac{\partial}{\partial \phi} \quad \text{at fixed } r.$$
$$\frac{\partial^2}{\partial y^2} = \left(\frac{\cos\phi}{r}\frac{\partial}{\partial \phi}\right)\left(\frac{\cos\phi}{r}\frac{\partial}{\partial \phi}\right)$$
$$= \frac{\cos^2\phi}{r^2}\frac{\partial^2}{\partial \phi^2} - \frac{\cos\phi\sin\phi}{r^2}\frac{\partial}{\partial \phi} \quad \text{at fixed } r.$$

$$\frac{\partial^2}{\partial x^2} + \frac{\partial^2}{\partial y^2} = \frac{\sin^2\phi}{r^2}\frac{\partial^2}{\partial\phi^2} + \frac{\sin\phi\cos\phi}{r^2}\frac{\partial}{\partial\phi} + \frac{\cos^2\phi}{r^2}\frac{\partial^2}{\partial\phi^2} - \frac{\cos\phi\sin\phi}{r^2}\frac{\partial}{\partial\phi}$$

$$= \frac{1}{r^2}\frac{\partial^2}{\partial\phi^2} \quad \text{at fixed } r.$$

So, $$\mathbf{H} = \frac{-\hbar^2}{2mr^2}\frac{\partial^2}{\partial\phi^2} \quad \text{(cylindrical coordinates, fixed } r\text{)}$$

$$= \frac{-\hbar^2}{2I}\frac{\partial^2}{\partial\phi^2}.$$

The Schrödinger equation for a particle on a ring then becomes

$$H\Psi = E\Psi,$$

$$\frac{-\hbar^2}{2I}\frac{\partial^2\Phi}{\partial\phi^2} = E\Phi,$$

$$\frac{\partial^2\Phi}{\partial\phi^2} = \left(\frac{-2IE}{\hbar^2}\right)\Phi.$$

The general solution to this equation is the now familiar expression:

$$\Phi(\phi) = C_1 e^{-im\phi} + C_2 e^{im\phi}, \quad \text{where} \quad m = \left(\frac{2IE}{\hbar^2}\right)^{\frac{1}{2}}.$$

Application of the cyclic boundary condition, $\Phi(\phi) = \Phi(\phi + 2\pi)$, results in the quantization of the energy expression: $E = \frac{m^2\hbar^2}{2I}$ where $m = 0, \pm1, \pm2, \pm3, \ldots$ It can be seen that the $\pm m$ values correspond to angular momentum of the same magnitude but opposite directions. Normalization of the wave function (over the region 0 to 2π) corresponding to + or $-m$ will result in a value of $(\frac{1}{2\pi})^{\frac{1}{2}}$ for the normalization constant.

$$\therefore \quad \Phi(\phi) = \left(\frac{1}{2\pi}\right)^{\frac{1}{2}} e^{im\phi}$$

— — $\frac{(\pm4)^2\hbar^2}{2I}$

— — $\frac{(\pm3)^2\hbar^2}{2I}$

— — $\frac{(\pm2)^2\hbar^2}{2I}$

↑↓ ↑↓ $\frac{(\pm1)^2\hbar^2}{2I}$

↑↓ $\frac{(0)^2\hbar^2}{2I}$

b.

$$\frac{\hbar^2}{2m} = 6.06 \times 10^{-28}\ \text{erg cm}^2,$$

$$\frac{\hbar^2}{2mr^2} = \frac{6.06 \times 10^{-28}\ \text{erg cm}^2}{(1.4 \times 10^{-8}\ \text{cm})^2}$$

$$= 3.09 \times 10^{-12}\ \text{erg},$$

$$\Delta E = (2^2 - 1^2)3.09 \times 10^{-12}\ \text{erg} = 9.27 \times 10^{-12}\ \text{erg}$$

$$\text{but} \quad \Delta E = h\nu = hc/\lambda, \quad \text{so } \lambda = hc/\Delta E.$$

$$\lambda = \frac{(6.63 \times 10^{-27} \text{ erg s})(3.00 \times 10^{10} \text{ cm s}^{-1})}{9.27 \times 10^{-12} \text{ erg}}$$

$$= 2.14 \times 10^{-5} \text{ cm} = 2.14 \times 10^{3} \text{Å}.$$

Sources of error in this calculation include:

(i) The attractive
force of the carbon nuclei is not included in the Hamiltonian.

(ii) The repulsive
force of the other π-electrons is not included in the Hamiltonian.

(iii) Benzene is not a ring.

(iv) Electrons move in three dimensions not one.

13

$$\Psi(\phi, 0) = \sqrt{\frac{4}{3\pi}} \cos^2 \phi.$$

This wave function needs to be expanded in terms of the eigenfunctions of the angular momentum operator, $(-i\hbar\frac{\partial}{\partial\phi})$. This is most easily accomplished by an exponential expansion of the cos function.

$$\Psi(\phi, 0) = \sqrt{\frac{4}{3\pi}} \left(\frac{e^{i\phi} + e^{-i\phi}}{2}\right) \left(\frac{e^{i\phi} + e^{-i\phi}}{2}\right)$$

$$= \left(\frac{1}{4}\right) \sqrt{\frac{4}{3\pi}} \left(e^{2i\phi} + e^{-2i\phi} + 2e^{(0)i\phi}\right).$$

The wave function is now written in terms of the eigenfunctions of the angular momentum operator, $(-i\hbar\frac{\partial}{\partial\phi})$, but they need to include their normalization constant, $\frac{1}{\sqrt{2\pi}}$:

$$\Psi(\phi, 0) = \left(\frac{1}{4}\right) \sqrt{\frac{4}{3\pi}} \sqrt{2\pi} \left(\frac{1}{\sqrt{2\pi}} e^{2i\phi} + \frac{1}{\sqrt{2\pi}} e^{-2i\phi} + 2\frac{1}{\sqrt{2\pi}} e^{(0)i\phi}\right)$$

$$= \left(\sqrt{\frac{1}{6}}\right) \left(\frac{1}{\sqrt{2\pi}} e^{2i\phi} + \frac{1}{\sqrt{2\pi}} e^{-2i\phi} + 2\frac{1}{\sqrt{2\pi}} e^{(0)i\phi}\right).$$

Once the wave function is written in this form (in terms of the normalized eigenfunctions of the angular momentum operator having $m\hbar$ as eigenvalues) the probabilities for observing angular momentums of $0\hbar$, $2\hbar$, and $-2\hbar$ can be easily identified as the squares of the coefficients of the corresponding eigenfunctions:

$$P_{2\hbar} = \left(\sqrt{\frac{1}{6}}\right)^2 = \frac{1}{6},$$

$$P_{-2\hbar} = \left(\sqrt{\frac{1}{6}}\right)^2 = \frac{1}{6},$$

$$P_{0\hbar} = \left(2\sqrt{\frac{1}{6}}\right)^2 = \frac{4}{6}.$$

14 a.

$$\frac{1}{2}mv^2 = 100 \text{ eV}\left(\frac{1.602 \times 10^{-12} \text{ erg}}{1 \text{ eV}}\right)$$
$$v^2 = \left(\frac{(2)1.602 \times 10^{-10} \text{ erg}}{9.109 \times 10^{-28} \text{ g}}\right)$$
$$v = 0.593 \times 10^9 \text{ cm s}^{-1}.$$

The length of the N_2 molecule is 2 Å $= 2 \times 10^{-8}$ cm.

$$v = \frac{d}{t}$$
$$t = \frac{d}{v} = \frac{2 \times 10^{-8} \text{ cm}}{0.593 \times 10^9 \text{ cm s}^{-1}} = 3.37 \times 10^{-17} \text{ s}.$$

b. The normalized ground-state harmonic oscillator can be written as

$$\Psi_0 = \left(\frac{\alpha}{\pi}\right)^{\frac{1}{4}} e^{-\alpha x^2/2}, \quad \text{where} \quad \alpha = \left(\frac{k\mu}{\hbar^2}\right)^{\frac{1}{2}} \quad \text{and} \quad x = r - r_e.$$

Calculating constants:

$$\alpha_{N_2} = \left(\frac{(2.294 \times 10^6 \text{ g s}^{-2})(1.1624 \times 10^{-23} \text{ g})}{(1.0546 \times 10^{-27} \text{ erg s})^2}\right)^{\frac{1}{2}}$$
$$= 0.489\,66 \times 10^{19} \text{ cm}^{-2} = 489.66 \text{ Å}^{-2}.$$

For N_2:

$$\Psi_0(r) = 3.533\,33 \text{ Å}^{-\frac{1}{2}} e^{-(244.83 \text{ Å}-2)(r-1.097\,69 \text{ Å})^2}$$
$$\alpha_{N_2^+} = \left(\frac{(2.009 \times 10^6 \text{ g s}^{-2})(1.1624 \times 10^{-23} \text{ g})}{(1.0546 \times 10^{-27} \text{ erg s})^2}\right)^{\frac{1}{2}}$$
$$= 0.458\,23 \times 10^{19} \text{ cm}^{-2} = 458.23 \text{ Å}^{-2}.$$

For N_2^+:

$$\Psi_0(r) = 3.475\,22 \text{ Å}^{-\frac{1}{2}} e^{-(229.113 \text{ Å}^{-2})(r-1.11642 \text{ Å})^2}.$$

c.

$$P(v = 0) = |\langle\Psi_{v=0}(N_2^+)|\Psi_{v=0}(N_2)\rangle|^2.$$

Let $P(v = 0) = I^2$, where I = integral:

$$I = \int_{-\infty}^{+\infty} \left(3.475\,22 \text{ Å}^{-\frac{1}{2}} e^{-(229.113 \text{ Å}^{-2})(r-1.116\,42 \text{ Å})^2}\right)$$
$$\times \left(3.533\,33 \text{ Å}^{-\frac{1}{2}} e^{-(244.830 \text{ Å}^{-2})(r-1.097\,69 \text{ Å})^2}\right) dr.$$

Let

$$C_1 = 3.475\,22\ \text{Å}^{-\frac{1}{2}}, \qquad C_2 = 3.533\,33\ \text{Å}^{-\frac{1}{2}},$$
$$A_1 = 229.113\ \text{Å}^{-2}, \qquad A_2 = 244.830\ \text{Å}^{-2},$$
$$r_1 = 1.116\,42\ \text{Å}, \qquad r_2 = 1.097\,69\ \text{Å},$$
$$I = C_1C_2 \int_{-\infty}^{+\infty} e^{-A_1(r-r_1)^2} e^{-A_2(r-r_2)^2}\, dr.$$

Focusing on the exponential:

$$\begin{aligned} -A_1(r-r_1)^2 - A_2(r-r_2)^2 &= -A_1\left(r^2 - 2r_1r + r_1^2\right) \\ &\quad -A_2\left(r^2 - 2r_2r + r_2^2\right) \\ &= -(A_1 + A_2)r^2 + (2A_1r_1 + 2A_2r_2)r \\ &\quad -A_1r_1^2 - A_2r_2^2. \end{aligned}$$

Let

$$\begin{aligned} A &= A_1 + A_2, \\ B &= 2A_1r_1 + 2A_2r_2, \\ C &= C_1C_2, \quad \text{and} \\ D &= A_1r_1^2 + A_2r_2^2. \\ I &= C\int_{-\infty}^{+\infty} e^{-Ar^2+Br-D}\, dr \\ &= C\int_{-\infty}^{+\infty} e^{-A(r-r_0)^2+D'}\, dr, \end{aligned}$$

where

$$\begin{aligned} -A(r-r_0)^2 + D' &= -Ar^2 + Br - D, \\ -A\left(r^2 - 2rr_0 + r_0^2\right) + D' &= -Ar^2 + Br - D, \end{aligned}$$

such that

$$\begin{aligned} 2Ar_0 &= B, \\ -Ar_0^2 + D' &= -D, \end{aligned}$$

and

$$\begin{aligned} r_0 &= \frac{B}{2A}, \\ D' &= Ar_0^2 - D = A\frac{B^2}{4A^2} - D = \frac{B^2}{4A} - D. \\ I &= C\int_{-\infty}^{+\infty} e^{-A(r-r_0)^2+D'}\, dr \\ &= Ce^{D'}\int_{-\infty}^{+\infty} e^{-Ay^2}\, dy \\ &= Ce^{D'}\sqrt{\frac{\pi}{A}}. \end{aligned}$$

Now back substituting all of these constants:

$$
\begin{aligned}
I &= C_1C_2\sqrt{\frac{\pi}{A_1+A_2}}\exp\left(\frac{(2A_1r_1+2A_2r_2)^2}{4(A_1+A_2)} - A_1r_1^2 - A_2r_2^2\right) \\
&= (3.475\,22)(3.533\,33)\sqrt{\frac{\pi}{(229.113)+(244.830)}} \\
&\quad \times \exp\left(\frac{(2(229.113)(1.116\,42)+2(244.830)(1.097\,69))^2}{4((229.113)+(244.830))}\right) \\
&\quad \times \exp(-(229.113)(1.116\,42)^2 - (244.830)(1.097\,69)^2) \\
&= 0.959.
\end{aligned}
$$

$P(v=0) = I^2 = 0.92$, so there is a 92% probability.

15 a.

$$
\begin{aligned}
E_v &= \left(\frac{\hbar^2 k}{\mu}\right)^{\frac{1}{2}}\left(v+\frac{1}{2}\right), \\
\Delta E &= E_{v+1} - E_v \\
&= \left(\frac{\hbar^2 k}{\mu}\right)^{\frac{1}{2}}\left\{v+1+\frac{1}{2}-v-\frac{1}{2}\right\} = \left(\frac{\hbar^2 k}{\mu}\right) \\
&= \left(\frac{(1.0546\times 10^{-27}\ \text{erg s})^2(1.87\times 10^6\ \text{g s}^{-2}}{6.857\ \text{g}/6.02\times 10^{23}}\right)^{\frac{1}{2}} \\
&= 4.27\times 10^{-13}\ \text{erg}, \\
\Delta E &= \frac{hc}{\lambda}, \\
\lambda &= \frac{hc}{\Delta E} = \frac{(6.626\times 10^{-27}\ \text{erg s})(3.00\times 10^{10}\ \text{cm s}^{-1})}{4.27\times 10^{-13}\ \text{erg}} \\
&= 4.66\times 10^{-4}\ \text{cm}, \\
\frac{1}{\lambda} &= 2150\ \text{cm}^{-1}.
\end{aligned}
$$

b.

$$
\begin{aligned}
\Psi_0 &= \left(\frac{\alpha}{\pi}\right)^{\frac{1}{4}} e^{-\alpha x^2/2}, \\
\langle x\rangle &= \langle \Psi_{v=0}|x|\Psi_{v=0}\rangle \\
&= \int_{-\infty}^{+\infty} \Psi_0^* x \Psi_0\, dx \\
&= \int_{-\infty}^{+\infty} \left(\frac{\alpha}{\pi}\right)^{\frac{1}{2}} x e^{-\alpha x^2}\, dx \\
&= \int_{-\infty}^{+\infty} \left(\frac{\alpha}{-\alpha^2\pi}\right)^{\frac{1}{2}} e^{-\alpha x^2}\, d(-\alpha x^2) \\
&= \left(\frac{-1}{\alpha\pi}\right)^{\frac{1}{2}} e^{-\alpha x^2}\Bigg|_{-\infty}^{+\infty} = 0, \\
\langle x^2\rangle &= \langle \Psi_{v=0}|x^2|\Psi_{v=0}\rangle \\
&= \int_{-\infty}^{+\infty} \Psi_0^* x^2 \Psi_0\, dx \\
&= \int_{-\infty}^{+\infty} \left(\frac{\alpha}{\pi}\right)^{\frac{1}{2}} x^2 e^{-\alpha x^2}\, dx
\end{aligned}
$$

$$= 2\left(\frac{\alpha}{\pi}\right)^{\frac{1}{2}} \int_{0}^{+\infty} x^2 e^{-\alpha x^2} dx$$

$$= 2\left(\frac{\alpha}{\pi}\right)^{\frac{1}{2}} \left(\frac{1}{2^{1+1}\alpha}\right)\left(\frac{\pi}{\alpha}\right)^{\frac{1}{2}}$$

$$= \left(\frac{1}{2\alpha}\right),$$

$$\Delta x = (\langle x^2\rangle - \langle x\rangle^{y})^{\frac{1}{2}} = \left(\frac{1}{2\alpha}\right)$$

$$= \left(\frac{\hbar}{2\sqrt{k\mu}}\right)^{\frac{1}{2}}$$

$$= \left(\frac{(1.0546 \times 10^{-27}\ \text{erg s})^2}{4(1.87 \times 10^6\ \text{g s}^{-2})(6.857\ \text{g}/6.02 \times 10^{23})}\right)^{\frac{1}{4}}$$

$$= 3.38 \times 10^{-10}\ \text{cm} = 0.0338\ \text{Å}.$$

c.

$$\Delta x = \left(\frac{\hbar}{2\sqrt{k\mu}}\right)^{\frac{1}{2}}.$$

The smaller k and μ become, the larger the uncertainty in the internuclear distance becomes. Helium has a small μ and small attractive force between atoms. This results in a very large Δx. This implies that it is extremely difficult for He atoms to "vibrate" with small displacement as a solid, even as absolute zero is approached.

16 a.

$$W = \int_{-\infty}^{\infty} \phi^* H \phi dx$$

$$= \left(\frac{2b}{\pi}\right)^{\frac{1}{2}} \int_{-\infty}^{\infty} e^{-bx^2}\left(-\frac{\hbar^2}{2m}\frac{d^2}{dx^2} + a|x|\right) e^{-bx^2} dx.$$

$$\frac{d^2}{dx^2} e^{-bx^2} = \frac{d}{dx}\left(-2bxe^{-bx^2}\right)$$

$$= (-2bx)\left(-2bxe^{-bx^2}\right) + \left(e^{-bx^2}\right)(-2b)$$

$$= \left(4b^2x^2e^{-bx^2}\right) + \left(-2be^{-bx^2}\right).$$

Making this substitution results in the following three integrals:

$$W = \left(\frac{2b}{\pi}\right)^{\frac{1}{2}}\left(-\frac{\hbar^2}{2m}\right)\int_{-\infty}^{\infty} e^{-bx^2} 4b^2x^2 e^{-bx^2} dx$$

$$+ \left(\frac{2b}{\pi}\right)^{\frac{1}{2}}\left(-\frac{\hbar^2}{2m}\right)\int_{-\infty}^{\infty} e^{-bx^2} - 2b\, e^{-bx^2} dx$$

$$+ \left(\frac{2b}{\pi}\right)^{\frac{1}{2}}\int_{-\infty}^{\infty} e^{-bx^2} a|x| e^{-bx^2} dx$$

$$= \left(\frac{2b}{\pi}\right)^{\frac{1}{2}}\left(-\frac{2b^2\hbar^2}{m}\right)\int_{-\infty}^{\infty} x^2 e^{-2bx^2} dx$$

$$+\left(\frac{2b}{\pi}\right)^{\frac{1}{2}}\left(\frac{b\hbar^2}{m}\right)\int_{-\infty}^{\infty}e^{-2bx^2}dx+\left(\frac{2b}{\pi}\right)^{\frac{1}{2}}a\int_{-\infty}^{\infty}|x|e^{-2bx^2}dx$$
$$=\left(\frac{2b}{\pi}\right)^{\frac{1}{2}}\left(-\frac{2b^2\hbar^2}{m}\right)2\left(\frac{1}{2^22b}\right)\sqrt{\frac{\pi}{2b}}$$
$$+\left(\frac{2b}{\pi}\right)^{\frac{1}{2}}\left(\frac{b\hbar^2}{m}\right)2\left(\frac{1}{2}\right)\sqrt{\frac{\pi}{2b}}+\left(\frac{2b}{\pi}\right)^{\frac{1}{2}}a\left(\frac{0!}{2b}\right)$$
$$=\left(-\frac{b\hbar^2}{m}\right)\left(\frac{1}{2}\right)+\left(\frac{b\hbar^2}{m}\right)+\left(\frac{2b}{\pi}\right)^{\frac{1}{2}}\left(\frac{a}{2b}\right)$$
$$W=\left(\frac{b\hbar^2}{2m}\right)+a\left(\frac{1}{2b\pi}\right)^{\frac{1}{2}}.$$

b. Optimize b by evaluating $\frac{dW}{db}=0$:

$$\frac{dW}{db}=\frac{d}{db}\left(\left(\frac{b\hbar^2}{2m}\right)+a\left(\frac{1}{2b\pi}\right)^{\frac{1}{2}}\right)$$
$$=\left(\frac{\hbar^2}{2m}\right)-\frac{a}{2}\left(\frac{1}{2\pi}\right)^{\frac{1}{2}}b^{-\frac{3}{2}}.$$

So,

$$\frac{a}{2}\left(\frac{1}{2\pi}\right)^{\frac{1}{2}}b^{-\frac{3}{2}}=\left(\frac{\hbar^2}{2m}\right)$$

or

$$b^{-\frac{3}{2}}=\left(\frac{\hbar^2}{2m}\right)\frac{2}{a}\left(\frac{1}{2\pi}\right)^{-\frac{1}{2}}=\left(\frac{\hbar^2}{ma}\right)\sqrt{2\pi},$$

and $b=\left(\frac{ma}{\sqrt{2\pi}\hbar^2}\right)^{\frac{2}{3}}$. Substituting this value of b into the expression for W gives

$$W=\left(\frac{\hbar^2}{2m}\right)\left(\frac{ma}{\sqrt{2\pi}\hbar^2}\right)^{\frac{2}{3}}+a\left(\frac{1}{2\pi}\right)^{\frac{1}{2}}\left(\frac{ma}{\sqrt{2\pi}\hbar^2}\right)^{-\frac{1}{3}}$$
$$=\left(\frac{\hbar^2}{2m}\right)\left(\frac{ma}{\sqrt{2\pi}\hbar^2}\right)^{\frac{2}{3}}+a\left(\frac{1}{2\pi}\right)^{\frac{1}{2}}\left(\frac{ma}{\sqrt{2\pi}\hbar^2}\right)^{-\frac{1}{3}}$$
$$=2^{-\frac{4}{3}}\pi^{-\frac{1}{3}}\hbar^{\frac{2}{3}}a^{\frac{2}{3}}m^{-\frac{1}{3}}+2^{-\frac{1}{3}}\pi^{-\frac{1}{3}}\hbar^{\frac{2}{3}}a^{\frac{2}{3}}m^{-\frac{1}{3}}$$
$$=\left(2^{-\frac{4}{3}}\pi^{-\frac{1}{3}}+2^{-\frac{1}{3}}\pi^{-\frac{1}{3}}\right)\hbar^{\frac{2}{3}}a^{\frac{2}{3}}m^{-\frac{1}{3}}=\frac{3}{2}(2\pi)^{-\frac{1}{3}}\hbar^{\frac{2}{3}}a^{\frac{2}{3}}m^{-\frac{1}{3}}$$
$$=0.812\,889\,106\,\hbar^{\frac{2}{3}}a^{\frac{2}{3}}m^{-\frac{1}{3}}\text{ which is in error by only 0.5284\% !!!!!}$$

17 a.

$$\mathbf{H}=-\frac{\hbar^2}{2m}\frac{d^2}{dx^2}+\frac{1}{2}kx^2,$$
$$\phi=\sqrt{\frac{15}{16}}a^{-\frac{5}{2}}(a^2-x^2)\qquad\text{for}\qquad -a<x<a,$$
$$\phi=0\qquad\text{for}\qquad |x|\geq a.$$

$$\int_{-\infty}^{+\infty} \phi^* \mathbf{H} \phi \, dx$$

$$= \int_{-a}^{+a} \sqrt{\frac{15}{16}} a^{-\frac{5}{2}} (a^2 - x^2) \left(-\frac{\hbar^2}{2m} \frac{d^2}{dx^2} + \frac{1}{2} k x^2 \right)$$
$$\times \sqrt{\frac{15}{16}} a^{-\frac{5}{2}} (a^2 - x^2) dx$$
$$= \left(\frac{15}{16}\right) a^{-5} \int_{-a}^{+a} (a^2 - x^2) \left(-\frac{\hbar^2}{2m} \frac{d^2}{dx^2} + \frac{1}{2} k x^2 \right) (a^2 - x^2) \, dx$$
$$= \left(\frac{15}{16}\right) a^{-5} \int_{-a}^{+a} (a^2 - x^2) \left(-\frac{\hbar^2}{2m} \right) \frac{d^2}{dx^2} (a^2 - x^2) \, dx$$
$$+ \left(\frac{15}{16}\right) a^{-5} \int_{-a}^{+a} (a^2 - x^2) \frac{1}{2} k x^2 (a^2 - x^2) \, dx$$
$$= \left(\frac{15}{16}\right) a^{-5} \int_{-a}^{+a} (a^2 - x^2) \left(-\frac{\hbar^2}{2m} \right) (-2) \, dx$$
$$+ \left(\frac{15}{32}\right) a^{-5} \int_{-a}^{+a} (k x^2)(a^4 - 2a^2 x^2 + x^4) \, dx$$
$$= \left(\frac{15\hbar^2}{16m}\right) a^{-5} \int_{-a}^{+a} (a^2 - x^2) \, dx$$
$$+ \left(\frac{15}{32}\right) a^{-5} \int_{-a}^{+a} a^4 k x^2 - 2a^2 k x^4 + k x^6 \, dx$$
$$= \left(\frac{15\hbar^2}{16m}\right) a^{-5} \left(a^2 x \Big|_{-a}^{a} - \frac{1}{3} x^3 \Big|_{-a}^{a} \right)$$
$$+ \left(\frac{15}{32}\right) a^{-5} \left(\frac{a^4 k}{3} x^3 \Big|_{-a}^{a} - \frac{2a^2 k}{5} x^5 \Big|_{-a}^{a} + \frac{k}{7} x^7 \Big|_{-a}^{a} \right)$$
$$= \left(\frac{15\hbar^2}{16m}\right) a^{-5} \left(2a^3 - \frac{2}{3} a^3 \right)$$
$$+ \left(\frac{15}{32}\right) a^{-5} \left(\frac{2a^7 k}{3} - \frac{4a^7 k}{5} + \frac{2k}{7} a^7 \right)$$
$$= \left(\frac{15}{16}\right) a^{-5} \left(\frac{4\hbar^2}{3m} a^3 + \frac{a^7 k}{3} - \frac{2a^7 k}{5} + \frac{k}{7} a^7 \right)$$
$$= \left(\frac{15}{16}\right) a^{-5} \left(\frac{4\hbar^2}{3m} a^3 + \left(\frac{k}{3} - \frac{2k}{5} + \frac{k}{7} \right) a^7 \right)$$
$$= \left(\frac{15}{16}\right) a^{-5} \left(\frac{4\hbar^2}{3m} a^3 + \left(\frac{35k}{105} - \frac{42k}{105} + \frac{15k}{105} \right) a^7 \right)$$
$$= \left(\frac{15}{16}\right) a^{-5} \left(\frac{4\hbar^2}{3m} a^3 + \left(\frac{8k}{105} \right) a^7 \right) = \frac{5\hbar^2}{4ma^2} + \frac{ka^2}{14}.$$

b. Substituting $a = b(\frac{\hbar^2}{km})^{\frac{1}{4}}$ into the above expression for E we obtain

$$E = \frac{5\hbar^2}{4b^2 m} \left(\frac{km}{\hbar^2} \right)^{\frac{1}{2}} + \frac{kb^2}{14} \left(\frac{\hbar^2}{km} \right)^{\frac{1}{2}}$$
$$= \hbar k^{\frac{1}{2}} m^{-\frac{1}{2}} \left(\frac{5}{4} b^{-2} + \frac{1}{14} b^2 \right).$$

c.

$$E = \frac{5\hbar^2}{4ma^2} + \frac{ka^2}{14},$$

$$\frac{dE}{da} = -\frac{10\hbar^2}{4ma^3} + \frac{2ka}{14} = -\frac{5\hbar^2}{2ma^3} + \frac{ka}{7} = 0,$$

$$\frac{5\hbar^2}{2ma^3} = \frac{ka}{7} \quad \text{and} \quad 35\hbar^2 = 2mka^4.$$

$$\text{So,} \quad a^4 = \frac{35\hbar^2}{2mk}, \quad \text{or} \quad a = \left(\frac{35\hbar^2}{2mk}\right)^{\frac{1}{4}}.$$

Therefore,

$$\phi_{\text{best}} = \sqrt{\frac{15}{16}}\left(\frac{35\hbar^2}{2mk}\right)^{-\frac{5}{8}}\left(\left(\frac{35\hbar^2}{2mk}\right)^{\frac{1}{2}} - x^2\right),$$

and

$$E_{\text{best}} = \frac{5\hbar^2}{4m}\left(\frac{2mk}{35\hbar^2}\right)^{\frac{1}{2}} + \frac{k}{14}\left(\frac{35\hbar^2}{2mk}\right)^{\frac{1}{2}} = \hbar k^{\frac{1}{2}} m^{-\frac{1}{2}}\left(\frac{5}{14}\right)^{\frac{1}{2}}.$$

d.

$$\frac{E_{\text{best}} - E_{\text{true}}}{E_{\text{true}}} = \frac{\hbar k^{\frac{1}{2}} m^{-\frac{1}{2}}\left(\left(\frac{5}{14}\right)^{\frac{1}{2}} - 0.5\right)}{\hbar k^{\frac{1}{2}} m^{-\frac{1}{2}} 0.5}$$

$$= \frac{\left(\frac{5}{14}\right)^{\frac{1}{2}} - 0.5}{0.5} = \frac{0.0976}{0.5} = 0.1952 = 19.52\%.$$

18 a.

$$\mathbf{H}_0 \psi_{lm}^{(0)} = \frac{\mathbf{L}^2}{2m_e r_0^2}\psi_{lm}^{(0)} = \frac{\mathbf{L}^2}{2m_e r_0^2} Y_{l,m}(\theta, \phi)$$

$$= \frac{1}{2m_e r_0^2}\hbar^2 l(l+1) Y_{l,m}(\theta, \phi),$$

$$E_{lm}^{(0)} = \frac{\hbar^2}{2m_e r_0^2} l(l+1).$$

b.

$$V = -e\varepsilon z = -e\varepsilon r_0 \cos\theta,$$

$$E_{00}^{(1)} = \langle Y_{00}|V|Y_{00}\rangle = \langle Y_{00}| - e\varepsilon r_0 \cos\theta|Y_{00}\rangle$$

$$= -e\varepsilon r_0 \langle Y_{00}|\cos\theta|Y_{00}\rangle.$$

Using the given identity this becomes:

$$E_{00}^{(1)} = -e\varepsilon r_0 \langle Y_{00} \mid Y_{10}\rangle \sqrt{\frac{(0+0+1)(0-0+1)}{(2(0)+1)(2(0)+3)}}$$

$$+ - e\varepsilon r_0 \langle Y_{00} \mid Y_{-10}\rangle \sqrt{\frac{(0+0)(0-0)}{(2(0)+1)(2(0)-1)}}.$$

The spherical harmonics are orthonormal, thus $\langle Y_{00}|Y_{10}\rangle = \langle Y_{00}|Y_{-10}\rangle = 0$, and $E_{00}^{(1)} = 0$.

$$E_{00}^{(2)} = \sum_{lm\neq 00} \frac{|\langle Y_{lm}|V|Y_{00}\rangle|^2}{E_{00}^{(0)} - E_{lm}^{(0)}},$$

$$\langle Y_{lm}|V|Y_{00}\rangle = -e\varepsilon r_0 \langle Y_{lm}|\cos\theta|Y_{00}\rangle.$$

Using the given identity this becomes:

$$\langle Y_{lm}|V|Y_{00}\rangle = -e\varepsilon r_0 \langle Y_{lm} \mid Y_{10}\rangle \sqrt{\frac{(0+0+1)(0-0+1)}{(2(0)+1)(2(0)+3)}}$$

$$+ - e\varepsilon r_0 \langle Y_{lm} \mid Y_{-10}\rangle \sqrt{\frac{(0+0)(0-0)}{(2(0)+1)(2(0)-1)}}$$

$$\langle Y_{lm}|V|Y_{00}\rangle = -\frac{e\varepsilon r_0}{\sqrt{3}} \langle Y_{lm} \mid Y_{10}\rangle.$$

This indicates that the only term contributing to the sum in the expression for $E_{00}^{(2)}$ is when $l = 1$, and $m = 0$, otherwise $\langle Y_{lm}|V|Y_{00}\rangle$ vanishes (from orthonormality). In quantum chemistry when using orthonormal functions it is typical to write the term $\langle Y_{lm}|Y_{10}\rangle$ as a delta function, for example $\delta_{lm,10}$, which only has values of 1 or 0; $\delta_{ij} = 1$ when $i = j$ and 0 when $i \neq j$. This delta function when inserted into the sum then eliminates the sum by "picking out" the non-zero component. For example,

$$\langle Y_{lm}|V|Y_{00}\rangle = -\frac{e\varepsilon r_0}{\sqrt{3}} \delta_{lm,10},$$

so

$$E_{00}^{(2)} = \sum_{lm\neq 00} = \frac{e^2\varepsilon^2 r_0^2}{3} \frac{\delta_{lm,10}^2}{E_{00}^{(0)} - E_{lm}^{(0)}} = \frac{e^2\varepsilon^2 r_0^2}{3} \frac{1}{E_{00}^{(0)} - E_{10}^{(0)}},$$

$$E_{00}^{(0)} = \frac{\hbar^2}{2m_e r_0^2} 0(0+1) = 0 \quad \text{and} \quad E_{10}^{(0)} = \frac{\hbar^2}{2m_e r_0^2} l(l+1) = \frac{\hbar^2}{m_e r_0^2}.$$

Inserting these energy expressions above yields:

$$E_{00}^{(2)} = -\frac{e^2\varepsilon^2 r_0^2}{3} \frac{m_e r_0^2}{\hbar^2} = -\frac{m_e e^2 \varepsilon^2 r_0^4}{3\hbar^2}.$$

c.

$$\begin{aligned} E_{00} &= E_{00}^{(0)} + E_{00}^{(1)} + E_{00}^{(2)} + \cdots \\ &= 0 + 0 - \frac{m_e e^2 \varepsilon^2 r_0^4}{3\hbar^2} \\ &= -\frac{m_e e^2 \varepsilon^2 r_0^4}{3\hbar^2}, \end{aligned}$$

$$\alpha = -\frac{\partial^2 E}{\partial^2 \varepsilon} = \frac{\partial^2}{\partial^2 \varepsilon}\left(\frac{m_e e^2 \varepsilon^2 r_0^4}{3\hbar^2}\right)$$

$$= \frac{2m_e e^2 r_0^4}{3\hbar^2}.$$

d.

$$\alpha = \frac{2(9.1095 \times 10^{-28}\ \text{g})\left(4.803\,24 \times 10^{-10}\ \text{g}^{\frac{1}{2}}\ \text{cm}^{\frac{3}{2}}\ \text{s}^{-1}\right)^2 r_0^4}{3(1.054\,59 \times 10^{-27}\ \text{g cm}^2\ \text{s}^{-1})^2}$$

$$= r_0^4\ 12\,598 \times 10^6\ \text{cm}^{-1} = r_0^4\ 1.2598\ \text{Å}^{-1},$$

$$\alpha_H = 0.0987\ \text{Å}^3,$$

$$\alpha_{Cs} = 57.57\ \text{Å}^3.$$

19

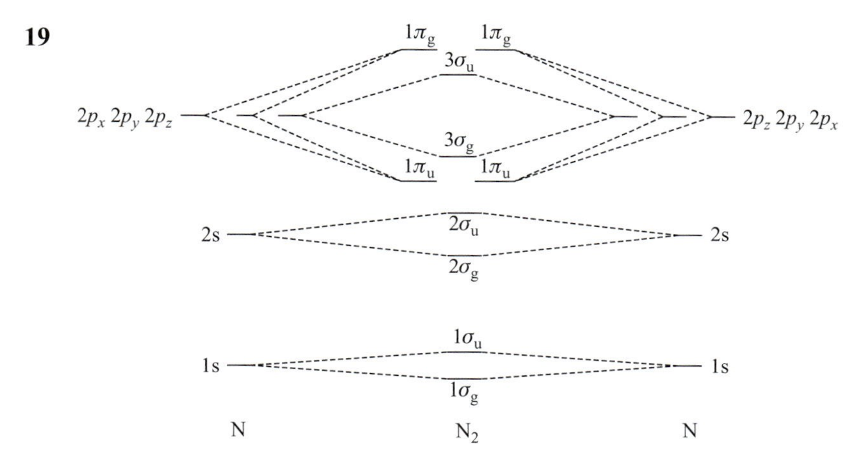

The above diagram indicates how the MOs are formed from the 1s, 2s, and 2p N atomic orbitals. It can be seen that there are $3\sigma_g$, $3\sigma_u$, $1\pi_{ux}$, $1\pi_{uy}$, $1\pi_{gx}$, and $1\pi_{gy}$ MOs. The Hamiltonian matrices (Fock matrices) are given. Each of these can be diagonalized to give the following MO energies:

$3\sigma_g$	-15.52, -1.45, and -0.54 (hartrees)
$3\sigma_u$	-15.52, -0.72, and 1.13
$1\pi_{ux}$	-0.58
$1\pi_{uy}$	-0.58
$1\pi_{gx}$	0.28
$1\pi_{gy}$	0.28

It can be seen that the $3\sigma_g$ orbitals are bonding, the $3\sigma_u$ orbitals are antibonding, the $1\pi_{ux}$ and $1\pi_{uy}$ orbitals are bonding, and the $1\pi_{gx}$ and $1\pi_{gy}$ orbitals are antibonding.

20 Using these approximate energies we can draw the following MO diagram:

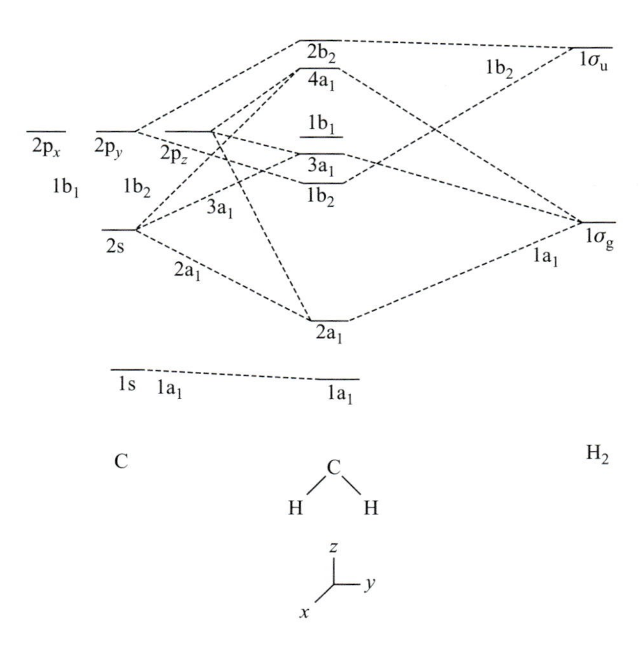

This MO diagram is not an orbital correlation diagram but can be used to help generate one. The energy levels on each side (C and H_2) can be "superimposed" to generate the reactant side of the orbital correlation diagram and the center CH_2 levels can be used to form the product side. Ignoring the core levels this generates the following orbital correlation diagram for the reaction $C + H_2 \longrightarrow CH_2$ (bent):

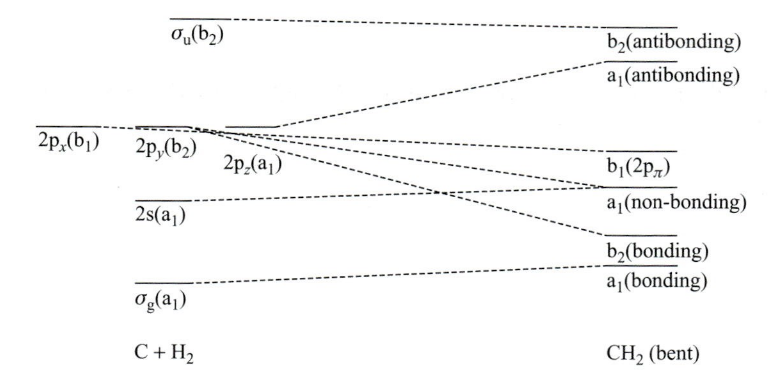

21

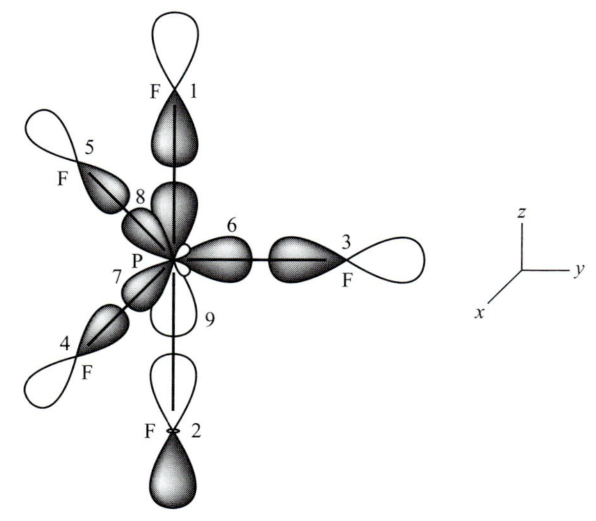

a. The two F p orbitals (top and bottom) generate the following reducible representation:

D_{3h}	E	$2C_3$	$3C_2$	σ_h	$2S_3$	$3\sigma_v$
Γ_p	2	2	0	0	0	2

This reducible representation reduces to $1A_1'$ and $1A_2''$ irreducible representations. Projectors may be used to find the symmetry-adapted AOs for these irreducible representations:

$$\phi_{a_1'} = \frac{1}{\sqrt{2}}(f_1 - f_2),$$
$$\phi_{a_2''} = \frac{1}{\sqrt{2}}(f_1 + f_2).$$

b. The three trigonal F p orbitals generate the following reducible representation:

D_{3h}	E	$2C_3$	$3C_2$	σ_h	$2S_3$	$3\sigma_v$
Γ_p	3	0	1	3	0	1

This reducible representation reduces to $1A_1'$ and $1E'$ irreducible representations. Projectors may be used to find the symmetry-adapted AOs for these irreducible representations (but they are exactly analogous to the previous few problems):

$$\phi_{a_1'} = \frac{1}{\sqrt{3}}(f_3 + f_4 + f_5),$$
$$\phi_{e'} = (1/6)^{-1/2}(2f_3 - f_4 - f_5),$$
$$\phi_{e'} = \frac{1}{\sqrt{2}}(f_4 - f_5).$$

c. The 3 P sp^2 orbitals generate the following reducible representation:

D_{3h}	E	$2C_3$	$3C_2$	σ_h	$2S_3$	$3\sigma_v$
Γ_{sp^2}	3	0	1	3	0	1

This reducible representation reduces to $1A_1'$ and $1E'$ irreducible representations. Again, projectors may be used to find the symmetry-adapted AOs for these irreducible representations:

$$\phi_{a1'} = \frac{1}{\sqrt{3}}(f_6 + f_7 + f_8),$$
$$\phi_{e'} = \frac{1}{\sqrt{6}}(2f_6 - f_7 - f_8),$$
$$\phi_{e'} = \frac{1}{\sqrt{2}}(f_7 - f_8).$$

The leftover P p$_z$ orbital generates the following irreducible representation:

D_{3h}	E	$2C_3$	$3C_2$	σ_h	$2S_3$	$3\sigma_v$
Γ_{pz}	1	1	-1	-1	-1	1

This irreducible representation is A_2'',

$$\phi_{a_2''} = f_9.$$

d, e. Drawing an energy level diagram using these AOs would result in the following:

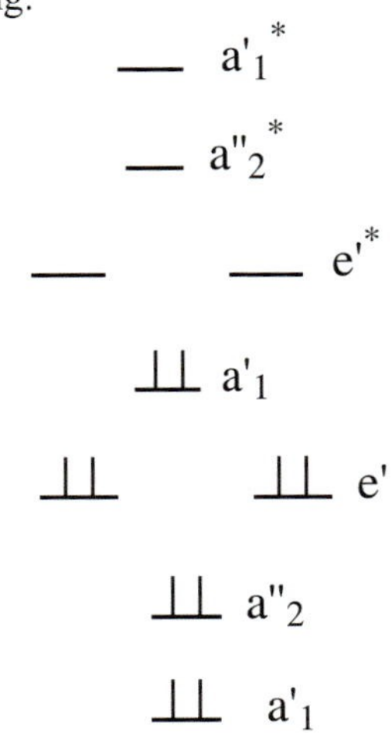

$a_1'^*$

$a_2''^*$

e'^*

a_1'

e'

a_2''

a_1'

22 a. For non-degenerate point groups, one can simply multiply the representations (since only one representation will be obtained):

$$a_1 \otimes b_1 = b_1.$$

Constructing a "box" in this case is unnecessary since it would only contain a single row. Two unpaired electrons will result in a singlet ($S = 0$, $M_S = 0$), and three triplets ($S = 1$, $M_S = 1$; $S = 1$, $M_S = 0$; $S = 1$, $M_S = -1$). The states will be: ${}^3B_1(M_S = 1)$, ${}^3B_1(M_S = 0)$, ${}^3B_1(M_S = -1)$, and ${}^1B_1(M_S = 0)$.

b. Remember that when coupling non-equivalent linear molecule angular momenta, one simple adds the individual L_z values and vector couples the electron spin. So, in this case ($1\pi_u^1 2\pi_u^1$), we have M_L values of $1 + 1$, $1 - 1$, $-1 + 1$, and $-1 - 1$ (2, 0, 0, and -2). The term symbol Δ is used to denote the spatially doubly degenerate level ($M_L = \pm 2$) and there are two distinct spatially non-degenerate levels denoted by the term symbol $\sum(M_L = 0)$. Again, two unpaired electrons will result in a singlet ($S = 0$, $M_S = 0$), and three triplets ($S = 1$, $M_S = 1$; $S = 1$, $M_S = 0$; $S = 1$, $M_S = -1$). The states generated are then:

${}^1\Delta(M_L = 2)$; one state ($M_S = 0$),
${}^1\Delta(M_L = -2)$; one state ($M_S = 0$),
${}^3\Delta(M_L = 2)$; three states ($M_S = 1, 0$, and -1),
${}^3\Delta(M_L = -2)$; three states ($M_S = 1, 0$, and -1),
${}^1\sum(M_L = 0)$; one state ($M_S = 0$),
${}^1\sum(M_L = 0)$; one state ($M_S = 0$),
${}^3\sum(M_L = 0)$; three states ($M_S = 1, 0$, and -1), and
${}^3\sum(M_L = 0)$; three states ($M_S = 1, 0$, and -1).

c. Constructing the "box" for two equivalent π electrons one obtains:

M_L / M_S	2	1	0
1			$\lvert\pi_1\alpha\pi_{-1}\alpha\rvert$
0	$\lvert\pi_1\alpha\pi_1\beta\rvert$		$\lvert\pi_1\alpha\pi_{-1}\beta\rvert$, $\lvert\pi_{-1}\alpha\pi_1\beta\rvert$

From this "box" one obtains six states:

${}^1\Delta(M_L = 2)$; one state ($M_S = 0$),
${}^1\Delta(M_L = -2)$; one state ($M_S = 0$),
${}^1\sum(M_L = 0)$; one state ($M_S = 0$),
${}^3\sum(M_L = 0)$; three states ($M_S = 1, 0$, and -1).

d. It is not necessary to construct a "box" when coupling non-equivalent angular momenta since vector coupling results in a range from the sum of the two individual angular momenta to the absolute value of their difference. In this case, $3d^14d^1$, $L = 4, 3, 2, 1, 0$, and $S = 1, 0$. The term symbols are: ^{3}G, ^{1}G, ^{3}F, ^{1}F, ^{3}D, ^{1}D, ^{3}P, ^{1}P, ^{3}S, and ^{1}S. The L and S angular momenta can be vector coupled to produce further splitting into levels:

$$J = L + S \cdots |L - S|.$$

Denoting J as a term symbol subscript one can identify all the levels and subsequent $(2J + 1)$ states:

$^3G_5$ (11 states),
$^3G_4$ (9 states),
$^3G_3$ (7 states),
$^1G_4$ (9 states),
$^3F_4$ (9 states),
$^3F_3$ (7 states),
$^3F_2$ (5 states),
$^1F_3$ (7 states),
$^3D_3$ (7 states),
$^3D_2$ (5 states),
$^3D_1$ (3 states),
$^1D_2$ (5 states),
$^3P_2$ (5 states),
$^3P_1$ (3 states),
$^3P_0$ (1 state),
$^1P_1$ (3 states),
$^3S_1$ (3 states), and
$^1S_0$ (1 state).

e. Construction of a "box" for the two equivalent d electrons generates (note the "box" has been turned sideways for convenience):

M_S / M_L	1	0
4		$\lvert d_2\alpha d_2\beta\rvert$
3	$\lvert d_2\alpha d_1\alpha\rvert$	$\lvert d_2\alpha d_1\beta\rvert$, $\lvert d_2\beta d_1\alpha\rvert$
2	$\lvert d_2\alpha d_0\alpha\rvert$	$\lvert d_2\alpha d_0\beta\rvert$, $\lvert d_2\beta d_0\alpha\rvert$, $\lvert d_1\alpha d_1\beta\rvert$
1	$\lvert d_1\alpha d_0\alpha\rvert$, $\lvert d_2\alpha d_{-1}\alpha\rvert$	$\lvert d_1\alpha d_0\beta\rvert$, $\lvert d_1\beta d_0\alpha\rvert$, $\lvert d_2\alpha d_{-1}\beta\rvert$, $\lvert d_2\beta d_{-1}\alpha\rvert$
0	$\lvert d_2\alpha d_{-2}\alpha\rvert$, $\lvert d_1\alpha d_{-1}\alpha\rvert$	$\lvert d_2\alpha d_{-2}\beta\rvert$, $\lvert d_2\beta d_{-2}\alpha\rvert$, $\lvert d_1\alpha d_{-1}\beta\rvert$, $\lvert d_1\beta d_{-1}\alpha\rvert$, $\lvert d_0\alpha d_0\beta\rvert$

The term symbols are: ^{1}G, ^{3}F, ^{1}D, ^{3}P, and ^{1}S. The L and S angular momenta can be vector coupled to produce further splitting into levels:

$^1G_4$ (9 states),
$^3F_4$ (9 states),
$^3F_3$ (7 states),
$^3F_2$ (5 states),
$^1D_2$ (5 states),
$^3P_2$ (5 states),
$^3P_1$ (3 states),
$^3P_0$ (1 state), and
$^1S_0$ (1 state).

23 a. Once the spatial symmetry has been determined by multiplication of the irreducible representations, the spin coupling gives the result:

$$\frac{1}{\sqrt{2}}(|3a_1\alpha 1b_1\beta| - |3a_1\beta 1b_1\alpha|).$$

b. There are three states here:

(i) $|3a_1\alpha 1b_1\alpha|$,
(ii) $\frac{1}{\sqrt{2}}(|3a_1\alpha 1b_1\beta| + |3a_1\beta 1b_1\alpha|)$, and
(iii) $|3a_1\beta 1b_1\beta|$.

c. $|3a_1\alpha 3a_1\beta|$.

24 a. All the Slater determinants have in common the $|1s\alpha 1s\beta 2s\alpha 2s\beta|$ "core" and hence this component will not be written out explicitly for each case.

$$\begin{aligned}
{}^3\mathrm{P}(M_L = 1, M_S = 1) &= |p_1\alpha p_0\alpha| \\
&= \left|\frac{1}{\sqrt{2}}(\mathrm{p}_x + i\mathrm{p}_y)\alpha(\mathrm{p}_z)\alpha\right| \\
&= \frac{1}{\sqrt{2}}(|\mathrm{p}_x\alpha \mathrm{p}_z\alpha| + i|\mathrm{p}_y\alpha p_z\alpha|),
\end{aligned}$$

$$\begin{aligned}
{}^3\mathrm{P}(M_L = 0, M_S = 1) &= |p_1\alpha p_{-1}\alpha| \\
&= \left|\frac{1}{\sqrt{2}}(\mathrm{p}_x + i\mathrm{p}_y)\alpha\frac{1}{\sqrt{2}}(\mathrm{p}_x - i\mathrm{p}_y)\alpha\right| \\
&= \frac{1}{2}(|\mathrm{p}_x\alpha\mathrm{p}_x\alpha| - i|\mathrm{p}_x\alpha\mathrm{p}_y\alpha| + i|\mathrm{p}_y\alpha\mathrm{p}_x\alpha| + |\mathrm{p}_y\alpha\mathrm{p}_y\alpha|) \\
&= \frac{1}{2}(0 - i|\mathrm{p}_x\alpha\mathrm{p}_y\alpha| - i|\mathrm{p}_x\alpha\mathrm{p}_y\alpha| + 0) \\
&= \frac{1}{2}(-2i|\mathrm{p}_x\alpha\mathrm{p}_y\alpha|) \\
&= -i|\mathrm{p}_x\alpha\mathrm{p}_y\alpha|,
\end{aligned}$$

$$
\begin{aligned}
{}^3\mathrm{P}(M_L = -1, M_S = 1) &= |\mathrm{p}_{-1}\alpha \mathrm{p}_0\alpha| \\
&= \left|\frac{1}{\sqrt{2}}(\mathrm{p}_x - i\mathrm{p}_y)\alpha(\mathrm{p}_z)\alpha\right| \\
&= \frac{1}{\sqrt{2}}(|\mathrm{p}_x\alpha\mathrm{p}_z\alpha| - i|\mathrm{p}_y\alpha\mathrm{p}_z\alpha|).
\end{aligned}
$$

As you can see, the symmetries of each of these states can not be labeled with a single irreducible representation of the C_{2v} point group. For example, $|\mathrm{p}_x\alpha\mathrm{p}_z\alpha|$ is $xz(\mathrm{B}_1)$ and $|\mathrm{p}_y\alpha\mathrm{p}_z\alpha|$ is $yz(\mathrm{B}_2)$ and hence the ${}^3\mathrm{P}(M_L = 1, M_S = 1)$ state is a combination of B_1 and B_2 symmetries. But, the three ${}^3\mathrm{P}(M_L, M_S = 1)$ functions are degenerate for the C atom and any combination of these three functions would also be degenerate. Therefore, we can choose new combinations that can be labeled with "pure" C_{2v} point group labels.

$$
\begin{aligned}
{}^3\mathrm{P}(xz, M_S = 1) &= |\mathrm{p}_x\alpha\mathrm{p}_z\alpha| \\
&= \frac{1}{\sqrt{2}}({}^3\mathrm{P}(M_L = 1, M_S = 1) + {}^3\mathrm{P}(M_L = -1, M_S = 1)) = {}^3\mathrm{B}_1, \\
{}^3\mathrm{P}(yx, M_S = 1) &= |\mathrm{p}_y\alpha\mathrm{p}_x\alpha| \\
&= \frac{1}{i}({}^3\mathrm{P}(M_L = 0, M_S = 1)) = {}^3\mathrm{A}_2, \\
{}^3\mathrm{P}(yz, M_S = 1) &= |\mathrm{p}_y\alpha\mathrm{p}_z\alpha| \\
&= \frac{1}{i\sqrt{2}}({}^3\mathrm{P}(M_L = 1, M_S = 1) - {}^3\mathrm{P}(M_L = -1, M_S = 1)) = {}^3\mathrm{B}_2.
\end{aligned}
$$

Now, we can do likewise for the five degenerate ${}^1\mathrm{D}$ states:

$$
\begin{aligned}
{}^1\mathrm{D}(M_L = 2, M_S = 0) &= |\mathrm{p}_1\alpha\mathrm{p}_1\beta| \\
&= \left|\frac{1}{\sqrt{2}}(\mathrm{p}_x + i\mathrm{p}_y)\alpha\frac{1}{\sqrt{2}}(\mathrm{p}_x + i\mathrm{p}_y)\beta\right| \\
&= \frac{1}{2}(|\mathrm{p}_x\alpha\mathrm{p}_x\beta| + i|\mathrm{p}_x\alpha\mathrm{p}_y\beta| + i|\mathrm{p}_y\alpha\mathrm{p}_x\beta| - |\mathrm{p}_y\alpha\mathrm{p}_y\beta|), \\
{}^1\mathrm{D}(M_L = -2, M_S = 0) &= |\mathrm{p}_{-1}\alpha\mathrm{p}_{-1}\beta| \\
&= \left|\frac{1}{\sqrt{2}}(\mathrm{p}_x - i\mathrm{p}_y)\alpha\frac{1}{\sqrt{2}}(\mathrm{p}_x - i\mathrm{p}_y)\beta\right| \\
&= \frac{1}{2}(|\mathrm{p}_x\alpha\mathrm{p}_x\beta| - i|\mathrm{p}_x\alpha\mathrm{p}_y\beta| - i|\mathrm{p}_y\alpha\mathrm{p}_x\beta| - |\mathrm{p}_y\alpha\mathrm{p}_y\beta|), \\
{}^1\mathrm{D}(M_L = 1, M_S = 0) &= \frac{1}{\sqrt{2}}(|\mathrm{p}_0\alpha\mathrm{p}_1\beta| - |\mathrm{p}_0\beta\mathrm{p}_1\alpha|) \\
&= \frac{1}{\sqrt{2}}\left(\left|(\mathrm{p}_z)\alpha\frac{1}{\sqrt{2}}(\mathrm{p}_x + i\mathrm{p}_y)\beta\right| - \left|(\mathrm{p}_z)\beta\frac{1}{\sqrt{2}}(\mathrm{p}_x + i\mathrm{p}_y)\alpha\right|\right) \\
&= \frac{1}{2}(|\mathrm{p}_z\alpha\mathrm{p}_x\beta| + i|\mathrm{p}_z\alpha\mathrm{p}_y\beta| - |\mathrm{p}_z\beta\mathrm{p}_x\alpha| - i|\mathrm{p}_z\beta\mathrm{p}_y\alpha|), \\
{}^1\mathrm{D}(M_L = -1, M_S = 0) &= \frac{1}{\sqrt{2}}(|\mathrm{p}_0\alpha\mathrm{p}_{-1}\beta| - |\mathrm{p}_0\beta\mathrm{p}_{-1}\alpha|) \\
&= \frac{1}{\sqrt{2}}\left(\left|(\mathrm{p}_z)\alpha\frac{1}{\sqrt{2}}(\mathrm{p}_x - i\mathrm{p}_y)\beta\right| - \left|(\mathrm{p}_z)\beta\frac{1}{\sqrt{2}}(\mathrm{p}_x - i\mathrm{p}_y)\alpha\right|\right) \\
&= \frac{1}{2}(|\mathrm{p}_z\alpha\mathrm{p}_x\beta| - i|\mathrm{p}_z\alpha\mathrm{p}_y\beta| - |\mathrm{p}_z\beta\mathrm{p}_x\alpha| + i|\mathrm{p}_z\beta\mathrm{p}_y\alpha|),
\end{aligned}
$$

$$
\begin{aligned}
{}^1\mathrm{D}(M_L=0,M_S=0) &= \frac{1}{\sqrt{6}}(2|\mathrm{p}_0\alpha\mathrm{p}_0\beta| + |\mathrm{p}_1\alpha\mathrm{p}_{-1}\beta| + |\mathrm{p}_{-1}\alpha\mathrm{p}_1\beta|) \\
&= \frac{1}{\sqrt{6}}\left(2\left|\mathrm{p}_z\alpha\mathrm{p}_z\beta| + |\frac{1}{\sqrt{2}}(\mathrm{p}_x + i\mathrm{p}_y)\alpha\frac{1}{\sqrt{2}}(\mathrm{p}_x - i\mathrm{p}_y)\beta\right|\right. \\
&\quad \left. + \left|\frac{1}{\sqrt{2}}(\mathrm{p}_x - i\mathrm{p}_y)\alpha\frac{1}{\sqrt{2}}(\mathrm{p}_x + i\mathrm{p}_y)\beta\right|\right) \\
&= \frac{1}{\sqrt{6}}\left(2|\mathrm{p}_z\alpha\mathrm{p}_z\beta|\right. \\
&\quad + \frac{1}{2}(|\mathrm{p}_x\alpha\mathrm{p}_x\beta| - i|\mathrm{p}_x\alpha\mathrm{p}_y\beta| + i|\mathrm{p}_y\alpha\mathrm{p}_x\beta| + |\mathrm{p}_y\alpha\mathrm{p}_y\beta|) \\
&\quad \left. + \frac{1}{2}(|\mathrm{p}_x\alpha\mathrm{p}_x\beta| + i|\mathrm{p}_x\alpha\mathrm{p}_y\beta| - i|\mathrm{p}_y\alpha\mathrm{p}_x\beta| + |\mathrm{p}_y\alpha\mathrm{p}_y\beta|)\right) \\
&= \frac{1}{\sqrt{6}}(2|\mathrm{p}_z\alpha\mathrm{p}_z\beta| + |\mathrm{p}_x\alpha\mathrm{p}_x\beta| + |\mathrm{p}_y\alpha\mathrm{p}_y\beta|).
\end{aligned}
$$

Analogous to the three ^{3}P states, we can also choose combinations of the five degenerate ^{1}D states which can be labeled with "pure" C_{2v} point group labels:

$$
\begin{aligned}
{}^1\mathrm{D}(xx-yy,M_S=0) &= |\mathrm{p}_x\alpha\mathrm{p}_x\beta| - |\mathrm{p}_y\alpha\mathrm{p}_y\beta| \\
&= ({}^1\mathrm{D}(M_L=2,M_S=0) + {}^1\mathrm{D}(M_L=-2,M_S=0)) \\
&= {}^1\mathrm{A}_1, \\
{}^1\mathrm{D}(yx,M_S=0) &= |\mathrm{p}_x\alpha\mathrm{p}_y\beta| + |\mathrm{p}_y\alpha\mathrm{p}_x\beta| \\
&= \frac{1}{i}({}^1\mathrm{D}(M_L=2,M_S=0) - {}^1\mathrm{D}(M_L=-2,M_S=0)) \\
&= {}^1\mathrm{A}_2, \\
{}^1\mathrm{D}(zx,M_S=0) &= |\mathrm{p}_z\alpha\mathrm{p}_x\beta| - |\mathrm{p}_z\beta\mathrm{p}_x\alpha| \\
&= ({}^1\mathrm{D}(M_L=1,M_S=0) + {}^1\mathrm{D}(M_L=-1,M_S=0)) \\
&= {}^1\mathrm{B}_1, \\
{}^1\mathrm{D}(zy,M_S=0) &= |\mathrm{p}_z\alpha\mathrm{p}_y\beta| - |\mathrm{p}_z\beta\mathrm{p}_y\alpha| \\
&= \frac{1}{i}({}^1\mathrm{D}(M_L=1,M_S=0) - {}^1\mathrm{D}(M_L=-1,M_S=0)) \\
&= {}^1\mathrm{B}_2, \\
{}^1\mathrm{D}(2zz+xx+yy,M_S=0) &= \frac{1}{\sqrt{6}}(2|\mathrm{p}_z\alpha\mathrm{p}_z\beta| + |\mathrm{p}_x\alpha\mathrm{p}_x\beta| + |\mathrm{p}_y\alpha\mathrm{p}_y\beta|)) \\
&= {}^1\mathrm{D}(M_L=0,M_S=0) = {}^1\mathrm{A}_1.
\end{aligned}
$$

The only state left is the ^{1}S:

$$
\begin{aligned}
{}^1\mathrm{S}(M_L=0,M_S=0) &= \frac{1}{\sqrt{3}}(|\mathrm{p}_0\alpha\mathrm{p}_0\beta| - |\mathrm{p}_1\alpha\mathrm{p}_{-1}\beta| - |\mathrm{p}_{-1}\alpha\mathrm{p}_1\beta|) \\
&= \frac{1}{\sqrt{3}}\left(|\mathrm{p}_z\alpha\mathrm{p}_z\beta| - \left|\frac{1}{\sqrt{2}}(\mathrm{p}_x + i\mathrm{p}_y)\alpha\frac{1}{\sqrt{2}}(\mathrm{p}_x - i\mathrm{p}_y)\beta\right|\right. \\
&\quad \left. - \left|\frac{1}{\sqrt{2}}(\mathrm{p}_x - i\mathrm{p}_y)\alpha\frac{1}{\sqrt{2}}(\mathrm{p}_x + i\mathrm{p}_y)\beta\right|\right)
\end{aligned}
$$

$$= \frac{1}{\sqrt{3}}\Bigg(|p_z\alpha p_z\beta|$$
$$-\frac{1}{2}(|p_x\alpha p_x\beta| - i|p_x\alpha p_y\beta| + i|p_y\alpha p_x\beta| + |p_y\alpha p_y\beta|)$$
$$-\frac{1}{2}(|p_x\alpha p_x\beta| + i|p_x\alpha p_y\beta| - i|p_y\alpha p_x\beta| + |p_y\alpha p_y\beta|)\Bigg)$$
$$= \frac{1}{\sqrt{3}}(|p_z\alpha p_z\beta| - |p_x\alpha p_x\beta| - |p_y\alpha p_y\beta|).$$

Each of the components of this state are A_1 and hence this state has A_1 symmetry.

b. Forming symmetry-adapted AOs from the C and H atomic orbitals would generate the following:

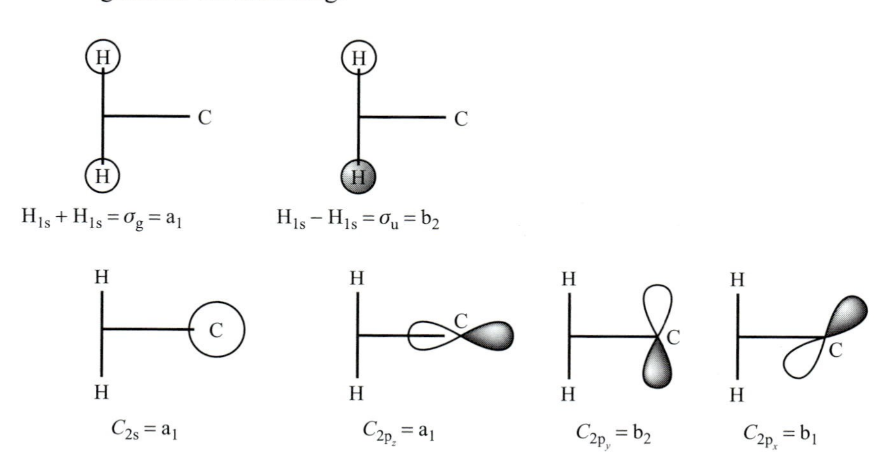

The bonding, non-bonding, and antibonding orbitals of CH_2 can be illustrated in the following manner:

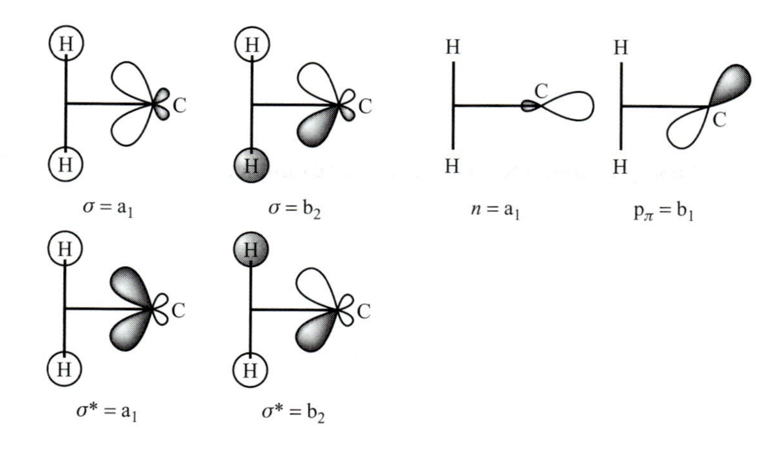

c. Orbital correlation diagram for the reaction $C + H_2 \longrightarrow CH_2$ (bent)

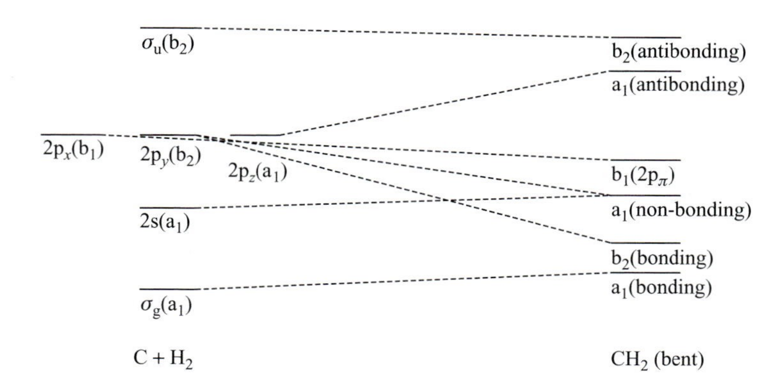

d,e. It is necessary to determine how the wave functions found in part (a) correlate with states of the CH_2 molecule:

$$
\begin{aligned}
&^3\mathrm{P}(xz, M_S = 1); {}^3\mathrm{B}_1 = \sigma_g^2 s^2 p_x p_z \longrightarrow \sigma^2 n^2 p_\pi \sigma^* \\
&^3\mathrm{P}(yx, M_S = 1); {}^3\mathrm{A}_2 = \sigma_g^2 s^2 p_x p_y \longrightarrow \sigma^2 n^2 p_\pi \sigma \\
&^3\mathrm{P}(yz, M_S = 1); {}^3\mathrm{B}_2 = \sigma_g^2 s^2 p_y p_z \longrightarrow \sigma^2 n^2 \sigma \sigma^* \\
&^1\mathrm{D}(xx - yy, M_S = 0); {}^1\mathrm{A}_1 \longrightarrow \sigma^2 n^2 p_\pi^2 - \sigma^2 n^2 \sigma^2 \\
&^1\mathrm{D}(yx, M_S = 0); {}^1\mathrm{A}_2 \longrightarrow \sigma^2 n^2 \sigma p_\pi \\
&^1\mathrm{D}(zx, M_S = 0); {}^1\mathrm{B}_1 \longrightarrow \sigma^2 n^2 \sigma^* p_\pi \\
&^1\mathrm{D}(zy, M_S = 0); {}^1\mathrm{B}_2 \longrightarrow \sigma^2 n^2 \sigma^* \sigma \\
&^1\mathrm{D}(2zz + xx + yy, M_S = 0); {}^1\mathrm{A}_1 \longrightarrow 2\sigma^2 n^2 \sigma^{*2} + \sigma^2 n^2 p_\pi^2 + \sigma^2 n^2 \sigma^2.
\end{aligned}
$$

Note, the $C + H_2$ state to which the lowest ${}^1A_1(\sigma^2 n^2 \sigma^2)$ CH_2 state decomposes would be $\sigma_g^2 s^2 p_y^2$. This state ($\sigma_g^2 s^2 p_y^2$) can not be obtained by a simple combination of the ^{1}D states. In order to obtain pure $\sigma_g^2 s^2 p_y^2$ it is necessary to combine ^{1}S with ^{1}D. For example,

$$
\sigma_g^2 s^2 p_y^2 = \frac{1}{6}(\sqrt{6}\,{}^1\mathrm{D}(0, 0) - 2\sqrt{3}\,{}^1\mathrm{S}(0, 0)) - \frac{1}{2}({}^1\mathrm{D}(2, 0) + {}^1\mathrm{D}(-2, 0)).
$$

This indicates that a configuration correlation diagram must be drawn with a barrier near the ^{1}D asymptote to represent the fact that ${}^1A_1$ CH_2 correlates with a mixture of ^{1}D and ^{1}S carbon plus hydrogen. The $C + H_2$

state to which the lowest $^3B_1$ $(\sigma^2 n\sigma^2 p_\pi)$ CH_2 state decomposes would be $\sigma_g^2 sp_y^2 p_x$.

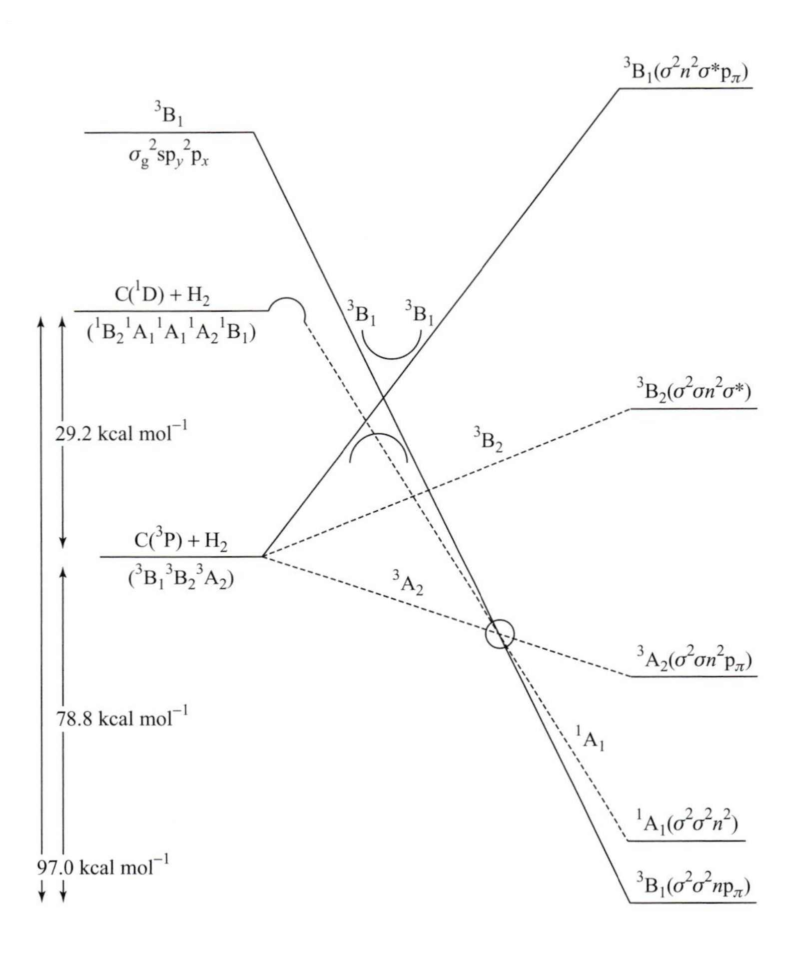

f. If you follow the $^3B_1$ component of the $C(^3P) + H_2$ (since it leads to the ground-state products) to $^3B_1$ CH_2 you must go over an approximately 20 kcal mol^{-1} barrier. Of course this path produces $^3B_1$ CH_2 product. Distortions away from C_{2v} symmetry, for example to C_s symmetry, would make the a_1 and b_2 orbitals identical in symmetry (a′). The b_1 orbitals would maintain their different symmetry going to a″ symmetry. Thus $^3B_1$ and $^3A_2$ (both $^3A''$ in C_s symmetry and odd under reflection through the molecular plane) can mix. The system could thus follow the $^3A_2$ component of the $C(^3P) + H_2$ surface to the place (marked with a circle on the CCD) where it crosses the $^3B_1$ surface upon which it then moves and continues to products. As a result, the barrier would be lowered.

You can estimate when the barrier occurs (late or early) using thermodynamic information for the reaction (i.e., slopes and asymptotic energies). For example, an early barrier would be obtained for a reaction with the characteristics:

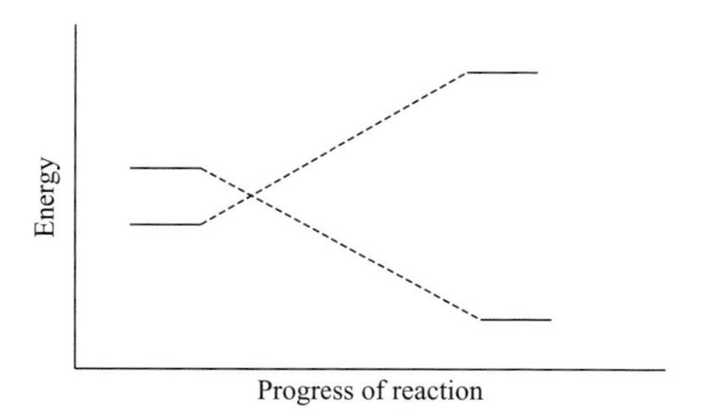

and a late barrier would be obtained for a reaction with the characteristics:

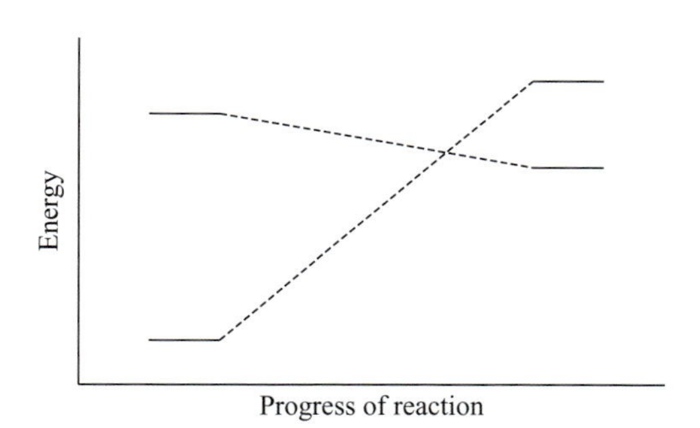

This relation between reaction endothermicity or exothermicity and the character of the transition state is known as the Hammond postulate. Note that the $C(^3P_1) + H_2 \rightarrow CH_2$ reaction of interest here has an early barrier.

g. The reaction $C(^1D) + H_2 \rightarrow CH_2(^1A_1)$ should have no symmetry barrier (this can be recognized by following the $^1A_1$ $(C(^1D) + H_2)$ reactants down to the $^1A_1$ (CH_2) products).

25 This problem in many respects is analogous to Problem 24.
The $^3B_1$ surface certainly requires a two configuration CI wavefunction; the $\sigma^2\sigma^2 n p_x$ ($\pi^2 p_y^2 sp_x$) and the $\sigma^2 n^2 p_x \sigma^*$ ($\pi^2 s^2 p_x p_z$). The $^1A_1$ surface could use the $\sigma^2\sigma^2 n^2$ ($\pi^2 s^2 p_y^2$) only but once again there is no combination of ^1D determinants which gives purely this configuration ($\pi^2 s^2 p_y^2$). Thus mixing of both ^1D and ^1S determinants is necessary to yield the required $\pi^2 s^2 p_y^2$ configuration. Hence even the $^1A_1$ surface would require a multiconfigurational wave function for adequate description.

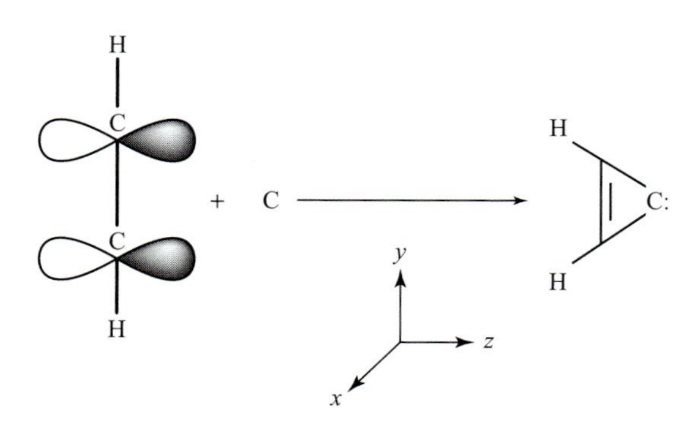

Orbital correlation diagram for the reaction $C_2H_2 + C \rightarrow C_3H_2$:

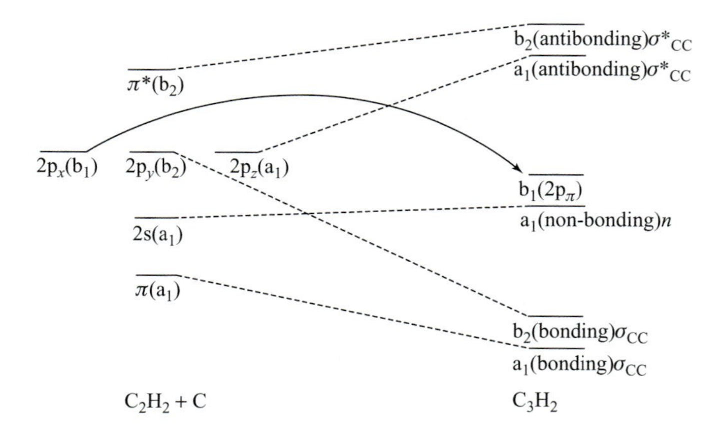

Configuration correlation diagram for the reaction $C_2H_2 + C \rightarrow C_3H_2$:

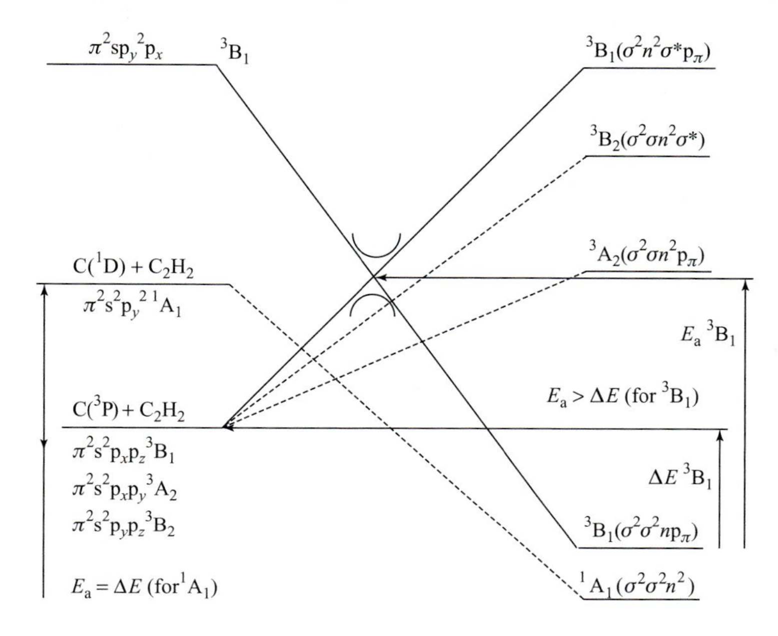

26 a. CCl_4 is tetrahedral and therefore is a spherical top. $CHCl_3$ has C_{3v} symmetry and therefore is a symmetric top. CH_2Cl_2 has C_{2v} symmetry and therefore is an asymmetric top.

b. CCl_4 has such high symmetry that it will not exhibit pure rotational spectra because it has no permanent dipole moment. $CHCl_3$ and CH_2Cl_2 will both exhibit pure rotational spectra.

27 NH_3 is a symmetric top (oblate). Use the given energy expression,

$$E = (A - B)K^2 + BJ(J + 1),$$

$A = 6.20\ \text{cm}^{-1}$, $B = 9.44\ \text{cm}^{-1}$, selection rules $\Delta J = \pm 1$, and the fact that μ_0 lies along the figure axis such that $\Delta K = 0$, to give

$$\Delta E = 2B(J + 1) = 2B, 4B, \quad \text{and} \quad 6B(J = 0, 1, \text{ and } 2).$$

So, lines are at $18.88\ \text{cm}^{-1}$, $37.76\ \text{cm}^{-1}$, and $56.64\ \text{cm}^{-1}$.

28 To convert between cm^{-1} and energy, multiply by $hc = (6.626\,18 \times 10^{-34}\ \text{J s})\,(2.997\,925 \times 10^{10}\ \text{cm s}^{-1}) = 1.9865 \times 10^{23}\ \text{J cm}$.
Let all quantities in cm^{-1} be designated with a bar, e.g. $\overline{B_e} = 1.78\ \text{cm}^{-1}$.

a.

$$hc\overline{B_e} = \frac{\hbar^2}{2\mu R_e^2},$$

$$R_e = \frac{\hbar}{\sqrt{2\mu hc\overline{B_e}}},$$

$$\mu = \frac{m_B m_O}{m_B + m_O} = \frac{(11)(16)}{(11+16)} \times 1.660\,56 \times 10^{-27}\ \text{kg}$$
$$= 1.0824 \times 10^{-26}\ \text{kg}.$$

$$hc\overline{B_e} = hc(1.78\ \text{cm}^{-1}) = 3.5359 \times 10^{-23}\,J,$$

$$R_e = \frac{1.054\,59 \times 10^{-34}\ \text{J s}}{\sqrt{(2)(1.0824 \times 10^{-26}\ \text{kg})(3.5359 \times 10^{-23}\ \text{J})}}$$
$$= 1.205 \times 10^{-10}\ \text{m} = 1.205\ \text{Å},$$

$$\omega_e x_e = \frac{\hbar\omega_e^2}{4D_e^0}, \quad \overline{\omega_e x_e} = \frac{\overline{\omega_e^2}}{4\overline{D_e^0}} = \frac{(1885\ \text{cm}^{-1})^2}{(4)(66\,782.2\ \text{cm}^{-1})} = 13.30\ \text{cm}^{-1},$$

$$D_0^0 = D_e^0 - \frac{\hbar\omega_e}{2} + \frac{\hbar\omega_e x_e}{4}, \quad \overline{D_0^0} = \overline{D_e^0} - \frac{\overline{\omega_e}}{2} + \frac{\overline{\omega_e x_e}}{4}$$
$$= 66\,782.2 - \frac{1885}{2} + \frac{13.3}{4}$$
$$= 65\,843.0\ \text{cm}^{-1} = 8.16\ \text{eV},$$

$$\alpha_e = \frac{-6B_e^2}{\hbar\omega_e} + \frac{6\sqrt{B_e^3\hbar\omega_e x_e}}{\hbar\omega_e},$$

$$\overline{\alpha_e} = \frac{-6\overline{B_e^2}}{\overline{\omega_e}} + \frac{6\sqrt{\overline{B_e^3}\,\overline{\omega_e x_e}}}{\overline{\omega_e}}$$
$$= \frac{(-6)(1.78)^2}{(1885)} + \frac{6\sqrt{(1.78)^3(13.3)}}{(1885)} = 0.0175\ \text{cm}^{-1},$$

$$B_0 = B_e - \alpha_e(1/2), \quad \overline{B_0} = \overline{B_e} - \overline{\alpha_e}(1/2) = 1.78 - 0.0175/2$$
$$= 1.77\ \text{cm}^{-1},$$

$$B_1 = B_e - \alpha_e(3/2), \quad \overline{B_1} = \overline{B_e} - \overline{\alpha_e}(3/2) = 1.78 - 0.0175(1.5)$$
$$= 1.75\ \text{cm}^{-1}.$$

b. The molecule has a dipole moment and so it should have a pure rotational spectrum. In addition, the dipole moment should change with R and so it should have a vibration–rotation spectrum.
The first three lines correspond to $J = 1 \to 0$, $J = 2 \to 1$, $J = 3 \to 2$:

$$E = \hbar\omega_e(v+1/2) - \hbar\omega_e x_e(v+1/2)^2 + B_v J(J+1) - D_e J^2(J+1)^2,$$

$$\Delta E = \hbar\omega_e - 2\hbar\omega_e x_e - B_0 J(J+1) + B_1 J(J-1) - 4D_e J^3,$$

$$\overline{\Delta E} = \overline{\omega_e} - 2\overline{\omega_e x_e} - \overline{B_0} J(J+1) + \overline{B_1} J(J-1) - 4\overline{D_e} J^3$$
$$= 1885 - 2(13.3) - 1.77J(J+1) + 1.75J(J-1) - 4(6.35 \times 10^{-6})J^3$$
$$= 1858.4 - 1.77J(J+1) + 1.75J(J-1) - 2.54 \times 10^{-5} J^3,$$

$$\overline{\Delta E}(J=1) = 1854.9\ \text{cm}^{-1},$$

$\overline{\Delta E}(J = 2) = 1851.3\ \text{cm}^{-1}$,
$\overline{\Delta E}(J = 3) = 1847.7\ \text{cm}^{-1}$.

29 The $C_2H_2Cl_2$ molecule has a σ_h plane of symmetry (plane of molecule), a C_2 axis ($\perp$ to the molecular plane), and inversion symmetry; this results in C_{2h} symmetry. Using C_{2h} symmetry, the modes can be labeled as follows: ν_1, ν_2, ν_3, ν_4, and ν_5 are a_g, ν_6 and ν_7 are a_u, ν_8 is b_g, and ν_9, ν_{10}, ν_{11}, and ν_{12} are b_u.

30

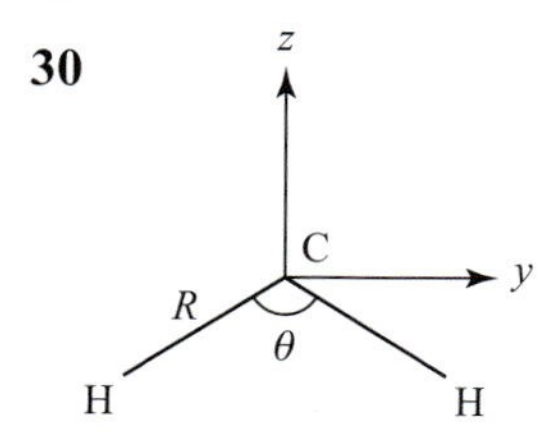

Molecule I
$R_{CH} = 1.121$ Å
$\angle_{HCH} = 104°$
$y_H = R\sin(\theta/2) = \pm 0.8834$
$z_H = R\cos(\theta/2) = -0.6902$

Molecule II
$R_{CH} = 1.076$ Å
$\angle_{HCH} = 136°$
$y_H = \pm 0.9976$
$z_H = -0.4031$

Center of mass (COM):

clearly, $X = Y = 0$,

$$Z = \frac{12(0) - 2R\cos(\theta/2)}{14} = -0.0986 \qquad Z = -0.0576$$

a.

$$I_{xx} = \sum_j m_j\left(y_j^2 + z_j^2\right) - M(Y^2 + Z^2)$$
$$I_{xy} = -\sum_j m_j x_j y_j - MXY$$

Molecule I:
$$I_{xx} = 2(1.121)^2 - 14(-0.0986)^2 = 2.377$$
$$I_{yy} = 2(0.6902)^2 - 14(-0.0986)^2 = 0.8167$$
$$I_{zz} = 2(0.8834)^2 = 1.561$$
$$I_{xz} = I_{yz} = I_{xy} = 0.$$

Molecule II:
$$I_{xx} = 2(1.076)^2 - 14(-0.0576)^2 = 2.269$$
$$I_{yy} = 2(0.4031)^2 - 14(-0.0576)^2 = 0.2786$$
$$I_{zz} = 2(0.9976)^2 = 1.990$$

b. Since the moment of inertia tensor is already diagonal, the principal moments of inertia have already been determined to be ($I_a < I_b < I_c$):

$I_{yy} < I_{zz} < I_{xx}$	$I_{yy} < I_{zz} < I_{xx}$
$0.8167 < 1.561 < 2.377$	$0.2786 < 1.990 < 2.269$

Using the formula:

$$A = \frac{h}{8\pi^2 c I_a} = \frac{6.626 \times 10^{-27}}{8\pi^2 (3 \times 10^{10}) I_a} \times \frac{6.02 \times 10^{23}}{(1 \times 10^{-8})^2}$$

$$A = \frac{16.84}{I_a}\ \text{cm}^{-1},$$

similarly,

$$B = \frac{16.84}{I_b}\ \text{cm}^{-1}, \quad \text{and} \quad C = \frac{16.84}{I_c}\ \text{cm}^{-1}.$$

So,

Molecule I	Molecule II
$y \Rightarrow A = 20.62$	$y \Rightarrow A = 60.45$
$z \Rightarrow B = 10.79$	$z \Rightarrow B = 8.46$
$x \Rightarrow C = 7.08$	$x \Rightarrow C = 7.42$

c. Averaging $B + C$:

$B = (B + C)/2 = 8.94$	$B = (B + C)/2 = 7.94$
$A - B = 11.68$	$A - B = 52.51$

Using the prolate top formula:

$E = (A - B)K^2 + BJ(J + 1)$,

Molecule I	Molecule II
$E = 11.68K^2 + 8.94J(J + 1)$	$E = 52.51K^2 + 7.94J(J + 1)$

Levels: $J = 0, 1, 2, \ldots$ and $K = 0, 1, \ldots J$

For a given level defined by J and K, there are M_J degeneracies given by:

$$(2J + 1) \times \left\{ \begin{array}{l} 1 \text{ for } K = 0 \\ 2 \text{ for } K \neq 0 \end{array} \right\}.$$

d. Molecule I Molecule II

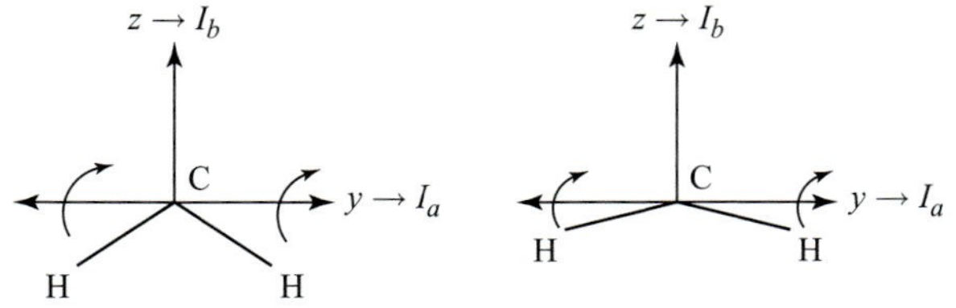

e. Assume molecule I is CH_2^- and molecule II is CH_2. Then,

$$\begin{aligned} \Delta E &= E_{J_j}(CH_2) - E_{J_i}(CH_2^-), \\ \text{where}\quad E(CH_2) &= 52.51K^2 + 7.94J(J+1), \\ \text{and}\quad E(CH_2^-) &= 11.68K^2 + 8.94J(J+1). \end{aligned}$$

For R-branches: $J_j = J_i + 1$, $\Delta K = 0$:

$$\begin{aligned} \Delta E_R &= E_{J_j}(CH_2) - E_{J_i}(CH_2^-) \\ &= 7.94(J_i+1)(J_i+1+1) - 8.94J_i(J_i+1) \\ &= (J_i+1)\{7.94(J_i+1+1) - 8.94J_i\} \\ &= (J_i+1)\{(7.94-8.94)J_i + 2(7.94)\} \\ &= (J_i+1)\{-J_i + 15.88\}. \end{aligned}$$

For P-branches: $J_j = J_i - 1$, $\Delta K = 0$:

$$\begin{aligned} \Delta E_P &= E_{J_j}(CH_2) - E_{J_i}(CH_2^-) \\ &= 7.94(J_i-1)(J_i-1+1) - 8.94J_i(J_i+1) \\ &= J_i\{7.94(J_i-1) - 8.94(J_i+1)\} \\ &= J_i\{(7.94-8.94)J_i - 7.94 - 8.94\} \\ &= J_i\{-J_i - 16.88\}. \end{aligned}$$

This indicates that the R-branch lines occur at energies which grow closer and closer together as J increases (since the $15.88 - J_i$ term will cancel). The P-branch lines occur at energies which lie more and more negative (i.e., to the left of the origin). So, you can predict that if molecule I is CH_2^- and molecule II is CH_2 then the R-branch has a band head and the P-branch does not. This is observed, therefore our assumption was correct: molecule I is CH_2^- and molecule II is CH_2.

f. The band head occurs when $\frac{d(\Delta E_R)}{dJ} = 0$.

$$\begin{aligned} \frac{d(\Delta E_R)}{dJ} &= \frac{d}{dJ}[(J_i+1)(-J_i+15.88)] \\ &= \frac{d}{dJ}\left(-J_i^2 - J_i + 15.88J_i + 15.88\right) \\ &= -2J_i + 14.88 = 0. \\ \therefore \qquad J_i &= 7.44, \quad \text{so} \quad J = 7 \text{ or } 8. \end{aligned}$$

At $J = 7.44$:

$$\begin{aligned} \Delta E_R &= (J+1)\{-J+15.88\} \\ &= (7.44+1)\{-7.44+15.88\} = (8.44)(8.44) = 71.2\ \text{cm}^{-1} \quad \text{above the origin.} \end{aligned}$$

31 a.

D_{6h}	E	$2C_6$	$2C_3$	C_2	$3C_2'$	$3C_2''$	i	$2S_3$	$2S_6$	σ_h	$3\sigma_d$	$3\sigma_v$		
A_{1g}	1	1	1	1	1	1	1	1	1	1	1	1		x^2+y^2, z^2
A_{2g}	1	1	1	1	−1	−1	1	1	1	1	−1	−1	R_z	
B_{1g}	1	−1	1	−1	1	−1	1	−1	1	−1	1	−1		
B_{2g}	1	−1	1	−1	−1	1	1	−1	1	−1	−1	1		
E_{1g}	2	1	−1	−2	0	0	2	1	−1	−2	0	0	R_x, R_y	(xz, yz)
E_{2g}	2	−1	−1	2	0	0	2	−1	−1	2	0	0		(x^2-y^2, xy)
A_{1u}	1	1	1	1	1	1	−1	−1	−1	−1	−1	−1		
A_{2u}	1	1	1	1	−1	−1	−1	−1	−1	−1	1	1	z	
B_{1u}	1	−1	1	−1	1	−1	−1	1	−1	1	−1	1		
B_{2u}	1	−1	1	−1	−1	1	−1	1	−1	1	1	−1		
E_{1u}	2	1	−1	−2	0	0	−2	−1	1	2	0	0	(x, y)	
E_{2u}	2	−1	−1	2	0	0	−2	1	1	−2	0	0		
Γ_{C-H}	6	0	0	0	0	2	0	0	0	6	2	0		

b. The number of irreducible representations may be found by using the following formula:

$$n_{\text{irrep}} = \frac{1}{g}\sum_R \chi_{\text{red}}(R)\chi_{\text{irrep}}(R),$$

where g = the order of the point group (24 for D_{6h}).

$$\begin{aligned}
n_{A_{1g}} &= \frac{1}{24}\sum_R \Gamma_{C-H}(R)\cdot A_{1g}(R) \\
&= \frac{1}{24}\{(1)(6)(1)+(2)(0)(1)+(2)(0)(1)+(1)(0)(1) \\
&\quad +(3)(0)(1)+(3)(2)(1)+(1)(0)(1)+(2)(0)(1) \\
&\quad +(2)(0)(1)+(1)(6)(1)+(3)(2)(1)+(3)(0)(1)\} \\
&= 1 \\
n_{A_{2g}} &= \frac{1}{24}\{(1)(6)(1)+(2)(0)(1)+(2)(0)(1)+(1)(0)(1) \\
&\quad +(3)(0)(-1)+(3)(2)(-1)+(1)(0)(1)+(2)(0)(1) \\
&\quad +(2)(0)(1)+(1)(6)(1)+(3)(2)(-1)+(3)(0)(-1)\} \\
&= 0 \\
n_{B_{1g}} &= \frac{1}{24}\{(1)(6)(1)+(2)(0)(-1)+(2)(0)(1)+(1)(0)(-1) \\
&\quad +(3)(0)(1)+(3)(2)(-1)+(1)(0)(1)+(2)(0)(-1) \\
&\quad +(2)(0)(1)+(1)(6)(-1)+(3)(2)(1)+(3)(0)(-1)\} \\
&= 0 \\
n_{B_{2g}} &= \frac{1}{24}\{(1)(6)(1)+(2)(0)(-1)+(2)(0)(1)+(1)(0)(-1) \\
&\quad +(3)(0)(-1)+(3)(2)(1)+(1)(0)(1)+(2)(0)(-1) \\
&\quad +(2)(0)(1)+(1)(6)(-1)+(3)(2)(-1)+(3)(0)(1)\} \\
&= 0
\end{aligned}$$

$$\begin{aligned}
n_{E_{1g}} &= \frac{1}{24}\{(1)(6)(2) + (2)(0)(1) + (2)(0)(-1) + (1)(0)(-2)\\
&+ (3)(0)(0) + (3)(2)(0) + (1)(0)(2) + (2)(0)(1)\\
&+ (2)(0)(-1) + (1)(6)(-2) + (3)(2)(0) + (3)(0)(0)\}\\
&= 0\\
n_{E_{2g}} &= \frac{1}{24}\{(1)(6)(2) + (2)(0)(-1) + (2)(0)(-1) + (1)(0)(2)\\
&+ (3)(0)(0) + (3)(2)(0) + (1)(0)(2) + (2)(0)(-1)\\
&+ (2)(0)(-1) + (1)(6)(2) + (3)(2)(0) + (3)(0)(0)\}\\
&= 1\\
n_{A_{1u}} &= \frac{1}{24}\{(1)(6)(1) + (2)(0)(1) + (2)(0)(1) + (1)(0)(1)\\
&+ (3)(0)(1) + (3)(2)(1) + (1)(0)(-1) + (2)(0)(-1)\\
&+ (2)(0)(-1) + (1)(6)(-1) + (3)(2)(-1) + (3)(0)(-1)\}\\
&= 0\\
n_{A_{2u}} &= \frac{1}{24}\{(1)(6)(1) + (2)(0)(1) + (2)(0)(1) + (1)(0)(1)\\
&+ (3)(0)(-1) + (3)(2)(-1) + (1)(0)(-1) + (2)(0)(-1)\\
&+ (2)(0)(-1) + (1)(6)(-1) + (3)(2)(1) + (3)(0)(1)\}\\
&= 0\\
n_{B_{1u}} &= \frac{1}{24}\{(1)(6)(1) + (2)(0)(-1) + (2)(0)(1) + (1)(0)(-1)\\
&+ (3)(0)(1) + (3)(2)(-1) + (1)(0)(-1) + (2)(0)(1)\\
&+ (2)(0)(-1) + (1)(6)(1) + (3)(2)(-1) + (3)(0)(1)\}\\
&= 0\\
n_{B_{2u}} &= \frac{1}{24}\{(1)(6)(1) + (2)(0)(-1) + (2)(0)(1) + (1)(0)(-1)\\
&+ (3)(0)(-1) + (3)(2)(1) + (1)(0)(-1) + (2)(0)(1)\\
&+ (2)(0)(-1) + (1)(6)(1) + (3)(2)(1) + (3)(0)(-1)\}\\
&= 1\\
n_{E_{1u}} &= \frac{1}{24}\{(1)(6)(2) + (2)(0)(1) + (2)(0)(-1) + (1)(0)(-2)\\
&+ (3)(0)(0) + (3)(2)(0) + (1)(0)(-2) + (2)(0)(-1)\\
&+ (2)(0)(1) + (1)(6)(2) + (3)(2)(0) + (3)(0)(0)\}\\
&= 1\\
n_{E_{2u}} &= \frac{1}{24}\{(1)(6)(2) + (2)(0)(-1) + (2)(0)(-1) + (1)(0)(2)\\
&+ (3)(0)(0) + (3)(2)(0) + (1)(0)(-2) + (2)(0)(1)\\
&+ (2)(0)(1) + (1)(6)(-2) + (3)(2)(0) + (3)(0)(0)\}\\
&= 0
\end{aligned}$$

We see that $\Gamma_{C-H} = A_{1g} \oplus E_{2g} \oplus B_{2u} \oplus E_{1u}$.

c. x and $y \rightarrow E_{1u}$, $z \Rightarrow A_{2u}$, so, the ground-state A_{1g} level can be excited to the degenerate E_{1u} level by coupling through the x or y transition dipoles. Therefore E_{1u} is infrared active and $\perp$ polarized.

d. $(x^2 + y^2, z^2) \rightarrow A_{1g}$, $(xz, yz) \rightarrow E_{1g}$, $(x^2 - y^2, xy) \rightarrow E_{2g}$, so the ground-state A_{1g} level can be excited to the degenerate E_{2g} level by coupling through the $x^2 - y^2$ or xy transitions or be excited to the degenerate A_{1g} level by coupling through the xz or yz transitions. Therefore A_{1g} and E_{2g} are Raman active.

e. The B_{2u} mode is not IR or Raman active.

32 a. Evaluate the z-component of μ_{fi}:

$$\mu_{fi} = \langle 2p_z | er\cos\theta | 1s \rangle,$$

where

$$\psi_{1s} = \frac{1}{\sqrt{\pi}}\left(\frac{Z}{a_0}\right)^{\frac{3}{2}} e^{\frac{-Zr}{a_0}}, \quad \text{and} \quad \psi_{2pz} = \frac{1}{4\sqrt{2\pi}}\left(\frac{Z}{a_0}\right)^{\frac{5}{2}} r\cos\theta e^{\frac{-Zr}{2a_0}}.$$

$$\begin{aligned}
\mu_{fi} &= \frac{1}{4\sqrt{2\pi}}\left(\frac{Z}{a_0}\right)^{\frac{5}{2}} \frac{1}{\sqrt{\pi}}\left(\frac{Z}{a_0}\right)^{\frac{3}{2}} \left\langle r\cos\theta e^{\frac{-Zr}{2a_0}} \middle| er\cos\theta \middle| e^{\frac{-Zr}{a_0}} \right\rangle \\
&= \frac{1}{4\pi\sqrt{2}}\left(\frac{Z}{a_0}\right)^4 \left\langle r\cos\theta e^{\frac{-Zr}{2a_0}} \middle| er\cos\theta \middle| e^{\frac{-Zr}{a_0}} \right\rangle \\
&= \frac{e}{4\pi\sqrt{2}}\left(\frac{Z}{a_0}\right)^4 \int_0^\infty r^2 dr \int_0^\pi \sin\theta d\theta \int_0^{2\pi} d\phi \left(r^2 e^{\frac{-Zr}{2a_0}} e^{\frac{-Zr}{a_0}}\right)\cos^2\theta \\
&= \frac{e}{4\pi\sqrt{2}} 2\pi \left(\frac{Z}{a_0}\right)^4 \int_0^\infty \left(r^4 e^{\frac{-3Zr}{2a_0}}\right) dr \int_0^{2\pi} \sin\theta\cos^2\theta d\theta \\
&= \frac{e}{4\pi\sqrt{2}} 2\pi \left(\frac{Z}{a_0}\right)^4 \frac{4!}{\left(\frac{3Z}{2a_0}\right)^5}\left(\frac{-1}{3}\right)\cos^3\theta\Big|_0^\pi \\
&= \frac{e}{4\pi\sqrt{2}} 2\pi \left(\frac{Z}{a_0}\right)^4 \frac{2^5 a_0^5 4!}{3^5 Z^5}\left(\frac{-1}{3}\right)((-1)^3 - (1)^3) \\
&= \frac{e}{\sqrt{2}}\frac{2^8 a_0}{3^5 Z} = \frac{ea_0}{Z}\frac{2^8}{\sqrt{2}3^5} = 0.7449\frac{ea_0}{Z}.
\end{aligned}$$

b. Examine the symmetry of the integrands for $\langle 2p_z | ex | 1s \rangle$ and $\langle 2p_z | ey | 1s \rangle$. Consider reflection in the xy plane:

Function	Symmetry
$2p_z$	-1
x	$+1$
1s	$+1$
y	$+1$

Under this operation, the integrand of $\langle 2p_z | ex | 1s \rangle$ is $(-1)(1)(1) = -1$ (it is antisymmetric) and hence $\langle 2p_z | ex | 1s \rangle = 0$.
Similarly, under this operation the integrand of $\langle 2p_z | ey | 1s \rangle$ is $(-1)(1)(1) = -1$ (it is also antisymmetric) and hence $\langle 2p_z | ey | 1s \rangle = 0$.

c.

$$\tau_R = \frac{3\hbar^4 c^3}{4(E_i - E_f)^3 |\mu_{fi}|^2},$$

$$E_i = E_{2p_z} = -\frac{1}{4} Z^2 \left(\frac{e^2}{2a_0}\right),$$

$$E_f = E_{1s} = -Z^2 \left(\frac{e^2}{2a_0}\right),$$

$$E_i - E_f = \frac{3}{8} \left(\frac{e^2}{a_0}\right) Z^2.$$

Making the substitutions for $E_i - E_f$ and $|\mu_{fi}|$ in the expression for τ_R we obtain:

$$\begin{aligned}
\tau_R &= \frac{3\hbar^4 c^3}{4\left[\frac{3}{8}\left(\frac{e^2}{a_0}\right) Z^2\right]^3 \left[\left(\frac{ea_0}{Z}\right) \frac{2^8}{\sqrt{2}3^5}\right]^2} \\
&= \frac{3\hbar^4 c^3}{4\frac{3^3}{8^3}\left(\frac{e^6}{a_0^3}\right) Z^6 \left(\frac{e^2 a_0^2}{Z^2}\right) \frac{2^{16}}{(2)3^{10}}} \\
&= \frac{\hbar^4 c^3 3^8 a_0}{e^8 Z^4 2^8}.
\end{aligned}$$

Inserting $e^2 = \frac{\hbar^2}{m_e a_0}$ we obtain

$$\begin{aligned}
\tau_R &= \frac{\hbar^4 c^3 3^8 a_0 m_e^4 a_0^4}{\hbar^8 Z^4 2^8} = \frac{3^8}{2^8} \frac{c^3 a_0^5 m_e^4}{\hbar^4 Z^4} \\
&= 25.6289 \frac{c^3 a_0^5 m_e^4}{\hbar^4 Z^4} \\
&= 25.6289 \left(\frac{1}{Z^4}\right) \\
&\quad \times \frac{(2.998 \times 10^{10}\ \text{cm s}^{-1})^3 (0.529\,177 \times 10^{-8}\ \text{cm})^5 (9.109 \times 10^{-28}\ \text{g})^4}{(1.0546 \times 10^{-27}\ \text{g cm}^2\ \text{sec}^{-1})^4} \\
&= 1.595 \times 10^{-9}\ \text{s} \times \left(\frac{1}{Z^4}\right).
\end{aligned}$$

So, for example:

Atom	τ_R
H	1.595 ns
He^+	99.7 ps
Li^{+2}	19.7 ps
Be^{+3}	6.23 ps
Ne^{+9}	159 fs

33 a. $H = H_0 + \lambda H'(t)$, $H'(t) = V\theta(t)$, $H_0\phi_k = E_k\phi_k$, $\omega_k = E_k/\hbar$, $i\hbar\frac{\partial\psi}{\partial t} = H\psi$.

Let $\psi(r,t) = i\hbar \sum_j c_j(t)\phi_j e^{-i\omega_j t}$ and insert into the Schrödinger equation:

$$i\hbar \sum_j [\dot{c}_j - i\omega_j c_j] e^{-i\omega_j t}\phi_j = i\hbar \sum_j c_j(t) e^{-i\omega_j t}(H_0 + \lambda H'(t))\phi_j$$

$$\sum_j [i\hbar\dot{c}_j + E_j c_j - c_j E_j - c_j \lambda H'] e^{-i\omega_j t}\phi_j = 0$$

$$\sum_j [i\hbar\dot{c}_j \langle m \mid j\rangle - c_j\lambda\langle m|H'|j\rangle] e^{-i\omega_j t} = 0$$

$$i\hbar\dot{c}_m e^{-i\omega_m t} = \sum_j c_j \lambda H'_{mj} e^{-i\omega j t}.$$

So,

$$\dot{c}_m = \frac{1}{i\hbar}\sum_j c_j\lambda H'_{mj} e^{-i\omega_{jm} t}.$$

Going back a few equations and multiplying from the left by ϕ_k instead of ϕ_m we obtain:

$$\sum_j [i\hbar\dot{c}_j \langle k \mid j\rangle - c_j\lambda\langle k|H'|j\rangle] e^{-i\omega j t} = 0,$$

$$i\hbar\dot{c}_k e^{-i\omega_k t} = \sum_j c_j\lambda H'_{kj} e^{-i\omega_j t}.$$

So,

$$\dot{c}_k = \frac{1}{i\hbar}\sum_j c_j\lambda H'_{kj} e^{-i\omega_{jk} t}.$$

Now, let

$$c_m = c_m^{(0)} + c_m^{(1)}\lambda + c_m^{(2)}\lambda^2 + \cdots$$
$$c_k = c_k^{(0)} + c_k^{(1)}\lambda + c_k^{(2)}\lambda^2 + \cdots$$

and substituting into the above we obtain:

$$\dot{c}_m^{(0)} + \dot{c}_m^{(1)}\lambda + \dot{c}_m^{(2)}\lambda^2 + \cdots = \frac{1}{i\hbar}\sum_j [c_j^{(0)} + c_j^{(1)}\lambda + c_j^{(2)}\lambda^2 + \cdots]\lambda H'_{mj} e^{-i\omega_{jm} t},$$

first order: $\dot{c}_m^{(0)} = 0 \rightarrow c_m^{(0)} = 1,$

second order: $\dot{c}_m^{(1)} = \frac{1}{i\hbar}\sum_j c_j^{(0)} H'_{mj} e^{-i\omega_{jm} t},$

$(n+1)$ order: $\dot{c}_m^{(n)} = \frac{1}{i\hbar}\sum_j c_j^{(n-1)} H'_{mj} e^{-i\omega_{jm} t}.$

Similarly:

first order: $\dot{c}_k^{(0)} = 0 \rightarrow c_{k\neq m}^{(0)} = 0,$

second order: $\dot{c}_k^{(1)} = \frac{1}{i\hbar}\sum_j c_j^{(0)} H'_{kj} e^{-i\omega_{jk} t},$

$(n+1)$ order: $\dot{c}_k^{(n)} = \frac{1}{i\hbar}\sum_j c_j^{(n-1)} H'_{kj} e^{-i\omega_{jk} t}.$

So,

$$\dot{c}_m^{(1)} = \frac{1}{i\hbar} c_m^{(0)} H'_{mm} e^{-i\omega_{mm}t} = \frac{1}{i\hbar} H'_{mm},$$
$$c_m^{(1)}(t) = \frac{1}{i\hbar}\int_0^t dt' V_{mm} = \frac{V_{mm}t}{i\hbar},$$

and similarly,

$$\dot{c}_k^{(1)} = \frac{1}{i\hbar} c_m^{(0)} H'_{km} e^{-i\omega_{mk}t} = \frac{1}{i\hbar} H'_{km} e^{-i\omega_{mk}t},$$
$$c_k^{(1)}(t) = \frac{1}{i\hbar} V_{km}\int_0^t dt' e^{-i\omega_{mk}t'} = \frac{V_{km}}{\hbar\omega_{mk}}[e^{-i\omega_{mk}t} - 1],$$
$$\dot{c}_m^{(2)} = \frac{1}{i\hbar}\sum_j c_j^{(1)} H'_{mj} e^{-i\omega_{jm}t},$$
$$\dot{c}_m^{(2)} = \sum_{j\neq m} \frac{1}{i\hbar}\frac{V_{jm}}{\hbar\omega_{mj}}[e^{-i\omega_{jm}t} - 1]H'_{mj}e^{-i\omega_{jm}t} + \frac{1}{i\hbar}\frac{V_{mm}t}{i\hbar}H'_{mm},$$
$$\dot{c}_m^{(2)} = \sum_{j\neq m} \frac{1}{i\hbar}\frac{V_{jm}V_{mj}}{\hbar\omega_{mj}}\int_0^t dt' e^{-i\omega_{jm}t'}[e^{-i\omega_{mj}t'} - 1] - \frac{V_{mm}V_{mm}}{\hbar^2}\int_0^t t'dt'$$
$$= \sum_{j\neq m} \frac{V_{jm}V_{mj}}{i\hbar^2\omega_{mj}}\int_0^t dt'[1 - e^{-i\omega_{jm}t'}] - \frac{|V_{mm}|^2}{\hbar^2}\frac{t^2}{2}$$
$$= \sum_{j\neq m} \frac{V_{jm}V_{mj}}{i\hbar^2\omega_{mj}}\left(t - \frac{e^{-i\omega_{jm}t} - 1}{-i\omega_{jm}}\right) - \frac{|V_{mm}|^2}{\hbar^2}\frac{t^2}{2}$$
$$= \sum_{j\neq m}{}' \frac{V_{jm}V_{mj}}{\hbar^2\omega_{mj}^2}(e^{-i\omega_{jm}t} - 1) + \sum_{j\neq m}{}' \frac{V_{jm}V_{mj}}{i\hbar^2\omega_{mj}}t - \frac{|V_{mm}|^2t^2}{2\hbar^2}.$$

Similarly,

$$\dot{c}_k^{(2)} = \frac{1}{i\hbar}\sum_j c_j^{(1)} H'_{kj} e^{-i\omega_{jk}t}$$
$$= \sum_{j\neq m} \frac{1}{i\hbar}\frac{V_{jm}}{\hbar\omega_{mj}}[e^{-i\omega_{mj}t} - 1]H'_{kj}e^{-i\omega_{jk}t} + \frac{1}{i\hbar}\frac{V_{mm}t}{i\hbar}H'_{km}e^{-i\omega_{mk}t},$$
$$c_k^{(2)}(t) = \sum_{j\neq m}{}' \frac{V_{jm}V_{kj}}{i\hbar^2\omega_{mj}}\int_0^t dt' e^{-i\omega_{jk}t'}[e^{-i\omega_{mj}t'} - 1] - \frac{V_{mm}V_{km}}{\hbar^2}\int_0^t t'dt' e^{-i\omega_{mk}t'}$$
$$= \sum_{j\neq m}{}' \frac{V_{jm}V_{kj}}{i\hbar^2\omega_{mj}}\left(\frac{e^{-i(\omega_{mj}+\omega_{jm})t} - 1}{-i\omega_{mk}} - \frac{e^{-i\omega_{jk}t} - 1}{-i\omega_{jk}}\right)$$
$$- \frac{V_{mm}V_{km}}{\hbar^2}\left[e^{-i\omega_{mk}t'}\left(\frac{t'}{-i\omega_{mk}} - \frac{1}{-(i\omega_{mk})^2}\right)\right]_0^t$$
$$= \sum_{j\neq m}{}' \frac{V_{jm}V_{kj}}{i\hbar^2\omega_{mj}}\left(\frac{e^{-i\omega_{mk}t} - 1}{\omega_{mk}} - \frac{e^{-i\omega_{jk}t} - 1}{i\omega_{jk}}\right)$$
$$+ \frac{V_{mm}V_{km}}{\hbar^2\omega_{mk}}\left[e^{-i\omega_{mk}t'}\left(\frac{t'}{i} - \frac{1}{\omega_{mk}}\right)\right]_0^t$$

$$= \sum_{j\neq m}{}' \frac{V_{jm}V_{kj}}{E_m - E_j}\left(\frac{e^{-i\omega_{mk}} - 1}{E_m - E_k} - \frac{e^{-i\omega_{jk}} - 1}{E_j - E_k}\right)$$

$$+ \frac{V_{mm}V_{km}}{\hbar(E_m - E_k)}\left[e^{-i\omega_{mk}t}\left(\frac{t}{i} - \frac{1}{\omega_{mk}}\right) + \frac{1}{\omega_{mk}}\right].$$

So, the overall amplitudes, cm and c_k, to second order are

$$c_m(t) = 1 + \frac{V_{mm}t}{i\hbar} + \sum_{j\neq m}{}' \frac{V_{jm}V_{mj}}{i\hbar(E_m - E_j)}t + \sum_{j\neq m}{}' \frac{V_{jm}V_{mj}}{\hbar^2(E_m - E_j)^2}(e^{-i\omega_{jm}t} - 1)$$
$$- \frac{|V_{mm}|^2 t^2}{2\hbar^2},$$

$$c_k(t) = \frac{V_{km}}{(E_m - E_k)}[e^{-i\omega_{mk}t} - 1] + \frac{V_{mm}V_{km}}{(E_m - E_k)^2}[1 - e^{-i\omega_{mk}t}]$$
$$+ \frac{V_{mm}V_{km}}{(E_m - E_k)}\frac{t}{\hbar i}e^{-i\omega_{mk}t} + \sum_{j\neq m}{}' \frac{V_{jm}V_{kj}}{E_m - E_j}\left(\frac{e^{-i\omega_{mk}t} - 1}{E_m - E_k} - \frac{e^{-i\omega_{jk}t} - 1}{E_j - E_k}\right).$$

b. The perturbation equations still hold:

$$\dot{c}_m^{(n)} = \frac{1}{i\hbar}\sum_j c_j^{(n-1)}H'_{mj}e^{-i\omega_{jm}t}; \qquad \dot{c}_k^{(n)} = \frac{1}{i\hbar}\sum_j c_j^{(n-1)}H'_{kj}e^{-i\omega_{jk}t}.$$

So, $c_m^{(0)} = 1$ and $c_k^{(0)} = 0$.

$$\dot{c}_m^{(1)} = \frac{1}{i\hbar}H'_{mm},$$

$$c_m^{(1)} = \frac{1}{i\hbar}V_{mm}\int_{-\infty}^{t} dt' e^{\eta t} = \frac{V_{mm}e^{\eta t}}{i\hbar\eta},$$

$$\dot{c}_k^{(1)} = \frac{1}{i\hbar}H'_{km}e^{-i\omega_{mk}t},$$

$$c_k^{(1)} = \frac{1}{i\hbar}V_{km}\int_{-\infty}^{t} dt' e^{-i(\omega_{mk}+\eta)t'} = \frac{V_{km}}{i\hbar(-i\omega_{mk} + \eta)}[e^{-i(\omega_{mk}+\eta)t}]$$

$$= \frac{V_{km}}{E_m - E_k + i\hbar\eta}[e^{-i(\omega_{mk}+\eta)t}],$$

$$\dot{c}_m^{(2)} = \sum_{j\neq m}{}' \frac{1}{i\hbar}\frac{V_{jm}}{E_m - E_j + i\hbar\eta}e^{-i(\omega_{mj}+\eta)t}V_{mj}e^{\eta t}e^{-i\omega_{jm}t} + \frac{1}{i\hbar}\frac{V_{mm}e^{\eta t}}{i\hbar\eta}V_{mm}e^{\eta t},$$

$$c_m^{(2)} = \sum_{j\neq m}{}' \frac{1}{i\hbar}\frac{V_{jm}V_{mj}}{E_m - E_j + i\hbar\eta}\int_{-\infty}^{t} e^{2\eta t'}dt' - \frac{|V_{mm}|^2}{\hbar^2\eta}\int_{-\infty}^{t} e^{2\eta t'}dt'$$

$$= \sum_{j\neq m}{}' \frac{V_{jm}V_{mj}}{i\hbar 2\eta(E_m - E_j + i\hbar\eta)}e^{2\eta t} - \frac{|V_{mm}|^2}{2\hbar^2\eta^2}e^{2\eta t},$$

$$\dot{c}_k^{(2)} = \sum_{j\neq m}{}' \frac{1}{i\hbar}\frac{1}{i\hbar}\frac{V_{jm}}{E_m - E_j + i\hbar\eta}e^{-i(\omega_{mj}+\eta)t}H'_{kj}e^{-i\omega_{jk}t} + \frac{1}{i\hbar}\frac{V_{mm}e^{\eta t}}{i\hbar\eta}H'_{km}e^{-i\omega_{mk}t},$$

$$c_k^{(2)} = \sum_{j \neq m}' \frac{1}{i\hbar} \frac{V_{jm}V_{kj}}{E_m - E_j + i\hbar\eta} \int_{-\infty}^{t} e^{-i(\omega_{mk}+2\eta)t} dt' - \frac{V_{mm}V_{km}}{\hbar^2\eta} \int_{-\infty}^{t} e^{-i(\omega_{mk}+2\eta)t'} dt'$$

$$= \sum_{j \neq m}' \frac{V_{jm}V_{kj}e^{-i(\omega_{mk}+2\eta)t'}}{(E_m - E_j + i\hbar\eta)(E_m - E_k + 2i\hbar\eta)} - \frac{V_{mm}V_{km}e^{-i(\omega_{mk}+2\eta)t}}{i\hbar\eta(E_m - E_k + 2i\hbar\eta)}.$$

Therefore, to second order:

$$c_m(t) = 1 + \frac{V_{mm}e^{\eta t}}{i\hbar\eta} + \sum_j \frac{V_{jm}V_{mj}}{i\hbar 2\eta(E_m - E_j + i\hbar\eta)} e^{2\eta t},$$

$$c_k(t) = \frac{V_{km}}{i\hbar(-i\omega_{mk} + \eta)}[e^{-i(\omega_{mk}+\eta)t}] + \sum_j \frac{V_{jm}V_{kj}e^{-i(\omega_{mk}+2\eta)t}}{(E_m - E_j + i\hbar\eta)(E_m - E_k + 2i\hbar\eta)}.$$

c. In part (a) the $c^{(2)}(t)$ grow linearly with time (for $V_{\mathrm{mm}} = 0$) while in part (b) they remain finite for $\eta > 0$. The result in part (a) is due to the sudden turning on of the field.

d.

$$|c_k(t)|^2 = \left| \sum_j \frac{V_{jm}V_{kj}e^{-i(\omega_{mk}+2\eta)t}}{(E_m - E_j + i\hbar\eta)(E_m - E_k + 2i\hbar\eta)} \right|^2$$

$$= \sum_{jj'} \frac{V_{kj}V_{kj'}V_{jm}V_{j'm}e^{-i(\omega_{mk}+2\eta)t}e^{i(\omega_{mk}+2\eta)t}}{(E_m - E_j + i\hbar\eta)(E_m - E_{j'} - i\hbar\eta)(E_m - E_k + 2i\hbar\eta)(E_m - E_k - 2i\hbar\eta)}$$

$$= \sum_{jj'} \frac{V_{kj}V_{kj'}V_{jm}V_{j'm}e^{4\eta t}}{[(E_m - E_j)(E_m - E_{j'}) + i\hbar\eta(E_j - E_{j'}) + \hbar^2\eta^2][(E_m - E_k)^2 + 4\hbar^2\eta^2)},$$

$$\frac{d}{dt}|c_k(t)|^2 = \sum_{jj'} \frac{4\eta V_{kj}V_{kj'}V_{jm}V_{j'm}}{[(E_m - E_j)(E_m - E_{j'}) + i\hbar\eta(E_j - E_{j'}) + \hbar^2\eta^2][(E_m - E_k)^2 + 4\hbar^2\eta^2)}.$$

Now, look at the limit as $\eta \to 0^+$:

$$\frac{d}{dt}|c_k(t)|^2 \neq 0 \quad \text{when} \quad E_m = E_k,$$

$$\eta \overset{\lim}{\to} 0^+ \frac{4\eta}{((E_m - E_k)^2 + 4\hbar^2\eta^2)} \alpha \delta(E_m - E_k).$$

So, the final result is the second order golden rule expression:

$$\frac{d}{dt}|c_k(t)|^2 = \frac{2\pi}{\hbar}\delta(E_m - E_k)_{\eta \overset{\lim}{\longrightarrow} 0^+} \left| \sum_j \frac{V_{jm}V_{kj}}{(E_j - E_m - i\hbar\eta)} \right|^2.$$

34 a.

$$T_{nm} \approx \frac{|\langle n|V|m\rangle|^2}{\hbar^2 \omega_{nm^2}}.$$

Evaluating $\langle 1\mathrm{s}|V|2\mathrm{s}\rangle$ (using only the radial portions of the 1s and 2s wave functions since the spherical harmonics will integrate to unity) where

$V = (e^2/r)$, the change in Coulomb potential when tritium becomes He:

$$\begin{aligned}
\langle 1s|V|2s\rangle &= \int 2\left(\frac{Z}{a_0}\right)^{\frac{3}{2}} e^{\frac{-Zr}{a_0}} \frac{1}{r}\frac{1}{\sqrt{2}}\left(\frac{Z}{a_0}\right)^{\frac{3}{2}}\left(1-\frac{Zr}{2a_0}\right)e^{\frac{-Zr}{2a_0}} r^2 dr \\
&= \frac{2}{\sqrt{2}}\left(\frac{Z}{a_0}\right)^3\left[\int re\frac{-3Zr}{2a_0}dr - \int \frac{Zr^2}{2a_0}e^{\frac{-3Zr}{2a_0}}dr\right] \\
&= \frac{2}{\sqrt{2}}\left(\frac{Z}{a_0}\right)^3\left[\frac{1}{\left(\frac{3Z}{2a_0}\right)^2} - \left(\frac{Z}{2a_0}\right)\frac{2}{\left(\frac{3Z}{2a_0}\right)^3}\right] \\
&= \frac{2}{\sqrt{2}}\left(\frac{Z}{a_0}\right)^3\left[\frac{2^2a_0^2}{3^2Z^2} - \frac{2^3a_0^2}{3^3Z^2}\right] \\
&= \frac{2}{\sqrt{2}}\left(\frac{Z}{a_0}\right)^3\left[\frac{(3)2^2a_0^2 - 2^3a_0^2}{3^3Z^2}\right] = \frac{8Z}{\sqrt{2}27a_0}.
\end{aligned}$$

Now,

$$E_n = -\frac{Z^2e^2}{n^2 2a_0}, \quad E_{1s} = -\frac{Z^2e^2}{2a_0}, \quad E_{2s} = -\frac{Z^2e^2}{8a_0}, \quad E_{2s} - E_{1s} = \frac{3Z^2e^2}{8a_0}.$$

So,

$$T_{nm} = \frac{\left(\frac{8Z}{\sqrt{2}27a_0}\right)^2}{\left(\frac{3Z^2}{8a_0}\right)^2} = \frac{2^6Z^2 2^6 a_0^2}{(2)3^8a_0^2Z^4} = \frac{2^{11}}{3^8Z^2} = 0.312 \quad (\text{for} Z = 1).$$

b.

$$\phi_m(r) = \phi_{1s} = 2\left(\frac{Z}{a_0}\right)^{\frac{3}{2}} e^{\frac{-Zr}{a_0}} Y_{00}.$$

The orthogonality of the spherical harmonics results in only s-states having non-zero values for A_{nm}. We can then drop the Y_{00} (integrating this term will only result in unity) in determining the value of $A_{1s,2s}$.

$$\psi_n(r) = \psi_{2s} = \frac{1}{\sqrt{2}}\left(\frac{Z}{a_0}\right)^{\frac{3}{2}}\left(1-\frac{Zr}{2a_0}\right)e^{\frac{-Zr}{2a_0}}.$$

Remember for $\phi_{1s} Z = 1$ and for $\psi_{2s} Z = 2$.

$$\begin{aligned}
A_{nm} &= \int 2\left(\frac{Z}{a_0}\right)^{\frac{3}{2}} e^{\frac{-Zr}{a_0}} \frac{1}{\sqrt{2}}\left(\frac{Z+1}{a_0}\right)^{\frac{3}{2}}\left(1-\frac{(Z+1)r}{2a_0}\right)e^{\frac{-(Z+1)r}{2a_0}} r^2 dr \\
&= \frac{2}{\sqrt{2}}\left(\frac{Z}{a_0}\right)^{\frac{3}{2}}\left(\frac{Z+1}{a_0}\right)^{\frac{3}{2}}\int e^{\frac{-(3Z+1)r}{2a_0}}\left(1-\frac{(Z+1)r}{2a_0}\right)r^2 dr \\
&= \frac{2}{\sqrt{2}}\left(\frac{Z}{a_0}\right)^{\frac{3}{2}}\left(\frac{Z+1}{a_0}\right)^{\frac{3}{2}}\left[\int r^2 e^{\frac{-(3Z+1)r}{2a_0}}dr - \int \frac{(Z+1)r^3}{2a_0}e^{\frac{-(3Z+1)r}{2a_0}}dr\right].
\end{aligned}$$

We obtain:

$$
\begin{aligned}
A_{nm} &= \frac{2}{\sqrt{2}}\left(\frac{Z}{a_0}\right)^{\frac{3}{2}}\left(\frac{Z+1}{a_0}\right)^{\frac{3}{2}}\left[\frac{2}{\left(\frac{3Z+1}{2a_0}\right)^3}-\left(\frac{Z+1}{2a_0}\right)\frac{(3)(2)}{\left(\frac{3Z+1}{2a_0}\right)^4}\right]\\
&= \frac{2}{\sqrt{2}}\left(\frac{Z}{a_0}\right)^{\frac{3}{2}}\left(\frac{Z+1}{a_0}\right)^{\frac{3}{2}}\left[\frac{2^4a_0^3}{(3Z+1)^3}-(Z+1)\frac{(3)2^4a_0^3}{(3Z+1)^4}\right]\\
&= \frac{2}{\sqrt{2}}\left(\frac{Z}{a_0}\right)^{\frac{3}{2}}\left(\frac{Z+1}{a_0}\right)^{\frac{3}{2}}\left[\frac{-2^5a_0^3}{(3Z+1)^4}\right]\\
&= -2\frac{[2^3Z(Z+1)]^{\frac{3}{2}}}{(3Z+1)^4}.
\end{aligned}
$$

The transition probability is the square of this amplitude:

$$
T_{nm} = \left(-2\frac{[2^3Z(Z+1)]^{\frac{3}{2}}}{(3Z+1)^4}\right)^2 = \frac{2^{11}Z^3(Z+1)^3}{(3Z+1)^8} = 0.25 \quad (\text{for } Z=1).
$$

The difference in these two results (parts (a) and (b)) will become negligible at large values of Z when the perturbation becomes less significant than in the case of $Z = 1$.

35 a. $\boldsymbol{\varepsilon}$ is along Z (lab fixed), and $\boldsymbol{\mu}$ is along z (the CI molecule fixed bond). The angle between Z and z is β:

$$
\boldsymbol{\varepsilon}\cdot\boldsymbol{\mu} = \varepsilon\mu\cos\beta = \varepsilon\mu D^{1*}_{00}(\alpha\beta\gamma).
$$

So,

$$
\begin{aligned}
I = \langle D^{J'}_{M'K'}|\boldsymbol{\varepsilon}\cdot\boldsymbol{\mu}|D^{J}_{MK}\rangle &= \int D^{J'}_{M'K'}\boldsymbol{\varepsilon}\cdot\boldsymbol{\mu}D^{J}_{MK}\sin\beta d\beta d\gamma d\alpha\\
&= \varepsilon\mu\int D^{J'}_{M'K'}D^{1*}_{00}D^{J}_{MK}\sin\beta d\beta d\gamma d\alpha.
\end{aligned}
$$

Now use

$$
D^{J'*}_{M'n'}D^{1}_{00}* = \sum_{jmn}\langle J'M'10\mid jm\rangle^* D^{j*}_{mn}\langle jn\mid J'K'10\rangle^*,
$$

to obtain

$$
I = \varepsilon\mu\sum_{jmn}\langle J'M'10\mid jm\rangle^*\langle jn\mid J'K'10\rangle^*\int D^{j*}_{mn}D^{J}_{MK}\sin\beta d\beta d\gamma d\alpha.
$$

Now use

$$
\int D^{j*}_{mn}D^{J}_{MK}\sin\beta d\beta d\gamma d\alpha = \frac{8\pi^2}{2J+1}\delta_{Jj}\delta_{Mm}\delta_{Kn},
$$

to obtain

$$
\begin{aligned}
I &= \varepsilon\mu\frac{8\pi^2}{2J+1}\sum_{jmn}\langle J'M'10\mid jm\rangle^*\langle jn\mid J'K'10\rangle^*\delta_{Jj}\delta_{Mm}\delta_{Kn}\\
&= \varepsilon\mu\frac{8\pi^2}{2J+1}\langle J'M'10\mid JM\rangle\langle JK\mid J'K'10\rangle.
\end{aligned}
$$

We use

$$\langle JK|J'K'10\rangle = \sqrt{2J+1}(-i)^{(J'-1+K)}\begin{pmatrix} J' & 1 & J \\ k' & 0 & K \end{pmatrix}$$

and

$$\langle J'M'10 \mid JM\rangle = \sqrt{2J+1}(-i)^{(J'-1+M)}\begin{pmatrix} J' & 1 & J \\ M & 0 & M \end{pmatrix}$$

to give

$$I = \varepsilon\mu\frac{8\pi^2}{2J+1}\sqrt{2J+1}(-i)^{(J'-1+M)}\begin{pmatrix} J' & 1 & J \\ M' & 0 & M \end{pmatrix}\sqrt{2J+1}(-i)^{(J'-1+K)}\begin{pmatrix} J' & 1 & J \\ K' & 0 & K \end{pmatrix}$$
$$= \varepsilon\mu 8\pi^2(-i)^{(J'-1+M+J'-1+K)}\begin{pmatrix} J' & 1 & J \\ M' & 0 & M \end{pmatrix}\begin{pmatrix} J' & 1 & J \\ K' & 0 & K \end{pmatrix}$$
$$= \varepsilon\mu 8\pi^2(-i)^{(M+K)}\begin{pmatrix} J' & 1 & J \\ M' & 0 & M \end{pmatrix}\begin{pmatrix} J' & 1 & J \\ K' & 0 & K \end{pmatrix}.$$

The $3-J$ symbols vanish unless $K'+0=K$ and $M'+0=M$.
So,

$$I = \varepsilon\mu 8\pi^2(-i)^{(M+K)}\begin{pmatrix} J' & 1 & J \\ M & 0 & M \end{pmatrix}\begin{pmatrix} J' & 1 & J \\ K & 0 & K \end{pmatrix}\delta_{M'M}\delta_{K'K}.$$

b. $\left(\begin{smallmatrix} J' & 1 & J \\ M & 0 & M \end{smallmatrix}\right)$ and $\left(\begin{smallmatrix} J' & 1 & J \\ K & 0 & K \end{smallmatrix}\right)$ vanish unless $J' = J+1, J, J-1$

$$\therefore \qquad \Delta J = \pm 1, 0.$$

The K quantum number can not change because the dipole moment lies along the molecule's C_3 axis and the light's electric field thus can exert no torque that twists the molecule about this axis. As a result, the light can not induce transitions that excite the molecule's spinning motion about this axis.

36 a. B atom: $1s^22s^22p^1$, ^2P ground state $L=1$, $S=\frac{1}{2}$, gives a degeneracy $((2L+1)(2S+1))$ of 6.
O atom: $1s^22s^22p^4$, ^3P ground state $L=1$, $S=1$, gives a degeneracy $((2L+1)(2S+1))$ of 9.
The total number of states formed is then $(6)(9)=54$.

b. We need only consider the p orbitals to find the low lying molecular states:

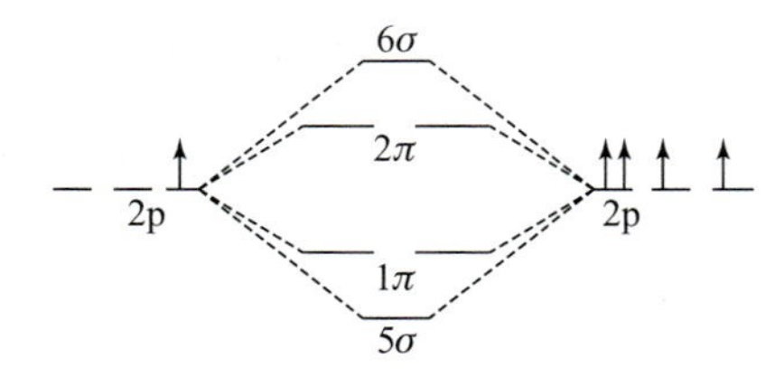

which in reality look like this:

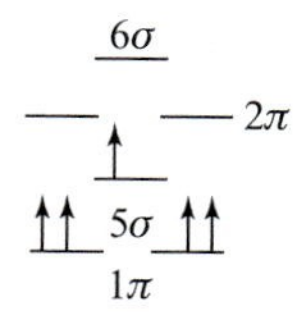

This is the correct ordering to give a $^2\sum^+$ ground state. The only low-lying electron configurations are $1\pi^3 5\sigma^2$ or $1\pi^4 5\sigma^1$. These lead to $^2\prod$ and $^2\sum^+$ states, respectively.

c. The bond orders in both states are 2.5.

d. The $^2\sum$ is + but g/u symmetry cannot be specified since this is a heteronuclear molecule.

e. Only one excited state, the $^2\prod$, is spin-allowed to radiate to the $^2\sum^+$. Consider symmetries of transition moment operators that arise in the electric dipole contributions to the transition rate $z \rightarrow \sum^+, x, y \rightarrow \prod$, $\therefore$ the $^2\prod \rightarrow {}^2\sum^+$ is electric dipole allowed via a perpendicular band.

f. Since ionization will remove a bonding electron, the BO^+ bond is weaker than the BO bond.

g. The ground state BO^+ is $^1\sum^+$ corresponding to a $1\pi^4$ electron configuration. An electron configuration of $1\pi^3 5\sigma^1$ leads to a $^3\prod$ and a $^1\prod$ state. The $^3\prod$ will be lower in energy. A $1\pi^2 5\sigma^2$ configuration will lead to higher lying states of $^3\sum^-$, $^1\Delta$, and $^1\sum^+$.

h. There should be three bands corresponding to formation of BO^+ in the $^1\sum^+$, $^3\prod$, and $^1\prod$ states. Since each of these involves removing a bonding electron, the Franck–Condon integrals will be appreciable for several vibrational levels, and thus a vibrational progression should be observed.

37 a. The bending (π) vibration is degenerate.

b. H—C≡N
  ⇑
  bending fundamental

c. H—C≡N
  ⇑
  stretching fundamental

d. CH stretch (ν_3 in figure) is σ, CN stretch is σ, and HCN bend (ν_2 in figure) is π.

e. Under $z(\sigma)$ light the CN stretch and the CH stretch can be excited, since $\psi_0 = \sigma$, $\psi_1 = \sigma$ and $z = \sigma$ provides coupling.

f. The bending vibration is active under (x, y) perpendicular polarized light. $\Delta J = 0, \pm 1$ are the selection rules for $\perp$ transitions. The CH stretching vibration is active under (z) $\parallel$ polarized light. $\Delta J = \pm 1$ are the selection rules for $\parallel$ transitions.

38 $$F\phi_i = \varepsilon_i\phi_j = h\phi_i + \sum_j [J_j - K_j]\phi_i.$$

Let the closed-shell Fock potential be written as

$$V_{ij} = \sum_k (2\langle ik \mid jk\rangle - \langle ik \mid kj\rangle),$$

and the $1e^-$ component as

$$h_{ij} = \langle\phi_i| - \frac{1}{2}\nabla^2 - \sum_A \frac{Z_A}{|r - R_A|}|\phi_j\rangle,$$

and the delta as

$$\delta_{ij} = \langle i \mid j\rangle,$$

so that

$$h_{ij} + V_{ij} = \delta_{ij}\varepsilon_i.$$

Using $\phi_i = \sum_\mu C_{\mu i}\chi_\mu$, $\phi_j = \sum_\nu C_{\nu j}\chi_\nu$, and $\phi_k = \sum_\gamma C_{\gamma k}\chi_\gamma$, and transforming from the MO to AO basis we obtain:

$$\begin{aligned} V_{ij} &= \sum_{k\mu\gamma\nu\kappa} C_{\mu i}C_{\gamma k}C_{\nu j}C_{\kappa k}(2\langle\mu\gamma \mid \nu\kappa\rangle - \langle\mu\gamma \mid \kappa\nu\rangle) \\ &= \sum_{k\mu\gamma\nu\kappa} (C_{\gamma k}C_{\kappa k})(C_{\mu i}C_{\nu j})(2\langle\mu\gamma \mid \nu\kappa\rangle - \langle\mu\gamma \mid \kappa\nu\rangle) \\ &= \sum_{\mu\nu}(C_{\mu i}C_{\nu j})V_{\mu\nu} \end{aligned}$$

where

$$V_{\mu\nu} = \sum_{\gamma\kappa} P_{\gamma\kappa}(2\langle\mu\gamma \mid \nu\kappa\rangle - \langle\mu\gamma \mid \kappa\nu\rangle), \quad \text{and} \quad P_{\gamma\kappa} = \sum_k (C_{\gamma k}C_{\kappa k}),$$

$$h_{ij} = \sum_{\mu\nu}(C_{\mu i}C_{\nu j})h_{\mu\nu},$$

where

$$h_{\mu\nu} = \langle\chi_\mu| - \frac{1}{2}\nabla^2 - \sum_A \frac{Z_A}{|r - R_A|}|\chi_\nu\rangle,$$

and

$$\delta_{ij} = \langle i \mid j\rangle = \sum_{\mu\nu}(C_{\mu i}S_{\mu\nu}C_{\nu j}).$$

So, $h_{ij} + V_{ij} = \delta_{ij}\varepsilon_j$ becomes:

$$\sum_{\mu\nu}(C_{\mu i}C_{\nu j})h_{\mu\nu} + \sum_{\mu\nu}(C_{\mu i}C_{\nu j})V_{\mu\nu} = \sum_{\mu\nu}(C_{\mu i}S_{\mu\nu}C_{\nu j})\varepsilon_j,$$

$$\sum_{\mu\nu}(C_{\mu i}S_{\mu\nu}C_{\nu j})\varepsilon_j - \sum_{\mu\nu}(C_{\mu i}C_{\nu j})h_{\mu\nu} - \sum_{\mu\nu}(C_{\mu i}C_{\nu j})V_{\mu\nu} = 0 \quad \text{for all } i, j,$$

$$\sum_{\mu\nu} C_{\mu i}[\varepsilon_j S_{\mu\nu} - h_{\mu\nu} - V_{\mu\nu}]C_{\nu j} = 0 \quad \text{for all } i, j.$$

Therefore,

$$\sum_{\nu}[h_{\mu\nu} + V_{\mu\nu} - \varepsilon_j S_{\mu\nu} -]C_{\nu j} = 0.$$

This is FC = SCE in the AO basis.

39 The Slater–Condon rule for zero (spin-orbital) difference with N electrons in N spin orbitals is

$$\begin{aligned} E = \langle |H + G| \rangle &= \sum_{i}^{N} \langle \phi_i | h | \phi_i \rangle + \sum_{i>j}^{N} (\langle \phi_i \phi_j | g | \phi_i \phi_j \rangle - \langle \phi_i \phi_j | g | \phi_j \phi_i \rangle) \\ &= \sum_i h_{ii} + \sum_{i>j} (g_{ijij} - g_{ijji}) \\ &= \sum_i h_{ii} + \frac{1}{2} \sum_{ij} (g_{ijij} - g_{ijji}). \end{aligned}$$

If all orbitals are doubly occupied and we carry out the spin integration we obtain:

$$E = 2 \sum_{i}^{\text{occ}} h_{ii} + \sum_{ij}^{\text{occ}} (2g_{ijij} - g_{ijji}),$$

where i and j now refer to orbitals (not spin-orbitals).

40 If the occupied orbitals obey $F\phi = \varepsilon_k \phi_k$, then the expression for E in Problem 39 can be rewritten as

$$E = \sum_{i}^{\text{occ}} \left(h_{ii} + \sum_{j}^{\text{occ}} (2g_{ijij} - g_{ijji}) \right) + \sum_{i}^{\text{occ}} h_{ii}.$$

We recognize the closed shell Fock operator expression and rewrite this as

$$E = \sum_{i}^{\text{occ}} F_{ii} + \sum_{i}^{\text{occ}} h_{ii} = \sum_{i}^{\text{occ}} (\varepsilon_i + h_{ii}).$$

41 I will use the *QMIC* software to do this problem. Let's just start from the beginning. Get the starting "guess" MO coefficients on disk. Using the program MOCOEFS it asks us for the first and second MO vectors. We input 1, 0 for the first MO (this means that the first MO is 1.0 times the He 1s orbital plus 0.0 times the H 1s orbital; this bonding MO is more likely to be heavily weighted on the atom having the higher nuclear charge) and 0, 1 for the second. Our beginning LCAO–MO array looks like: $\begin{bmatrix} 1.0 & 0.0 \\ 0.0 & 1.0 \end{bmatrix}$ and is placed on disk in a file we choose to call "mocoefs.dat". We also put the AO integrals on disk using the program RW_INTS. It asks for the unique one- and two-electron integrals and places a canonical list of these on disk in a file we choose to call "ao_integrals.dat". At this point it is useful for us to step back and look at the set of equations which we wish to solve: $FC = SCE$. The *QMIC* software does not provide us with a so-called generalized eigenvalue solver (one that contains an overlap matrix; or metric), so in order to use the diagonalization program that is provided we must

transform this equation ($FC = SCE$) to one that looks like ($F'C' = C'E$). We do that in the following manner:
Since S is symmetric and positive definite we can find an $S^{-\frac{1}{2}}$ such that $S^{-\frac{1}{2}}S^{+\frac{1}{2}} = 1$, $S^{-\frac{1}{2}}S = S^{+\frac{1}{2}}$, etc.
Rewrite $FC = SCE$ by inserting unity between FC and multiplying the whole equation on the left by $S^{-\frac{1}{2}}$. This gives:

$$S^{-\frac{1}{2}}FS^{-\frac{1}{2}}S^{+\frac{1}{2}}C = S^{-\frac{1}{2}}SCE = S^{+\frac{1}{2}}CE.$$

Letting:

$$F' = S^{-\frac{1}{2}}FS^{-\frac{1}{2}},$$
$$C' = S^{+\frac{1}{2}}C, \quad \text{and inserting these expressions above gives}$$
$$F'C' = C'E.$$

Note that to get the next iteration's MO coefficients we must calculate C from C':
$C' = S^{+\frac{1}{2}}C$, so, multiplying through on the left by $S^{-\frac{1}{2}}$ gives

$$S^{-\frac{1}{2}}C' = S^{-\frac{1}{2}}S^{+\frac{1}{2}}C = C.$$

This will be the method we will use to solve our Fock equations.
Find $S^{-\frac{1}{2}}$ by using the program FUNCT_MAT (this program generates a function of a matrix). This program will ask for the elements of the S array and write to disk a file (name of your choice – a good name might be "shalf") containing the $S^{-\frac{1}{2}}$ array. Now we are ready to begin the iterative Fock procedure.

a. Calculate the Fock matrix, F, using program FOCK which reads in the MO coefficients from "mocoefs.dat" and the integrals from "ao_integrals.dat" and writes the resulting Fock matrix to a user-specified file (a good filename to use might be something like "fock1").
b. Calculate $F' = S^{-\frac{1}{2}}FS^{-\frac{1}{2}}$ using the program UTMATU which reads in F and $S^{-\frac{1}{2}}$ from files on the disk and writes F' to a user-specified file (a good filename to use might be something like "fock1p"). Diagonalize F' using the program DIAG. This program reads in the matrix to be diagonalized from a user-specified filename and writes the resulting eigenvectors to disk using a user-specified filename (a good filename to use might be something like "coef1p"). You may wish to choose the option to write the eigenvalues (Fock orbital energies) to disk in order to use them at a later time in program FENERGY. Calculate C by using $C = S^{-\frac{1}{2}}C'$. This is accomplished by using the program MATXMAT which reads in two matrices to be multiplied from user specified files and writes the product to disk using a user-specified filename (a good filename to use might be something like "mocoefs.dat").

c. The *QMIC* program FENERGY calculates the total energy:

$$\sum_{kl} 2\langle k|h|k\rangle + 2\langle kl \mid kl\rangle - \langle kl \mid lk\rangle + \sum_{\mu>\nu} \frac{Z_\mu Z_\nu}{R_{\mu\nu}},$$

and

$$\sum_{k} \varepsilon_k + \langle k|h|k\rangle + \sum_{\mu>\nu} \frac{Z_\mu Z_\nu}{R_{\mu\nu}}.$$

This is the conclusion of one iteration of the Fock procedure – you may continue by going back to part (a) and proceeding onward.

d. and e. Results for the successful convergence of this system using the supplied *QMIC* software are as follows (this data is provided to give the student assurance that they are on the right track; alternatively one could switch to the *QMIC* program SCF and allow that program to iteratively converge the Fock equations):

The one-electron AO integrals: $\begin{bmatrix} -2.644\,200 & -1.511\,300 \\ -1.511\,300 & -1.720\,100 \end{bmatrix}$

The two-electron AO integrals:

1 1 1 1	1.054 700
2 1 1 1	0.474 4000
2 1 2 1	0.566 4000
2 2 1 1	0.246 9000
2 2 2 1	0.350 4000
2 2 2 2	0.625 0000

The "initial" MO–AO coefficients: $\begin{bmatrix} 1.000\,000 & 0.000\,000 \\ 0.000\,000 & 1.000\,000 \end{bmatrix}$

AO overlap matrix (S): $\begin{bmatrix} 1.000\,000 & 0.578\,400 \\ 0.578\,400 & 1.000\,000 \end{bmatrix}$

$S^{-\frac{1}{2}}$ $\begin{bmatrix} 1.168\,032 & -0.372\,070\,9 \\ -0.372\,070\,9 & 1.168\,031 \end{bmatrix}$

ITERATION 1

The charge bond order matrix $\begin{bmatrix} 1.000\,000 & 0.000\,000\,0 \\ 0.000\,000\,0 & 0.000\,000\,0 \end{bmatrix}$

The Fock matrix (F) $\begin{bmatrix} -1.589\,500 & -1.036\,900 \\ -1.036\,900 & -0.834\,200\,1 \end{bmatrix}$

$S^{-\frac{1}{2}}FS^{-\frac{1}{2}}$ $\begin{bmatrix} -1.382\,781 & -0.504\,867\,9 \\ -0.504\,867\,8 & -0.456\,888\,3 \end{bmatrix}$

The eigenvalues of this matrix (Fock orbital energies) are

$$[-1.604\,825 \quad -0.234\,845\,0]$$

Their corresponding eigenvectors ($C' = S^{+\frac{1}{2}}*C$) are

$$\begin{bmatrix} -0.915\,380\,9 & -0.402\,588\,8 \\ -0.402\,588\,8 & 0.915\,381\,0 \end{bmatrix}$$

The "new" MO–AO coefficients ($C = S^{-\frac{1}{2}}*C'$)

$$\begin{bmatrix} -0.919\,402\,2 & -0.810\,823\,1 \\ -0.129\,649\,8 & 1.218\,985 \end{bmatrix}$$

The one-electron MO integrals

$$\begin{bmatrix} -2.624\,352 & -0.164\,433\,6 \\ -0.164\,433\,6 & -1.306\,845 \end{bmatrix}$$

The two-electron MO integrals

1 1 1 1	0.977 933 1
2 1 1 1	0.192 462 3
2 1 2 1	0.597 207 5
2 2 1 1	0.117 083 8
2 2 2 1	−0.000 794 519 4
2 2 2 2	0.615 732 3

The closed-shell Fock energy from formula

$$\sum_{kl} 2\langle k|h|k\rangle + 2\langle kl \mid kl\rangle - \langle kl \mid lk\rangle + \sum_{\mu>\nu} \frac{Z_\mu Z_\nu}{R_{\mu\nu}} = -2.842\,199\,33,$$

from formula

$$\sum_{k} \varepsilon_k + \langle k|h|k\rangle + \sum_{\mu>\nu} \frac{Z_\mu Z_\nu}{R_{\mu\nu}} = -2.80060530,$$

the difference is −0.04159403.

ITERATION 2

The charge bond order matrix $\begin{bmatrix} 0.845\,300\,5 & 0.119\,200\,3 \\ 0.119\,200\,3 & 0.016\,809\,06 \end{bmatrix}$

The Fock matrix $\begin{bmatrix} -1.624\,673 & -1.083\,623 \\ -1.083\,623 & -0.877\,207\,1 \end{bmatrix}$

$S^{-\frac{1}{2}} F S^{-\frac{1}{2}}$ $\begin{bmatrix} -1.396\,111 & -0.541\,103\,7 \\ -0.541\,103\,7 & -0.479\,821\,3 \end{bmatrix}$

The eigenvalues of this matrix (Fock orbital energies) are

$$[-1.646\,972 \quad -0.228\,959\,9]$$

Their corresponding eigenvectors ($C' = S^{+\frac{1}{2}} {*} C$) are

$$\begin{bmatrix} -0.907\,242\,7 & -0.420\,607\,4 \\ -0.420\,607\,4 & 0.907\,242\,7 \end{bmatrix}$$

The "new" MO-AO coefficients ($C = S^{-\frac{1}{2}} {*} C'$)

$$\begin{bmatrix} -0.903\,192\,3 & -0.828\,841\,3 \\ -0.153\,724\,0 & 1.216\,184 \end{bmatrix}$$

The one-electron MO integrals

$$\begin{bmatrix} -2.617\,336 & -0.190\,347\,5 \\ -0.190\,347\,5 & -1.313\,861 \end{bmatrix}$$

The two-electron MO integrals

1 1 1 1	0.962 607 0
2 1 1 1	0.194 982 8
2 1 2 1	0.604 814 3
2 2 1 1	0.124 690 7
2 2 2 1	0.003 694 540
2 2 2 2	0.615 843 7

The closed-shell Fock energy from formula

$$\sum_{kl} 2\langle k|h|k\rangle + 2\langle kl \mid kl\rangle - \langle kl \mid lk\rangle + \sum_{\mu>\nu} \frac{Z_\mu Z_\nu}{R_{\mu\nu}} = -2.843\,492\,98,$$

from formula

$$\sum_k \varepsilon_k + \langle k|h|k\rangle + \sum_{\mu>\nu} \frac{Z_\mu Z_\nu}{R_{\mu\nu}} = -2.835\,736\,75,$$

the difference is $-0.007\,756\,23$.

ITERATION 3

The charge bond order matrix $\begin{bmatrix} 0.815\,756\,3 & 0.138\,842\,3 \\ 0.138\,842\,3 & 0.023\,631\,07 \end{bmatrix}$

The Fock matrix $\begin{bmatrix} -1.631\,153 & -1.091\,825 \\ -1.091\,825 & -0.885\,351\,4 \end{bmatrix}$

$S^{-\frac{1}{2}}FS^{-\frac{1}{2}}$ $\begin{bmatrix} -1.398\,951 & -0.547\,073\,1 \\ -0.547\,073\,0 & -0.484\,700\,7 \end{bmatrix}$

The eigenvalues of this matrix (Fock orbital energies) are

$$[-1.654\,745 \quad -0.228\,907\,8]$$

Their corresponding eigenvectors ($C' = S^{+\frac{1}{2}} * C$) are

$$\begin{bmatrix} -0.905\,870\,9 & -0.423\,554\,6 \\ -0.423\,554\,5 & 0.905\,870\,6 \end{bmatrix}$$

The "new" MO–AO coefficients ($C = S^{-\frac{1}{2}} * C'$)

$$\begin{bmatrix} -0.900\,493\,5 & -0.831\,773\,3 \\ -0.157\,676\,7 & 1.215\,678 \end{bmatrix}$$

The one-electron MO integrals $\begin{bmatrix} -2.616\,086 & -0.194\,581\,1 \\ -0.194\,581\,1 & -1.315\,112 \end{bmatrix}$

The two-electron MO integrals

1 1 1 1 0.960 070 7
2 1 1 1 0.195 325 5
2 1 2 1 0.606 057 2
2 2 1 1 0.125 933 2
2 2 2 1 0.004 475 587
2 2 2 2 0.615 897 2

The closed-shell Fock energy from formula

$$\sum_{kl} 2\langle k|h|k\rangle + 2\langle kl \mid kl\rangle - \langle kl \mid lk\rangle + \sum_{\mu>\nu} \frac{Z_\mu Z_\nu}{R_{\mu\nu}} = -2.843\,530\,18,$$

from formula

$$\sum_{k} \varepsilon_k + \langle k|h|k\rangle + \sum_{\mu>\nu} \frac{Z_\mu Z_\nu}{R_{\mu\nu}} = -2.842\,259\,41,$$

the difference is $-0.001\,270\,77$.

ITERATION 4

The charge bond order matrix $\begin{bmatrix} 0.810\,888\,5 & 0.141\,986\,9 \\ 0.141\,986\,9 & 0.024\,861\,94 \end{bmatrix}$

The Fock matrix $\begin{bmatrix} -1.632\,213 & -1.093\,155 \\ -1.093\,155 & -0.886\,690\,9 \end{bmatrix}$

$S^{-\frac{1}{2}}FS^{-\frac{1}{2}}$ $\begin{bmatrix} -1.399\,426 & -0.548\,028\,7 \\ -0.548\,028\,7 & -0.485\,519\,1 \end{bmatrix}$

The eigenvalues of this matrix (Fock orbital energies) are

$$[-1.656\,015 \quad -0.228\,930\,8]$$

Their corresponding eigenvectors ($C' = S^{+\frac{1}{2}} * C$) are

$$\begin{bmatrix} -0.905\,649\,4 & -0.424\,027\,1 \\ -0.424\,027\,1 & 0.905\,649\,5 \end{bmatrix}$$

The "new" MO–AO coefficients ($C = S^{-\frac{1}{2}} * C'$)

$$\begin{bmatrix} -0.900\,058\,9 & -0.832\,242\,8 \\ -0.158\,311\,1 & 1.215\,595 \end{bmatrix}$$

The one-electron MO integrals $\begin{bmatrix} -2.615\,881 & -0.195\,259\,4 \\ -0.195\,259\,4 & -1.315\,315 \end{bmatrix}$

The two-electron MO integrals

1 1 1 1	0.959 661 5
2 1 1 1	0.195 378 1
2 1 2 1	0.606 255 7
2 2 1 1	0.126 132 1
2 2 2 1	0.004 601 604
2 2 2 2	0.615 906 5

The closed-shell Fock energy from formula

$$\sum_{kl} 2\langle k|h|k\rangle + 2\langle kl \mid kl\rangle - \langle kl \mid lk\rangle + \sum_{\mu>\nu} \frac{Z_\mu Z_\nu}{R_{\mu\nu}} = -2.843\,529\,22,$$

from formula

$$\sum_k \varepsilon_k + \langle k|h|k\rangle + \sum_{\mu>\nu} \frac{Z_\mu Z_\nu}{R_{\mu\nu}} = -2.843\,324\,18,$$

the difference is $-0.000\,205\,04$.

ITERATION 5

The charge bond order matrix $\begin{bmatrix} 0.810\,106\,0 & 0.142\,489\,3 \\ 0.142\,489\,3 & 0.025\,062\,41 \end{bmatrix}$

The Fock matrix $\begin{bmatrix} -1.632\,385 & -1.093\,368 \\ -1.093\,368 & -0.886\,906\,6 \end{bmatrix}$

$S^{-\frac{1}{2}}FS^{-\frac{1}{2}}$ $\begin{bmatrix} -1.399\,504 & -0.548\,181\,2 \\ -0.548\,181\,3 & -0.485\,651\,6 \end{bmatrix}$

The eigenvalues of this matrix (Fock orbital energies) are

$$[-1.656\,219 \quad -0.228\,936\,0]$$

Their corresponding eigenvectors ($C' = S^{+\frac{1}{2}}*C$) are

$$\begin{bmatrix} -0.905\,613\,8 & -0.424\,102\,6 \\ -0.424\,102\,8 & 0.905\,614\,1 \end{bmatrix}$$

The "new" MO–AO coefficients ($C = S^{-\frac{1}{2}}*C'$)

$$\begin{bmatrix} -0.899\,989\,2 & -0.832\,317\,9 \\ -0.158\,412\,7 & 1.215\,582 \end{bmatrix}$$

The one-electron MO integrals $\begin{bmatrix} -2.615\,847 & -0.195\,367\,4 \\ -0.195\,367\,4 & -1.315\,348 \end{bmatrix}$

The two-electron MO integrals

1 1 1 1	0.959 595 6
2 1 1 1	0.195 386 2
2 1 2 1	0.606 287 2
2 2 1 1	0.126 163 9
2 2 2 1	0.004 621 811
2 2 2 2	0.615 907 8

The closed-shell Fock energy from formula

$$\sum_{kl} 2\langle k|h|k\rangle + 2\langle kl \mid kl\rangle - \langle kl \mid lk\rangle + \sum_{\mu>\nu} \frac{Z_\mu Z\nu}{R_{\mu\nu}} = -2.843\,527\,79,$$

from formula

$$\sum_k \varepsilon_k + \langle k|h|k\rangle + \sum_{\mu>\nu} \frac{Z_\mu Z_\nu}{R_{\mu\nu}} = -2.843\,494\,89,$$

the difference is $-0.000\,032\,90$.

ITERATION 6

The charge bond order matrix $\begin{bmatrix} 0.809\,980\,5 & 0.142\,569\,8 \\ 0.142\,569\,8 & 0.025\,094\,60 \end{bmatrix}$

The Fock matrix $\begin{bmatrix} -1.632\,412 & -1.093\,402 \\ -1.093\,402 & -0.886\,941\,3 \end{bmatrix}$

$S^{-\frac{1}{2}}FS^{-\frac{1}{2}}$ $\begin{bmatrix} -1.399\,517 & -0.548\,205\,6 \\ -0.548\,205\,6 & -0.485\,673\,0 \end{bmatrix}$

The eigenvalues of this matrix (Fock orbital energies) are

$$[-1.656\,253 \quad -0.228\,937\,5]$$

Their corresponding eigenvectors ($C' = S^{+\frac{1}{2}} * C$) are

$$\begin{bmatrix} -0.905\,608\,5 & -0.424\,114\,4 \\ -0.424\,114\,4 & 0.905\,608\,6 \end{bmatrix}$$

The "new" MO–AO coefficients ($C = S^{-\frac{1}{2}} * C'$)

$$\begin{bmatrix} -0.899\,978\,6 & -0.832\,329\,6 \\ -0.158\,428\,3 & 1.215\,580 \end{bmatrix}$$

The one-electron MO integrals $\begin{bmatrix} -2.615\,843 & -0.195\,384\,6 \\ -0.195\,384\,6 & -1.315\,353 \end{bmatrix}$

The two-electron MO integrals

1 1 1 1	0.959 585 9
2 1 1 1	0.195 387 8
2 1 2 1	0.606 292 5
2 2 1 1	0.126 169 0
2 2 2 1	0.004 625 196
2 2 2 2	0.615 908 3

The closed-shell Fock energy from formula

$$\sum_{kl} 2\langle k \mid h \mid k\rangle + 2\langle kl \mid kl\rangle - \langle kl \mid lk\rangle + \sum_{\mu>\nu} \frac{Z_\mu Z_\nu}{R_{\mu\nu}} = -2.843\,528\,27,$$

from formula

$$\sum_k \varepsilon_k + \langle k|h|k\rangle + \sum_{\mu>\nu} \frac{Z_\mu Z_\nu}{R_{\mu\nu}} = -2.843\,523\,98,$$

the difference is $-0.000\,004\,29$.

ITERATION 7

The charge bond order matrix: $\begin{bmatrix} 0.809\,961\,6 & 0.142\,582\,1 \\ 0.142\,582\,1 & 0.025\,099\,52 \end{bmatrix}$

The Fock matrix: $\begin{bmatrix} -1.632\,416 & -1.093\,407 \\ -1.093\,407 & -0.886\,946\,4 \end{bmatrix}$

$$S^{-\frac{1}{2}}FS^{-\frac{1}{2}} \quad \begin{bmatrix} -1.399\,519 & -0.548\,209\,3 \\ -0.548\,209\,2 & -0.485\,676\,1 \end{bmatrix}$$

The eigenvalues of this matrix (Fock orbital energies) are

$$[-1.656\,257 \quad -0.228\,937\,4]$$

Their corresponding eigenvectors ($C' = S^{+\frac{1}{2}}*C$) are

$$\begin{bmatrix} -0.905\,607\,6 & -0.424\,116\,4 \\ -0.424\,116\,4 & 0.905\,607\,7 \end{bmatrix}$$

The "new" MO–AO coefficients ($C = S^{-\frac{1}{2}}*C'$)

$$\begin{bmatrix} -0.899\,977\,0 & -0.832\,331\,7 \\ -0.158\,431\,0 & 1.215\,580 \end{bmatrix}$$

The one-electron MO integrals $\begin{bmatrix} -2.615\,843 & -0.195\,387\,6 \\ -0.195\,387\,6 & -1.315\,354 \end{bmatrix}$

The two-electron MO integrals

1 1 1 1	0.959 584 9
2 1 1 1	0.195 388 1
2 1 2 1	0.606 293 6
2 2 1 1	0.126 169 7
2 2 2 1	0.004 625 696
2 2 2 2	0.615 908 3

The closed-shell Fock energy from formula

$$\sum_{kl} 2\langle k|h|k\rangle + 2\langle kl \mid kl\rangle - \langle kl \mid lk\rangle + \sum_{\mu>\nu} \frac{Z_\mu Z_\nu}{R_{\mu\nu}} = -2.843\,529\,22,$$

from formula

$$\sum_k \varepsilon_k + \langle k|h|k\rangle + \sum_{\mu>\nu} \frac{Z_\mu Z_\nu}{R_{\mu\nu}} = -2.843\,528\,27,$$

the difference is $-0.000\,000\,95$.

ITERATION 8

The charge bond order matrix $\begin{bmatrix} 0.809\,958\,5 & 0.142\,584\,2 \\ 0.142\,584\,2 & 0.025\,100\,37 \end{bmatrix}$

The Fock matrix $\begin{bmatrix} -1.632\,416 & -1.093\,408 \\ -1.093\,408 & -0.886\,947\,0 \end{bmatrix}$

$$S^{-\frac{1}{2}}FS^{-\frac{1}{2}} \quad \begin{bmatrix} -1.399\,518 & -0.548\,210\,3 \\ -0.548\,210\,2 & -0.485\,676\,1 \end{bmatrix}$$

The eigenvalues of this matrix (Fock orbital energies) are

$$[-1.656\,258 \quad -0.228\,936\,8]$$

Their corresponding eigenvectors ($C' = S^{+\frac{1}{2}}*C$) are

$$\begin{bmatrix} -0.905\,607\,4 & -0.424\,116\,8 \\ -0.424\,116\,8 & 0.905\,607\,5 \end{bmatrix}$$

The "new" MO–AO coefficients ($C = S^{-\frac{1}{2}}*C'$)

$$\begin{bmatrix} -0.899\,976\,5 & -0.832\,332\,0 \\ -0.158\,431\,5 & 1.215\,579 \end{bmatrix}$$

The one-electron MO integrals $\begin{bmatrix} -2.615\,842 & -0.195\,388\,2 \\ -0.195\,388\,2 & -1.315\,354 \end{bmatrix}$

The two-electron MO integrals

1 1 1 1	0.959 584 1
2 1 1 1	0.195 388 1
2 1 2 1	0.606 293 4
2 2 1 1	0.126 170 0
2 2 2 1	0.004 625 901
2 2 2 2	0.615 908 1

The closed shell Fock energy from formula

$$\sum_{kl} 2\langle k|h|k\rangle + 2\langle kl \mid kl\rangle - \langle kl \mid lk\rangle + \sum_{\mu>\nu} \frac{Z_\mu Z_\nu}{R_{\mu\nu}} = -2.843\,528\,27,$$

from formula

$$\sum_{k} \varepsilon_k + \langle k|h|k\rangle + \sum_{\mu>\nu} \frac{Z_\mu Z_\nu}{R_{\mu\nu}} = -2.843\,528\,27,$$

the difference is 0.000 000 00.

f. In looking at the energy convergence we see the following:

Iter	Formula 1	Formula 2
1	−2.842 199 33	−2.800 605 30
2	−2.843 492 98	−2.835 736 75
3	−2.843 530 18	−2.842 259 41
4	−2.843 529 22	−2.843 324 18
5	−2.843 527 79	−2.843 494 89
6	−2.843 528 27	−2.843 523 98
7	−2.843 52922	−2.843 528 27
8	−2.843 528 27	−2.843 528 27

If you look at the energy differences (SCF at iteration n – SCF converged) and plot this data versus iteration number, and do a fifth order polynomial fit, we see the following:

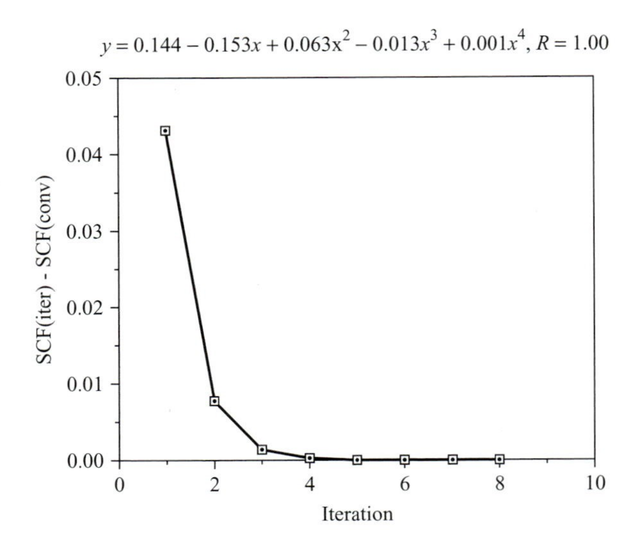

In looking at the polynomial fit we see that the convergence is primarily linear since the coefficient of the linear term is much larger than those of the cubic and higher terms.

g. The converged SCF total energy calculated using the result of Problem 40 is an upper bound to the ground-state energy, but, during the iterative procedure it is not. Only at convergence does the expectation value of the Hamiltonian for the Hartree–Fock determinant become equal to that given by the equation in Problem 40.

h. Yes, the $1\sigma^2$ configuration does dissociate properly because at $R \longrightarrow \infty$ the lowest energy state is $He + H^+$, which also has a $1\sigma^2$ orbital occupancy (i.e., $1s^2$ on He and $1s^0$ on H^+).

42 At convergence the MO coefficients are

$$\phi_1 = \begin{bmatrix} -0.899\,976\,5 \\ -0.158\,431\,5 \end{bmatrix} \qquad \phi_2 = \begin{bmatrix} -0.832\,332\,0 \\ 1.215\,579 \end{bmatrix},$$

and the integrals in this MO basis are

$$\begin{array}{lll} h_{11} = -2.615\,842 & h_{21} = -0.195\,388\,2 & h_{22} = -1.315\,354 \\ g_{1111} = 0.959\,584\,1 & g_{2111} = 0.195\,388\,1 & g_{2121} = 0.606\,293\,4 \\ g_{2211} = 0.126\,170\,0 & g_{2221} = 004\,625\,901 & g_{2222} = 0.615\,908\,1 \end{array}$$

a.
$$H = \begin{bmatrix} \langle 1\sigma^2|H|1\sigma^2\rangle & \langle 1\sigma^2|H|2\sigma^2\rangle \\ \langle 2\sigma^2|H|1\sigma^2\rangle & \langle 2\sigma^2|H|2\sigma^2\rangle \end{bmatrix} = \begin{bmatrix} 2h_{11} + g_{1111} & g_{1122} \\ g_{1122} & 2h_{22} + g_{2222} \end{bmatrix}$$
$$= \begin{bmatrix} 2^* - 2.615\,842 + 0.959\,584\,1 & 0.126\,170\,0 \\ 0.126\,170\,0 & 2^* - 1.315\,354 + 0.615\,908\,1 \end{bmatrix}$$
$$= \begin{bmatrix} -4.272\,100 & 0.126\,170 \\ 0.126\,170 & -2.014\,800 \end{bmatrix}.$$

b. The eigenvalues are $E_1 = -4.279\,131$ and $E_2 = -2.007\,770$. The corresponding eigenvectors are

$$C_1 = \begin{bmatrix} -0.998\,451\,23 \\ 0.055\,634\,39 \end{bmatrix}, \qquad C_2 = \begin{bmatrix} 0.055\,634\,38 \\ 0.998\,451\,40 \end{bmatrix}.$$

c.
$$\frac{1}{2}\Big[\Big|\left(a^{\frac{1}{2}}\phi_1 + b^{\frac{1}{2}}\phi_2\right)_\alpha \left(a^{\frac{1}{2}}\phi_1 - b^{\frac{1}{2}}\phi_2\right)_\beta\Big|$$
$$+ \Big|\left(a^{\frac{1}{2}}\phi_1 - b^{\frac{1}{2}}\phi_2\right)_\alpha \left(a^{\frac{1}{2}}\phi_1 + b^{\frac{1}{2}}\phi_2\right)_\beta\Big|\Big]$$
$$= \frac{1}{2\sqrt{2}}\Big[\left(a^{\frac{1}{2}}\phi_1 + b^{\frac{1}{2}}\phi_2\right)\left(a^{\frac{1}{2}}\phi_1 - b^{\frac{1}{2}}\phi_2\right)$$
$$+ \left(a^{\frac{1}{2}}\phi_1 - b^{\frac{1}{2}}\phi_2\right)\left(a^{\frac{1}{2}}\phi_1 + b^{\frac{1}{2}}\phi_2\right)\Big]_{(\alpha\beta - \beta\alpha)}$$
$$= \frac{1}{\sqrt{2}}(a\phi_1\phi_1 - b\phi_2\phi_2)(\alpha\beta - \beta\alpha)$$
$$= a|\phi_1\alpha\phi_1\beta| - b|\phi_2\alpha\phi_2\beta|.$$

d. The third configuration $|1\sigma 2\sigma| = \frac{1}{\sqrt{2}}[|1\alpha 2\beta| - |1\beta 2\alpha|]$. Adding this configuration to the previous 2×2 CI results in the following 3×3 "full" CI:

$$H = \begin{bmatrix} \langle 1\sigma^2|H|1\sigma^2\rangle & \langle 1\sigma^2|H|2\sigma^2\rangle & \langle 1\sigma^2|H|1\sigma 2\sigma\rangle \\ \langle 2\sigma^2|H|1\sigma^2\rangle & \langle 2\sigma^2|H|2\sigma^2\rangle & \langle 2\sigma^2|H|1\sigma 2\sigma\rangle \\ \langle 1\sigma 2\sigma|H|1\sigma^2\rangle & \langle 2\sigma^2|H|1\sigma 2\sigma\rangle & \langle 1\sigma 2\sigma|H|1\sigma 2\sigma\rangle \end{bmatrix}$$
$$= \begin{bmatrix} 2h_{11} + g_{1111} & g_{1122} & \frac{1}{\sqrt{2}}[2h_{12} + 2g_{2111}] \\ g_{1122} & 2h_{22} + g_{2222} & \frac{1}{\sqrt{2}}[2h_{12} + 2g_{2221}] \\ \frac{1}{\sqrt{2}}[2h_{12} + 2g_{2111}] & \frac{1}{\sqrt{2}}[2h_{12} + 2g_{2221}] & h_{11} + h_{22} + g_{2121} + g_{2211} \end{bmatrix}.$$

Evaluating the new matrix elements:

$$H_{13} = H_{31} = \sqrt{2} * (-0.195\,388\,2 + 0.195\,388\,1) = 0.0$$
$$H_{23} = H_{32} = \sqrt{2} * (-0.195\,388\,2 + 0.004\,626) = -0.269\,778$$

$$H_{33} = -2.615\,842 - 1.315\,354 + 0.606\,293 + 0.126\,170$$
$$= -3.198\,733$$
$$= \begin{bmatrix} -4.272\,100 & 0.126\,170 & 0.0 \\ 0.126\,170 & -2.014\,800 & -0.269\,778 \\ 0.0 & -0.269\,778 & -3.198\,733 \end{bmatrix}.$$

e. The eigenvalues are $E_1 = -4.279\,345$, $E_2 = -3.256\,612$ and $E_3 = -1.949\,678$. The corresponding eigenvectors are

$$C_1 = \begin{bmatrix} -0.998\,252\,80 \\ 0.057\,322\,90 \\ 0.014\,310\,85 \end{bmatrix}, \quad C_2 = \begin{bmatrix} -0.026\,053\,43 \\ -0.209\,692\,83 \\ -0.977\,420\,00 \end{bmatrix}, \quad C_3 = \begin{bmatrix} -0.053\,027\,67 \\ -0.976\,085\,40 \\ 0.210\,820\,04 \end{bmatrix}.$$

f. We need the non-vanishing matrix elements of the dipole operator in the MO basis. These can be obtained by calculating them by hand. They are more easily obtained by using the TRANS program. Put the $1e^-$ AO integrals on disk by running the program RW_INTS. In this case you are inserting $z_{11} = 0.0$, $z_{21} = 0.2854$, and $z_{22} = 1.4$ (insert 0.0 for all the $2e^-$ integrals) – call the output file "ao_dipole.ints" for example. The converged MO–AO coefficients should be in a file ("mocoefs.dat" is fine). The transformed integrals can be written to a file (name of your choice) for example "mo_dipole.ints". These matrix elements are

$$z_{11} = 0.116\,526\,90, \quad z_{21} = -0.544\,209\,90, \quad z_{22} = 1.491\,173\,20.$$

The excitation energies are $E_2 - E_1 = -3.256\,612 - (-4.279\,345) = 1.022\,733$, and $E_3 - E_1 = -1.949\,678 - (-4.279\,345) = 2.329\,667$. Using the Slater–Condon rules to obtain the matrix elements between configurations we obtain:

$$H_z = \begin{bmatrix} \langle 1\sigma^2|z|1\sigma^2\rangle & \langle 1\sigma^2|z|2\sigma^2\rangle & \langle 1\sigma^2|z|1\sigma 2\sigma\rangle \\ \langle 2\sigma^2|z|1\sigma^2\rangle & \langle 2\sigma^2|z|2\sigma^2\rangle & \langle 2\sigma^2|z|1\sigma 2\sigma\rangle \\ \langle 1\sigma 2\sigma|z|1\sigma^2\rangle & \langle 2\sigma^2|z|1\sigma 2\sigma\rangle & \langle 1\sigma 2\sigma|z|1\sigma 2\sigma\rangle \end{bmatrix}$$

$$= \begin{bmatrix} 2z_{11} & 0 & \frac{1}{\sqrt{2}}[2z_{12}] \\ 0 & 2z_{22} & \frac{1}{\sqrt{2}}[2z_{12}] \\ \frac{1}{\sqrt{2}}[2z_{12}] & \frac{1}{\sqrt{2}}[2z_{12}] & z_{11} + z_{22} \end{bmatrix}$$

$$= \begin{bmatrix} 0.233\,054 & 0 & -0.769\,629 \\ 0 & 2.982\,346 & -0.769\,629 \\ -0.769\,629 & -0.769\,629 & 1.607\,700 \end{bmatrix}.$$

Now, $\langle\Psi_1|z|\Psi_2\rangle = C_1{}^T H_z C_2$ (this can be accomplished with the program UTMATU)

$$= \begin{bmatrix} -0.998\,252\,80 \\ 0.057\,322\,90 \\ 0.014\,310\,85 \end{bmatrix}^T \begin{bmatrix} 0.233\,054 & 0 & -0.769\,629 \\ 0 & 2.982\,346 & -0.769\,629 \\ -0.769\,629 & -0.769\,629 & 1.607\,700 \end{bmatrix} \begin{bmatrix} -0.026\,053\,43 \\ -0.209\,692\,83 \\ -0.977\,420\,00 \end{bmatrix}$$

$= -0.757\,494$

and $\langle \Psi_1 | z | \Psi_3 \rangle = C_1{}^T H_z C_3$

$$= \begin{bmatrix} -0.998\,252\,80 \\ 0.057\,322\,90 \\ 0.014\,310\,85 \end{bmatrix}^T \begin{bmatrix} 0.233\,054 & 0 & -0.769\,629 \\ 0 & 2.982\,346 & -0.769\,629 \\ -0.769\,629 & -0.769\,629 & 1.607\,700 \end{bmatrix} \begin{bmatrix} -0.053\,027\,67 \\ -0.976\,085\,40 \\ 0.210\,820\,04 \end{bmatrix}$$

$= 0.014\,322.$

g. Using the converged coefficients the orbital energies obtained from solving the Fock equations are $\varepsilon_1 = -1.656\,258$ and $\varepsilon_2 = -0.228\,938$. The resulting expression for the PT first order wavefunction becomes

$$\begin{aligned} |1\sigma^2\rangle^{(1)} &= -\frac{g_{2211}}{2(\varepsilon_2 - \varepsilon_1)} |2\sigma^2\rangle \\ &= -\frac{0.126\,170}{2(-0.228\,938 + 1.656\,258)} |2\sigma^2\rangle \\ &= -0.044\,198\,2 |2\sigma^2\rangle. \end{aligned}$$

h. As you can see from part (c), the matrix element $\langle 1\sigma^2 | H | 1\sigma 2\sigma \rangle = 0$ (this is also a result of the Brillouin theorem) and hence this configuration does not enter into the first order wave function.

i. $|0\rangle = |1\sigma^2\rangle - 0.044\,198\,2 |2\sigma^2\rangle$. To normalize we divide by

$$\sqrt{[1 + (0.044\,198\,2)^2]} = 1.000\,976\,2,$$
$$|0\rangle = 0.999\,025 |1\sigma^2\rangle - 0.044\,155 |2\sigma^2\rangle.$$

In the 2×2 CI we obtained

$$|0\rangle = 0.998\,451\,23 |1\sigma^2\rangle - 0.055\,634\,39 |2\sigma^2\rangle.$$

j. The expression for the 2nd order RSPT is

$$\begin{aligned} E^{(2)} &= -\frac{|g_{2211}|^2}{2(\varepsilon_2 - \varepsilon_1)} = -\frac{0.126\,170^2}{2(-0.228\,938 + 1.656\,258)} \\ &= -0.005\,576 \text{ au}. \end{aligned}$$

Comparing the 2×2 CI energy obtained to the SCF result we have

$$-4.279\,131 - (-4.272\,102) = -0.007\,029 \text{ au}.$$

43

STO total energy	−2.843 528 3
STO3G total energy	−2.834 056 1
3-21G total energy	−2.886 440 5

The STO3G orbitals were generated as a best fit of three primitive Gaussians (giving one CGTO) to the STO. So, STO3G can at best reproduce the STO result. The 3-21G orbitals are more flexible since there are two CGTOs per atom. This gives four orbitals (more parameters to optimize) and a lower total energy.

44

R	HeH^+ energy	H_2 energy
1.0	−2.812 787 056	−1.071 953 297
1.2	−2.870 357 513	−1.113 775 015
1.4	−2.886 440 516	−1.122 933 507
1.6	−2.886 063 576	−1.115 567 684
1.8	−2.880 080 938	−1.099 872 589
2.0	−2.872 805 595	−1.080 269 098
2.5	−2.856 760 263	−1.026 927 710
10.0	−2.835 679 293	−0.736 170 530 3

Plotting total energy vs. geometry for HeH^+:

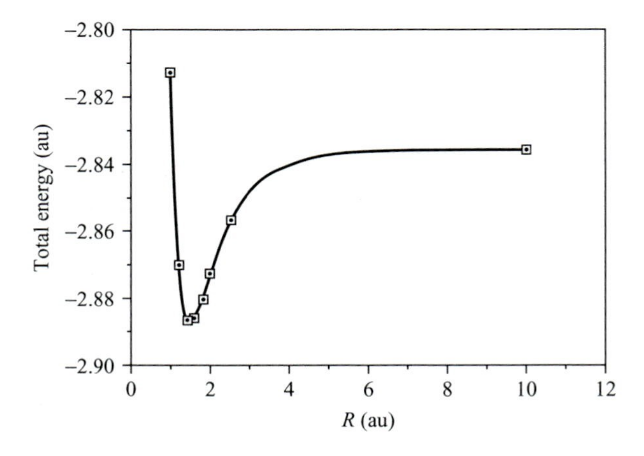

Plotting total energy vs. geometry for H_2:

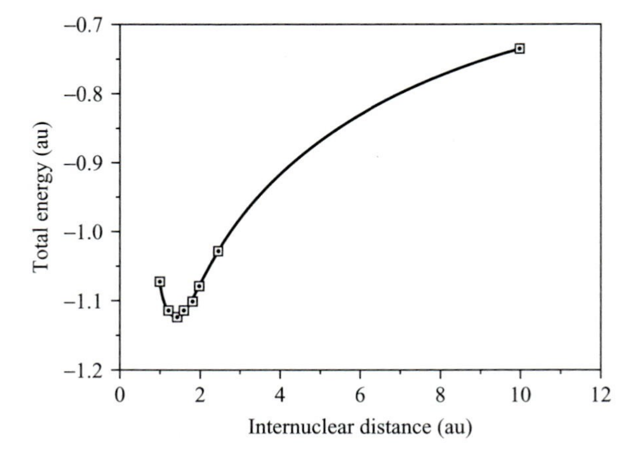

For HeH^+ at $R = 10.0$ au, the eigenvalues of the converged Fock matrix and the corresponding converged MO–AO coefficients are:

−0.1003571E + 01	−0.4961988E + 00	0.5864846E + 00	0.1981702E + 01
0.4579189E + 00	−0.8245406E − 05	0.1532163E − 04	0.1157140E + 01
0.6572777E + 00	−0.4580946E − 05	−0.6822942E − 05	−0.1056716E + 01
−0.1415438E − 05	0.3734069E + 00	0.1255539E + 01	−0.1669342E − 04
0.1112778E − 04	0.7173244E + 00	−0.1096019E + 01	0.2031348E − 04

Notice that this indicates that orbital 1 is a combination of the s functions on He only (dissociating properly to He + H^+).
For H_2 at $R = 10.0$ au, the eigenvalues of the converged Fock matrix and the corresponding converged MO–AO coefficients are:

−0.2458041E + 00	−0.1456223E + 00	0.1137235E + 01	0.1137825E + 01
0.1977649E + 00	−0.1978204E + 00	0.1006458E + 01	−0.7903225E + 00
0.5632566E + 00	−0.5628273E + 00	−0.8179120E + 00	0.6424941E + 00
0.1976312E + 00	0.1979216E + 00	0.7902887E + 00	0.1006491E + 01
0.5629326E + 00	0.5631776E + 00	−0.6421731E + 00	−0.8181460E + 00

Notice that this indicates that orbital 1 is a combination of the s functions on both H atoms (dissociating improperly; equal probabilities of H_2 dissociating to two neutral atoms or to a proton plus hydride ion).

45 The H_2 CI result:

R	${}^1\Sigma_g^+$	${}^3\Sigma_u^+$	${}^1\Sigma_u^+$	${}^1\Sigma_g^+$
1.0	−1.074 970	−0.532 342 9	−0.399 741 2	0.384 167 6
1.2	−1.118 442	−0.645 077 8	−0.489 880 5	0.176 301 8
1.4	−1.129 904	−0.722 178 1	−0.544 034 6	0.015 191 3
1.6	−1.125 582	−0.778 732 8	−0.578 442 8	−0.114 007 4
1.8	−1.113 702	−0.822 116 6	−0.601 385 5	−0.219 014 4
2.0	−1.098 676	−0.856 255 5	−0.617 276 1	−0.304 495 6
2.5	−1.060 052	−0.914 196 8	−0.638 455 7	−0.453 064 5
5.0	−0.983 588 6	−0.979 054 5	−0.587 966 2	−0.580 244 7
7.5	−0.980 623 8	−0.980 579 5	−0.524 741 5	−0.524 664 6
10.0	−0.980 598	−0.980 598 2	−0.491 405 8	−0.491 353 2

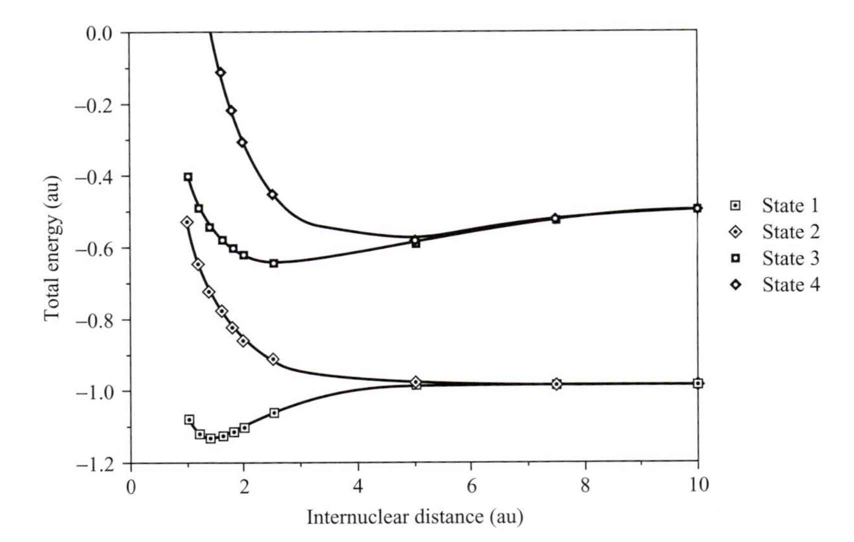

For H_2 at $R = 1.4$ au, the eigenvalues of the Hamiltonian matrix and the corresponding determinant amplitudes are:

Determinant	−1.129 904	−0.722 178	−0.544 035	0.015 191
$\lvert 1\sigma_g\alpha 1\sigma_g\beta \rvert$	0.996 95	0.000 00	0.000 00	0.078 02
$\lvert 1\sigma_g\beta 1\sigma_u\alpha \rvert$	0.000 00	0.707 11	0.707 11	0.000 00
$\lvert 1\sigma_g\alpha 1\sigma_u\beta \rvert$	0.000 00	0.707 11	−0.707 11	0.000 00
$\lvert 1\sigma_u\alpha 1\sigma_u\beta \rvert$	−0.078 02	0.000 00	0.000 00	0.996 95

This shows, as expected, the mixing of the first ${}^1\sum_g^+(1\sigma_g^2)$ and the second ${}^1\sum_g^+(1\sigma_u^2)$ determinants in the first and fourth states, and the

$$^3\sum_u^+ = \left[\frac{1}{\sqrt{2}}(\lvert 1\sigma_g\beta 1\sigma_u\alpha\rvert + \lvert 1\sigma_g\alpha 1\sigma_u\beta\rvert)\right],$$

$$\text{and} \quad ^1\sum_u^+ = \left[\frac{1}{\sqrt{2}}(\lvert 1\sigma_g\beta 1\sigma_u\alpha\rvert - \lvert 1\sigma_g\alpha 1\sigma_u\beta\rvert)\right]$$

states as the second and third states.
Also notice that the first ${}^1\sum_g{}^+$ state has coefficients (0.996 95 − 0.078 02) (note specifically the $+-$ combination) and the second ${}^1\sum_g{}^+$ state has the opposite coefficients with the same signs (note specifically the $++$ combination). The $++$ combination always gives a higher energy than the $+-$ combination.

46 F atoms have $1s^22s^22p^5\ {}^2P$ ground electronic states that are split by spin–orbit coupling into ${}^2P_{3/2}$ and ${}^2P_{1/2}$ states that differ by only 0.05 eV in energy.

a. The degeneracy of a state having a given J is $2J + 1$, and the $J = 3/2$ state is lower in energy because the 2p orbital shell is more than half filled

(I learned this in inorganic chemistry class), so

$$q_{\rm el} = 4\exp(-0/kT) + 2\exp(-0.05\ {\rm eV}/kT).$$

0.05 eV is equivalent to k(500 K), so $0.05/kT = 500/T$, hence

$$q_{\rm el} = 4\exp(-0/kT) + 2\exp(-500/T).$$

b. $$Q = q^N/N!$$

so, $\ln Q = N \ln q - \ln N!$

$$E = kT^2\ \partial \ln Q/\partial T = NkT^2 \partial \ln q/\partial T$$
$$= Nk\{1000\exp(-500/T)/[4 + 2\exp(-500/T)]\}.$$

c. Using the fact that $kT = 0.03$ eV at $T = 300$ K, make a (qualitative) graph of $\bar{E}/N$ vs. T for T ranging from 100 K to 3000 K.

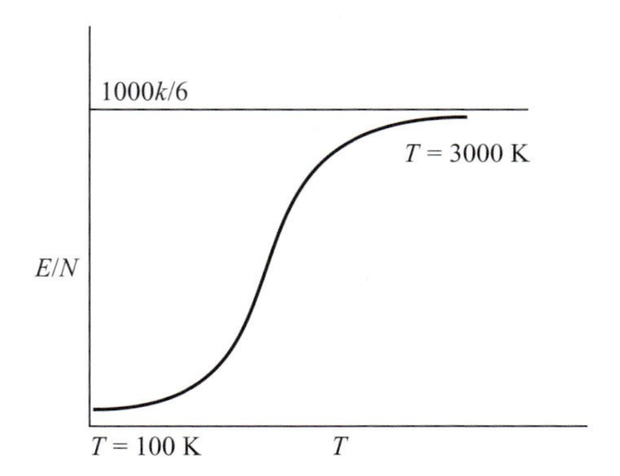

At $T = 100$ K, E/N is small and equal to $1000k\exp(-5)/(4 + 2\exp(-5))$.
At $T = 3000$ K, E/N has grown to $1000k\exp(-1/6)/(4 + 2\exp(-1/6))$ which is approximately $1000k/6$.

47 a. The difference between a linear and bent transition state would arise in the vibrational and rotational partition functions. For the linear TS, one has $3N - 6$ vibrations (recall that one loses one vibration as a reaction coordinate), but for the bent TS, one has $3N - 7$ vibrations. For the linear TS, one has two rotational axes, and for the bent TS, one has three.
So the ratio of rate constants will reduce to ratios of vibration and rotation partition functions. In particular, one will have

$$k_{\rm linear}/k_{\rm bent} = (q_{\rm vib}^{3N-6} q_{\rm rot}^2 / q_{\rm vib}^{3N-7} q_{\rm rot}^3) = (q_{\rm vib}/q_{\rm rot}).$$

b. Using

$$q_t \sim 10^8, \qquad q_r \sim 10^2, \qquad q_v \sim 1,$$

I would expect $k_{\text{linear}}/k_{\text{bent}}$ to be of the order of $1/10^2 = 10^{-2}$.

48 Constructing the Slater determinant corresponding to the "state" $1\text{s}(\alpha)1\text{s}(\alpha)$ with the rows labeling the orbitals and the columns labeling the electron gives

$$\begin{aligned}|1\text{s}\alpha 1\text{s}\alpha| &= \frac{1}{\sqrt{2!}}\begin{vmatrix}1\text{s}\alpha(1) & 1\text{s}\alpha(2)\\ 1\text{s}\alpha(1) & 1\text{s}\alpha(2)\end{vmatrix}\\ &= \frac{1}{\sqrt{2}}(1\text{s}\alpha(1)1\text{s}\alpha(2) - 1\text{s}\alpha(1)1\text{s}\alpha(2))\\ &= 0.\end{aligned}$$

49 Starting with the $M_S = 1\ {}^3\text{S}$ state (which in a "box" for this $M_L = 0$, $M_S = 1$ case would contain only one product function; $|1\text{s}\alpha 2\text{s}\alpha|$) and applying S_- gives

$$\begin{aligned}S_-{}^3S(S=1, M_S=1) &= \sqrt{1(1+1) - 1(1-1)}\hbar\,{}^3S(S=1, M_S=0)\\ &= \hbar\sqrt{2}\,{}^3S(S=1, M_S=0)\\ &= (S_-(1) + S_-(2))|1\text{s}\alpha 2\text{s}\alpha|\\ &= S_-(1)|1\text{s}\alpha 2\text{s}\alpha| + S_-(2)|1\text{s}\alpha 2\text{s}\alpha|\\ &= \hbar\sqrt{\frac{1}{2}\left(\frac{1}{2}+1\right) - \frac{1}{2}\left(\frac{1}{2}-1\right)}|1\text{s}\beta 2\text{s}\alpha|\\ &\quad +\hbar\sqrt{\frac{1}{2}\left(\frac{1}{2}+1\right) - \frac{1}{2}\left(\frac{1}{2}-1\right)}|1\text{s}\alpha 2\text{s}\beta|\\ &= \hbar(|1\text{s}\beta 2\text{s}\alpha| + |1\text{s}\alpha 2\text{s}\beta|).\end{aligned}$$

So,

$$\begin{aligned}\hbar\sqrt{2}\,{}^3S(S=1, M_S=0) &= \hbar(|1\text{s}\beta 2\text{s}\alpha| + |1\text{s}\alpha 2\text{s}\beta|)\\ {}^3S(S=1, M_S=0) &= \frac{1}{\sqrt{2}}(|1\text{s}\beta 2\text{s}\alpha| + |1\text{s}\alpha 2\text{s}\beta|).\end{aligned}$$

The three triplet states are then

$$\begin{aligned}{}^3S(S=1, M_S=1) &= |1\text{s}\alpha 2\text{s}\alpha|,\\ {}^3S(S=1, M_S=0) &= \frac{1}{\sqrt{2}}(|1\text{s}\beta 2\text{s}\alpha| + |1\text{s}\alpha 2\text{s}\beta|),\\ \text{and}\quad {}^3S(S=1, M_S=-1) &= |1\text{s}\beta 2\text{s}\beta|.\end{aligned}$$

The singlet state which must be constructed orthogonal to the three singlet states (and in particular to the ${}^3S(S=1, M_S=0)$ state) can be seen to be

$${}^1S(S=0, M_S=0) = \frac{1}{\sqrt{2}}(|1\text{s}\beta 2\text{s}\alpha| - |1\text{s}\alpha 2\text{s}\beta|).$$

Applying S^2 and S_z to each of these states gives:

$$\begin{aligned}
S_z|1s\alpha 2s\alpha| &= (S_z(1)+S_z(2))|1s\alpha 2s\alpha| \\
&= S_z(1)|1s\alpha 2s\alpha| + S_z(2))|1s\alpha 2s\alpha| \\
&= \hbar\left(\frac{1}{2}\right)|1s\alpha 2s\alpha| + \hbar\left(\frac{1}{2}\right)|1s\alpha 2s\alpha| \\
&= \hbar|1s\alpha 2s\alpha|. \\
S^2|1s\alpha 2s\alpha| &= (S_-S_+ + S_z^2 + \hbar S_z)|1s\alpha 2s\alpha| \\
&= S_-S_+|1s\alpha 2s\alpha| + S_z^2|1s\alpha 2s\alpha| + \hbar S_z|1s\alpha 2s\alpha| \\
&= 0 + \hbar^2|1s\alpha 2s\alpha| + \hbar^2|1s\alpha 2s\alpha| \\
&= 2\hbar^2|1s\alpha 2s\alpha|. \\
S_z\frac{1}{\sqrt{2}}(|1s\beta 2s\alpha| + |1s\alpha 2s\beta|) &= (S_z(1)+S_z(2))\frac{1}{\sqrt{2}}(|1s\beta 2s\alpha| + |1s\alpha 2s\beta|) \\
&= \frac{1}{\sqrt{2}}(S_z(1)+S_z(2))|1s\beta 2s\alpha| \\
&\quad + \frac{1}{\sqrt{2}}(S_z(1)+S_z(2))|1s\alpha 2s\beta| \\
&= \frac{1}{\sqrt{2}}\left(\hbar\left(-\frac{1}{2}\right)+\hbar\left(\frac{1}{2}\right)\right)|1s\beta 2s\alpha| \\
&\quad + \frac{1}{\sqrt{2}}\left(\hbar\left(\frac{1}{2}\right)+\hbar\left(-\frac{1}{2}\right)\right)|1s\alpha 2s\beta| \\
&= 0\hbar\frac{1}{\sqrt{2}}(|1s\beta 2s\alpha| + |1s\alpha 2s\beta|). \\
S^2\frac{1}{\sqrt{2}}(|1s\beta 2s\alpha| + |1s\alpha 2s\beta|) &= (S_-S_+ + S_z^2 + \hbar S_z)\frac{1}{\sqrt{2}}(|1s\beta 2s\alpha| + |1s\alpha 2s\beta|) \\
&= S_-S_+\frac{1}{\sqrt{2}}(|1s\beta 2s\alpha| + |1s\alpha 2s\beta|) \\
&= \frac{1}{\sqrt{2}}(S_-(S_+(1)+S_+(2))|1s\beta 2s\alpha|) \\
&\quad + S_-(S_+(1)+S_+(2))|1s\alpha 2s\beta|) \\
&= \frac{1}{\sqrt{2}}(S_-\hbar|1s\alpha 2s\alpha| + S_-\hbar|1s\alpha 2s\alpha|) \\
&= 2\hbar\frac{1}{\sqrt{2}}((S_-(1)+S_-(2))|1s\alpha 2s\alpha|) \\
&= 2\hbar\frac{1}{\sqrt{2}}(\hbar|1s\beta 2s\alpha| + \hbar|1s\alpha 2s\beta|) \\
&= 2\hbar^2\frac{1}{\sqrt{2}}(|1s\beta 2s\alpha| + |1s\alpha 2s\beta|). \\
S_z|1s\beta 2s\beta| &= (S_z(1)+S_z(2))|1s\beta 2s\beta| \\
&= S_z(1)|1s\beta 2s\beta| + S_z(2))|1s\beta 2s\beta| \\
&= \hbar\left(-\frac{1}{2}\right)|1s\beta 2s\beta| + \hbar\left(-\frac{1}{2}\right)|1s\beta 2s\beta| \\
&= -\hbar|1s\beta 2s\beta|.
\end{aligned}$$

$$
\begin{aligned}
S^2|1s\beta 2s\beta| &= (S_+S_- + S_z^2 - \hbar S_z)|1s\beta 2s\beta| \\
&= S_+S_-|1s\beta 2s\beta| + S_z^2|1s\beta 2s\beta| - \hbar S_z|1s\beta 2s\beta| \\
&= 0 + \hbar^2|1s\beta 2s\beta| + \hbar^2|1s\beta 2s\beta| \\
&= 2\hbar^2|1s\beta 2s\beta|.
\end{aligned}
$$

$$
\begin{aligned}
S_z\frac{1}{\sqrt{2}}(|1s\beta 2s\alpha| - |1s\alpha 2s\beta|) &= (S_z(1) + S_z(2))\frac{1}{\sqrt{2}}(|1s\beta 2s\alpha| - |1s\alpha 2s\beta|) \\
&= \frac{1}{\sqrt{2}}(S_z(1) + S_z(2))|1s\beta 2s\alpha| \\
&\quad - \frac{1}{\sqrt{2}}(S_z(1) + S_z(2))|1s\alpha 2s\beta| \\
&= \frac{1}{\sqrt{2}}\left(\hbar\left(-\frac{1}{2}\right) + \hbar\left(\frac{1}{2}\right)\right)|1s\beta 2s\alpha| \\
&\quad - \frac{1}{\sqrt{2}}\left(\hbar\left(\frac{1}{2}\right) + \hbar\left(-\frac{1}{2}\right)\right)|1s\alpha 2s\beta| \\
&= 0\hbar\frac{1}{\sqrt{2}}(|1s\beta 2s\alpha| - |1s\alpha 2s\beta|).
\end{aligned}
$$

$$
\begin{aligned}
S^2\frac{1}{\sqrt{2}}(|1s\beta 2s\alpha| - |1s\alpha 2s\beta|) &= (S_-S_+ + S_z^2 + \hbar S_z)\frac{1}{\sqrt{2}}(|1s\beta 2s\alpha| - |1s\alpha 2s\beta|) \\
&= S_-S_+\frac{1}{\sqrt{2}}(|1s\beta 2s\alpha| - |1s\alpha 2s\beta|) \\
&= \frac{1}{\sqrt{2}}(S_-(S_+(1) + S_+(2))|1s\beta 2s\alpha| \\
&\quad -S_-(S_+(1) + S_+(2))|1s\alpha 2s\beta|) \\
&= \frac{1}{\sqrt{2}}(S_-\hbar|1s\alpha 2s\alpha| - S_-\hbar|1s\alpha 2s\alpha|) \\
&= 0\hbar\frac{1}{\sqrt{2}}((S_-(1) + S_-(2))|1s\alpha 2s\alpha|) \\
&= 0\hbar\frac{1}{\sqrt{2}}(\hbar|1s\beta 2s\alpha| - \hbar|1s\alpha 2s\beta|) \\
&= 0\hbar^2\frac{1}{\sqrt{2}}(|1s\beta 2s\alpha| - |1s\alpha 2s\beta|).
\end{aligned}
$$

50 As shown in Problem 22c, for two equivalent π electrons one obtains six states:

$$
\begin{aligned}
&{}^1\Delta(M_L = 2);\ \text{one state}\ (M_S = 0), \\
&{}^1\Delta(M_L = -2);\ \text{one state}\ (M_S = 0), \\
&{}^1\textstyle\sum(M_L = 0);\ \text{one state}\ (M_S = 0),\ \text{and} \\
&{}^3\textstyle\sum(M_L = 0);\ \text{three states}\ (M_S = 1, 0,\ \text{and}\ -1).
\end{aligned}
$$

By inspecting the "box" in Problem 22c, it should be fairly straightforward to write down the wave functions for each of these:

$$
\begin{aligned}
&{}^1\Delta(M_L = 2); |\pi_1\alpha\pi_1\beta| \\
&{}^1\Delta(M_L = -2); |\pi_{-1}\alpha\pi_{-1}\beta|
\end{aligned}
$$

$$^1\sum(M_L = 0); \frac{1}{\sqrt{2}}(|\pi_1\beta\pi_{-1}\alpha| - |\pi_1\alpha\pi_{-1}\beta|)$$
$$^3\sum(M_L = 0, M_S = 1); |\pi_1\alpha\pi_{-1}\alpha|$$
$$^3\sum(M_L = 0, M_S = 0); \frac{1}{\sqrt{2}}(|\pi_1\beta\pi_{-1}\alpha| + |\pi_1\alpha\pi_{-1}\beta|)$$
$$^3\sum(M_L = 0, M_S = -1); |\pi_1\beta\pi_{-1}\beta|.$$

51 We can conveniently couple another s electron to the states generated from the $1s^1 2s^1$ configuration:

$$^3S(L = 0, S = 1) \text{ with } 3s^1 \left(L = 0, S = \frac{1}{2}\right) \text{ giving:}$$
$$L = 0, S = \frac{3}{2}, \frac{1}{2}; \ ^4S(4 \text{ states}) \text{ and } ^2S(2 \text{ states}).$$
$$^1S(L = 0, S = 0) \text{ with } 3s^1 \left(L = 0, S = \frac{1}{2}\right) \text{ giving:}$$
$$L = 0, S = \frac{1}{2}; \ ^2S(2 \text{ states}).$$

Constructing a "box" for this case would yield:

M_L / M_S	0
$\frac{3}{2}$	$\lvert 1s\alpha 2s\alpha 3s\alpha\rvert$
$\frac{1}{2}$	$\lvert 1s\alpha 2s\alpha 3s\beta\rvert$, $\lvert 1s\alpha 2s\beta 3s\alpha\rvert$, $\lvert 1s\beta 2s\alpha 3s\alpha\rvert$

One can immediately identify the wave functions for two of the quartets (they are single entries):

$$^4S\left(S = \frac{3}{2}, M_S = \frac{3}{2}\right) : |1s\alpha 2s\alpha 3s\alpha|,$$
$$^4S\left(S = \frac{3}{2}, M_S = -\frac{3}{2}\right) : |1s\beta 2s\beta 3s\beta|.$$

Applying S_- to $^4S\left(S = \frac{3}{2}, M_S = \frac{3}{2}\right)$ yields

$$S_-^4S\left(S = \frac{3}{2}, M_S = \frac{3}{2}\right) = \hbar\sqrt{\frac{3}{2}\left(\frac{3}{2}+1\right) - \frac{3}{2}\left(\frac{3}{2}-1\right)}\ ^4S\left(S = \frac{3}{2}, M_S = \frac{1}{2}\right)$$
$$= \hbar\sqrt{3}\ ^4S\left(S = \frac{3}{2}, M_S = \frac{1}{2}\right),$$
$$S_-|1s\alpha 2s\alpha 3s\alpha| = \hbar(|1s\beta 2s\alpha 3s\alpha| + |1s\alpha 2s\beta 3s\alpha| + |1s\alpha 2s\alpha 3s\beta|).$$

So, $$^4S\left(S = \frac{3}{2}, M_S = \frac{1}{2}\right) = \frac{1}{\sqrt{3}}(|1s\beta 2s\alpha 3s\alpha| + |1s\alpha 2s\beta 3s\alpha| + |1s\alpha 2s\alpha 3s\beta|).$$

Applying S_+ to ^{4}S $\left(S = \frac{3}{2}, M_S = -\frac{3}{2}\right)$ yields

$$S_+^4\text{S}\left(S = \frac{3}{2}, M_S = -\frac{3}{2}\right) = \hbar\sqrt{\frac{3}{2}\left(\frac{3}{2}+1\right) - \left[-\frac{3}{2}\left(-\frac{3}{2}+1\right)\right]}$$

$$\times {}^4\text{S}\left(S = \frac{3}{2}, M_S = -\frac{1}{2}\right)$$

$$= \hbar\sqrt{3}\,{}^4\text{S}\left(S = \frac{3}{2}, M_S = -\frac{1}{2}\right),$$

$$S_+|1s\beta 2s\beta 3s\beta| = \hbar(|1s\alpha 2s\beta 3s\beta| + |1s\beta 2s\alpha 3s\beta| + |1s\beta 2s\beta 3s\alpha|).$$

So, $${}^4\text{S}\left(S = \frac{3}{2}, \text{M}_\text{S} = -\frac{1}{2}\right) = \frac{1}{\sqrt{3}}(|1s\alpha 2s\beta 3s\beta| + |1s\beta 2s\alpha 3s\beta| + |1s\beta 2s\beta 3s\alpha|).$$

It only remains to construct the doublet states which are orthogonal to these quartet states. Recall that the orthogonal combinations for systems having three equal components (for example when symmetry adapting the 3 sp^2 hybrids in C_{2v} or D_{3h} symmetry) give results of $+++$, $+2--$, and $0+-$. Notice that the quartets are the $+++$ combinations and therefore the doublets can be recognized as

$${}^2\text{S}\left(S = \frac{1}{2}, \text{M}_\text{S} = \frac{1}{2}\right) = \frac{1}{\sqrt{6}}(|1s\beta 2s\alpha 3s\alpha| + |1s\alpha 2s\beta 3s\alpha| - 2|1s\alpha 2s\alpha 3s\beta|),$$

$${}^2\text{S}\left(S = \frac{1}{2}, \text{M}_\text{S} = \frac{1}{2}\right) = \frac{1}{\sqrt{2}}(|1s\beta 2s\alpha 3s\alpha| - |1s\alpha 2s\beta 3s\alpha| + 0|1s\alpha 2s\alpha 3s\beta|),$$

$${}^2\text{S}\left(S = \frac{1}{2}, \text{M}_\text{S} = -\frac{1}{2}\right) = \frac{1}{\sqrt{6}}(|1s\alpha 2s\beta 3s\beta| + |1s\beta 2s\alpha 3s\beta| - 2|1s\beta 2s\beta 3s\alpha|),$$

$${}^2\text{S}\left(S = \frac{1}{2}, \text{M}_\text{S} = -\frac{1}{2}\right) = \frac{1}{\sqrt{3}}(|1s\alpha 2s\beta 3s\beta| - |1s\beta 2s\alpha 3s\beta| + 0|1s\beta 2s\beta 3s\alpha|).$$

52 As illustrated in Problem 24, a p^2 configuration (two equivalent p electrons) gives rise to the term symbols: ^{3}P, ^{1}D, and ^{1}S. Coupling an additional electron (3d^1) to this p^{2}configuration will give the desired 1s^{2}2s^{2}2p^{2}3d^1 term symbols:

^{3}P ($L = 1, S = 1$) with ^{2}D $\left(L = 2, \text{S} = \frac{1}{2}\right)$ generates

$L = 3, 2, 1,$ and $S = \frac{3}{2}, \frac{1}{2}$ with term symbols ^{4}F, ^{2}F, ^{4}D, ^{2}D, ^{4}P, and ^{2}P;

^{1}D ($L = 2, S = 0$) with ^{2}D $\left(L = 2, S = \frac{1}{2}\right)$ generates

$L = 4, 3, 2, 1, 0,$ and $S = \frac{1}{2}$ with term symbols ^{2}G, ^{2}F, ^{2}D, ^{2}P, and ^{2}S;

^{1}S ($L = 0, S = 0$) with ^{2}D $\left(L = 2, S = \frac{1}{2}\right)$ generates

$L = 2$ and $S = \frac{1}{2}$ with term symbol ^{2}D.

53 The notation used for the Slater–Condon rules will be as follows:

(i) zero (spin-orbital) difference:

$$\begin{aligned}\langle|F+G|\rangle &= \sum_i \langle\phi_i|f|\phi_i\rangle + \sum_{i>j}(\langle\phi_i\phi_j|g|\phi_i\phi_j\rangle - \langle\phi_i\phi_j|g|\phi_j\phi_i\rangle)\\ &= \sum_i f_{ii} + \sum_{i>j}(g_{ijij} - g_{ijji}).\end{aligned}$$

(ii) one (spin-orbital) difference ($\phi_p \neq \phi_{p'}$):

$$\begin{aligned}\langle|F+G|\rangle &= \langle\phi_p|f|\phi_{p'}\rangle + \sum_{j\neq p;p'}(\langle\phi_p\phi_j|g|\phi_{p'}\phi_j\rangle - \langle\phi_p\phi_j|g|\phi_j\phi_{p'}\rangle)\\ &= f_{pp'} + \sum_{j\neq p;p'}(g_{pjp'j} - g_{pjjp'}).\end{aligned}$$

(iii) two (spin-orbital) differences ($\phi_p \neq \phi_{p'}$ and $\phi_q \neq \phi_{q'}$):

$$\begin{aligned}\langle|F+G|\rangle &= \langle\phi_p\phi_q|g|\phi_{p'}\phi_{q'}\rangle - \langle\phi_p\phi_q|g|\phi_{q'}\phi_{p'}\rangle\\ &= g_{pqp'q'} - g_{pqq'p'}.\end{aligned}$$

(iv) three or more (spin-orbital) differences:

$$\langle|F+G|\rangle = 0.$$

a. $^3\mathrm{P}(M_L = 1, M_S = 1) = |\mathrm{p}_1\alpha\mathrm{p}_0\alpha|\langle|\mathrm{p}_1\alpha\mathrm{p}_0\alpha|H|\mathrm{p}_1\alpha\mathrm{p}_0\alpha|\rangle = \langle|10|H|10|\rangle$.
Using the Slater–Condon rule (i) above (I will denote these SCi–SCiv):

$$\langle|10|H|10|\rangle = f_{11} + f_{00} + g_{1010} - g_{1001}.$$

b. $^3\mathrm{P}(M_L = 0, M_S = 0) = \frac{1}{\sqrt{2}}(|\mathrm{p}_1\alpha\mathrm{p}_{-1}\beta| + |\mathrm{p}_1\beta\mathrm{p}_{-1}\alpha|)$ $\quad\langle^3\mathrm{P}(M_L = 0, M_S = 0)|H|^3\mathrm{P}(M_L = 0, M_S = 0)\rangle$

$$\begin{aligned}&= \frac{1}{2}(\langle|\mathrm{p}_1\alpha\mathrm{p}_{-1}\beta|H|\mathrm{p}_1\alpha\mathrm{p}_{-1}\beta|\rangle + \langle|\mathrm{p}_1\alpha\mathrm{p}_{-1}\beta|H|\mathrm{p}_1\beta\mathrm{p}_{-1}\alpha|\rangle\\ &\quad + \langle|\mathrm{p}_1\beta\mathrm{p}_{-1}\alpha|H|\mathrm{p}_1\alpha\mathrm{p}_{-1}\beta|\rangle + \langle|\mathrm{p}_1\beta\mathrm{p}_{-1}\alpha|H|\mathrm{p}_1\beta\mathrm{p}_{-1}\alpha|\rangle).\end{aligned}$$

Evaluating each matrix element gives

$$\begin{aligned}\langle|\mathrm{p}_1\alpha\mathrm{p}_{-1}\beta|H|\mathrm{p}_1\alpha\mathrm{p}_{-1}\beta|\rangle &= f_{1\alpha1\alpha} + f_{-1\beta-1\beta} + g_{1\alpha-1\beta1\alpha-1\beta} - g_{1\alpha-1\beta-1\beta1\alpha} \quad \text{(SCi)}\\ &= f_{11} + f_{-1-1} + g_{1-11-1} - 0,\\ \langle|\mathrm{p}_1\alpha\mathrm{p}_{-1}\beta|H|\mathrm{p}_1\beta\mathrm{p}_{-1}\alpha|\rangle &= g_{1\alpha-1\beta1\beta-1\alpha} - g_{1\alpha-1\beta-1\alpha1\beta} \quad \text{(SCiii)}\\ &= 0 - g_{1-1-11},\\ \langle|\mathrm{p}_1\beta\mathrm{p}_{-1}\alpha|H|\mathrm{p}_1\alpha\mathrm{p}_{-1}\beta|\rangle &= g_{1\beta-1\alpha1\alpha-1\beta} - g_{1\beta-1\alpha-1\beta1\alpha} \quad \text{(SCiii)}\\ &= 0 - g_{1-1-11},\\ \langle|\mathrm{p}_1\beta\mathrm{p}_{-1}\alpha|H|\mathrm{p}_1\beta\mathrm{p}_{-1}\alpha|\rangle &= f_{1\beta1\beta} + f_{-1\alpha-1\alpha} + g_{1\beta-1\alpha1\beta-1\alpha} - g_{1\beta-1\alpha-1\alpha1\beta} \quad \text{(SCi)}\\ &= f_{11} + f_{-1-1} + g_{1-11-1} - 0.\end{aligned}$$

Substitution of these expressions gives:

$$\begin{aligned}
&\langle {}^3\mathrm{P}(M_L = 0, M_S = 0)|H|{}^3\mathrm{P}(M_L = 0, M_S = 0)\rangle \\
&= \frac{1}{2}(f_{11} + f_{-1-1} + g_{1-11-1} - g_{1-1-11} - g_{1-1-11} + f_{11} + f_{-1-1} + g_{1-11-1}) \\
&= f_{11} + f_{-1-1} + g_{1-11-1} - g_{1-1-11}.
\end{aligned}$$

c. ${}^1\mathrm{S}(M_L = 0, M_S = 0) = \frac{1}{\sqrt{3}}(|\mathrm{p}_0\alpha\mathrm{p}_0\beta| - |\mathrm{p}_1\alpha\mathrm{p}_{-1}\beta| - |\mathrm{p}_{-1}\alpha\mathrm{p}_1\beta|)$
$\langle {}^1\mathrm{S}(M_L = 0, M_S = 0)|H|{}^1\mathrm{S}(M_L = 0, M_S = 0)\rangle$

$$\begin{aligned}
&= \frac{1}{3}(\langle|\mathrm{p}_0\alpha\mathrm{p}_0\beta|H|\mathrm{p}_0\alpha\mathrm{p}_0\beta|\rangle - \langle|\mathrm{p}_0\alpha\mathrm{p}_0\beta|H|\mathrm{p}_1\alpha\mathrm{p}_{-1}\beta|\rangle \\
&\quad -\langle|\mathrm{p}_0\alpha\mathrm{p}_0\beta|H|\mathrm{p}_{-1}\alpha\mathrm{p}_1\beta|\rangle - \langle|\mathrm{p}_1\alpha\mathrm{p}_{-1}\beta|H|\mathrm{p}_0\alpha\mathrm{p}_0\beta|\rangle \\
&\quad +\langle|\mathrm{p}_1\alpha\mathrm{p}_{-1}\beta|H|\mathrm{p}_1\alpha\mathrm{p}_{-1}\beta|\rangle + \langle|\mathrm{p}_1\alpha\mathrm{p}_{-1}\beta|H|\mathrm{p}_{-1}\alpha\mathrm{p}_1\beta|\rangle \\
&\quad -\langle|\mathrm{p}_{-1}\alpha\mathrm{p}_1\beta|H|\mathrm{p}_0\alpha\mathrm{p}_0\beta|\rangle + \langle|\mathrm{p}_{-1}\alpha p_1\beta|H|\mathrm{p}_1\alpha\mathrm{p}_{-1}\beta|\rangle \\
&\quad +\langle|\mathrm{p}_{-1}\alpha\mathrm{p}_1\beta|H|\mathrm{p}_{-1}\alpha\mathrm{p}_1\beta|\rangle).
\end{aligned}$$

Evaluating each matrix element gives

$$\begin{aligned}
\langle|\mathrm{p}_0\alpha\mathrm{p}_0\beta|H|\mathrm{p}_0\alpha p_0\beta|\rangle &= f_{0\alpha0\alpha} + f_{0\beta0\beta} + g_{0\alpha0\beta0\alpha0\beta} - g_{0\alpha0\beta0\beta0\alpha} \quad \text{(SCi)} \\
&= f_{00} + f_{00} + g_{0000} - 0, \\
\langle|\mathrm{p}_0\alpha\mathrm{p}_0\beta|H|\mathrm{p}_1\alpha\mathrm{p}_{-1}\beta|\rangle &= \langle|\mathrm{p}_1\alpha\mathrm{p}_{-1}\beta|H|\mathrm{p}_0\alpha\mathrm{p}_0\beta|\rangle \\
&= g_{0\alpha0\beta1\alpha-1\beta} - g_{0\alpha0\beta-1\beta1\alpha} \quad \text{(SCiii)} \\
&= g_{001-1} - 0, \\
\langle|\mathrm{p}_0\alpha\mathrm{p}_0\beta|H|\mathrm{p}_{-1}\alpha\mathrm{p}_1\beta|\rangle &= \langle|\mathrm{p}_{-1}\alpha\mathrm{p}_1\beta|H|\mathrm{p}_0\alpha\mathrm{p}_0\beta|\rangle \\
&= g_{0\alpha0\beta-1\alpha1\beta} - g_{0\alpha0\beta1\beta-1\alpha} \quad \text{(SCiii)} \\
&= g_{00-11} - 0, \\
\langle|\mathrm{p}_1\alpha\mathrm{p}_{-1}\beta|H|\mathrm{p}_1\alpha\mathrm{p}_{-1}\beta|\rangle &= f_{1\alpha1\alpha} + f_{-1\beta-1\beta} + g_{1\alpha-1\beta1\alpha-1\beta} - g_{1\alpha-1\beta-1\beta1\alpha} \quad \text{(SCi)} \\
&= f_{11} + f_{-1-1} + g_{1-11-1} - 0, \\
\langle|\mathrm{p}_1\alpha\mathrm{p}_{-1}\beta|H|\mathrm{p}_{-1}\alpha\mathrm{p}_1\beta|\rangle &= \langle|\mathrm{p}_{-1}\alpha\mathrm{p}_1\beta|H|\mathrm{p}_1\alpha\mathrm{p}_{-1}\beta|\rangle \\
&= g_{1\alpha-1\beta-1\alpha1\beta} - g_{1\alpha-1\beta1\beta-1\alpha} \quad \text{(SCiii)} \\
&= g_{1-1-11} - 0, \\
\langle|\mathrm{p}_{-1}\alpha\mathrm{p}_1\beta|H|\mathrm{p}_{-1}\alpha\mathrm{p}_1\beta|\rangle &= f_{-1\alpha-1\alpha} + f_{1\beta1\beta} + g_{-1\alpha1\beta-1\alpha1\beta} - g_{-1\alpha1\beta1\beta-1\alpha} \quad \text{(SCi)} \\
&= f_{-1-1} + f_{11} + g_{-11-11} - 0.
\end{aligned}$$

Substitution of these expressions gives

$$\begin{aligned}
&\langle {}^1\mathrm{S}(M_L = 0, M_S = 0)|H|{}^1\mathrm{S}(M_L = 0, M_S = 0)\rangle \\
&= \frac{1}{3}(f_{00} + f_{00} + g_{0000} - g_{001-1} - g_{00-11} - g_{001-1} + f_{11} + f_{-1-1} \\
&\quad + g_{1-11-1} + g_{1-1-11} - g_{00-11} + g_{1-1-11} + f_{-1-1} + f_{11} + g_{-11-11}) \\
&= \frac{1}{3}(2f_{00} + 2f_{11} + 2f_{-1-1} + g_{0000} - 4g_{001-1} + 2g_{1-11-1} + 2g_{1-1-11}).
\end{aligned}$$

d. ${}^1\mathrm{D}(M_L = 0,\ M_S = 0) = \frac{1}{\sqrt{6}}(2|\mathrm{p}_0\alpha\mathrm{p}_0\beta| + |\mathrm{p}_1\alpha\mathrm{p}_{-1}\beta| + |\mathrm{p}_{-1}\alpha\mathrm{p}_1\beta|)$
Evaluating $\langle {}^1\mathrm{D}(M_L = 0, M_S = 0)|H|{}^1\mathrm{D}(M_L = 0, M_S = 0)\rangle$ we note that all the Slater–Condon matrix elements generated are the same as

those evaluated in part (c) (the signs for the wave function components and the multiplicative factor of two for one of the components, however, are different).

$$\begin{aligned}
&\langle {}^1\mathrm{D}(M_L = 0, M_S = 0)|H|{}^1\mathrm{D}(M_L = 0, M_S = 0)\rangle \\
&= \frac{1}{6}(4f_{00} + 4f_{00} + 4g_{0000} + 2g_{001-1} + 2g_{00-11} + 2g_{001-1} + f_{11} + f_{-1-1} \\
&\quad + g_{1-11-1} + g_{1-1-11} + 2g_{00-11} + g_{1-1-11} + f_{-1-1} + f_{11} + g_{-11-11}) \\
&= \frac{1}{6}(8f_{00} + 2f_{11} + 2f_{-1-1} + 4g_{0000} + 8g_{001-1} + 2g_{1-11-1} + 2g_{1-1-11}).
\end{aligned}$$

54 a. $$\begin{aligned}
{}^1\Delta(M_L = 2, M_S = 0) = |\pi_1\alpha\pi_1\beta| \langle {}^1\Delta(M_L = 2, M_S = 0)|H|{}^1 \\
\times \Delta(M_L = 2, M_S = 0)\rangle \\
&= \langle|\pi_1\alpha\pi_1\beta|H|\pi_1\alpha\pi_1\beta|\rangle \\
&= f_{1\alpha1\alpha} + f_{1\beta1\beta} + g_{1\alpha1\beta1\alpha1\beta} - g_{1\alpha1\beta1\beta1\alpha} \qquad \text{(SCi)} \\
&= f_{11} + f_{11} + g_{1111} - 0 \\
&= 2f_{11} + g_{1111}.
\end{aligned}$$

b.

$$\begin{aligned}
{}^1\sum(M_L = 0, M_S = 0) &= \frac{1}{\sqrt{2}}(|\pi_1\alpha\pi_{-1}\beta| - |\pi_1\beta\pi_{-1}\alpha|) \\
&\times \left\langle {}^3\sum(M_L = 0, M_S = 0)\right| H \left|{}^3\sum(M_L = 0, M_S = 0)\right\rangle \\
&= \frac{1}{2}(\langle|\pi_1\alpha\pi_{-1}\beta|H|\pi_1\alpha\pi_{-1}\beta|\rangle - \langle|\pi_1\alpha\pi_{-1}\beta|H|\pi_1\beta\pi_{-1}\alpha|\rangle \\
&\quad - \langle|\pi_1\beta\pi_{-1}\alpha|H|\pi_1\alpha\pi_{-1}\beta|\rangle + \langle|\pi_1\beta\pi_{-1}\alpha|H|\pi_1\beta\pi_{-1}\alpha|\rangle).
\end{aligned}$$

Evaluating each matrix element gives:

$$\begin{aligned}
\langle|\pi_1\alpha\pi_{-1}\beta|H|\pi_1\alpha\pi_{-1}\beta|\rangle &= f_{1\alpha1\alpha} + f_{-1\beta-1\beta} + g_{1\alpha-1\beta1\alpha-1\beta} - g_{1\alpha-1\beta-1\beta1\alpha} \quad \text{(SCi)} \\
&= f_{11} + f_{-1-1} + g_{1-11-1} - 0, \\
\langle|\pi_1\alpha\pi_{-1}\beta|H|\pi_1\beta\pi_{-1}\alpha|\rangle &= g_{1\alpha-1\beta1\beta-1\alpha} - g_{1\alpha-1\beta-1\alpha1\beta} \quad \text{(SCiii)} \\
&= 0 - g_{1-1-11}, \\
\langle|\pi_1\beta\pi_{-1}\alpha|H|\pi_1\alpha\pi_{-1}\beta|\rangle &= g_{1\beta-1\alpha1\alpha-1\beta} - g_{1\beta-1\alpha-1\beta1\alpha} \quad \text{(SCiii)} \\
&= 0 - g_{1-1-11}, \\
\langle|\pi_1\beta\pi_{-1}\alpha|H|\pi_1\beta\pi_{-1}\alpha|\rangle &= f_{1\beta1\beta} + f_{-1\alpha-1\alpha} + g_{1\beta-1\alpha1\beta-1\alpha} - g_{1\beta-1\alpha-1\alpha1\beta} \quad \text{(SCa)} \\
&= f_{11} + f_{-1-1} + g_{1-11-1} - 0.
\end{aligned}$$

Substitution of these expressions gives:

$$\begin{aligned}
&\left\langle {}^3\sum(M_L = 0, M_S = 0)\right| H \left|{}^3\sum(M_L = 0, M_S = 0)\right\rangle \\
&= \frac{1}{2}(f_{11} + f_{-1-1} + g_{1-11-1} + g_{1-1-11} + g_{1-1-11} + f_{11} + f_{-1-1} + g_{1-11-1}) \\
&= f_{11} + f_{-1-1} + g_{1-11-1} + g_{1-1-11}.
\end{aligned}$$

c.

$$
\begin{aligned}
{}^3\sum(M_L = 0, M_S = 0) &= \frac{1}{\sqrt{2}}(|\pi_1\alpha\pi_{-1}\beta| + |\pi_1\beta\pi_{-1}\alpha|) \\
&\quad \times \left\langle {}^3\sum(M_L = 0, M_S = 0)\right| H \left|{}^3\sum(M_L = 0, M_S = 0)\right\rangle \\
&= f_{11} + f_{-1-1} + g_{1-11-1} - 0, \\
\langle|\pi_1\alpha\pi_{-1}\beta|H|\pi_1\beta\pi_{-1}\alpha|\rangle &= g_{1\alpha-1\beta1\beta-1\alpha} - g_{1\alpha-1\beta-1\alpha1\beta} \qquad \text{(SCiii)} \\
&= 0 - g_{1-1-11}, \\
\langle|\pi_1\beta\pi_{-1}\alpha|H|\pi_1\alpha\pi_{-1}\beta|\rangle &= g_{1\beta-1\alpha1\alpha-1\beta} - g_{1\beta-1\alpha-1\beta1\alpha} \qquad \text{(SCiii)} \\
&= 0 - g_{1-1-11}, \\
\langle|\pi_1\beta\pi_{-1}\alpha|H|\pi_1\beta\pi_{-1}\alpha|\rangle &= f_{1\beta1\beta} + f_{-1\alpha-1\alpha} + g_{1\beta-1\alpha1\beta-1\alpha} - g_{1\beta-1\alpha-1\alpha1\beta} \quad \text{(SCa)} \\
&= f_{11} + f_{-1-1} + g_{1-11-1} - 0.
\end{aligned}
$$

Substitution of these expressions gives

$$
\begin{aligned}
&\left\langle {}^3\sum(M_L = 0, M_S = 0)\right| H \left|{}^3\sum(M_L = 0, M_S = 0)\right\rangle \\
&= \frac{1}{2}(f_{11} + f_{-1-1} + g_{1-11-1} - g_{1-1-11} - g_{1-1-11} + f_{11} + f_{-1-1} + g_{1-11-1}) \\
&= f_{11} + f_{-1-1} + g_{1-11-1} - g_{1-1-11}.
\end{aligned}
$$

55 The order of the answers is J, I, G, K, B, D, E, A, C, H, F.

56

$$
\begin{aligned}
\frac{Vp}{kTN} &= \frac{1}{1 - \frac{N}{V}b} - \frac{N}{V}\frac{a}{kT} \\
&= 1 + \frac{N}{V}b + \left(\frac{N}{V}\right)^2 b^2 + \cdots - \frac{N}{V}\frac{a}{kT},
\end{aligned}
$$

so

$$B_2 = b - \frac{a}{kT}.$$

57 a. MD because you need to keep track of how far the molecule moves as a function of time and MC does not deal with time.

b. MC is capable of doing this although MD is also. However, MC requires fewer computational steps, so I would prefer to use it.

c. MC can do this, as could MD. Again, because MC needs fewer computational steps, I'd use it.

Suppose you are carrying out a Monte-Carlo simulation involving 1000 Ar atoms. Further suppose that the potentials are pairwise additive and that your computer requires approximately 50 floating point operations (FPOs) (e.g., multiply, add, divide, etc.) to compute the interaction potential between any pair of atoms.

d. For each MC move, we must compute only the change in potential energy. To do this, we need to compute only the change in the pair energies that involve the atom that was moved. This will require 999×50 FPOs (the 99 being the number of atoms other than the one that moved). So, for a million MC steps, I would need $10^6 \times 999 \times 50$ FPOs. At 100×10^6 FPOs per second, this will require 495 seconds, or a little over eight minutes.

e. Because the statistical fluctuations in MC calculations are proportional to $(1/N)^{1/2}$, where N is the number of steps taken, I will have to take four times as many steps to cut the statistical errors in half. So, this will require 4×495 seconds or 1980 seconds.

f. If we have one million rather than one thousand atoms, the 495 second calculation of part (d) would require

$$999\,999/999$$

times as much time. This ratio arises because the time to compute the change in potential energy accompanying a MC move is proportional to the number of other atoms. So, the calculation would take $495 \times (999\,999/999)$ seconds or about 500 000 seconds or about 140 hours.

g. We would be taking 10^{-9} s/(10^{-15} s per step) $= 10^6$ MD steps. Each step requires that we compute all forces $(-\partial V \partial R_{I,J})$ between all pairs of atoms. There are $1000 \times 999/2$ such pairs. So, to compute all the forces would require $(1000 \times 999/2) \times 50$ FPOs $= 2.5 \times 10^7$ FPOs. So, we will need 2.5×10^7 FPOs/step $\times 10^6$ steps/(100 FPOs per second) $= 2.5 \times 10^5$ seconds or about 70 hours.

h. The graduate student is 10^8 times slower than the 100 Mflop computer, so it will take her/him 10^8 times as long, so 495×10^8 seconds or about 1570 years.

58 First, Na has a ^{2}S ground-state term symbol whose degeneracy is $2S + 1 = 2$.
Na$_2$ has a $^1\sum$ ground state whose degeneracy is 1.
The symmetry number for Na$_2$ is $\sigma = 2$.
The D_0 value given is 17.3 kcal mol^{-1}.
The K_p equilibrium constant would be given in terms of partial pressures as (and then using $pV = NkT$)

$$K_p = \frac{p_{\text{Na}}^2}{p_{\text{Na}_2}} = kT \left(\frac{q_{\text{Na}}}{V}\right)^2 \Big/ \left(\frac{q_{\text{Na}_2}}{V}\right)$$

in terms of the partition functions.

a.
$$q_{\text{Na}} = \left(\frac{2\pi mkT}{h^2}\right)^{3/2} V q_{\text{el}},$$
$$q_{\text{NA}_2} = \left(\frac{2\pi m'kT}{h^2}\right)^{3/2} V \left(\frac{8\pi^2 IkT}{h^2}\right) \frac{1}{2} \exp\left(\frac{-h\nu}{2kT}\right)$$
$$\times \left(1 - \exp\left(\frac{-h\nu}{kT}\right)\right)^{-1} \exp\left(\frac{D_e}{kT}\right).$$

We can combine the D_e and the $-h\nu/2kT$ to obtain the D_0 which is what we were given.

b. For Na (I will use cgs units in all cases):

$$q/V = [(2\pi \times 23)(1.66 \times 10^{-24})(1.38 \times 10^{-16})1000/(6.626 \times 10^{-27})^2]^{3/2}2$$
$$= (6.54 \times 10^{26}) \times 2 = 1.31 \times 10^{27}.$$

For Na_2:

$$q/V = 2^{3/2} \times (6.54 \times 10^{26})(1000/0.221)(1/2)(1 - \exp(-229/1000))^{-1} \exp(D_0/kT)$$
$$= 1.85 \times 10^{27}(2.26 \times 10^3)(4.88)(5.96 \times 10^3)$$
$$= 1.22 \times 10^{35}.$$

So,

$$K_p = \frac{(1.72 \times 10^{54})(1.38 \times 10^{-16})1000}{1.22 \times 10^{35}}$$
$$= 1.95 \times 10^6 \text{ dynes cm}^{-2}$$
$$= 1.93 \text{ atm}.$$

59 The differences in k_{rate} will arise from differences in the number of translational, rotational, and vibrational partition functions arising in the adsorbed and gas-phase species. Recall that

$$k_{\text{rate}} = \left(\frac{kT}{h}\right) \exp\left(-\frac{E*}{kT}\right) \left[\frac{q^{\text{TS}}}{V}\right] \left[\left(\frac{q_{\text{NO}}}{V}\right)\left(\frac{q_{\text{Cl}_2}}{V}\right)\right].$$

In the gas phase,

NO has three translations, two rotations, and one vibration
Cl_2 has three translations, two rotations, and one vibration
$NOCl_2$ TS, which is bent, has three translations, three rotations, and five vibrations (recall that one vibration is missing and is the reaction coordinate).

In the adsorbed state,

NO has two translations, one rotation, and three vibrations
Cl_2 has two translations, one rotation, and three vibrations
$NOCl_2$ TS, which is bent, has two translations, one rotation, and eight vibrations (again, one vibration is missing and is the reaction coordinate).

So, in computing the partition function ratio:

$$\frac{[q^{\mathrm{TS}}/V]}{[(q_{\mathrm{NO}}/V)(q_{\mathrm{Cl_2}}/V)]}$$

for the adsorbed and gas-phase cases, one does not obtain the same number of translational, rotational, and vibrational factors. In particular, the ratio of these factors for the adsorbed and gas-phase cases gives the ratio of rate constants as follows:

$$\frac{k_{\mathrm{ad}}}{k_{\mathrm{gas}}} = \frac{(q_{\mathrm{trans}}/V)}{q_{\mathrm{vib}}}$$

which should be of the order of 10^8 (using the ratio of partition functions as given). Notice that this result suggests that reaction rates can be altered by constraining the reacting species to move freely in lower dimensions even if one does not alter the energetics (e.g., activation energy or thermochemistry).

Appendix

Character tables

C_1	E
A	1

C_s	E	σ_h		
A′	1	1	x,y,R_z	x^2,y^2,z^2,xy
A″	1	−1	z,R_x,R_y	yz,xz

C_i	E	i		
A_g	1	1	R_x,R_y,R_z	x^2,y^2,z^2,xy,xz,yz
A_u	1	−1	x,y,z	

C_2	E	C_2		
A	1	1	z,R_z	x^2,y^2,z^2,xy
B	1	−1	x,y,R_x,R_y	yz,xz

D_2	E	$C_2(z)$	$C_2(y)$	$C_2(x)$		
A	1	1	1	1		x^2, y^2, z^2
B_1	1	1	−1	−1	z, R_z	xy
B_2	1	−1	1	−1	y, R_y	xz
B_3	1	−1	−1	1	x, R_x	yz

D_3	E	$2C_3$	$3C_2$		
A_1	1	1	1		$x^2 + y^2, z^2$
A_2	1	1	−1	z, R_z	
E	2	−1	0	$(x, y)(R_x, R_y)$	$(x^2 - y^2, xy)(xz, yz)$

D_4	E	$2C_4$	$C_2(= C_4^2)$	$2C_2'$	$2C_2''$		
A_1	1	1	1	1	1		$x^2 + y^2, z^2$
A_2	1	1	1	−1	−1	z, R_z	
B_1	1	−1	1	1	−1		$x^2 - y^2$
B_2	1	−1	1	−1	1		xy
E	2	0	−2	0	0	$(x, y)(R_x, R_y)$	(xz, yz)

C_{2v}	E	C_2	$\sigma_v(xz)$	$\sigma_v'(yz)$		
A_1	1	1	1	1	z	x^2, y^2, z^2
A_2	1	1	−1	−1	R_z	xy
B_1	1	−1	1	−1	x, R_y	xz
B_2	1	−1	−1	1	y, R_x	yz

C_{3v}	E	$2C_3$	$3\sigma_v$		
A_1	1	1	1	z	$x^2 + y^2, z^2$
A_2	1	1	−1	R_z	
E	2	−1	0	$(x, y)(R_x, R_y)$	$(x^2 - y^2, xy)(xz, yz)$

C_{4v}	E	$2C_4$	C_2	$2\sigma_v$	$2\sigma_d$		
A_1	1	1	1	1	1	z	x^2+y^2,z^2
A_2	1	1	1	−1	−1	R_z	
B_1	1	−1	1	1	−1		x^2-y^2
B_2	1	−1	1	−1	1		xy
E	2	0	−2	0	0	$(x,y)(R_x,R_y)$	(xz,yz)

C_{2h}	E	C_2	i	σ_h		
A_g	1	1	1	1	R_z	x^2,y^2,z^2,xy
B_g	1	−1	1	−1	R_x,R_y	xz,yz
A_u	1	1	−1	−1	z	
B_u	1	−1	−1	1	x,y	

D_{2h}	E	$C_2(z)$	$C_2(y)$	$C_2(x)$	i	$\sigma(xy)$	$\sigma(xz)$	$\sigma(yz)$		
A_g	1	1	1	1	1	1	1	1		x^2,y^2,z^2
B_{1g}	1	1	−1	−1	1	1	−1	−1	R_z	xy
B_{2g}	1	−1	1	−1	1	−1	1	−1	R_y	xz
B_{3g}	1	−1	−1	1	1	−1	−1	1	R_x	yz
A_u	1	1	1	1	−1	−1	−1	−1		
B_{1u}	1	1	−1	−1	−1	−1	1	1	z	
B_{2u}	1	−1	1	−1	−1	1	−1	1	y	
B_{3u}	1	−1	−1	1	−1	1	1	−1	x	

D_{3h}	E	$2C_3$	$3C_2$	σ_h	$2S_3$	$3\sigma_v$		
A_1'	1	1	1	1	1	1		x^2+y^2,z^2
A_2'	1	1	−1	1	1	−1	R_z	
E'	2	−1	0	2	−1	0	(x,y)	(x^2-y^2,xy)
A_1''	1	1	1	−1	−1	−1		
A_2''	1	1	−1	−1	−1	1	z	
E''	2	−1	0	−2	1	0	(R_x,R_y)	(xz,yz)

D_{4h}	E	$2C_4$	C_2	$2C_2'$	$2C_2''$	i	$2S_4$	σ_h	$2\sigma_v$	$2\sigma_d$		
A_{1g}	1	1	1	1	1	1	1	1	1	1		x^2+y^2, z^2
A_{2g}	1	1	1	−1	−1	1	1	1	−1	−1	R_z	
B_{1g}	1	−1	1	1	−1	1	−1	1	1	−1		x^2-y^2
B_{2g}	1	−1	1	−1	1	1	−1	1	−1	1		xy
E_g	2	0	−2	0	0	2	0	−2	0	0	(R_x, R_y)	(xz, yz)
A_{1u}	1	1	1	1	1	−1	−1	−1	−1	−1		
A_{2u}	1	1	1	−1	−1	−1	−1	−1	1	1	z	
B_{1u}	1	−1	1	1	−1	−1	1	−1	−1	1		
B_{2u}	1	−1	1	−1	1	−1	1	−1	1	−1		
E_u	2	0	−2	0	0	−2	0	2	0	0	(x, y)	

D_{6h}	E	$2C_6$	$2C_3$	C_2	$3C_2'$	$3C_2''$	i	$2S_3$	$2S_6$	σ_h	$3\sigma_d$	$3\sigma_v$		
A_{1g}	1	1	1	1	1	1	1	1	1	1	1	1		x^2+y^2, z^2
A_{2g}	1	1	1	1	−1	−1	1	1	1	1	−1	−1	R_z	
B_{1g}	1	−1	1	−1	1	−1	1	−1	1	−1	1	−1		
B_{2g}	1	−1	1	−1	−1	1	1	−1	1	−1	−1	1		
E_{1g}	2	1	−1	−2	0	0	2	1	−1	−2	0	0	(R_x, R_y)	(xz, yz)
E_{2g}	2	−1	−1	2	0	0	2	−1	−1	2	0	0		(x^2-y^2, xy)
A_{1u}	1	1	1	1	1	1	−1	−1	−1	−1	−1	−1		
A_{2u}	1	1	1	1	−1	−1	−1	−1	−1	−1	1	1	z	
B_{1u}	1	−1	1	−1	1	−1	−1	1	−1	1	−1	1		
B_{2u}	1	−1	1	−1	−1	1	−1	1	−1	1	1	−1		
E_{1u}	2	1	−1	−2	0	0	−2	−1	1	2	0	0	(x, y)	
E_{2u}	2	−1	−1	2	0	0	−2	1	1	−2	0	0		

D_{2d}	E	$2S_4$	C_2	$2C_2'$	$2\sigma_d$		
A_1	1	1	1	1	1		x^2+y^2, z^2
A_2	1	1	1	−1	−1	R_z	
B_1	1	−1	1	1	−1		x^2-y^2
B_2	1	−1	1	−1	1	z	xy
E	2	0	−2	0	0	$(x, y)(R_x, R_y)$	(xz, yz)

D_{3d}	E	$2C_3$	$3C_2$	i	$2S_6$	$3\sigma_d$		
A_{1g}	1	1	1	1	1	1		x^2+y^2,z^2
A_{2g}	1	1	−1	1	1	−1	R_z	
E_g	2	−1	0	2	−1	0	(R_x,R_y)	$(x^2-y^2,xy)(xz,yz)$
A_{1u}	1	1	1	−1	−1	−1		
A_{2u}	1	1	−1	−1	−1	1	z	
E_u	2	−1	0	−2	1	0	(x,y)	

S_4	E	S_4	C_2	S_4^3		
A	1	1	1	1	R_z	x^2+y^2,z^2
B	1	−1	1	−1	z	x^2-y^2,xy
E	1	i	−1	−i	$(x,y)\ (R_x,R_y)$	(xz,yz)
	1	−i	−1	i		

T	E	$4C_3$	$4C_3^2$	$3C_2$		$\varepsilon=\exp(2\pi i/3)$
A	1	1	1	1		$x^2+y^2+z^2$
E	1	ε	ε^*	1		$(2z^2-x^2-y^2,$
	1	ε^*	ε	−1		$x^2-y^2)$
T	3	0	0	−1	$(R_x,R_y,R_z)(x,y,z)$	(xy,xz,yz)

T_h	E	$4C_3$	$4C_3^2$	$3C_2$	i	$4S_6$	$4S_6^5$	$3\sigma_h$		$\varepsilon=\exp(2\pi i/3)$
A_g	1	1	1	1	1	1	1	1		$x^2+y^2+z^2$
A_u	1	1	1	1	−1	−1	−1	−1		
E_g	1	ε	ε^*	1	1	ε	ε^*	1		$(2z^2-x^2-y^2,$
	1	ε^*	ε	1	1	ε^*	ε	1		$x^2-y^2)$
E_u	1	ε	ε^*	1	−1	$-\varepsilon$	$-\varepsilon^*$	−1		
	1	ε^*	ε	1	−1	$-\varepsilon^*$	$-\varepsilon$	−1		
T_g	3	0	0	−1	1	0	0	−1	(R_x,R_y,R_z)	
T_u	3	0	0	−1	−1	0	0	1	(x,y,z)	(xy,xz,yz)

T_d	E	$8C_3$	$3C_2$	$6S_4$	$6\sigma_d$		
A_1	1	1	1	1	1		$x^2+y^2+z^2$
A_2	1	1	1	−1	−1		
E	2	−1	2	0	0		$(2z^2-x^2-y^2, x^2-y^2)$
T_1	3	0	−1	1	−1	(R_x,R_y,R_z)	
T_2	3	0	−1	−1	1	(x,y,z)	(xy,xz,yz)

O	E	$6C_4$	$3C_2(=C_4^2)$	$8C_3$	$6C_2$		
A_1	1	1	1	1	1		$x^2+y^2+z^2$
A_2	1	−1	1	1	−1		
E	2	0	2	−1	0		$(2z^2-x^2-y^2, x^2-y^2)$
T_1	3	1	−1	0	−1	$(R_x,R_y,R_z)(x,y,z)$	
T_2	3	−1	−1	0	1		(xy,xz,yz)

O_h	E	$8C_3$	$6C_2$	$6C_4$	$3C_2$ $(=C_4^2)$	i	$6S_4$	$8S_6$	$3\sigma_h$	$6\sigma_d$		
A_{1g}	1	1	1	1	1	1	1	1	1	1		$x^2+y^2+z^2$
A_{2g}	1	1	−1	−1	1	1	−1	1	1	−1		
E_g	2	−1	0	0	2	2	0	−1	2	0		$(2z^2-x^2-y^2, x^2-y^2)$
T_{1g}	3	0	−1	1	−1	3	1	0	−1	−1	(R_x,R_y,R_z)	
T_{2g}	3	0	1	−1	−1	3	−1	0	−1	1		(xy,xz,yz)
A_{1u}	1	1	1	1	1	−1	−1	−1	−1	−1		
A_{2u}	1	1	−1	−1	1	−1	1	−1	−1	1		
E_u	2	−1	0	0	2	−2	0	1	−2	0		
T_{1u}	3	0	−1	1	−1	−3	−1	0	1	1	(x,y,z)	
T_{2u}	3	0	1	−1	−1	−3	1	0	1	−1		

$C_{\infty v}$	E	$2C_{\infty}^{\Phi}$	$\cdots$	$\infty\sigma_v$		
$A_1 \equiv \sum^+$	1	1	$\cdots$	1	z	$x^2 + y^2, z^2$
$A_2 \equiv \sum^-$	1	1	$\cdots$	-1	R_z	
$E_1 \equiv \prod$	2	$2\cos\Phi$	$\cdots$	0	$(x,y)(R_x,R_y)$	(xz,yz)
$E_2 \equiv \Delta$	2	$2\cos 2\Phi$	$\cdots$	0		$(x^2 - y^2, xy)$
$E_3 \equiv \Phi$	2	$2\cos 3\Phi$	$\cdots$	0		
$\cdots$	$\cdots$	$\cdots$	$\cdots$	$\cdots$		

$D_{\infty h}$	E	$2C_{\infty}^{\Phi}$	$\cdots$	$\infty\sigma_v$	i	$2S_{\infty}^{\Phi}$	$\cdots$	∞C_2		
$\sum_g^+$	1	1	$\cdots$	1	1	1	$\cdots$	1		$x^2 + y^2, z^2$
$\sum_g^-$	1	1	$\cdots$	-1	1	1	$\cdots$	-1	R_z	
$\prod_g$	2	$2\cos\Phi$	$\cdots$	0	2	$-2\cos\Phi$	$\cdots$	0	(R_x,R_y)	(xz,yz)
Δ_g	2	$2\cos 2\Phi$	$\cdots$	0	2	$2\cos 2\Phi$	$\cdots$	0		$(x^2 - y^2, xy)$
$\cdots$	$\cdots$	$\cdots$	$\cdots$	$\cdots$	$\cdots$	$\cdots$	$\cdots$	$\cdots$		
$\sum_u^+$	1	1	$\cdots$	1	-1	-1	$\cdots$	-1	z	
$\sum_u^-$	1	1	$\cdots$	-1	-1	-1	$\cdots$	1		
$\prod_u$	2	$2\cos\Phi$	$\cdots$	0	-2	$2\cos\Phi$	$\cdots$	0	(x,y)	
Δ_u	2	$2\cos 2\Phi$	$\cdots$	0	-2	$-2\cos 2\Phi$	$\cdots$	0		
$\cdots$	$\cdots$	$\cdots$	$\cdots$	$\cdots$	$\cdots$	$\cdots$	$\cdots$	$\cdots$		

Index